Handbook of Measurements

Benchmarks for Systems Accuracy and Precision

Industrial Innovation Series

Series Editor

Adedeji B. Badiru

Air Force Institute of Technology (AFIT) – Dayton, Ohio

PUBLISHED TITLES

Quality Management in Construction Projects, *Abdul Razzak Rumane*

Quality Tools for Managing Construction Projects, *Abdul Razzak Rumane*

Social Responsibility: Failure Mode Effects and Analysis, *Holly Alison Duckworth & Rosemond Ann Moore*

Statistical Techniques for Project Control, *Adedeji B. Badiru & Tina Agustiady*

STEP Project Management: Guide for Science, Technology, and Engineering Projects, *Adedeji B. Badiru*

Sustainability: Utilizing Lean Six Sigma Techniques, *Tina Agustiady & Adedeji B. Badiru*

Systems Thinking: Coping with 21st Century Problems, *John Turner Boardman & Brian J. Sauser*

Techonomics: The Theory of Industrial Evolution, *H. Lee Martin*

Total Project Control: A Practitioner's Guide to Managing Projects as Investments, Second Edition, *Stephen A. Devaux*

Triple C Model of Project Management: Communication, Cooperation, Coordination, *Adedeji B. Badiru*

FORTHCOMING TITLES

3D Printing Handbook: Product Development for the Defense Industry, *Adedeji B. Badiru & Vhance V. Valencia*

Company Success in Manufacturing Organizations: A Holistic Systems Approach, *Ana M. Ferreras & Lesia L. Crumpton-Young*

Design for Profitability: Guidelines to Cost Effectively Management the Development Process of Complex Products, *Salah Ahmed Mohamed Elmoselhy*

Essentials of Engineering Leadership and Innovation, *Pamela McCauley-Bush & Lesia L. Crumpton-Young*

Global Manufacturing Technology Transfer: Africa-USA Strategies, Adaptations, and Management, *Adedeji B. Badiru*

Guide to Environment Safety and Health Management: Developing, Implementing, and Maintaining a Continuous Improvement Program, *Frances Alston & Emily J. Millikin*

Handbook of Construction Management: Scope, Schedule, and Cost Control, *Abdul Razzak Rumane*

Handbook of Measurements: Benchmarks for Systems Accuracy and Precision, *Adedeji B. Badiru & LeeAnn Racz*

Introduction to Industrial Engineering, Second Edition, *Avraham Shtub & Yuval Cohen*

Manufacturing and Enterprise: An Integrated Systems Approach, *Adedeji B. Badiru, Oye Ibidapo-Obe & Babatunde J. Ayeni*

Project Management for Research: Tools and Techniques for Science and Technology, *Adedeji B. Badiru, Vhance V. Valencia & Christina Rusnock*

Project Management Simplified: A Step-by-Step Process, *Barbara Karten*

A Six Sigma Approach to Sustainability: Continual Improvement for Social Responsibility, *Holly Allison Duckworth & Andrea Hoffmeier Zimmerman*

Total Productive Maintenance: Strategies and Implementation Guide, *Tina Agustiady & Elizabeth A. Cudney*

Handbook of
Measurements

Benchmarks for Systems Accuracy and Precision

Edited by
Adedeji B. Badiru and LeeAnn Racz

CRC Press
Taylor & Francis Group
Boca Raton London New York

CRC Press is an imprint of the
Taylor & Francis Group, an **informa** business

CRC Press
Taylor & Francis Group
6000 Broken Sound Parkway NW, Suite 300
Boca Raton, FL 33487-2742

First issued in paperback 2017

© 2016 by Taylor & Francis Group, LLC
CRC Press is an imprint of Taylor & Francis Group, an Informa business

No claim to original U.S. Government works

ISBN-13: 978-1-4822-2522-8 (hbk)
ISBN-13: 978-1-138-74940-5 (pbk)

Library of Congress Cataloging-in-Publication Data

Handbook of measurements : benchmarks for systems accuracy and precision / editors, Adedeji B. Badiru and LeeAnn Racz.
    pages cm. -- (Industrial innovation series)
    Includes bibliographical references and index.
    ISBN 978-1-4822-2522-8
    1. Measurement--Handbooks, manuals, etc. 2. Weights and measures--Handbooks, manuals, etc.
I. Badiru, Adedeji Bodunde, 1952- II. Racz, LeeAnn.

T50.H265 2016
530.8'1--dc23   2015012623

Visit the Taylor & Francis Web site at
http://www.taylorandfrancis.com

and the CRC Press Web site at
http://www.crcpress.com

To Deleeville,
for the inspiration and motivation for this work

Contents

Preface

Planning, measurement, and attention to detail form the basis for success in engineering operations. Measurements pervade everything we do and must be viewed from a systems perspective. Using a systems framework, *The Handbook of Measurements: Benchmarks for Systems Accuracy*, presents a comprehensive guide to everything about measurement. Its technically rigorous approach to systems linking of measurements sets it apart from other handbooks. The broad approach of the handbook covers both qualitative and quantitative topics of measurement. Of particular benefit is the inclusion of human-centric measurements such as measurement of personnel productivity and contractor performance. The handbook opens with a chapter on the fundamentals of measurement. It is well understood that humans cannot manage anything that cannot be measured. All elements involved in our day-to-day decision making involve some forms of measurement. Measuring an attribute of a system and then analyzing it against some standard, specification, best practice, or benchmark empowers a decision maker to take appropriate and timely actions.

Fundamentally, measurement is the act, or the result of a quantitative comparison between a predefined standard and an unknown magnitude. This handbook uses a systems view of measurement to link all aspects. For example, one chapter in the handbook addresses systems interoperability measurement, which further illustrates how the elements of any large complex system interoperate to accomplish a desired end goal. Other chapters in the handbook include human factors measurements for work system analysis, measurements of environmental health, measurement of environmental contamination, measurements of land, measuring building performance, energy systems measurements, economic systems measurements, measurement and quantum mechanics, social science measurement, a measurement model to quantify health, large-scale dimensional metrology, performance-based logistics, data processing, visualization of big data, maintenance metrics analysis, measuring success of services contracts, measurement of personnel productivity, measurement risk analysis, data modeling for forecasting, measurements from archival observational data, reduction of measurement imprecision, measurements in ergonomics studies, metrics to manage contractor performance, and a low-clutter method for bistatic radar cross-section measurements. The handbook concludes with three appendices on measurement references, conversion factors, equations, formulas, and statistics for measurement.

Adedeji B. Badiru and LeeAnn Racz

Acknowledgments

We gratefully acknowledge the contributions and support of all those who played a part in the writing of this book. Special thanks go to Ms. Annabelle Sharp for her responsiveness and dedication to the needs and challenges of typing and organizing the complex manuscript. We also appreciate the editorial and organizational support provided by Richard D. Cook and Thomas M. Dickey in the initial draft of the book. We gratefully acknowledge the extraordinary contributions of Ms. Anna E. Maloney, who pored over the galley proofs to ensure that all final t's are crossed and all final i's are dotted. Her keen sense of technical review positively impacted the overall quality of the handbook.

Editors

Adedeji B. Badiru is the dean and senior academic officer for the Graduate School of Engineering and Management at the Air Force Institute of Technology (AFIT). He is responsible for planning, directing, and controlling all operations related to granting doctoral and master's degrees, professional continuing cyber education, and research and development programs. Previously, Deji Badiru was professor and head of Systems Engineering and Management at AFIT (Air Force Institute of Technology), professor and head of the Department of Industrial & Information Engineering at the University of Tennessee in Knoxville, and professor of industrial engineering and dean of University College at the University of Oklahoma, Norman. He is a registered professional engineer (PE), a certified project management professional (PMP), a Fellow of the Institute of Industrial Engineers, and a Fellow of the Nigerian Academy of Engineering. He holds a BS in industrial engineering, an MS in mathematics, and an MS in industrial engineering from Tennessee Technological University, and a PhD in industrial engineering from the University of Central Florida. His areas of interest include mathematical modeling, systems efficiency analysis, and high-tech product development. He is the author of over 30 books, 35 book chapters, 75 technical journal articles, and 115 conference proceedings and presentations. He has also published 30 magazine articles and 20 editorials and periodicals to his credit. He is a member of several professional associations and scholastic honor societies. Dr. Badiru has won several awards for his teaching, research, and professional accomplishments.

LeeAnn Racz is a bioenvironmental engineering flight commander at RAF Lakenheath, United Kingdom. She previously served as an assistant professor of environmental engineering and director of the Graduate Environmental Engineering and Science Program in the Systems & Engineering Management at the Air Force Institute of Technology. She currently holds the rank of Lieutenant Colonel in the US Air Force. Her other assignments have taken her to the US Air Force School of Aerospace Medicine, San Antonio, Texas; the US Air Force Academy, Colorado Springs, Colorado; Peterson Air Force Base, Colorado Springs, Colorado; Osan Air Base, South Korea; and Cannon Air Force Base, Clovis, New Mexico. She is a registered professional engineer (PE), certified industrial hygienist, and board-certified environmental engineer. She holds a BS in environmental engineering from California Polytechnic State University (San Luis Obispo), an MS in biological and agricultural engineering from the University of Idaho, and a PhD in civil and environmental engineering from the University of Utah. Her areas of interest include characterizing the fate of chemical warfare agents and pollutants of emerging concern in the natural and engineered environments, environmental health issues, and using biological reactors to treat industrial waste. Dr. Racz has authored dozens of refereed journal articles, conference proceedings, magazine articles, and presentations, and one handbook. She is a member of several professional associations and honor societies. Dr. Racz has received numerous awards such as the 2014 Air Education and Training Command Military Educator of the Year Award, the 2014 Military Officers Association of America Educator of the Year Award, the 2012 Southwestern Ohio Council for Higher Education Teaching Excellence Award, and was the 2011 Faculty Scholar of the Year for the Department of Systems and Engineering Management. She is also the 2014 recipient of the Air Force Meritorious Service Medal (one oak leaf cluster) and Air Force Commendation Medal (three oak leaf clusters).

Contributors

Dario Assante
University Uninettuno
Rome, Italy

Adedeji B. Badiru
Department of Systems Engineering
  and Management
Air Force Institute of Technology
Dayton, Ohio

Peter J. Collins
Department of Electrical and Computer
  Engineering
Air Force Institute of Technology
Dayton, Ohio

Livio Conti
University Uninettuno
Rome, Italy

Justin D. Delorit
Civil Engineer School
Air Force Institute of Technology
Dayton, Ohio

Kenneth Doerr
Graduate School of Business and Public
  Policy
Naval Postgraduate School
Monterey, California

Isaac J. Donaldson
U.S. Navy
Naval Computer and Telecommunications
  Station Naples
Naples, Italy

Donald R. Eaton
Graduate School of Business and Public
  Policy
Naval Postgraduate School
Monterey, California

John J. Elshaw
Department of Systems Engineering
  and Management
Air Force Institute of Technology
Dayton, Ohio

R. Marshall Engelbeck
Graduate School of Business and Public
  Policy
Naval Postgraduate School
Monterey, California

Thomas C. Ford
Department of Systems Engineering
  and Management
Air Force Institute of Technology
Dayton, Ohio

Fiorenzo Franceschini
Politecnico di Torino
Department of Management
  and Production Engineering
University of Turin
Turin, Italy

Ronald E. Giachetti
Department of Systems Engineering
Naval Postgraduate School
Monterey, California

Patrick Hagan
Graduate School of Business and Public
    Policy
Naval Postgraduate School
Monterey, California

Adrian S. Hanley
U.S. Environmental Protection Agency
Office of Water
Washington, DC

Sandra C. Hom
Department of Information Science
Naval Postgraduate School
Monterey, California

Thomas Housel
Department of Information Science
Graduate School of Operational
    and Information Sciences
Naval Postgraduate School
Monterey, California

Paul F. M. Krabbe
Department of Epidemiology
University of Groningen
Groningen, the Netherlands

Hai-Woong Lee
Department of Physics
Korea Advanced Institute of Science
    and Technology (KAIST)
Daejeon, Korea

Ira A. Lewis
Graduate School of Business and Public
    Policy
Naval Postgraduate School
Monterey, California

Matthew L. Magnuson
U.S. Environmental Protection Agency
Washington, DC

Domenico Maisano
Politecnico di Torino
Department of Management
    and Production Engineering
University of Turin
Turin, Italy

Farman A. Moayed
Department of the Built
    Environment
Indiana State University
Terre Haute, Indiana

Johnathan Mun
Department of Information Science
Naval Postgraduate School
Monterey, California

David W. Nehrkorn
Redwood City, California

Olufemi A. Omitaomu
Computational Sciences and Engineering
    Division
Oak Ridge National Laboratory
Oak Ridge, Tennessee

Paola Pedone
Department of Calibration Laboratories
Accredia
Turin, Italy

LeeAnn Racz
Department of Systems Engineering
    and Management
Air Force Institute of Technology
Dayton, Ohio

Steven J. Schuldt
Department of Systems Engineering
    and Management
Air Force Institute of Technology
Dayton, Ohio

Vittorio Sgrigna
Department of Mathematics and Physics
University of Rome Tre
Rome, Italy

Trent Silkey
Graduate School of Business and Public
    Policy
Naval Postgraduate School
Monterey, California

Joseph Spede
Graduate School of Business and Public
    Policy
Naval Postgraduate School
Monterey, California

Trisha Sutton
Graduate School of Business and Public
    Policy
Naval Postgraduate School
Monterey, California

David E. Weeks
Department of Engineering Physics
Air Force Institute of Technology
Dayton, Ohio

Stuart A. Willison
U.S. Environmental Protection Agency
Washington, DC

Dirk Yamamoto
Department of Systems Engineering
    and Management
Air Force Institute of Technology
Dayton, Ohio

David Zilpimiani
National Institute of Geophysics
Georgian Academy Sciences
Tbilisi, Georgia

chapter one

Fundamentals of measurement

Adedeji B. Badiru

Contents

1.1   Introduction

Throughout history, humans have strived to come up with better tools, techniques, and instruments for measurement. From the very ancient times to the present fast-paced society, our search for more precise, more convenient, and more accessible measuring devices has led to new developments over the years. The following quotes confirm the importance and efficacy of measurements in our lives. Appendix A presents the most common measurement conversion factors.

> Measure twice, cut once
>
> **English Proverb**

> Where there is no Standard there can be no Kaizen (improvement)
>
> **Taiichi Ohno**

> Where there is no measurement, there can be no standard
>
> **Adedeji Badiru**

> Measurement mitigates mess-ups
>
> **Adedeji Badiru**

Measurement pervades everything we do; this applies to technical, management, and social activities and requirements. Even in innocuous situations such as human leisure, the importance of measurement comes to the surface; for example, how much, how far, how good, how fast, and how long are typical expressions of some sort of measurement. Consider a possible newspaper-classified advertisement that provides the following measurement disclaimer:

> The acceptable units of measure for all firewood advertisement are cord or fraction of a cord. The units of measurements for a cord are 4″ by 4″ by 8″. The terms face cord, rack, pile, rick, truckload or similar terms are not acceptable.

Who would have thought that firewood had such a serious measurement constraint and guideline? Social, psychological, physical, economic, cognitive, and metabolic attributes, as well as other human characteristics, are all amenable to measurement systems just as the mechanical devices around us.

1.2   What is measurement?

It is well understood that we cannot manage anything if we cannot measure it. All elements involved in our day-to-day decision making involve some form of measurement. Measuring an attribute of a system and then analyzing it against some standard, some best practice, or some benchmark empowers a decision maker to take appropriate and timely actions.

Fundamentally, measurement is the act or the result of a quantitative comparison between a predefined standard and an unknown magnitude. Beckwith and Buck (1965), Shillito and De Marle (1992), Morris (1997), Badiru et al. (2012), and Badiru and Kovach (2012) all address concepts, tools, and techniques of measurement systems. If the result is to be generally meaningful, two requirements must be met in the act of measurement:

1. The standard, which is used for comparison, must be accurately known and commonly accepted.
2. The procedure and instrument employed for obtaining the comparison must be provable and repeatable.

The first requirement is that there should be an accepted standard of comparison. A weight cannot simply be heavy. It can only be proportionately as heavy as something else, namely the standard. A comparison must be made, and unless it is made relative to something generally recognized as standard, the measurement can only have a limited meaning; this holds for any quantitative measurement we may wish to make. In general, the comparison is of magnitude, and a numerical result is presupposed. The quantity in question may be twice or 1.4 times as large as the standard, or in some other ratio, but a numerical comparison must be made for it to be meaningful. The typical characteristics of a measurement process include the following

- Precision
- Accuracy
- Correlation
- Stability
- Linearity
- Type of data

1.3   The dawn and present of measurement

Weights and measures may be ranked among the necessaries of life to every individual of human society. They enter into the economical arrangements and daily concerns of every family. They are necessary to every occupation of human industry; to the distribution and security of every species of property; to every transaction of trade and commerce; to the labors of the husbandman; to the ingenuity of the artificer; to the studies of the philosopher; to the researches of the antiquarian; to the navigation of the mariner, and the marches of the soldier; to all

the exchanges of peace, and all the operations of war. The knowledge
of them, as in established use, is among the first elements of education,
and is often learned by those who learn nothing else, not even to read
and write. This knowledge is riveted in the memory by the habitual
application of it to the employments of men throughout life.

John Quincy Adams
Report to the Congress, 1821

The historical accounts of measurements presented in this section are based mostly on
NIST (1974), HistoryWorld (2014), and TANF (2014). Weights and measures were among the
earliest tools invented by man. Primitive societies needed rudimentary measures for many
tasks, such as house and road construction and commerce of raw materials. Man, in early
years, used parts of the human body and the natural surroundings as device-measuring
standards. Early Babylonian and Egyptian records and the Bible indicate that length was
first measured with the foot, forearm, hand, or finger; time was measured by the periods of
the sun, moon, and other heavenly bodies. When it was necessary to compare the capacities of containers such as gourds, clay, or metal vessels, they were filled with plant seeds
that were then counted to measure volumes. With the development of scales as a means
of weighing, seeds and stones served as standards. For instance, the "carat," still used as a
mass unit for gems, is derived from the carob seed.

As societies evolved, measurements became more complex. The invention of number-
ing systems and the science of mathematics made it possible to create whole systems of
measurement units suitable for trade and commerce, land division, taxation, and scientific
research. For these more sophisticated uses, it was necessary to not only weigh and measure more complex items, but it was also necessary to do it accurately and repeatedly at
different locations. With the limited international exchange of goods and communication
of ideas in ancient times, different measuring systems evolved for the same measures and
became established in different parts of the world. In ancient times, different parts of the
same country might use different measuring systems for the same purpose. Historical
records indicate that early measurement systems evolved locally in Africa to take advantage of the African natural environment. For example, common early units of measure, in
some parts of Africa relied on standardizations based on the cocoa bean sizes and weights.

1.3.1   *The English system*

The measurement system commonly used in the United States today is nearly the same
as that brought by the American colony settlers from England. These measures had their
origins in a variety of cultures, including Babylonian, Egyptian, Roman, Anglo-Saxon,
and Nordic French. The ancient "digit," "palm," "span," and "cubic" units of length slowly
lost preference to the length units, "inch," "foot," and "yard." Roman contributions include
the use of 12 as a base number and the words from which we derive many of the modern
names of measurement units. For example, the 12 divisions of the Roman "pes," or foot
were called "unciae." The "foot," as a unit of measuring length is divided into 12 in. The
common words "inch" and "ounce" are both derived from Latin words. The "yard" as
a measure of length can be traced back to early Saxon kings. They wore a sash or girdle
around the waist that could be removed and used as a convenient measuring device. Thus,
the word "yard" comes from the Saxon word "gird," which represents the circumference
of a person's waist, preferably a "standard person," such as a king.

Evolution and standardization of measurement units often had interesting origins. For example, it was recorded that King Henry I decreed that a yard should be the distance from the tip of his nose to the end of his outstretched thumb. The length of a furlong (or furrow-long) was established by early Tudor rulers as 220 yards, this led Queen Elizabeth I to declare in the sixteenth century that the traditional Roman mile of 5000 ft would be replaced by one of the 5280 ft, making the mile exactly 8 furlongs and providing a convenient relationship between the furlong and the mile. To this day, there are 5280 ft in 1 mile, which is 1760 yards. Thus, through royal edicts, England by the eighteenth century had achieved a greater degree of standardization than other European countries. The English units were well suited to commerce and trade because they had been developed and refined to meet commercial needs. Through English colonization and its dominance of world commerce during the seventeenth, eighteenth, and nineteenth centuries, the English system of measurement units became established in many parts of the world, including the American colonies. The early 13 American colonies, however, had undesirable differences with respect to measurement standards for commerce. The need for a greater uniformity led to clauses in the Articles of Confederation (ratified by the original colonies in 1781) and the Constitution of the United States (ratified in 1788) that gave Congress the power to fix uniform standards for weights and measures across the colonies. Today, standards provided by the U.S. National Institute of Standards and Technology (NIST) ensure uniformity of measurement units throughout the country.

1.3.2 The metric system

The need for a single worldwide coordinated measurement system was recognized over 300 years ago. In 1670, Gabriel Mouton, vicar of St. Paul's Church in Lyons, France, and an astronomer, proposed a comprehensive decimal measurement system based on the length of 1 arc-min of a great circle of the Earth. Mouton also proposed the swing length of a pendulum with a frequency of 1 beat/s as the unit of length. A pendulum with this beat would have been fairly easily reproducible, thus facilitating the widespread distribution of uniform standards.

In 1790, in the midst of the French Revolution, the National Assembly of France requested the French Academy of Sciences to "deduce an invariable standard for all the measures and all the weights." The Commission appointed by the Academy created a system that was, at once, simple and scientific. The unit of length was to be a portion of the Earth's circumference. Measures for capacity (volume) and mass were to be derived from the unit of length, thus relating the basic units of the system to each other and nature. Furthermore, larger and smaller multiples of each unit were to be created by multiplying or dividing the basic units by 10 and powers of 10. This feature provided a great convenience to users of the system by eliminating the need for such calculations as dividing by 16 (to convert ounces to pounds) or by 12 (to convert inches to feet). Similar calculations in the metric system could be performed simply by shifting the decimal point. Thus, the metric system is a "base-10" or "decimal" system.

The Commission assigned the name metre (i.e., meter in English) to the unit of length. This name was derived from the Greek word metron, meaning "a measure." The physical standard representing the meter was to be constructed so that it would equal one 10 millionth of the distance from the North Pole to the equator along the meridian running near Dunkirk in France and Barcelona in Spain. The initial metric unit of mass, the "gram," was defined as the mass of 1 cm^3 (a cube that is 0.01 m on each side) of water at its temperature of maximum density. The cubic decimeter (a cube 0.1 m on each side) was chosen as the unit of capacity. The fluid volume measurement for the cubic decimeter was given the name "liter." Although the metric system was not accepted with much enthusiasm at first, adoption by other nations occurred steadily after France made its use compulsory in 1840. The standardized structure and decimal features of the metric system made it well suited for scientific and engineering work. Consequently, it is not surprising that the rapid spread of the system coincided with an age of rapid technological development. In the United States, by Act of Congress in 1866, it became "lawful throughout the United States of America to employ the weights and measures of the metric system in all contracts, dealings or court proceedings." However, the United States has remained a hold-out with respect to a widespread adoption of the metric system. Previous attempts to standardize to the metric system in the United States have failed. Even today, in some localities of the United States, both English and metric systems are used side by side.

As an illustration of dual usage of measuring systems, a widespread news report in late September 1999 reported how the National Aeronautics and Space Administration (NASA) lost a $125 million Mars orbiter in a crash onto the surface of Mars because a Lockheed Martin engineering team used the English units of measurement while the agency's team used the more conventional metric system for a key operation of the spacecraft. The unit's mismatch prevented navigation information from transferring between the Mars Climate Orbiter spacecraft team at Lockheed Martin in Denver and the flight team at NASA's Jet Propulsion Laboratory in Pasadena, California. Therefore, even at such a high-stakes scientific endeavor, nonstandardization of measuring units can create havoc.

Getting back to history, the late 1860s saw the need for even better metric standards to keep pace with scientific advances. In 1875, an international agreement, known as the Meter Convention, set up well-defined metric standards for length and mass and established permanent mechanisms to recommend and adopt further refinements in the metric system. This agreement, commonly called the "Treaty of the Meter" in the United States, was signed by 17 countries, including the United States. As a result of the treaty, metric standards were constructed and distributed to each nation that ratified the Convention. Since 1893, the internationally adopted metric standards have served as the fundamental measurement standards of the United States, at least, in theory if not in practice.

By 1900, a total of 35 nations, including the major nations of continental Europe and most of South America, had officially accepted the metric system. In 1960, the General Conference on Weights and Measures, the diplomatic organization made up of the signatory nations to the Meter Convention, adopted an extensive revision and simplification of the system. The following seven units were adopted as the base units for the metric system:

1. Meter (for length)
2. Kilogram (for mass)
3. Second (for time)
4. Ampere (for electric current)
5. Kelvin (for thermodynamic temperature)
6. Mole (for amount of substance)
7. Candela (for luminous intensity)

1.3.3 The SI system

Based on the general standardization described above, the name Système International d'Unités (International System of Units), with the international abbreviation SI, was adopted for the modern metric system. Throughout the world, measurement science research and development continue to develop more precise and easily reproducible ways of defining measurement units. The working organizations of the General Conference on Weights and Measures coordinate the exchange of information about the use and refinement of the metric system and make recommendations concerning improvements in the system and its related standards. Our daily lives are mostly ruled or governed by the measurements of length, weight, volume, and time. These are briefly described in the following sections.

1.3.4   Length

The measurement of distance, signified as length, is the most ubiquitous measurement in our world. The units of length represent how we conduct everyday activities and transactions. The two basic units of length measurement are the British units (inch, foot, yard, and mile) and the metric system (meters, kilometers). In its origin, anatomically, the inch is a thumb. The foot logically references the human foot. The yard relates closely to a human pace, but also derives from two cubits (the measure of the forearm). The mile originates from the Roman *mille passus*, which means a thousand paces. The ancient Romans defined a pace as two steps. Therefore, approximately, a human takes two paces within one yard. The average human walking speed is about 5 kilometers per hour (km/h), or about 3.1 miles per hour (mph).

For the complex measuring problems of ancient civilization—surveying the land to register property rights, or selling a commodity by length—a more precise unit was required. The solution was a rod or bar of an exact length, which was kept in a central public place. From this "standard" other identical rods could be copied and distributed through the community. In Egypt and Mesopotamia, these standards were kept in temples. The basic unit of length in both civilizations was the cubit, based on the measurement of a forearm from the elbow to the tip of the middle finger. When a length such as this is standardized, it is usually the king's dimension, which is first taken as the norm.

1.3.5   Weight

For measurements of weight, the human body does not provide convenient approximations as for length. Compared to other grains, grains of wheat are reasonably standard in size. Weight can be expressed with some degree of accuracy in terms of a number of grains of wheat. The use of grains to convey weight is still used today in the measurement of precious metals, such as gold, by jewelers. As with measurements of length, a block of metal was kept in the temples as an official standard for a given number of grains. Copies of this were cast and weighed in the balance for perfect accuracy. However, imperfect human integrity in using scales made it necessary to have an inspectorate of weights and measures for practical adjudication of measurements in the olden days.

1.3.6   Volume

From the ancient time of trade and commerce to the present day, a reliable standard of volume is one of the hardest to accomplish. However, we improvise by using items from nature and art. Items such as animal skins, baskets, sacks, or pottery jars could be made to approximately consistent sizes, such that they were sufficient for measurements in ancient measurement transactions. Where the exact amount of any commodity needs to be known, weight is the measure more likely to be used instead of volume.

1.3.7   Time

Time is a central aspect of human life. Throughout human history, time has been appreciated in very precise terms. Owing to the celestial preciseness of day and night, the day and the week are easily recognized and recorded. However, an accurate calendar for a year is more complicated to achieve universally. The morning time before midday (forenoon) is easily distinguishable from the time after midday (afternoon), provided the sun is shining, and the position of the sun in the landscape can reveal roughly how much of the day has passed. In contrast, the smaller units of time, such as hours, minutes, and seconds, were initially (in ancient times) unmeasurable and unneeded. Unneeded because the ancient man had big blocks of time to accomplish whatever was needed to be done. Microallocation of time was, thus, not essential. However, in our modern society, the minute (tiny) time measurements of seconds and minutes are very essential. The following reprinted poem by the coeditor conveys the modern appreciation of the passage of time:

The Flight of Time

> What is the speed and direction of Time?
> Time flies; but it has no wings.
> Time goes fast; but it has no speed.
> Where has time gone? But it has no destination.
> Time goes here and there; but it has no direction.
> Time has no embodiment. It neither flies, walks, nor goes anywhere.
> Yet, the passage of time is constant.

Adedeji Badiru
2006

1.3.7.1   *Sundial and water clock*

The sundial and water clock originated in the second millennium BC. The movement of the sun through the sky makes possible a simple estimate of time, from the length and position of a shadow cast by a vertical stick. If marks were made where the sun's shadow fell, the time of day could be recorded in a consistent manner. The result was the invention of sundial. An Egyptian example survives from about 800 BC, but records indicate that the principle was familiar to astronomers of earlier times. Practically, it is difficult to measure time precisely on a sundial because the sun's path through the sky changes with the seasons. Early attempts at precision in timekeeping, relied on a different principle known as the water clock. The water clock, known from a Greek word as the clepsydra, attempted to measure time by the amount of water that dripped from a tank. This would have been a reliable form of the clock if the flow of water could be controlled perfectly. In practice, at that time, it could not. The hourglass, using sand on the same principle, had an even longer history and utility. It was a standard feature used in the eighteenth-century pulpits in Britain to ensure a sermon of standard and sufficient duration.

1.3.7.2   *Origin of the hour*

The hour, as a unit of time measurement, originated in the fourteenth century. Until the arrival of clockwork, in the fourteenth century AD, an hour was a variable concept. It is a practical division of the day into 12 segments (12 being the most convenient number for dividing into fractions, since it is divisible by 2, 3, and 4). For the same reason 60, divisible by 2, 3, 4, and 5, has been a larger framework of measurement ever since the Babylonian times. The traditional concept of the hour, as one-twelfth of the time between dawn and dusk, was useful in terms of everyday timekeeping. Approximate appointments could be made easily, at times that could be easily sensed. Noon is always the sixth hour. Half way through the afternoon is the ninth hour. This is famous as the time of the death of Jesus on the Cross. The trouble with the traditional hour is that it differs in length from day to day. In addition, a daytime hour is different from one in the night (also divided into 12 equal hours). A clock cannot reflect this variation, but it can offer something more useful. It can provide every day something, which occurs naturally only twice a year, at the spring and autumn equinox, when the 12 h of the day and the 12 h of the night are of the same lengths. In the fourteenth century, coinciding with the first practical clocks, the meaning of an hour gradually changed. It became a specific amount of time, one twenty-fourth of a full solar cycle from dawn to dawn. Today, the day is recognized as 24 h, although it still features on clock faces as two twelves.

1.3.7.3 Minutes and seconds: Fourteenth to sixteenth century

Minutes and seconds, as we know them today, originated in the fourteenth to the sixteenth centuries. Even the first clocks could measure periods less than an hour, but soon striking the quarter-hours seemed insufficient. With the arrival of dials for the faces of clocks, in the fourteenth century, something like a minute was required. The Middle Ages inherited a scale of scientific measurement based on 60 from Babylon. In Medieval Latin, the unit of one-sixtieth is *pars minuta prima* ("first very small part"), and a sixtieth of that is *pars minute secunda* ("second very small part"). Thus, based on a principle that is 3000 years old, minutes and seconds find their way into our modern time. Minutes were mentioned from the fourteenth century onward, but clocks were not precise enough for "seconds" of time to be needed until two centuries later.

1.3.8 Hero's *Dioptra*

Hero's *Dioptra* was written in the first century AD. One of the surviving books of Hero of Alexandria, titled *On the Dioptra*, describes a sophisticated technique, which he had developed for surveying land. Plotting the relative position of features in a landscape, essential for any accurate map, is a more complex task than simply measuring distances. It is necessary to discover accurate angles in both the horizontal and vertical planes. To make this possible, a surveying instrument must somehow maintain both planes consistently in different places, to take readings of the deviation in each plane between one location and another. This is what Hero achieved with the instrument mentioned in his title, the *dioptra*, which approximately means, the "spyhole," through which the surveyor looks when pinpointing the target in order to read the angles. For his device, Hero adapted an instrument long used by Greek astronomers (e.g., Hipparchus) for measuring the angle of stars in the sky. In his days, Hero achieved his device without the convenience of two modern inventions, the compass and the telescope.

1.3.9 Barometer and atmospheric pressure

Barometer and atmospheric pressure originated between 1643 and 1646. Like many significant discoveries, the principle of the barometer was observed by accident. Evangelista Torricelli, assistant to Galileo at the end of his life, was interested in knowing why it is more difficult to pump water from a well in which the water lies far below ground level. He suspected that the reason might be the weight of the extra column of air above the water, and he devised a way of testing this theory. He filled a glass tube with mercury; submerging it in a bath of mercury and raising the sealed end to a vertical position, he found that the mercury slipped a little way down the tube. He reasoned that the weight of air on the mercury in the bath was supporting the weight of the column of mercury in the tube. If this was true, then the space in the glass tube above the mercury column must be a vacuum. This rushed him into controversy with traditional scientists of the day, who believed nature abhorred a vacuum. However, it also encouraged von Guericke, in the next decade, to develop the vacuum pump. The concept of variable atmospheric pressure occurred to Torricelli when he noticed, in 1643, that the height of his column of mercury sometimes varied slightly from its normal level, which was 760 mm above the mercury level in the bath. Observation suggested that these variations related closely to changes in the weather. This was the origin of the barometer. With the concept thus establishing that the air had weight, Torricelli was able to predict that there must be less atmospheric pressure at higher altitudes. In 1646, Blaise Pascal, aided by his brother-in-law, carried a barometer to different levels of the 4000-feet mountain Puy de Dôme, near Clermont, to take readings. The confirmation was that atmospheric pressure varied with altitude.

1.3.10   Mercury thermometer

The mercury thermometer originated ca. between 1714 and 1742. Gabriel Daniel Fahrenheit, a German glass blower and instrument maker working in Holland, was interested in improving the design of thermometer that had been in use for half a century. Known as the Florentine thermometer, because it was developed in the 1650s in Florence's Accademia del Cimento, this pioneering instrument depended on the expansion and contraction of alcohol within a glass tube. Alcohol expands rapidly with a rise in temperature but not at an entirely regular speed of expansion. This made accurate readings difficult, as also did the sheer technical problem of blowing glass tubes with very narrow and entirely consistent bores. By 1714, Fahrenheit had made great progress on the technical front, creating two separate alcohol thermometers, which agreed precisely in their reading of temperature. In that year, he heard about the research of a French physicist, Guillaume Amontons, into the thermal properties of mercury. Mercury expands less than alcohol (about 7 times less for the same rise in temperature), but it does so in a more regular manner. Fahrenheit saw the advantage of this regularity, and he had the glass-making skills to accommodate the smaller rate of expansion. He constructed the first mercury thermometer that, subsequently, became a standard. There remained the problem of how to calibrate the thermometer to show degrees of temperature. The only practical method was to choose two temperatures, which could be established independently, mark them on the thermometer, and divide the intervening length of tube into a number of equal degrees. In 1701, Sir Isaac Newton had proposed the freezing point of water for the bottom of the scale and the temperature of the human body for the top end. Fahrenheit, accustomed to Holland's cold winters, wanted to include temperatures below the freezing point of water. He, therefore, accepted blood temperature for the top of his scale but adopted the freezing point of salt water for the lower extreme.

Measurement is conventionally done in multiples of 2, 3, and 4, therefore, Fahrenheit split his scale into 12 sections, each of them divided into 8 equal parts. This gave him a total of 96°, zero being the freezing point of brine and 96° as an inaccurate estimate of the average temperature of the human blood. Actual human body temperature is 98.6°. With his thermometer calibrated on these two points, Fahrenheit could take a reading for the freezing point (32°) and boiling point (212°) of water. In 1742, a Swede, Anders Celsius, proposed an early example of decimalization. His centigrade scale took the freezing and boiling temperatures of water as 0° and 100°, respectively. In English-speaking countries, this less complicated system took more than two centuries to be embraced. Yet even today, the Fahrenheit unit of temperature is more prevalent in some countries, such as the United States.

1.3.11   The chronometer

The chronometer was developed ca. 1714–1766. Two centuries of ocean travel made it important for ships on naval or merchant business to be able to calculate their positions accurately in any of the oceans in the world. With the help of the simple and ancient astrolabe, the stars would reveal latitude. However, on a revolving planet, longitude is harder. It was essential to know what time it was before it could be determined what place it was. The importance of this was made evident in 1714 when the British government set up a Board of Longitude and offered a massive prize of £20,000, at that time, to any inventor who could produce a clock capable of keeping accurate time at sea. The terms were demanding. To win the prize, a chronometer must be sufficiently accurate to calculate longitude within thirty nautical miles at the end of a journey to the West Indies. This meant that in rough seas, damp salty conditions, and sudden changes of temperature, the instrument must lose or gain not more than three seconds a day. This was a level of accuracy unmatched at the time by the best clocks. The challenge appealed to John Harrison, who was at the time a 21-year-old Lincolnshire carpenter with an interest in clocks. It was nearly 60 years before he could win the money. Luckily, he lived long enough to collect it. By 1735, Harrison had built the first chronometer, which he believed approached the necessary standard. Over the next quarter-century, he replaced it with three improved models before formally undergoing the government's test. His innovations included bearings that reduce friction, weighted balances interconnected by coiled springs to minimize the effects of movement, and the use of two metals in the balance spring to cope with expansion and contraction caused by changes of temperature. Harrison's first "sea clock," in 1735, weighed 72 lb and was 3 ft in all dimensions. His fourth, in 1759, was more like a watch, being circular and 5 in. in diameter. It was this version that underwent the sea trials. Harrison was at that time 67 years old. Therefore, his son, took the chronometer on its test journey to Jamaica in 1761. It was 5 s slow at the end of the voyage. The government argued that this might be a fluke and offered Harrison only £2500. After further trials and the successful building of a Harrison chronometer by another craftsman (at the huge cost of £450), the inventor was finally paid the full prize money in 1773.

Harrison had proved in 1761 what was possible, but his chronometer was an elaborate and expensive way of achieving the purpose. It was in France, where a large prize was also on offer from the Académie des Sciences, that the practical chronometer of the future was developed. The French trial, open to all comers, took place in 1766 on a voyage from Le Havre in a specially commissioned yacht, the *Aurore*. The only chronometer ready for the test was designed by Pierre Le Roy. At the end of 46 days, his machine was accurate to within 8 s. Le Roy's timepiece was larger than Harrison's final model, but it was much easier to construct. It provided the pattern of the future. With further modifications from various sources over the next two decades, the marine chronometer emerged before the end of the eighteenth century. Using it in combination with the sextant, explorers traveling the world's oceans could then bring back accurate information of immense value to the makers of maps and charts of the world.

1.3.12   Sextant

The sextant originated between 1731 and 1757. The eighteenth-century search for a way of discovering longitude was accompanied by refinements in the ancient method of establishing latitude. This had been possible since the second century BC by means of the astrolabe. From the beginning of the European voyages in the fifteenth century, practical improvements had been made to the astrolabe, mainly by providing more convenient cali-brated arcs on which the user could read the number of degrees of the sun or a star above the horizon. The size of these arcs was defined in relation to the full circle. A quadrant (a quarter of the circle) showed 90°, a sextant 60°, and an octant 45°. The use of such arcs in conjunction with the traditional astrolabe is evident from a text dating back to 1555 that reported about voyaging to the West Indies. The author wrote on "quadrant and astro-labe, instruments of astronomy." The important development during the eighteenth cen-tury was the application of optical devices (mirrors and lenses) to the task of working out angles above the horizon. Slightly differing solutions, by instrument makers in Europe and America, competed during the early decades of the century. The one that prevailed, mainly because it was more convenient at sea, was designed as an octant in 1731 by John Hadley, an established English maker of reflecting telescopes. Hadley's instrument, like others designed by his contemporaries, used mirrors to bring any two points into align-ment in the observer's sight line. For the navigator at that time, these two points would usually be the sun and the horizon. To read the angle of the sun, the observer looked through the octant's eyepiece at the horizon and then turned an adjusting knob until the reflected sphere of the sun (through a darkened glass) was brought down to the same level. The double reflection meant that the actual angle of the sun above the horizon was twice that on the octant's arc of 45°. Therefore, Hadley's instrument could read angles up to 90°. In 1734, Hadley added an improvement, which became the standard, by installing an additional level so that the horizontal could be found even if the horizon was not visible. In 1757, after Hadley's death, a naval captain proposed that the arc in the instrument can be extended from 45° to 60°, making a reading up to 120° possible. With this, Hadley's octant became a sextant, and the instrument has been in general use since then.

1.3.13   Ancient measurement systems in Africa

Africa is home to the world's earliest known use of measuring and calculation, confirming the continent as the origin of both basic and advanced mathematics. Thousands of years ago, while parallel developments were going on in Europe, Africans were using rudimen-tary numerals, algebra, and geometry in daily life. This knowledge spread throughout the entire world after a series of migrations out of Africa, beginning around 30,000 BC and later following a series of invasions of Africa by Europeans and Asians (1700 BC-present). It is historically documented that early man migrated out of Africa to Europe and Asia. This feat of early travel and navigation could have been facilitated by the indigenous measure-ment systems of ancient Africa. The following sections recount measuring and counting in ancient Africa.

1.3.13.1   Lebombo bone (35,000 BC)

The oldest known mathematical instrument is the Lebombo bone, a baboon fibula used as a measuring device and so named after its location of discovery in the Lebombo moun-tains of Swaziland. The device is at least 35,000 years old. Judging from its 29 distinct markings, it could have been either used to track lunar cycles or used as a measuring stick. It is rather interesting to note the significance of the 29 markings (roughly the same num-ber as lunar cycle, i.e., 29.531 days) on the baboon fibula because it is the oldest indication that the baboon, a primate indigenous to Africa, was symbolically linked to Khonsu, who was also associated with time. The Kemetic god, Djehuty ("Tehuti" or "Toth"), was later depicted as a baboon or an ibis, which is a large tropical wading bird with a long neck and long legs. This animal symbolism is usually associated with the moon, math, writing, and science. The use of baboon bones as measuring devices had continued throughout Africa, suggesting that Africans always held the baboon as sacred and associated it with the moon, math, and time.

1.3.13.2 Ishango bone (20,000 BC)

The world's oldest evidence of advanced mathematics was also a baboon fibula that was discovered in the present-day Democratic Republic of Congo and dates to at least 20,000 BC. The bone is now housed in the Museum of Natural Sciences in Brussels. The Ishango bone is not merely a measuring device or tally stick as some people erroneously suggest. The bone's inscriptions are clearly separated into clusters of markings that represent various quantities. When the markings are counted, they are all odd numbers with the left column containing all prime numbers between 10 and 20 and the right column containing added and subtracted numbers. When both columns are calculated, they add up to 60 (nearly double the length of the lunar cycle). We recall that the number 60 also featured prominently in the development of early measuring devices in Europe.

1.3.13.3 Gebet'a or "Mancala" game (700 BC-present)

Although the oldest known evidence of the ancient counting board game, Gebet'a or "Mancala" as it is more popularly known, comes from Yeha (700 BC) in Ethiopia, it was probably used in Central Africa many years prior to that. The game forced players to strategically capture a greater number of stones than one's opponent. The game usually consists of a wooden board with two rows of six holes each and two larger holes at either end. However, in antiquity, the holes were more likely to be carved into stone, clay, or mud. More advanced versions found in Central and East Africa, such as the Omweso, Igisoro, and Bao, usually involve four rows of eight holes each. A variant of this counting game still exists today in the Yoruba culture of Nigeria. It is called "Ayo" game, which tests the counting and tracking ability of players. A photograph of a modern Yoruba Ayo game board from coeditor Adedeji Badiru's household is shown in Figure 1.1. Notice the row of six holes on each player's side, with a master counting hole above the row.

Figure 1.1 Nigerian Yoruba Ayo game board (for counting). (Photo courtesy of Adedeji Badiru family, 2014.)

This example and other similar artifacts demonstrated the handed-down legacy of ancient counting in Africa.

1.3.14 *"Moscow" papyrus (2000 BC)*

Housed in Moscow's Pushkin State Museum of Fine Arts, what is known as "Moscow" papyrus was purchased by Vladimir Golenishchev sometime in the 1890s. Written in hieratic from perhaps the 13th dynasty in Kemet, the name of ancient Egypt, the papyrus is one of the world's oldest examples of the use of geometry and algebra. The document contains approximately 25 mathematical problems, including how to calculate the length of a ship's rudder, the surface area of a basket, the volume of a frustum (a truncated pyramid), and various ways of solving for unknowns.

1.3.15 *"Rhind" mathematical papyrus (1650 BC)*

Purchased by Alexander Rhind in 1858 AD, the so-called "Rhind" Mathematical Papyrus dates to approximately 1650 BC and is presently housed in the British Museum. Although some Egyptologists link this to the foreign Hyksos, this text was found during excavations at the Ramesseum in Waset (Thebes) in Southern Egypt, which never came under Hyksos' rule. The first page contains 20 arithmetic problems, including addition and multiplication of fractions, and 20 algebraic problems, including linear equations. The second page shows how to calculate the volume of rectangular and cylindrical granaries, with pi (Π) estimated at 3.1605. There are also calculations for the area of triangles (slopes of a pyramid) and an octagon. The third page continues with 24 problems, including the multiplication of algebraic fractions, among others.

1.3.16 *Timbuktu mathematical manuscripts (1200s AD)*

Timbuktu in Mali is well known as a hub of commerce in ancient times. Timbuktu is home to one of the world's oldest universities, Sankore, which had libraries full of manuscripts mainly written in Ajami (African languages, such as Hausa in this case, written in a script similar to "Arabic") in the 1200s AD. When Europeans and Western Asians began visiting and colonizing Mali from the 1300s to 1800s AD, Malians began to hide the manuscripts. Many of the scripts were mathematical and astronomical in nature. In recent years, as many as 700,000 scripts have been rediscovered and attest to the continuous knowledge of advanced mathematics, science, and measurements in Africa, well before European colonization.

1.3.17 *Fundamental scientific equations*

This section presents some of the seminal and fundamental theoretical scientific equations that have emerged over the centuries. Perhaps the most quoted and recognized in the modern scientific literature is Einstein's equation.

Einstein's equation

$$E = mc^2 \tag{1.1}$$

The fundamental relationship connecting energy, mass, and the speed of light emerges from Einstein's theory of special relativity, published in 1905. Showing the equivalence of mass and energy, it may be the most famous and beautiful equation in all of modern science. Its power was graphically demonstrated less than four decades later with the discovery of nuclear fission, a process in which a small amount of mass is converted to a very large amount of energy, precisely in accord with this equation.

Einstein's field equation

$$R_{\mu\nu} - \frac{1}{2}g_{\mu\nu}R + \Lambda g_{\mu\nu} = 8\pi GT_{\mu\nu} \tag{1.2}$$

Einstein's elegant equation published in 1916 is the foundation of his theory of gravity, the theory of general relativity. The equation relates the geometrical curvature of space–time with the energy density of matter. The theory constructs an entirely new picture of space and time, out of which gravity emerges in the form of geometry and from which Newton's theory of gravity emerges as a limiting case. Einstein's field equation explains many features of modern cosmology, including the expansion of the universe and the bending of star light by matter, and it predicts black holes and gravitational waves. He introduced a cosmological constant in the equation, which he called his greatest blunder, but that quantity may be needed if, as recent observations suggest, the expansion of the universe is accelerating. A remaining challenge for physicists in the twenty-first century is to produce a fundamental theory uniting gravitation and quantum mechanics.

Heisenberg's uncertainty principle

$$\Delta x \Delta p \geq \frac{h}{2} \tag{1.3}$$

In 1927, Werner Heisenberg's matrix formulation of quantum mechanics led him to discover that an irreducible uncertainty exists when measuring the position and momentum of an object simultaneously. Unlike classical mechanics, quantum mechanics requires that the more accurately the position of an object is known, the less accurately its momentum is known, and vice versa. The magnitude of that irreducible uncertainty is proportional to Planck's constant.

Schrödinger equation

$$i\hbar\frac{\partial\Psi}{\partial t} = H\Psi \tag{1.4}$$

In 1926, Erwin Schrödinger derived his nonrelativistic wave equation for the quantum mechanical motion of particles such as electrons in atoms. The probability density of finding a particle at a particular position in space is the square of the absolute value of the complex wave function, which is calculated from Schrödinger's equation. This equation accurately predicts the allowed energy levels for the electron in the hydrogen atom. With the use of modern computers, generalizations of this equation predict the properties of larger molecules and the behavior of electrons in complex materials.

Dirac equation

$$i\hbar\frac{\partial\Psi}{\partial t} = \left[c\vec{\alpha}\cdot(\vec{p} - \vec{A}) + \beta mc^2 + e\Phi\right]\Psi \tag{1.5}$$

In 1928, Paul Dirac derived a relativistic generalization of Schrödinger's wave equation for the quantum mechanical motion of a charged particle in an electromagnetic field.

His marvelous equation predicts the magnetic moment of the electron and the existence of antimatter.

Maxwell's equations

$$\vec{\nabla} \cdot \vec{D} = p \tag{1.6}$$

$$\vec{\nabla} \times \vec{H} = \vec{J} + \frac{\partial \vec{D}}{\partial t} \tag{1.7}$$

$$\vec{\nabla} \times \vec{E} + \frac{\partial \vec{B}}{\partial t} = 0 \tag{1.8}$$

$$\vec{\nabla} \cdot \vec{B} = 0 \tag{1.9}$$

The fundamental equations explaining classical electromagnetism were developed over many years by James Clerk Maxwell and were completed in his famous treatise published in 1873. His classical field theory provides an elegant framework for understanding electricity, magnetism, and propagation of light. Maxwell's theory was a major achievement of nineteenth-century physics, and it contained one of the clues that was used years later by Einstein to develop special relativity. Classical field theory was also the springboard for the development of quantum filed theory.

Boltzmann's equation for entropy

$$S = k \ln W \tag{1.10}$$

Ludwig Boltzmann, one of the founders of statistical mechanics in the late nineteenth century, proposed that the probability for any physical state of macroscopic system is proportional to the number of ways in which the internal state of that system can be rearranged without changing the system's external properties. When more arrangements are possible, the system is more disordered. Boltzmann showed that the logarithm of the multiplicity of states of a system, or its disorder, is proportional to its entropy, and the constant of proportionality is Boltzmann's constant k. The second law of thermodynamics states that the total entropy of a system and its surroundings always increase as time elapses. Boltzmann's equation for entropy is carved on his grave.

Planck–Einstein equation

$$E = hv \tag{1.11}$$

The simple relation between the energy of a light quantum and the frequency of the associated light wave first emerged in a formula discovered in 1900 by Max Planck. He was examining the intensity of electromagnetic radiation emitted by the atoms in the walls of an enclosed cavity (a blackbody) at fixed temperature. He found that he could fit the experimental data by assuming that the energy associated with each mode of the electromagnetic field is an integral multiple of some minimum energy that is proportional to the frequency. The constant of proportionality, h, is known as Planck's constant. It is one of the most important fundamental numbers in physics. In 1905, Albert Einstein recognized that Planck's equation implies that light is absorbed or emitted in discrete quanta, explaining the photoelectric effect and igniting the quantum mechanical revolution.

Planck's blackbody radiation formula

$$u = \frac{8\pi h}{c^3} v^3 \left[e^{\frac{hv}{kT}} - 1 \right]^{-1} \tag{1.12}$$

In studying the energy density of radiation in a cavity, Max Planck compared two approximate formulas, one for low frequency and another for high frequency. In 1900, using an ingenious extrapolation, he found his equation for the energy density of black-body radiation, which reproduced experimental results. Seeking to understand the significance of his formula, he discovered the relation between energy and frequency known as Planck–Einstein equation.

Hawking equation for black hole temperature

$$T_{BH} = \frac{hc^3}{8\pi GMk} \tag{1.13}$$

Using insights from thermodynamics, relativist quantum mechanics, and Einstein's gravitational theory, Stephen Hawking predicted in 1974 the surprising result that gravitational black holes, which are predicted by general relativity, would radiate energy. His formula for the temperature of the radiating black hole depends on the gravitational constant, Planck's constant, the speed of light, and Boltzmann's constant. While Hawking radiation remains to be observed, his formula provides a tempting glimpse of the insights that will be uncovered in a unified theory combining quantum mechanics and gravity.

Navier–Stokes equation for a fluid

$$\rho \frac{\partial \vec{v}}{\partial t} + \rho(\vec{v} \cdot \vec{\nabla})\vec{v} = -\vec{\nabla}p + \mu \nabla^2 \vec{v} + (\lambda + \mu)\vec{\nabla}(\vec{\nabla} \cdot \vec{v}) + \rho \vec{g} \tag{1.14}$$

The Navier–Stokes equation was derived in the nineteenth century from Newtonian mechanics to model viscous fluid flow. Its nonlinear properties make it extremely difficult to solve, even with modern analytic and computational technique. However, its solutions describe a rich variety of phenomena including turbulence.

Lagrangian for quantum chromodynamics

$$L_{QDC} = -\frac{1}{4} F_a^{\mu v} \cdot F_{a\mu v} + \sum_f \overline{\Psi}_f [i\slashed{\nabla} - g\slashed{A}_a t_a - m_f] \Psi_f \tag{1.15}$$

Relativistic quantum field theory had its first great success with quantum electro-dynamics, which explains the interaction of charged particles with the quantized electromagnetic field. Exploration of non-Abelian gauge theories led next to the spectacular unification of the electromagnetic and weak interactions. Then, with insights developed from the quark model, quantum chromodynamics was developed to explain the strong interactions. This theory predicts that quarks are bound more tightly together as their separation increases, which explains why individual quarks are not seen directly in experiments. The standard model, which incorporates strong, weak, and electromagnetic interactions into a single quantum field theory, describes the interaction of quarks, gluons, and leptons and has achieved remarkable success in predicting experimental results in elementary particle physics.

Bardeen–Cooper–Schrieffer equation for superconductivity

$$T_c = 1.13\Theta e^{-\frac{1}{N(0)V}} \tag{1.16}$$

Superconductors are materials that exhibit no electrical resistance at low temperatures. In 1957, John Bardeen, Leon N. Cooper, and J. Robert Schrieffer applied quantum field theory with an approximate effective potential to explain this unique behavior of electrons in a superconductor. The electrons were paired and move collectively without resistance in the crystal lattice of the superconducting material. The BCS theory and its later generalizations predict a wide variety of phenomena that agree with experimental observations and have many practical applications. John Bardeen's contributions to solid-state physics also include inventing the transistor, made from semiconductors, with Walter Brattain and William Shockley in 1947.

Josephson effect

$$\frac{d(\Delta\varphi)}{dt} = \frac{2eV}{h} \tag{1.17}$$

In 1962, Brian Josephson made the remarkable prediction that electric current could flow between two thin pieces of superconducting material separated by a thin piece of insulating material without application of a voltage. Using the BCS theory of superconductivity, he also predicted that if a voltage difference were maintained across the junction, there would be an alternating current with a frequency related to the voltage and Planck's constant. The presence of magnetic fields influences the Josephson effect, allowing it to be used to measure very weak magnetic fields approaching the microscopic limit set by quantum mechanics.

Fermat's last theorem

$$x^n + y^n = z^n \tag{1.18}$$

While studying the properties of whole numbers, or integers, the French mathematician Pierre de Fermat wrote in 1637 that it is impossible for the cube of an integer to be written as the sum of the cubes of two other integers. More generally, he stated that it is impossible to find such a relation between three integers for any integral power greater than two. He went on to write a tantalizing statement in the margin of his copy of a Latin translation of Diophantus's *Arithemetica*: "I have a truly marvelous demonstration of this proposition, which this margin is too narrow to contain." It took over 350 years to prove Fermat's simple conjecture. The feat was achieved by Andrew Wiles in 1994 with a "tour de force" proof of many pages using newly developed techniques in number theory. It is noteworthy that many researchers, mathematicians, and scholars toiled for almost four centuries before a credible proof of Fermat's last theorem was found. Indeed, the lead editor of this handbook, as a mathematics graduate student in the early 1980s, was introduced to the problem during his advanced calculus studies under Professor Reginald Mazeres at Tennessee Technological University in 1980. Like many naïve researchers before him, he struggled with the problem as a potential thesis topic for 6 months before abandoning it to pursue a more doable topic in predictive time series modeling.

1.4   Fundamental methods of measurement

There are two basic methods of measurement:

1. *Direct comparison* with either a primary or a secondary standard.
2. *Indirect comparison* with a standard with the use of a calibrated system.

1.4.1   Direct comparison

How do you measure the length of a cold-rolled bar? You probably use a steel tape. You compare the bar's length with a standard. The bar is so many feet long because that many units on your standard have the same length as the bar. You have determined this by making a direct comparison. Although you do not have access to the primary standard defining the unit, you manage very well with a secondary standard. Primary measurement standards have the least amount of uncertainty compared to the certified value and are traceable directly to the SI. Secondary standards, on the other hand, are derived by assigning value by comparison to a primary standard.

In some respect, measurement by direct comparison is quite common. Many length measurements are made in this way. In addition, time of the day is usually determined by comparison, with a watch used as a secondary standard. The watch goes through its double-cycle, in synchronization with the earth's rotation. Although, in this case, the primary standard is available to everyone, the watch is more convenient because it works on cloudy days, indoors, outdoors, in daylight, and in the dark (at night). It is also more precise. That is, its resolution is better. In addition, if well regulated, the watch is more accurate because the earth does not rotate at a uniform speed. It is seen, therefore, that in some cases, a secondary standard is actually more useful than the primary standard.

Measuring by direct comparison implies stripping the measurement problem to its bare essentials. However, the method is not always the most accurate or the best. The human senses are not equipped to make direct comparisons of all quantities with equal facility. In many cases, they are not sensitive enough. We can make direct length comparisons using a steel rule with a level of precision of about 0.01 in. Often we wish for a greater accuracy, in which case we must call for additional assistance from some calibrated measuring system.

1.4.2   Indirect comparison

While we can do a reasonable job through direct comparison of length, how well can we compare masses, for example? Our senses enable us to make rough comparisons. We can lift a pound of meat and compare its effect with that of some unknown mass. If the unknown is about the same weight, we may be able to say that it is slightly heavier, or perhaps, not quite as heavy as our "standard" pound, but we could never be certain that the two masses were the same, even say within one ounce. Our ability to make this comparison is not as good as it is for the displacement of the mass. Our effectiveness in coming close to the standard is related to our ability to "gage" the relative impacts of mass on our ability to displace the mass. This brings to mind the common riddle, "Which weighs more? A pound of feathers or a pound of stones?" Of course, both weigh the same with respect to the standard weight of "pound."

In making most engineering measurements, we require the assistance of some form of the measuring system, and measurement by direct comparison is less general than measurement by indirect comparison.

1.5   Generalized mechanical measuring system

Most mechanical measurement systems (Beckwith and Buck, 1965) fall within the framework of a generalized arrangement consisting of three stages, as follows:

Stage I: A detector–transducer stage
Stage II: An intermediate modifying stage
Stage III: The terminating stage, consisting of one or a combination of an indicator, a recorder, or some form of the controller.

Each stage is made up of a distinct component or grouping of components, which perform required and definite steps in the measurement. These may be termed basic elements, whose scope is determined by their functioning rather than their construction. First stage detector–transducer: The prime function of the first stage is to detect or to sense the input signal. This primary device must be sensitive to the input quantity. At the same time, ideally it should be insensitive to every other possible input. For instance, if it is a pressure pickup, it should not be sensitive to, say, acceleration; if it is a strain gauge, it should be insensitive to temperature; or if a linear accelerometer, it should be insensitive to angular acceleration, and so on. Unfortunately, it is very rare indeed to find a detecting device that is completely selective. As an example of a simple detector–transducer device, consider an automobile tire pressure gauge. It consists of a cylinder and a piston, a spring resisting the piston movement, and a stem with scale divisions. As the air pressure bears against the piston, the resulting force compresses the spring until the spring and air forces are balanced. The calibrated stem, which remains in place after the spring returns the piston, indicates the applied pressure. In this case, the piston–cylinder combination along with the spring makes up the detector–transducer. The piston and cylinder form one basic element, while the spring is another basic element. The piston–cylinder combination, serving as a force-summing device, senses the pressure effect, and the spring transduces it into the displacement. Realistically, not all measurements we encounter in theory and practice are of transduceable mechanical settings. Measurements, thus, can take more generic paths of actualization. Figure 1.2 shows a generic measurement loop revolving around variable

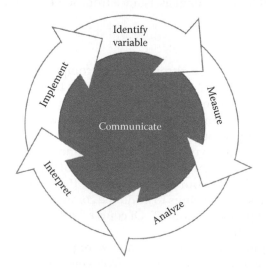

Figure 1.2 Generic measurement loop.

identification, actual measurement, analyzing the measurement result, interpreting the measuring in the context of the prevailing practical application, and implementing the measurement for actionable decisions. In each stage of the loop, communication is a central requirement. Communication can be in the form of a pictorial display, a verbal announcement, or a written dissemination. A measurement is not usable unless it is communicated in an appropriate form and at the appropriate time.

1.6   Data types and measurement scales

Every decision requires data collection, measurement, and analysis. In practice, we encounter different types of measurement scales depending on the particular items of interest. Data may need to be collected on decision factors, costs, performance levels, outputs, and so on. The different types of data measurement scales that are applicable are presented below.

1.6.1   Nominal scale of measurement

The nominal scale is the lowest level of measurement scales. It classifies items into categories. The categories are mutually exclusive and collectively exhaustive. That is, the categories do not overlap, and they cover all possible categories of the characteristics being observed. For example, in the analysis of the critical path in a project network, each job is classified as either critical or not critical. Gender, type of industry, job classification, and color are examples of measurements on a nominal scale.

1.6.2   Ordinal scale of measurement

An ordinal scale is distinguished from a nominal scale by the property of order among the categories. An example is the process of prioritizing project tasks for resource allocation. We know that first is above second, but we do not know how far above. Similarly, we know that better is preferred to good, but we do not know by how much. In quality control, the ABC classification of items based on the Pareto distribution is an example of a measurement on an ordinal scale.

1.6.3   Interval scale of measurement

An interval scale is distinguished from an ordinal scale by having equal intervals between the units of measurement. The assignment of priority ratings to project objectives on a scale of 0–10 is an example of a measurement on an interval scale. Even though an objective may have a priority rating of zero, it does not mean that the objective has absolutely no significance to the project team. Similarly, the scoring of zero on an examination does not imply that a student knows absolutely nothing about the materials covered by the examination. Temperature is a good example of an item that is measured on an interval scale. Even though there is a zero point on the temperature scale, it is an arbitrary relative measure. Other examples of interval scale are IQ measurements and aptitude ratings.

1.6.4   Ratio scale measurement

A ratio scale has the same properties of an interval scale, but with a true zero point. For example, an estimate of zero time units for the duration of a task is a ratio scale measurement. Other examples of items measured on a ratio scale are cost, time, volume, length,

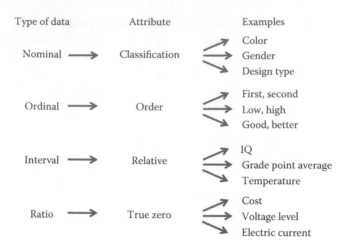

Figure 1.3 Four primary types of data.

height, weight, and inventory level. Many of the items measured in engineering systems will be on a ratio scale.

Another important aspect of measurement involves the classification scheme used. Most systems will have both quantitative and qualitative data. Quantitative data require that we describe the characteristics of the items being studied numerically. Qualitative data, on the other hand, are associated with attributes that are not measured numerically. Most items measured on the nominal and ordinal scales will normally be classified into the qualitative data category while those measured on the interval and ratio scales will normally be classified into the quantitative data category. The implication for engineering system control is that qualitative data can lead to bias in the control mechanism because qualitative data are subjected to the personal views and interpretations of the person using the data. As much as possible, data for an engineering systems control should be based on a quantitative measurement. Figure 1.3 illustrates the four different types of data classification. Notice that the temperature is included in the "relative" category rather the "true zero" category. Even though there are zero temperature points on the common temperature scales (i.e., Fahrenheit, Celsius, and Kelvin), those points are experimentally or theoretically established. They are not true points as one might find in a counting system.

1.7   Common units of measurements

Some common units of measurement include the following:

Acre: An area of 43,560 ft^2.
Agate: 1/14 in. (used in printing for measuring column length).
Ampere: Unit of electric current.
Astronomical (A.U.): 93,000,000 miles; the average distance of the earth from the sun (used in astronomy).
Bale: A large bundle of goods. In the United States, approximate weight of a bale of cotton is 500 lbs. The weight of a bale may vary from country to country.
Board foot: 144 in.^{3} (12 × 12 × 1 used for lumber).
Bolt: 40 yards (used for measuring cloth).

Btu: British thermal unit; the amount of heat needed to increase the temperature of 1 lb of water by 1°F (252 cal).

Carat: 200 mg or 3086 troy; used for weighing precious stones (originally the weight of a seed of the carob tree in the Mediterranean region). *See also Karat.*

Chain: 66 ft; used in surveying (1 mile = 80 chains).

Cubit: 18 in. (derived from the distance between elbow and tip of the middle finger).

Decibel: Unit of relative loudness.

Freight Ton: 40 ft^3 of merchandise (used for cargo freight).

Gross: 12 dozen (144).

Hertz: Unit of measurement of electromagnetic wave frequencies (measures cycles per second).

Hogshead: Two liquid barrels or 14,653 in.^{3}

Horsepower: The power needed to lift 33,000 lbs a distance of 1 ft in 1 min (about 1 1/2 times the power an average horse can exert); used for measuring the power of mechanical engines.

Karat: A measure of the purity of gold. It indicates how many parts out of 24 are pure. 18-karat gold is 3/4 pure gold.

Knot: Rate of the speed of 1 nautical mile/h; used for measuring the speed of ships (not distance).

League: Approximately 3 miles.

Light-year: 5,880,000,000,000 miles; distance traveled by light in 1 year at the rate of 186,281.7 miles/s; used for measurement of interstellar space.

Magnum: Two-quart bottle; used for measuring wine.

Ohm: Unit of electrical resistance.

Parsec: Approximately 3.26 light-years of 19.2 trillion miles; used for measuring interstellar distances.

Pi (π): 3.14159265+; the ratio of the circumference of a circle to its diameter.

Pica: 1/6 in. or 12 points; used in printing for measuring column width.

Pipe: Two hogsheads; used for measuring wine and other liquids.

Point: 0.013837 (~1/72 in. or 1/12 pica); used in printing for measuring type size.

Quintal: 100,000 g or 220.46 lbs avoirdupois.

Quire: 24 or 25 sheets; used for measuring paper (20 quires is one ream).

Ream: 480 or 500 sheets; used for measuring paper.

Roentgen: Dosage unit of radiation exposure produced by x-rays.

Score: 20 units.

Span: 9 in. or 22.86 cm; derived from the distance between the end of the thumb and the end of the little finger when both are outstretched.

Square: 100 ft^2; used in building.

Stone: 14 lbs avoirdupois in Great Britain.

Therm: 100,000 Btus.

Township: U.S. land measurement of almost 36 square miles; used in surveying.

Tun: 252 gallons (sometimes larger); used for measuring wine and other liquids.

Watt: Unit of power.

1.7.1   Common constants

Speed of light: 2.997,925 × 10^{10} cm/s (983.6 × 10^6 ft/s; 186,284 miles/s)

Velocity of sound: 340.3 m/s (1116 ft/s)

Gravity (acceleration): 9.80665 m/s^2 (32.174 ft/s^2; 386.089 in./s^2)

1.7.2   Measurement numbers and exponents

Exponentiation is essential in measurements both small and large numbers. The standard exponentiation numbers and prefixes are presented as follows:

yotta (10^{24}):	1 000 000 000 000 000 000 000 000
zetta (10^{21}):	1 000 000 000 000 000 000 000
exa (10^{18}):	1 000 000 000 000 000 000
peta (10^{15}):	1 000 000 000 000 000
tera (10^{12}):	1 000 000 000 000
giga (10^{9}):	1 000 000 000
mega (10^{6}):	1 000 000
kilo (10^{3}):	1 000
hecto (10^{2}):	100
deca (10^{1}):	10
deci (10^{-1}):	0.1
centi (10^{-2}):	0.01
milli (10^{-3}):	0.001
micro (10^{-6}):	0.000 001
nano (10^{-9}):	0.000 000 001
pico (10^{-12}):	0.000 000 000 001
femto (10^{-15}):	0.000 000 000 000 001
atto (10^{-18}):	0.000 000 000 000 000 001
zepto (10^{-21}):	0.000 000 000 000 000 000 001
yocto (10^{-24}):	0.000 000 000 000 000 000 000 001

1.8   Patterns of numbers in measurements

Numbers are the basis for any measurement. They have many inherent properties that are fascinating and these should be leveraged in measurement systems. Some interesting number patterns relevant to measurement systems are shown as follows:

$1 \times 8 + 1 = 9$
$12 \times 8 + 2 = 98$
$123 \times 8 + 3 = 987$
$1234 \times 8 + 4 = 9876$
$12,345 \times 8 + 5 = 98,765$
$123,456 \times 8 + 6 = 987,654$
$1,234,567 \times 8 + 7 = 9,876,543$
$12,345,678 \times 8 + 8 = 98,765,432$
$123,456,789 \times 8 + 9 = 987,654,321$

$1 \times 9 + 2 = 11$
$12 \times 9 + 3 = 111$
$123 \times 9 + 4 = 1111$
$1234 \times 9 + 5 = 11,111$
$12,345 \times 9 + 6 = 111,111$
$123,456 \times 9 + 7 = 1,111,111$
$1,234,567 \times 9 + 8 = 11,111,111$
$12,345,678 \times 9 + 9 = 111,111,111$
$123,456,789 \times 9 + 10 = 1,111,111,111$

$9 \times 9 + 7 = 88$
$98 \times 9 + 6 = 888$
$987 \times 9 + 5 = 8888$
$9876 \times 9 + 4 = 88,888$
$98,765 \times 9 + 3 = 888,888$
$987,654 \times 9 + 2 = 8,888,888$
$9,876,543 \times 9 + 1 = 88,888,888$
$98,765,432 \times 9 + 0 = 888,888,888$

$1 \times 1 = 1$
$11 \times 11 = 121$
$111 \times 111 = 12,321$
$1111 \times 1111 = 1,234,321$
$11,111 \times 11,111 = 123,454,321$
$111,111 \times 111111 = 12,345,654,321$
$1,111,111 \times 1,111,111 = 1,234,567,654,321$
$11,111,111 \times 11,111,111 = 123,456,787,654,321$
$111,111,111 \times 111,111,111 = 12,345,678,987,654,321$
$111,111,111 \times 111,111,111 = 12,345,678,987,654,321$

$$1 \times 8 + 1 = 9$$
$$12 \times 8 + 2 = 98$$
$$123 \times 8 + 3 = 987$$
$$1234 \times 8 + 4 = 9876$$
$$12,345 \times 8 + 5 = 98,765$$
$$123,456 \times 8 + 6 = 987,654$$
$$1,234,567 \times 8 + 7 = 9,876,543$$
$$12,345,678 \times 8 + 8 = 98,765,432$$
$$123,456,789 \times 8 + 9 = 987,654,321$$

$$1 \times 9 + 2 = 11$$
$$12 \times 9 + 3 = 111$$
$$123 \times 9 + 4 = 1111$$
$$1234 \times 9 + 5 = 11,111$$
$$12,345 \times 9 + 6 = 111,111$$
$$123,456 \times 9 + 7 = 1,111,111$$
$$1,234,567 \times 9 + 8 = 11,111,111$$
$$12,345,678 \times 9 + 9 = 111,111,111$$
$$123,456,789 \times 9 + 10 = 1,111,111,111$$

$$9 \times 9 + 7 = 88$$
$$98 \times 9 + 6 = 888$$
$$987 \times 9 + 5 = 8888$$
$$9876 \times 9 + 4 = 88,888$$
$$98,765 \times 9 + 3 = 888,888$$
$$987,654 \times 9 + 2 = 8,888,888$$
$$9,876,543 \times 9 + 1 = 88,888,888$$
$$98,765,432 \times 9 + 0 = 888,888,888$$

$$1 \times 1 = 1$$
$$11 \times 11 = 121$$
$$111 \times 111 = 12,321$$
$$1111 \times 1111 = 1,234,321$$
$$11,111 \times 11111 = 123,454,321$$
$$111,111 \times 111,111 = 12,345,654,321$$
$$1,111,111 \times 1,111,111 = 1,234,567,654,321$$
$$11,111,111 \times 11,111,111 = 123,456,787,654,321$$
$$111,111,111 \times 111,111,111 = 12,345,678,987,654,321$$

1.9   Statistics in measurement

Statistical data management is essential for measurement with respect to analyzing and interpreting measurement outputs. In this section, a project control scenario is used to illustrate data management for measurement of project performance.

Transient data is defined as a volatile set of data that is used for one-time decision making and is not needed again. As an example, the number of operators that show up at a job site on a given day; unless there is some correlation between the day-to-day attendance records of operators, this piece of information will be relevant only for that given day. The project manager can make his decision for that day based on that day's attendance record. Transient data need not be stored in a permanent database unless it may be needed for future analysis or uses (e.g., forecasting, incentive programs, and performance review).

Recurring data refers to the data that is encountered frequently enough to necessitate storage on a permanent basis. An example is a file containing contract due dates. This file will need to be kept at least through the project life cycle. Recurring data may be further categorized into *static data* and *dynamic data*. A recurring data that is static will retain its original parameters and values each time it is retrieved and used. A recurring data that is dynamic has the potential for taking on different parameters and values each time it is retrieved and used. Storage and retrieval considerations for project control should address the following questions:

1. What is the origin of the data?
2. How long will the data be maintained?
3. Who needs access to the data?
4. What will the data be used for?
5. How often will the data be needed?
6. Is the data for reference purposes only (i.e., no printouts)?
7. Is the data for reporting purposes (i.e., generate reports)?
8. In what format is the data needed?
9. How fast will the data need to be retrieved?
10. What security measures are needed for the data?

1.10   Data determination and collection

It is essential to determine what data to collect for project control purposes. Data collection and analysis are basic components of generating information for project control. The requirements for data collection are discussed next.

Choosing the data. This involves selecting data based on their relevance, the level of likelihood that they will be needed for future decisions, and whether or not they contribute to making the decision better. The intended users of the data should also be identified.

Collecting the data. This identifies a suitable method of collecting the data as well as the source from which the data will be collected. The collection method depends on the particular operation being addressed. The common methods include manual tabulation, direct keyboard entry, optical character reader, magnetic coding, electronic scanner, and more recently, voice command. An input control may be used to confirm the accuracy of collected data. Examples of items to control when collecting data are the following:

Relevance check. This checks if the data are relevant to the prevailing problem. For example, data collected on personnel productivity may not be relevant for decision-involving marketing strategies.

Limit check. This checks to ensure that the data are within known or acceptable limits. For example, an employee overtime claim amounting to over 80 h/week for several weeks in a row is an indication of a record well beyond ordinary limits.

Critical value. This identifies a boundary point for data values. Values below or above a critical value fall in different data categories. For example, the lower specification limit for a given characteristic of a product is a critical value that determines whether the product meets quality requirements.

Coding the data. This refers to the technique used for representing data in a form useful for generating information. This should be done in a compact and yet meaningful format. The performance of information systems can be greatly improved if effective data formats and coding are designed into the system right from the beginning.

Processing the data. Data processing is the manipulation of data to generate useful information. Different types of information may be generated from a given data set depending on how it is processed. The processing method should consider how the information will be used, who will be using it, and what caliber of system response time is desired. If possible, processing controls should be used.

Control total. Check for the completeness of the processing by comparing accumulated results to a known total. An example of this is the comparison of machine throughput to a standard production level or the comparison of cumulative project budget depletion to a cost accounting standard.

Consistency check. Check if the processing is producing the same results for similar data. For example, an electronic inspection device that suddenly shows a measurement that is 10 times higher than the norm warrants an investigation of both the input and the processing mechanisms.

Scales of measurement. For numeric scales, specify units of measurement, increments, the zero point on the measurement scale, and the range of values.

Using the information. Using information involves people. Computers can collect data, manipulate data, and generate information, but the ultimate decision rests with people, and decision making starts when information becomes available. Intuition, experience, training, interest, and ethics are just a few of the factors that determine how people use information. The same piece of information that is positively used to further the progress of a project in one instance may also be used negatively in another instance. To assure that data and information are used appropriately, computer-based security measures can be built into the information system. Project data may be obtained from several sources. Some potential sources are

- Formal reports
- Interviews and surveys

- Regular project meetings
- Personnel time cards or work schedules

The timing of data is also very important for project control purposes. The contents, level of detail, and frequency of data can affect the control process. An important aspect of project management is the determination of the data required to generate the information needed for project control. The function of keeping track of the vast quantity of rapidly changing and interrelated data about project attributes can be very complicated. The major steps involved in data analysis for project control are

- Data collection
- Data analysis and presentation
- Decision making
- Implementation of action

Data is processed to generate information. Information is analyzed by the decision maker to make the required decisions. Good decisions are based on timely and relevant information, which in turn is based on reliable data. Data analysis for project control may involve the following functions:

- Organizing and printing computer-generated information in a form usable by managers
- Integrating different hardware and software systems to communicate in the same project environment
- Incorporating new technologies such as expert systems into data analysis
- Using graphics and other presentation techniques to convey project information

Proper data management will prevent misuse, misinterpretation, or mishandling. Data is needed at every stage in the lifecycle of a project from the problem identification stage through the project phase-out stage. The various items for which data may be needed are project specifications, feasibility study, resource availability, staff size, schedule, project status, performance data, and phase-out plan. The documentation of data requirements should cover the following

- *Data summary*. A data summary is a general summary of the information and decisions for which the data is required as well as the form in which the data should be prepared. The summary indicates the impact of the data requirements on the organizational goals.
- *Data-processing environment*. The processing environment identifies the project for which the data is required, the user personnel, and the computer system to be used in processing the data. It refers to the project request or authorization and relationship to other projects and specifies the expected data communication needs and mode of transmission.
- *Data policies and procedures*. Data handling policies and procedures describe policies governing data handling, storage, and modification and the specific procedures for implementing changes to the data. Additionally, they provide instructions for data collection and organization.
- *Static data*. A static data description describes that portion of the data that is used mainly for reference purposes and it is rarely updated.

- *Dynamic data.* A dynamic data description defines the portion of the data that is frequently updated based on the prevailing circumstances in the organization.
- *Data frequency.* The frequency of data update specifies the expected frequency of data change for the dynamic portion of the data, for example, quarterly. This data change frequency should be described in relation to the frequency of processing.
- *Data constraints.* Data constraints refer to the limitations on the data requirements. Constraints may be procedural (e.g., based on corporate policy), technical (e.g., based on computer limitations), or imposed (e.g., based on project goals).
- *Data compatibility.* Data compatibility analysis involves ensuring that data collected for project control needs will be compatible with future needs.
- *Data contingency.* A data contingency plan concerns data security measures in case of accidental or deliberate damage or sabotage affecting hardware, software, or personnel.

1.11   Data analysis and presentation

Data analysis refers to the various mathematical and graphical operations that can be performed on data to elicit the inherent information contained in the data. The manner in which project data is analyzed and presented can affect how the information is perceived by the decision maker. The examples presented in this section illustrate how basic data analysis techniques can be used to convey important information for project control.

In many cases, data are represented as the answer to direct questions such as, when is the project deadline? Who are the people assigned to the first task? How many resource units are available? Are enough funds available for the project? What are the quarterly expenditures on the project for the past two years? Is personnel productivity low, average, or high? Who is the person in charge of the project? Answers to these types of questions constitute data of different forms or expressed on different scales. The resulting data may be qualitative or quantitative. Different techniques are available for analyzing the different types of data. This section discusses some of the basic techniques for data analysis. The data presented in Table 1.1 is used to illustrate the data analysis techniques.

1.11.1   Raw data

Raw data consists of ordinary observations recorded for a decision variable or factor. Examples of factors for which data may be collected for decision making are revenue, cost, personnel productivity, task duration, project completion time, product quality, and resource availability. Raw data should be organized into a format suitable for visual review and computational analysis. The data in Table 1.1 represents the quarterly revenues from projects A, B, C, and D. For example, the data for quarter 1 indicates that project C yielded

Table 1.1 Quarterly revenue from four projects (in $1000s)

Project	Quarter 1	Quarter 2	Quarter 3	Quarter 4	Row total
A	3000	3200	3400	2800	12,400
B	1200	1900	2500	2400	8000
C	4500	3400	4600	4200	16,700
D	2000	2500	3200	2600	10,300
Total	10,700	11,000	13,700	12,000	47,400

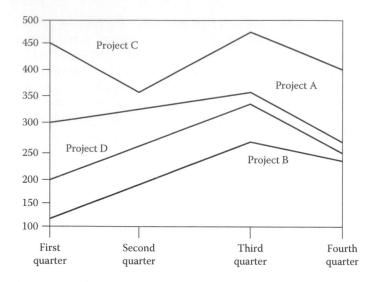

Figure 1.4 Line graph of quarterly project revenues.

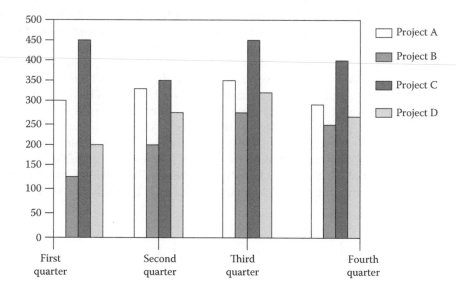

Figure 1.5 Multiple bar chart of quarterly project revenues.

the highest revenue of $4,500,000 while project B yielded the lowest revenue of $1,200,000. Figure 1.4 presents the raw data of project revenue as a line graph. The same information is presented as a multiple bar chart in Figure 1.5.

1.11.2   *Total revenue*

A total or sum is a measure that indicates the overall effect of a particular variable. If X_1, X_2, X_3,..., X_n represent a set of n observations (e.g., revenues), then the total is computed as

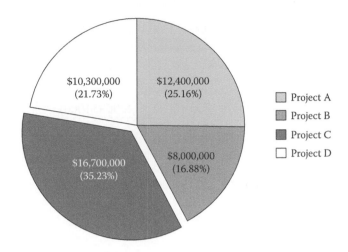

Figure 1.6 Pie chart of total revenue per project.

$$T = \sum_{i=1}^{n} X_i \qquad (1.19)$$

For the data in Table 1.1, the total revenue for each project is shown in the last column. The totals indicate that project C brought in the largest total revenue over the four quarters under consideration while project B produced the lowest total revenue. The last row of the table shows the total revenue for each quarter. The totals reveal that the largest revenue occurred in the third quarter. The first quarter brought in the lowest total revenue. The grand total revenue for the four projects over the four quarters is shown as $47,400,000 in the last cell in the table. The total revenues for the four projects over the four quarters are shown in a pie chart in Figure 1.6. The percentage of the overall revenue contributed by each project is also shown on the pie chart.

1.11.3 Average revenue

Average is one of the most used measures in data analysis. Given n observations (e.g., revenues), $X_1, X_2, X_3, \ldots, X_n$, the average of the observations is computed as

$$\bar{X} = \frac{\sum_{i=1}^{n} X_i}{n} = \frac{T_x}{n} \qquad (1.20)$$

where T_x is the sum of n revenues. For our sample data, the average quarterly revenues for the four projects are

$$\bar{X}_A = \frac{(3000 + 3200 + 3400 + 2800)(\$1000)}{4}$$

$$= \$3,100,000$$

$$\bar{X}_B = \frac{(1200 + 1900 + 2500 + 2400)(\$1000)}{4}$$

$$= \$2,000,000$$

$$\bar{X}_C = \frac{(4500 + 3400 + 4600 + 4200)(\$1000)}{4}$$

$$= \$4,175,000$$

$$\bar{X}_D = \frac{(2000 + 2500 + 3200 + 2600)(\$1000)}{4}$$

$$= 2,575,000$$

Similarly, the expected average revenues per project for the four quarters are

$$\bar{X}_1 = \frac{(3000 + 1200 + 4500 + 2000)(\$1000)}{4}$$

$$= \$2,675,000$$

$$\bar{X}_2 = \frac{(3200 + 1900 + 3400 + 2500)(\$1000)}{4}$$

$$= \$2,750,000$$

$$\bar{X}_3 = \frac{(3400 + 2500 + 4600 + 3200)(\$1000)}{4}$$

$$= \$3,425,000$$

$$\bar{X}_4 = \frac{(2800 + 2400 + 4200 + 2600)(\$1000)}{4}$$

$$= \$3,000,000$$

The above values are shown in a bar chart in Figure 1.7. The average revenue from any of the four projects in any given quarter is calculated as the sum of all the observations divided by the number of observations. That is,

$$\bar{\bar{X}} = \frac{\sum_{i=1}^{N}\sum_{j=1}^{M} X_{ij}}{K} \tag{1.21}$$

where
   N = number of projects
   M = number of quarters
   K = total number of observations ($K = NM$)

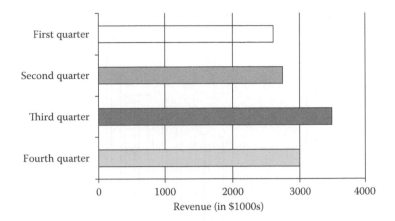

Figure 1.7 Average revenue per project for each quarter.

The overall average per project per quarter is

$$\bar{\bar{X}} = \frac{\$47,400,000}{16}$$
$$= \$2,962,500$$

As a cross check, the sum of the quarterly averages should be equal to the sum of the project revenue averages, which is equal to the grand total divided by 4

$$(2675 + 2750 + 3425 + 3000)(\$1000) = (3100 + 2000 + 4175 + 2575)(\$1000)$$
$$= \$11,800,000$$
$$= \$47,400,000/4$$

The cross check procedure above works because we have a balanced table of observations. That is, we have four projects and four quarters. If there were only three projects, for example, the sum of the quarterly averages would not be equal to the sum of the project averages.

1.11.4 *Median revenue*

The median is the value that falls in the middle of a group of observations arranged in order of magnitude. One-half of the observations are above the median, and the other half are below the median. The method of determining the median depends on whether or not the observations are organized into a frequency distribution. For unorganized data, it is necessary to arrange the data in an increasing or decreasing order before finding the median. Given K observations (e.g., revenues), X_1, X_2, X_3, ..., X_K, arranged in increasing or decreasing order, the median is identified as the value in position $(K + 1)/2$ in the data arrangement if K is an odd number. If K is an even number, then the average of the two middle values is considered to be the median. If the sample data were arranged in increasing order, we would get the following

1200, 1900, 2000, 2400, 2500, 2500, 2600, 2800, 3000, 3200, 3200, 3400, 3400, 4200, 4500, and 4600.

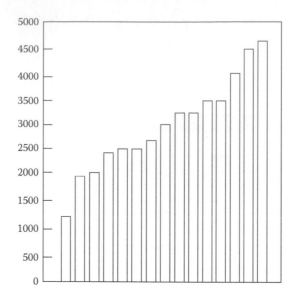

Figure 1.8 Ordered bar chart.

The median is then calculated as (2800 + 3000)/2 = 2900. Half of the recorded revenues are expected to be above $2,900,000 while half are expected to be below that amount.
Figure 1.8 presents a bar chart of the revenue data arranged in an increasing order. The
median is anywhere between the eighth and ninth values in the ordered data.

1.11.5   Quartiles and percentiles

The median is a position measure because its value is based on its position in a set of observations. Other measures of position are *quartiles* and *percentiles*. There are three quartiles,
which divide a set of data into four equal categories. The first quartile, denoted Q_1, is the
value below which one-fourth of all the observations in the data set fall. The second quartile, denoted, Q_2, is the value below which two-fourths or one-half of all the observations
in the data set fall. The third quartile, denoted Q_3, is the value below which three-fourths
of the observations fall. The second quartile is identical to the median. It is technically
incorrect to talk of the fourth quartile because that will imply that there is a point within
the data set below which all the data points fall: a contradiction! A data point cannot lie
within the range of the observations and at the same time exceed all the observations,
including itself.

The concept of percentiles is similar to the concept of quartiles except that reference is
made to percentage points. There are 99 percentiles that divide a set of observations into
100 equal parts. The X percentile is the value below which X percent of the data fall. The
99 percentile refers to the point below which 99% of the observations fall. The three quartiles discussed previously are regarded as the 25th, 50th, and 75th percentiles. It would be
technically incorrect to talk of the 100th percentile. In performance ratings, such as on an
examination or product quality level, the higher the percentile of an individual or product,
the better. In many cases, recorded data are classified into categories that are not indexed
by numerical measures. In such cases, other measures of central tendency or position will
be needed. An example of such a measure is the mode.

1.11.6   The mode

The mode is defined as the value that has the highest frequency in a set of observations. When the recorded observations can be classified only into categories, the mode can be particularly helpful in describing the data. Given a set of K observations (e.g., revenues), X_1, X_2, X_3, ..., X_K, the mode is identified as that value that occurs more than any other value in the set. Sometimes, the mode is not unique in a set of observations. For example, in Table 1.2, $2500, $3200, and $3400 all have the same number of occurrences. Each of them is a mode of the set of revenue observations. If there is a unique mode in a set of observations, then the data is said to be unimodal. The mode is very useful in expressing the central tendency for observations with qualitative characteristics such as color, marital status, or state of origin.

1.11.7   Range of revenue

The range is determined by the two extreme values in a set of observations. Given K observations (e.g., revenues), X_1, X_2, X_3,..., X_K, the range of the observations is simply the difference between the lowest and the highest observations. This measure is useful when the analyst wants to know the extent of extreme variations in a parameter. The range of the revenues in our sample data is ($4,600,000 – $1,200,000) = $3,400,000. Because of its dependence on only two values, the range tends to increase as the sample size increases. Furthermore, it does not provide a measurement of the variability of the observations relative to the center of the distribution. This is why the standard deviation is normally used as a more reliable measure of dispersion than the range.

The variability of a distribution is generally expressed in terms of the deviation of each observed value from the sample average. If the deviations are small, the set of data is said to have low variability. The deviations provide information about the degree of dispersion in a set of observations. A general formula to evaluate the variability of data cannot be based on the deviations. This is because some of the deviations are negative while some are positive, and the sum of all the deviations is equal to zero. One possible solution to this is to compute the average deviation.

1.11.8   Average deviation

The average deviation is the average of the absolute values of the deviations from the sample average. Given K observations (e.g., revenues), X_1, X_2, X_3,..., X_K, the average deviation of the data is computed as

$$\bar{D} = \frac{\sum_{i=1}^{K} |X_i - \bar{X}|}{K} \tag{1.22}$$

Table 1.2 shows how the average deviation is computed for our sample data. One aspect of the average deviation measure is that the procedure ignores the sign associated with each deviation. Despite this disadvantage, its simplicity and ease of computation make it useful. In addition, knowledge of the average deviation helps in understanding the standard deviation, which is the most important measure of dispersion available.

Table 1.2 Average deviation, standard deviation, and variance

Observation number (i)	Recorded observation X_i	Deviation from average $X_i - \bar{X}$	Absolute value $\lvert X_i - \bar{X} \rvert$	Square of deviation $(X_i - \bar{X})^2$
1	3000	37.5	37.5	1406.25
2	1200	−1762.5	1762.5	3,106,406.30
3	4500	1537.5	1537.5	2,363,906.30
4	2000	−962.5	962.5	926,406.25
5	3200	237.5	237.5	56,406.25
6	1900	−1062.5	1062.5	1,128,906.30
7	3400	437.5	437.5	191,406.25
8	2500	−462.5	462.5	213,906.25
9	3400	437.5	437.5	191,406.25
10	2500	−462.5	462.5	213,906.25
11	4600	1637.5	1637.5	2,681,406.30
12	3200	237.5	237.5	56,406.25
13	2800	−162.5	162.5	26,406.25
14	2400	−562.5	562.5	316,406.25
15	4200	1237.5	1237.5	1,531,406.30
16	2600	−362.5	362.5	131,406.25
Total	47,400.0	0.0	11,600.0	13,137,500.25
Average	2962.5	0.0	725.0	821,093.77
Square root	–	–	–	906.14

1.11.9   *Sample variance*

The sample variance is the average of the squared deviations computed from a set of observations. If the variance of a set of observations is large, the data is said to have a large variability. For example, a large variability in the levels of productivity of a project team may indicate a lack of consistency or improper methods in the project functions. Given K observations (e.g., revenues), $X_1, X_2, X_3,..., X_K$, the sample variance of the data is computed as

$$s^2 = \frac{\sum_{i=1}^{K}(X_i - \bar{X})^2}{K - 1} \tag{1.23}$$

The variance can also be computed by the following alternate formulas:

$$s^2 = \frac{\sum_{i=1}^{K}(X_i^2 - (1/K)) \left[\sum_{i=1}^{K} X_i \right]^2}{K - 1}$$

$$s^2 = \frac{\sum_{i=1}^{K} X_i^2 - K(\bar{X}^2)}{K - 1}$$

Using the first formula, the sample variance of the data in Table 1.2 is calculated as

$$s^2 = \frac{13,137,500.25}{16-1}$$
$$= 875,833.33$$

The average calculated in the last column of Table 1.1 is obtained by dividing the total for that column by 16 instead of $16 - 1 = 15$. That average is not the correct value of the sample variance. However, as the number of observations gets very large, the average, as computed in the table, will become a close estimate for the correct sample variance. Analysts make a distinction between the two values by referring to the number calculated in the table as the population variance when K is very large, and referring to the number calculated by the formulas above as the sample variance, particularly, when K is small. For our example, the population variance is given by

$$\sigma^2 = \frac{\sum_{i=1}^{K}(X_i - \bar{X})^2}{K}$$
$$= \frac{13,137,500.25}{16}$$
$$= 821,093.77 \tag{1.24}$$

while the sample variance, as shown previously for the same data set, is given by

$$\sigma^2 = \frac{\sum_{i=1}^{K}(X_i - \bar{X})^2}{K-1}$$
$$= \frac{13,137,500.25}{(16-1)} = 875,833.33 \tag{1.25}$$

1.11.10   Standard deviation

The sample standard deviation of a set of observations is the positive square root of the sample variance. The use of variance as a measure of variability has some drawbacks. For example, the knowledge of the variance is helpful only when two or more sets of observations are compared. Because of the squaring operation, the variance is expressed in square units rather than the original units of the raw data. To get a reliable feel for the variability in the data, it is necessary to restore the original units by performing the square root operation on the variance. This is why the standard deviation is a widely recognized measure of variability. Given K observations (e.g., revenues), $X_1, X_2, X_3,..., X_K$, the sample standard deviation of the data is computed as

$$s = \sqrt{\frac{\sum_{i=1}^{K}(X_i - \bar{X})^2}{K-1}} \tag{1.26}$$

As in the case of the sample variance, the sample standard deviation can also be computed by the following alternate formulas:

$$s = \sqrt{\frac{\sum_{i=1}^{K} X_i^2 - \left(\frac{1}{K}\right)\left[\sum_{i=1}^{K} X_i\right]^2}{K-1}} \tag{1.27}$$

$$s = \sqrt{\frac{\sum_{i=1}^{K} X_i^2 - K(\bar{X})^2}{K-1}} \tag{1.28}$$

Using the first formula, the sample standard deviation of the data is calculated as

$$s = \sqrt{\frac{13,137,500.25}{(16-1)}} = \sqrt{875,833.33} = 935.8597$$

We can say that the variability in the expected revenue per project per quarter is $935,859.70. The population sample standard deviation is given by the following

$$\sigma = \sqrt{\frac{\sum_{i=1}^{K} (X_i - \bar{X})^2}{K}}$$

$$= \sqrt{\frac{13,137,500.25}{16}}$$

$$= \sqrt{821,093.77}$$

$$= 906.1423 \tag{1.29}$$

The sample standard deviation is given by the following expression:

$$s = \sqrt{\frac{\sum_{i=1}^{K} (X_i - \bar{X})^2}{K-1}}$$

$$= \sqrt{\frac{13,137,500.25}{(16-1)}}$$

$$= 935.8597 \tag{1.30}$$

The results of data analysis can be reviewed directly to determine where and when project control actions may be needed. The results can also be used to generate control charts.

1.12 Conclusion

Human interfaces with measurement systems and tools date back to ancient times. Even in the modern times, we must recognize the past approaches to measurements, appreciate the present practices of measurements, and be prepared for the future developments

Figure 1.9 Ancient Korean measurement tool (Angbuilgu Sundial). (Photographed by A. B. Badiru in South Korea, May 31, 2006.)

of measurement. Figure 1.9 illustrates an ancient Korean Angbuilgu Sundial. Figure 1.10 shows a time-correction table used with the Angbuilgu.

The Angbuilgu is a sundial that was used during the Josean Dynasty in Korea. The name "Angbuilgu" means "upward-looking kettle that catches the shadow of the sun." The thirteen horizontal lines mark the 24 periods of seasonal change from the winter solstice to the summer solstice and allow the season of the year to be determined. The vertical lines are time lines. The device is aligned to face the North Star, and it is marked with pictures of animals, rather than letters, for the sake of the common people, who could not

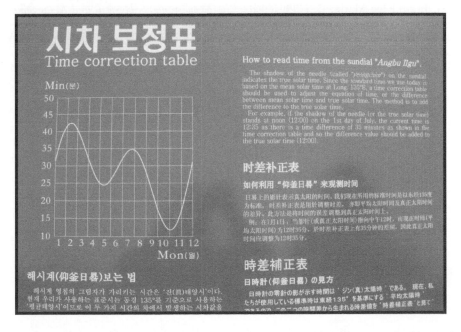

Figure 1.10 Time measurement correction table for Angbuilgu Sundial. (Photographed by A. B. Badiru in South Korea, May 31, 2006.)

read. The sundial tool was first made in 1434, but the later Angbuilgu was created in the latter half of the seventeenth century.

Linking the past, present, and future of measurements, the presentations in this chapter demonstrate that it is a world of systems, where elements are linked through measurement systems. A system is a collection of interrelated elements that are linked through synergy such that the collective output of the system is higher than the sum of the individual outputs of the subsystems. This is, indeed, what happens in the systems and subsystems making up our world. Measurement scales and systems make the interfaces of these elements possible. The cover of this book illustrates this fact aptly. Chapter 11 in this handbook presents systems interoperability measurement techniques, which further illustrate how the elements of any large complex system inter-operate to accomplish a desired end goal.

References

Badiru, A. B., Oye Ibidapo-Obe, and B. J. Ayeni 2012. *Industrial Control Systems: Mathematical and Statistical Models and Techniques.* Taylor & Francis CRC Press, Boca Raton, FL.

Badiru, A. B. and Tina Kovach 2012. *Statistical Techniques for Project Control.* Taylor & Francis CRC Press, Boca Raton, FL.

Beckwith, T. G. and Buck, N. L. 1965. *Mechanical Measurements.* Addison-Wesley Publishing Company, Reading, MA.

HistoryWorld 2014. http://www.historyworld.net/wrldhis/PlainTextHistories.asp?historyid=ac07 (accessed December 24, 2014).

Morris, A. S. 1997. *Measurement and Calibration Requirements for Quality Assurance to ISO 9000.* John Wiley & Sons, Inc., New York, NY.

NIST 1974. NBS Special Publication; 304A (Stock Number 003-003-03501-7), U.S. Department of Commerce Technical Administration National Institute of Standards and Technology, Washington, DC, September 1974.

Shillito, M. L. and De Marle, D. J. 1992. *Value: Its Measurement, Design, and Management.* John Wiley & Sons, Inc., New York, NY.

TANF 2014. Restoring Africa's Lost Legacy. http://www.taneter.org/math.html, TA Neter Foundation (accessed December 23, 2014).

chapter two

Human factors measurement

Farman A. Moayed

Contents

2.1   Introduction

One of the primary goals of human factors analysis is to improve human performance in the work environment. Industrial engineers (IEs) can study human performance in the workplace by creating different models and analyzing the interaction between workers and their environment. In general, "Human-at-Work" is considered a complex system that

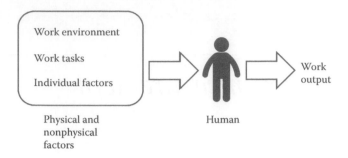

Figure 2.1 Block diagram of human factors analysis.

is affected by numerous inputs, and the human factors analysis is concerned with the synergistic effects of such inputs to optimize human performance at the workplace.

There are three types of input factors that can affect human performance in the workplace: (i) work environment, (ii) work tasks, and (iii) individual factors. Figure 2.1 demonstrates how the three types of input factors can determine the physical and nonphysical factors of working conditions affecting human performance, and as a result the work output. In this chapter, human performance is defined as the pattern of activities carried out to achieve an objective according to certain qualitative and/or quantitative work standards—in other words, performance is measured as the degree of accomplishing work standards that can be defined in the following different ways:

- Output quantity
- Output productivity
- Output quality
- Work safety

2.2   *Applications of human factors analysis*

Regardless of which method or model is selected for analysis, human factors analysis can be used and applied in numerous single or interdisciplinary fields such as industrial engineering, business, and medicine. The following is a list of human factors analysis applications in different fields:

1. *Design of work and improvement of work tasks/processes by IEs, business managers, supervisors, and teams*: Human factors analysis (also known as work system analysis) can serve as a diagnostic tool to point out the deficiencies in a work system (content and context of work). Consequently, the most appropriate and relevant design guidelines can be utilized to improve the work tasks and work environment. It can also be used as a monitoring and evaluating tool for work improvement efforts (Hackman and Oldham, 1980).
2. *Planning and scheduling work by managers, IEs, and supervisors:* Human factors analysis can provide detailed information regarding the difficulty, frequency, duration, and variability of work demands (including work content and context). This information can lead to a precise depiction of job attributes, which in turn, improves the accuracy of predicting project duration and efficiency in the coordination of work activities (McCormick, 1979).

3. *Documentation of work demands and loads by human resource specialists for personnel purposes such as selection and compensation*: Work demands and loads can provide a written record of any job's task demands and ability requirements. This information can be used as a screening tool for job placement by matching potential candidates to job requirements. The listing of job demands and loads can also assist in the compensation process by providing a list of criteria for job performance evaluation (Borman and Peterson, 1982; McCormick, 1982; Karwowski and Rodrick, 2001).

4. *Design of the most appropriate medical examinations by occupational physicians:* A database of work demands and loads can be created and used as a checklist of functional physical capacity for essential job tasks. Occupational physicians and nurses can evaluate an individual's physical capabilities for a specific job (as demonstrated in a pre-employment health screening) according to such a database. An inventory of work demands and loads can also be utilized to identify workers who participate in hazardous tasks and, thus, alert medical specialists to monitor such workers for adverse health effects (Fraser, 1992).

5. *Design of rehabilitation and return-to-work programs by health practitioners:* The human factors analysis provides information to classify which tasks are essential and which are nonessential regarding job execution. That means critical work demands and loads can be used to provide input to physical therapists as recommendations for therapy guidelines. Later, the critical demands and loads can serve as the performance criteria for determining whether the worker possesses the sufficient capability to return to work (Fraser, 1992).

6. *Decision making regarding the work restriction and provisions of reasonable accommodations by occupational physicians, human resource specialists, and managers:* In recent years, there has been considerable government attention regarding the needs of handicapped and disabled people in the active workforce. This has put considerable scrutiny over medical selection procedures for job candidates. A worker's limitations with respect to a given job can be evaluated using a detailed job description that characterizes the nature of the crucial tasks. Human factors analysis provides the means for collecting and classifying this information (Fraser, 1992).

A typical functional capacity assessment model is presented in Figure 2.2 (modeled after Fraser, 1992), which shows how human factors analysis or work systems analysis can integrate different sets of information to assess workers' capacity according to their work demands. In such model, a worker's physical and mental capacity are assessed against the information about the physical work environment and work content to determine if the worker is capable of performing the job effectively and safely. In case of any imbalance between the worker's capabilities and work demands, the output of the job will be negatively affected in terms of safety, quality, productivity, and quantity.

2.3 Elements of human-at-work model

Any human-at-work model is developed based on the fact that "people who work in organizations share a common point that they are each *somebody* doing *something, somewhere*." That means that there are three elements to human-at-work models:

1. The *activity* or *task*: It covers every physical, mental, or social aspect of any work including any required tools or equipment that a worker has to use to fulfill the activity.

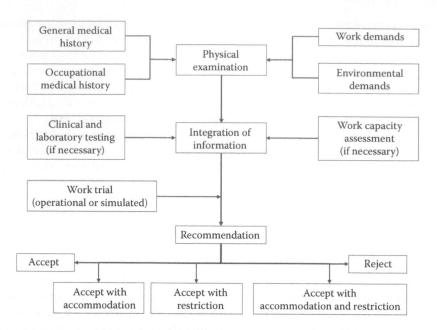

Figure 2.2 Model of functional capacity assessment.

2. The *context* in which an activity is performed: This includes the physical and socio-organizational work environment that affects workers' interactions with others and the environment.

3. The general state or condition of the *human*: This consists of an individual's abilities, needs, personality, etc., which affect the worker's performance and decision-making process.

In a human-at-work model (Figure 2.3, modeled after Rummler and Brache, 1995), a worker can have the optimum performance and output if the individual is physically and mentally capable (Factor 6), if he/she is well trained and has the necessary knowledge and skills (Factor 5), if the goals and expectations are clearly set and determined (Factor 1), if the individual has the appropriate task supports (Factor 2), if meaningful reinforcement mechanisms exist (Factor 3), and if appropriate and constructive feedback is provided to the worker (Factor 4).

Experts believe and studies have shown that the highest percentage of performance improvement opportunities can be found in the environment (Factors 1–4) in which individuals work. While the figure varies somewhat in different jobs, industries, and countries, overall 80% of performance improvement opportunities can be found in the work environment and management, and usually, 15%–20% of the opportunities are in the workers' skills and knowledge area. Less than 1% of performance problems result from individual capacity deficiencies. W. Edwards Deming maintained that only 15% of performance problems are worker problems, and 85% are management problems (Rummler and Brache, 1995).

Humans have to go through four basic operational functions involved in work activities. These functions are common to all jobs; they are: (i) sensing, (ii) information storage, (iii) information processing and decision making, and (iv) action. As demonstrated in Figure 2.4 (modeled after McCormick, 1979), in many typical situations, the individual

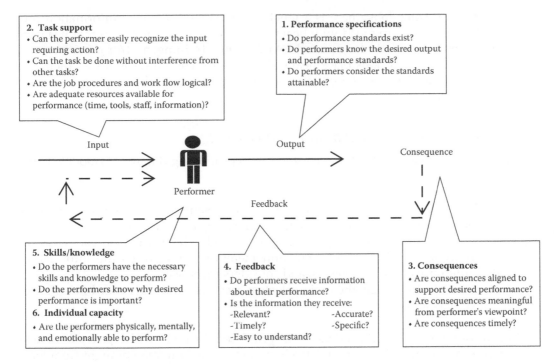

2. **Task support**
• Can the performer easily recognize the input requiring action?
• Can the task be done without interference from other tasks?
• Are the job procedures and work flow logical?
• Are adequate resources available for performance (time, tools, staff, information)?

1. **Performance specifications**
• Do performance standards exist?
• Do performers know the desired output and performance standards?
• Do performers consider the standards attainable?

Input   Output

Consequence

Performer

Feedback

5. **Skills/knowledge**
• Do the performers have the necessary skills and knowledge to perform?
• Do the performers know why desired performance is important?
6. **Individual capacity**
• Are the performers physically, mentally, and emotionally able to perform?

4. **Feedback**
• Do performers receive information about their performance?
• Is the information they receive:
-Relevant?   -Accurate?
-Timely?   -Specific?
-Easy to understand?

3. **Consequences**
• Are consequences aligned to support desired performance?
• Are consequences meaningful from performer's viewpoint?
• Are consequences timely?

Figure 2.3 Human-at-work model.

receives stimuli from the work environment that serve as information input to him/her. Such information interacts with the stored information (i.e., the memory) and usually triggers some form of information processing that leads to a decision followed by some type of action or response. The action may be a physical act or a communication act. Although each of these operational functions are involved in virtually every job, the degree and nature of that involvement vary widely from job to job.

In repetitive jobs, in which the worker performs a predetermined action, the load for information processing and decision between the information input and the action function is minimal since it resembles a person programmed to do the action. On the other hand, in some jobs, there may be a premium on the use of psychomotor skills in executing physical actions that must be done carefully (e.g., engraving work), in a certain sequence (e.g., keypunching), or with precise time control (e.g., playing piano).

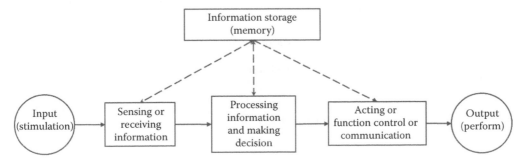

Figure 2.4 Four basic operational functions involved in all human work activities.

In addition, in some jobs, there is a much greater involvement of information processing and decision-making load with heavy dependence on information that has been stored. Such jobs may involve long periods of education, training, and/or previous experience. Meanwhile, certain jobs place heavy demands on communications as their output—written or oral. Jobs involving oral communications may also require associated interpersonal skills.

2.4   Model of human information processing

There are at least three stages of human information processing (Figure 2.5, modeled after Wickens et al., 1998).

1. *Perceptual stage*: Information is brought through the senses (sensation) and compared with knowledge from the memory to give it meaning (perception).
2. *Cognitive stage*: In this central processing or thought stage, a person compares the new information with current goals and memories, transforms the information, makes inferences, solves problems, and considers responses.
3. *Action stage*: In this stage, the brain selects a response and then coordinates/sends motor signals for action.

Any information receiving attention is processed further in the *perception* stage. Perception compares the input with relatively permanent information brought from long-term memory and adds meaning to the information. As a result, many stimuli are assigned to a single perceptual category. Once the meaning is added to the sensory information, a person either reacts to the perceptions with some type of response or sends information onto *working memory* for further processing. Working memory is a term for both the

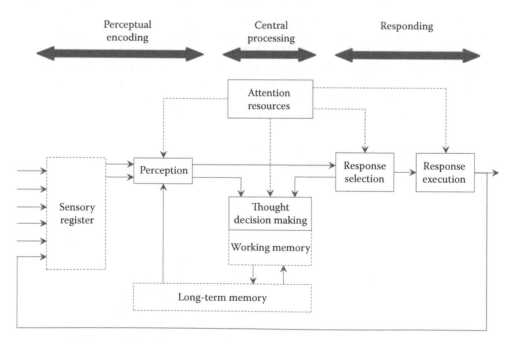

Figure 2.5 Generic model of human information processing.

short-term storage of whatever information is currently active in central processing and also a kind of workbench of consciousness in which a person compares, evaluates, and transforms cognitive representations. The information in working memory decays rapidly unless it is rehearsed (with repetition) to keep it there. This activity is maintained until response selection and execution processes occur; then, under certain circumstances, information is encoded into *long-term memory* for later use. Most cognitive processes require some allocation of *attentional resources* to proceed efficiently. The two exceptions are the sensory register and maintenance of material in the long-term memory (Wickens et al., 1998).

There are multiple models of human information processing. According to one such model, humans process information at two different levels: automatic level and conscious level. Human performance at the automatic level is considered a skilled performance and can only take place after the skill for that particular activity or task has been developed. Information processing at this level takes place without conscious control and in many cases without conscious awareness. It is fast, and more than one piece of information can be processed in parallel. Automatic processing is virtually unlimited (Bailey, 1996).

On the other hand, the conscious-level information processing is serial or sequential in nature. One piece of information can be processed at a time, requiring more time to complete a task compared to automatic processing. This type of information processing has limited capacity. Figure 2.6 (modeled after Bailey, 1996) is a block diagram of automatic and conscious levels of information processing.

Another model expands the two-level model of human information processing to three levels (Figure 2.7, modeled after Rasmussen, 1983), that is, cognitive control, computational processing, and emotional or effective processing. At the emotional information-processing level, the information reaching human senses is processed at a lower level (compared to cognitive control) over which a person has no control and of which he/she is hardly conscious. Only the output of emotional processing reaches our consciousness. It is not logical and does not follow formal rules, and the individual is unable to stop or start the processing of effective information (Rasmussen, 1983).

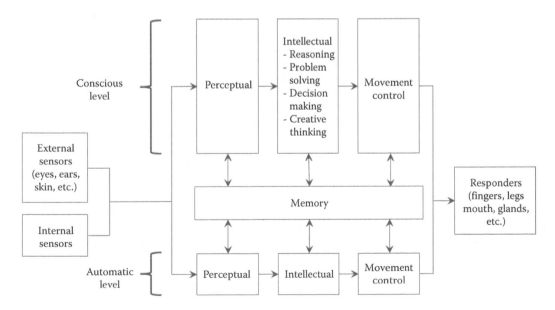

Figure 2.6 Two-level model of human information processing.

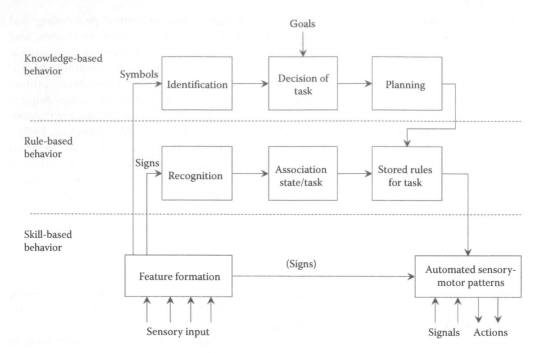

Figure 2.7 Three levels of performance of skilled human operators.

At the cognitive control level of information processing, which is a higher mechanism, performance is evaluated and parameters are provided as the basis for computational processing activities. The cognitive control evaluates whether a certain level of task performance is attained. On the basis of knowledge of results, the person may change his/her strategy with which the task is executed (Rasmussen, 1983).

Finally, at the computational information-processing level, the information is processed in two ways: (i) controlled processing (serial, slow, and requiring attention and therefore effort) and (ii) automatic processing (parallel, fast, and taking place at an unconscious level). The computational information-processing level can be adopted only after the person has received training for some time for a particular activity or task (Rasmussen, 1983).

2.5   *Evolution of human factors analysis efforts*

To understand how human factors analysis has developed over time, it is necessary to look at the history of industrial developments, which can be divided into multiple eras. Each era represents specific characteristics that played important roles in the development of certain types of analysis.

The era of general craftsmanship: Prior to the industrial revolution in England, the general craftsmanship approach served as the dominant manufacturing method for producing goods. This approach resulted in good quality of output. However, volume was low in comparison to the mass production concept in today's economy. The general craftsmanship method was typically utilized in small shops where each worker or a small group of workers was responsible for producing a whole product. The fundamental strength of this approach was that each craftsperson, being accountable for the product as a whole, was driven by a great deal of satisfaction from seeing the output of his/her work (Genaidy et al., 2002).

The era of specialized craftsmanship: As the time progressed, technology also advanced, and the world population increased. This created the need to produce large volumes and varied types of goods; hence, specialized craftsmanship grew from the roots of the industrial revolution in England. To meet these needs, the efficient utilization of resources was required to produce cost-effective products. As such, the general craftsmanship approach gradually evolved into the era of specialized craftsmanship (Genaidy et al., 2002).

Classical theory of work analysis and design: The era of specialized craftsmanship evolved into what is now known as the "Classical Theory of Work Analysis and Design." The theory centers around the division of labor, a concept put into practice for thousands of years—for example, jobs were based on specialization and grouped by function during the construction of the pyramids in ancient Egypt. An increase in work efficiency from job specialization was primarily achieved through time reductions for manufacturing and an increase in skill due to repetition, fewer tool changes, and decreased waste (Genaidy et al., 1999).

Principles of scientific management (PSM): Frederick Winslow Taylor's PSM popularized this concept, and job specialization spread through the American industry and eventually to other countries. The basis of PSM was to select the best worker for each particular task, and then, to train, teach, and develop the worker. Taylor cautioned that for the PSM to be successful, it is necessary to establish good relations between the management and workers with shared responsibilities (Genaidy et al., 1999).

The PSM formed the cornerstone of assembly line automobile manufacturing perfected by Henry Ford. Gradually, "Motion Study" or "Work Methods," "Time Study" or "Work Measurement," and "Motion and Time Study" or "Work Design" grew out of PSM to describe standards for work performance and guidelines for improving work methods (Genaidy et al., 1999).

"Time Study" determined the standard number of minutes that a qualified, properly trained, and experienced person should take to perform a task, and "Motion Study" advocated the study of body motions used in performing an operation to optimize the task by eliminating unnecessary motions and discovery of the most favorable motion sequence for maximum efficiency. Motion Study and Time Study merged to form what is now known as Motion and Time Study, the systematic study of work systems for the purposes of (i) developing the preferred method, usually the one with the lowest cost, (ii) standardizing such system and method, (iii) determining the time required by a qualified and properly trained person working at a normal pace to do a specific task or operation, and (iv) assisting an organization for training workers about the preferred method (Genaidy et al., 1999).

Deviation from the original PSM: The promise of improvement from the practical application of the PSM created a demand for efficiency experts, engineers, and consultants who encouraged narrow views of the workplace and treating employees as replaceable machine parts. As a result, increasingly, large numbers of workers objected to these practices and Taylor's caution to establish good relations and divide work into equal shares between the management and workers was lost during this period (Genaidy et al., 1999).

Additional factors played an important role in deviation from the original PSM such as regulation of working conditions through organized labor groups and government agencies (e.g., Occupational Safety and Health Administration [OSHA]). The advancements in technology also changed and are still changing working conditions (e.g., automation). Studies such as Hawthorne experiments (illumination experiments and relay experiments) made the investigators conclude that human relations and attention to worker needs are important factors for work improvement, confirming Taylor's earlier notion of this requirement for successful implementation of scientific management (Genaidy et al., 1999).

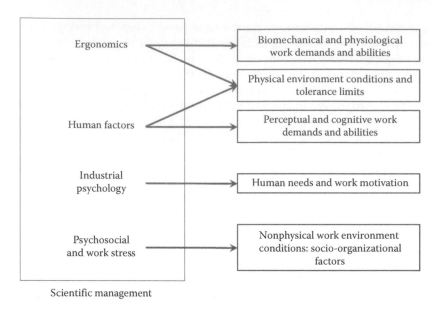

Figure 2.8 Evolution of human factors analysis theories and methods.

New directions: The recent approach to human factors analysis involves different aspects of human nature. As demonstrated in Figure 2.8, experts in human factors analysis tend to integrate different models for work systems analysis. In the early days of ergonomics development in England, an emphasis was placed on matching the physical task demands with the physical strength and endurance to minimize fatigue and risk of injuries and illnesses. Meanwhile, in the United States, human factors efforts were mainly concerned with the fitting of the mental task demands and the perceptual/cognitive capabilities to minimize the likelihood of human error leading to industrial accidents. Both efforts emphasize the study of physical environment conditions such as noise and vibration to ensure the physical environment conditions are within the tolerance limits of workers (Genaidy et al., 2002).

Industrial psychology models tend to put emphasis on optimizing employee motivation by satisfying the higher needs of humans, while in psychosocial and work stress models, emphasis was made on the reduction of exposure to harmful nonphysical work factors (i.e., mental, social, and organizational). The physical factors were also considered but not with equal importance. The goal was to balance the perceived work demands with perceived capabilities that could reduce work stress; human relations models have been used to maximize personal and team effectiveness (Genaidy et al., 1999).

2.6   *Evaluation of physical environment conditions*

The physical environment conditions are made of three major elements that are chemical, physical, and biological agents. The evaluation of such conditions (as described below) falls under the industrial hygiene field.

- Chemical agents: Toxic and nontoxic chemicals (e.g., dust and fumes)
- Physical agents: Harmful physical properties of the environment/materials (e.g., noise and radiation)
- Biological agents: Harmful microbial living organism (e.g., bacteria and viruses)

Most, if not all, physical environmental conditions may be considered physical hazards. Therefore, specialized equipment has been developed to precisely measure the extent of physical hazards such as toxicity of chemicals. Industrial hygienists can implement different methods to monitor, evaluate, and even control the exposure of workers to such agents. Different organizations and agencies such as OSHA have established standards and requirements regarding the permissible exposure level (PEL) to most of the agents, which can be harmful if exceeded.

Within the context of human factors analysis, experts are concerned with the effect of dealing with physical hazards on human effort and fatigue (i.e., physical, mental and emotional effort, and fatigue) that can lead to an added risk of musculoskeletal disorders (MSDs) as well as human error (consequently, industrial accidents). Overall, no universal tool is available to assess the global workload imposed upon the worker due to the effect of dealing with different physical environmental conditions.

2.7   Human limitations

The human body has limitations that can determine its capacity in performing different tasks from various aspects. These limitations are different from one individual to another, and they change over time and personal experience. Human limitations can be classified into five different groups:

Physiological: Refers to human physical capabilities to maintain stability and life, such as physical strength and endurance as well as the capacity to maintain stable internal physiological conditions under adverse situations. These physiological limitations vary from person to person and can affect his/her work performance. In addition, if these limitations prevent a person from meeting the physiological demand of his/her work, the risk of accidents will increase. It is important to have pre-employment and/or periodical physical checkups by occupational physicians to make sure that employees are physiologically capable of fulfilling their jobs.

Psychological: Refers to human mental capabilities to learn and use acquired knowledge such as learning capacity, skills, performance capability, tolerance of adverse conditions, and motivation. Individuals have different levels of psychological capabilities, and the content of their jobs should match their limitations to optimize their performance and reduce the risk of injury. Therefore, mental evaluations of employees help to ensure that their job content is suitable to their psychological capabilities.

Anthropometric: Refers to the shape and structural features of the human body, such as fixed morphology, tissue structure, size and shape of the work envelope, and postural requirements. Anthropometric parameters can determine how workers fit in their work environment. Anthropometric measurements must be considered in workstation design/setups, tool design, and equipment layout and setup to enhance the man–machine interaction and reduce the risk of fatigue and accidents.

Nutritional: Refers to a human's capability to metabolize food intake (solid or liquid) to generate energy and discharge the waste. This limitation can be assessed based on the need for maintenance of appropriate food, water intake, and requirements for eliminations. The metabolic rate can vary in people and can change with aging. Some tasks demand higher levels of physical and/or mental activities that lead to higher metabolic rates. If the metabolic rate cannot keep pace with the physical and/or mental demands of work, there may be unsafe conditions for the worker.

Clinical: Refers to a human's status of health and the overall performance of his/her internal organs, such as the person's state of health, presence of disease, aging, race/

ethnicity/gender, familial history of diseases, and genetics. These parameters, among others, can put a person at risk for occupational illnesses or injuries if not considered for work system design.

Knowing and understanding the limits of employees in all aspects, engineers and safety experts can make adjustments to the existing working conditions to either eliminate hazards or minimize the risk of accidents. The information about human limits regarding the employees can be obtained by a combination of direct clinical examinations, surveys, direct observations, and even measurements.

2.8   Ergonomics efforts

Ergonomics efforts rest on the premise that the physical demands of work should be balanced with worker endurance and strength to minimize fatigue and the risk of MSDs. Researchers have developed a variety of assessment methods. The following methods are briefly described in this section:

- National Institute for Occupational Safety and Health (NIOSH) lifting equation
- Strain index (SI)
- Rapid upper limb assessment (RULA)
- Arbeitswissenschaftliches Erhebungsverfahren zur Tatigkeitsanalyse (AET) (ergonomic job analysis procedure)

2.8.1   NIOSH lifting equation

The NIOSH developed the lifting equation. The objective of this method is to estimate and recommend a weight limit for the lifting task by utilizing a mathematical equation that includes major biomechanical, physiological, and psychophysical parameters. The NIOSH lifting equation is defined as (Waters et al., 1994)

$$RWL = LC \times HM \times VM \times DM \times AM \times FM \times CM \qquad (2.1)$$

where
　　RWL = Recommended weight limit
　　LC = Load constant
　　HM = Horizontal distance multiplier
　　VM = Vertical multiplier
　　DM = Distance multiplier
　　AM = Asymmetric multiplier
　　FM = Frequency multiplier
　　CM = Coupling multiplier

Equation 2.1 can be implemented by using both metric systems and U.S. standard units. The load constant (LC) is the basic weight limit that is adjusted by other multipliers according to the following measurements (Waters et al., 1994):

1. Horizontal distance of hands from the midpoint between the ankles
2. Vertical travel distance of the hands from the floor
3. Vertical travel distance between the origin and the destination of the lift
4. Angular displacement of the load from the sagittal plane

5. Average frequency rate of lifting measured in lifts per minute for a given task duration T; that is, $T \le 1$ h, $1 \le T \le 2$ h, or $2 \le T \le 8$ h
6. Coupling condition

Once the recommended weight limit (RWL) is estimated, the lifting index (LI) can be calculated as the ratio of the weight of actual load to the RWL, and the value of this index can be used to assess the risk of developing MSDs. The higher the value of LI, the higher the risk of developing MSDs is. Although not stated by NIOSH, the generally accepted rule for LI interpretation is

- If LI ≤ 1, the lifting conditions are considered safe to most of the workforce.
- If LI ≥ 3, the lifting conditions are considered unsafe to most of the workforce.
- If $1 < $ LI $ < 3$, the lifting conditions are considered safe to some and unsafe to others.

There are several limitations to this method. First, the NIOSH lifting equation can be applied only to a manual lifting method in which the worker uses both hands to lift rigid objects. This assessment method should not be used if the worker is employing any tool to lift the load such as a wheelbarrow, if the load is bulky such as a bag of liquid, or if the worker is using only one hand to lift the load. Second, this method can only be used for healthy workers. That means, if a worker has a history of back disorders, this equation cannot be used for risk assessment. Third, this method cannot be used if the lifting task is combined with other activities such as pushing or pulling the load (Waters et al., 1994).

2.8.2   Strain index

The SI is another assessment method used for hand and wrist movements in manual tasks. The objective of this assessment method is to assess the risk of developing MSDs or cumulative trauma disorders (CTDs) such as carpal tunnel syndrome in hands and wrists by developing and applying a mathematical equation that accounts for the major contributing factors. The SI equation is defined as (Moore and Garg, 1995)

$$SI = IM \times DM \times EM \times PM \times SM \times DDM \tag{2.2}$$

where
    SI = Strain index
    IM = Intensity of exertion multiplier
    DM = Duration of exertion multiplier
    EM = Exertions per minute multiplier
    PM = Posture multiplier
    SM = Speed of work multiplier
    DDM = Duration per day multiplier

SI can be estimated by using Equation 2.2 as each parameter is given a rating from one to five based on required measurements or from predetermined tables. A multiplier is then assigned to every SI parameter depending on the rating given, and eventually, SI can be calculated as the product of all the parameter multipliers. SI parameters are defined as (Moore and Garg, 1995)

1. *Intensity of exertion:* This is an estimate of the strength required to perform the task one at a time.

2. *Duration of exertion:* This is calculated by measuring the duration of all exertions during an observation period, then dividing the measured duration of exertion by the total observation time, and multiplying by 100.
3. *Exertions per minute:* This is measured by counting the number of exertions that occur during an observation period and then dividing the number of exertions by the duration of the observation period.
4. *Hand/wrist posture:* This is an estimate of the position of the hand or wrist relative to the neutral position.
5. *Speed of work:* This is an estimate of how fast the employee is working.
6. *Duration of task per day:* This is either measured or obtained from plant personnel.

Once the SI is calculated, the result can be interpreted as (Moore and Garg, 1995)

- If SI ≤ 3, the job is probably safe.
- If SI > 5, the job will be associated with distal upper extremity disorders.
- If SI ≥ 7, the job is hazardous.

There are several limitations to this method. First, it is only applied to hands and wrists; elbow and shoulder postures are not considered in this model. Second, not only are two of the rating criteria subjective and determined by the analyst (i.e., *hand/wrist posture* and *speed of work*), but also the person who is performing the job might not agree with their corresponding multipliers. Finally, the rating of *exertions per minute* does not differentiate whether the exertions occur in equal intervals, which can affect the time the muscles rest between exertions.

2.8.3   Rapid upper limb assessment

RULA is a coding system used to generate an action list that indicates the level of intervention required to reduce the risks of injury due to physical workload on workers. The action levels are developed on the basis of a rating score that ranges from 1 to 7. The final RULA score represents the overall predicted muscular effort associated with the working posture, exerting force, and performing static or repetitive work. The action levels are (McAtamney and Corlett, 1993)

- Level one: A score of 1 or 2 indicates that the posture is acceptable if it is not maintained or repeated for long periods.
- Level two: A score of 3 or 4 indicates that further investigation is needed and changes may be required.
- Level three: A score of 5 or 6 indicates that investigation and changes are required soon.
- Level four: A score of 7 indicates that investigation and changes are required immediately.

As shown in Figure 2.9 (adapted from McAtamney and Corlett, 1993), the final RULA score is calculated based on the postural score "A" and "B" obtained from different tables and are functions of (i) the categories of the upper extremity postures defined around the shoulder, elbow, and wrist, and (ii) the categories of the neck, trunk, and lower extremity postures, respectively. Once scores "A" and "B" are adjusted according to the level of muscle use and applied on force or load, the final RULA score is obtained from the third table (McAtamney and Corlett, 1993).

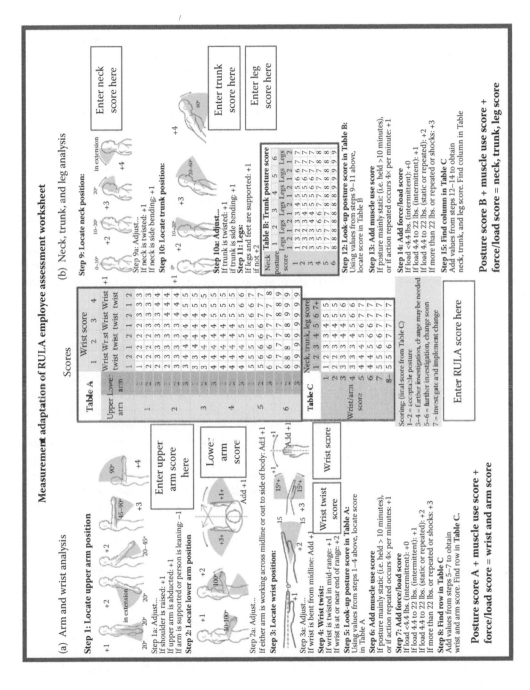

Figure 2.9 Sample of RULA worksheet. (Adapted from McAtamney, L. and Corlett, E. N. 1993. *Applied Ergonomics*, 24(2), 91–99.)

The RULA method has its own limitations. First, this method does not consider some of the factors such as twisting, lateral bending, and abduction of hands and arms that can contribute to and increase the risk of developing MSDs. Second, different postures of body parts are weighted equally and receive the same score. For example, the angle of 90° and 120° for the upper arm is considered equivalent with a score of 4, while in reality, they can be different. Third, RULA does not differentiate between left and right body parts, although it is very likely that during actual work, left and right arms or legs have different postures and could affect the way scores are assigned, and the final RULA score.

2.8.4 *Arbeitswissenschaftliches Erhebungsverfahren zur Tatigkeitsanalyse*

The AET method also known as "Arbeitswissenschaftliches Erhebungsverfahren zur Tatigkeitsanalyse" originally started as a study ordered by the German government to investigate wage discrimination against women at work. This method developed gradually and continuously over the years and has been used in different industrial situations. This method is structured in three parts: Part A is used for work system analysis (objects, equipment, and work environment), Part B is for task analysis, and Part C is for job demand analysis. Each part is made of series of items regarding properties and characteristics of the work system, which are assessed by the analysis (through observation or interview) by using different codes to indicate the significance, exclusive, duration, and frequency of the property with respect to the entire system, and other properties (Rohmert and Landau, 1983).

The AET method can be utilized to assess the temporal distribution and location of physical stresses (Part C of the AET instrument) due to postural work (postures/body position), static work, heavy dynamic work, active light work, and its application of forces and frequency of motion (Rohmert and Landau, 1983).

Physical stress is directly associated with metabolic processes that take place during muscular work. There are two types of chemical reactions to generate energy for muscular work (muscle contraction): anaerobic reaction (without the presence of oxygen) and aerobic reaction (with the presence of oxygen). During these metabolic processes, adenosine triphosphate (ATP) in the muscles is broken down to create energy for muscle contraction. ATP stored in cells breaks down to adenosine diphosphate (ADP) and phosphate, and generates energy that is enough for about 1 s of muscle contraction. Another source of ATP is the ATP–creatine phosphate (CP) system. In this system, CP breaks down to creatine and phosphate releasing energy, and later, the phosphate reacts with ADP to produce additional ATP, which is enough for 3–4 s of muscle contraction. Neither of these two sources of energy are enough to help the human body to perform muscular work for longer than 5 s (Chaffin et al., 2006; Tayyari and Smith, 1997).

There are two more sources of energy to support muscular contraction for a longer period of time: anaerobic glycolysis and aerobic glycolysis. Anaerobic glycolysis is a metabolic process that breaks down the glucose stored in cells without oxygen to generate ATP. The human body uses this process for highly intense physical activities such as sprint racing in a track-and-field competition. The anaerobic glycolysis process produces lactic acid, and the human body reaches its maximum tolerance to it in 35–45 s. That is why this process cannot sustain for a long period of time (Chaffin et al., 2006; Tayyari and Smith, 1997). During aerobic glycolysis, glucose (glycogen in fat or proteins) is broken down with oxygen to generate ATP. This process can continue and generate energy from a few seconds to several hours. In contrast to sprint race runners, marathoners use this process as the source of their energy (Chaffin et al., 2006; Tayyari and Smith, 1997).

Stress due to postural work: Postural work leads to strain due to the need to maintain a certain body position without any external forces applied. Postural stresses are evaluated in terms of the amount of time spent in the given body positions/postures such as sitting (normal), sitting (bent), standing (normal), kneeling (normal), crouching, standing/stooping, and standing considerably stooped.

Stress due to static work: Static work implies a long-term (more than 4 s) muscular effort that does not result in body movement. Muscular effort can take place in static work not only as a result of the exertion of an external force but also because of the effort required to maintain the weight of the extremities. Stress due to static work is evaluated in terms of the amount of time spent on different body parts such as finger/hand/forearm, arm/shoulder/back, and leg/foot.

Stress due to heavy dynamic work: Heavy dynamic work is a class of work such as walking, running, climbing, or crawling that requires the use of large muscle groups and results in an increased consumption of energy. It is evaluated in terms of the proportion of shift/time spent in activities that require the use of both arms with support of muscles in the upper part of the body and both legs with support of the pelvic muscles.

Stress due to active light work: Active light work is a class of work that requires the use of one or several small muscle groups with an active mass smaller than 1/7 of the overall muscle mass of the body and whose frequency is higher than 15 exertions per minute. It is evaluated in terms of the proportion of shift/time spent in activities that require the use of finger/hand, hand/arm, and foot/leg.

Applications of forces and frequency of motion: The extent of force (in relation to the region of the body involved) is evaluated for static work, heavy dynamic work, and light active work in terms of different levels (e.g., low or high). However, the frequency of motion is evaluated for active light work in terms of graded levels (e.g., low or high). For example, 120 or more motions per minute for fingers, 90 or more motions per minute for the hand and forearm, and 60 or more motions per minute for the upper arm, foot, and legs are considered as "high" frequency (Rohmert and Landau, 1983).

The AET is the only method that evaluates the different elements of physical task demands but lacks a way to integrate the different information in a global index such as the NIOSH LI, SI, or RULA.

2.9   Human factors efforts

Human factors methods emphasize balancing mental task demands with the perceptual and cognitive abilities of workers to reduce fatigue and to minimize the likelihood of human error leading to industrial accidents. Here, we explain four of several assessment methods in this field.

2.9.1   Cooper–Harper scales

This method was initially designed to measure aircraft-handling characteristics. It used a decision tree with multidimensional descriptors of handling difficulty, and the final result was a one-dimensional rating scale of 1–10 in which a larger number represents a higher level of mental workload (Donmez et al., 2008).

The Cooper–Harper scale has been modified and adjusted to measure workload in both aviation and other settings. As shown in Figure 2.10 (modeled after Wierwille and Casali, 1983), this method focuses on the consequences of any decision made by the operator. It starts with the frequency or magnitude of errors; if the errors can cause the system

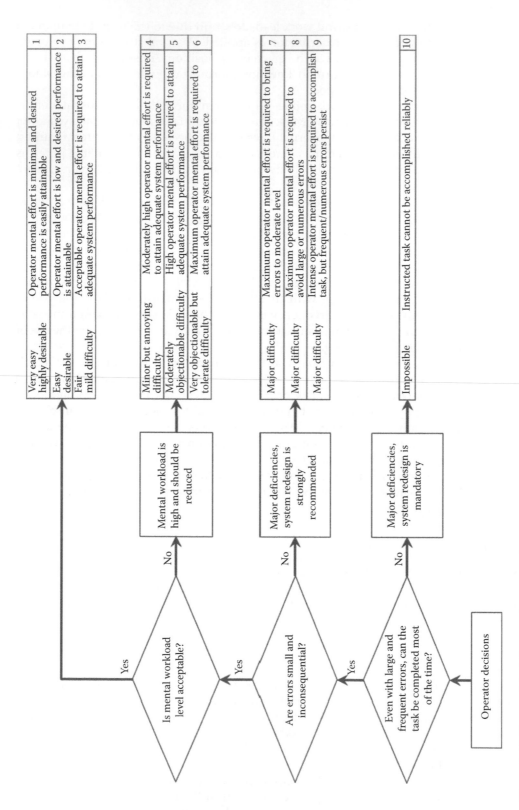

Figure 2.10 The modified Cooper–Harper scale.

to fail, then the rating is 10. If the operator needs to spend a significantly high level of mental effort to avoid or reduce the severity of consequences of errors, then, it is rated from seven to nine. Similarly, if the operator needs to spend the moderate level of mental effort to control the system performance, then the Cooper–Harper scale is rated from four to six. If the level of mental workload is acceptable and the operator can easily maintain the performance of a system at adequate or acceptable conditions, then, the scale is rated from one to three. It has been concluded that the modified method is statistically reliable and an acceptable indicator of the overall mental workload (Wierwille and Casali, 1983).

2.9.2 NASA task load index

The NASA-task load index (NASA-TLX) was developed based on the belief that workload is a subjective experience caused by the interaction between the requirement of a task, the circumstances under which it is performed (external factors), and the skills, behaviors, and perceptions of the operators (internal factors). The goal of the NASA-TLX was to develop a workload-rating scale that provides a sensitive summary of workload variations within and between tasks (Hart and Staveland, 1988). Unlike the Cooper–Harper scale (one dimensional), the NASA-TLX is a multidimensional scale that integrates six parameters (i.e., mental demand, physical demand, temporal demand, performance, effort, and frustration level). These six parameters of NASA-TLX incorporate three scale categories (Hart and Staveland, 1988):

1. Task-related scales (mental demand and physical demand).
2. Behavior-related scales (performance and effort)—Provide subjective evaluations of the effort that subjects exert to satisfy task requirements and opinions about how successful they are in doing so.
3. Subject-related scales (frustration)—Focus on the psychological impact of task demands, behavior, and performance of the subjects.

For any given task, each of the six parameters is evaluated based on two types of information: (i) the subjective importance of the parameter, (ii) the magnitude of the parameter. The importance of parameters is determined by comparing them in pairs and assigning a weight (from zero to five). Also, the magnitude of each parameter is determined by assigning a number ranging from 1 to 100. Eventually, an average of the six scales, weighted to reflect the contribution of each parameter to the workload of a specific activity from the perspective of the operator (rater) is proposed in the NASA-TLX as an integrated measure of the overall workload (Hart and Staveland, 1988). The overall measure is referred to as the weighted workload (WWL), and the creators of the NASA-TLX suggest that their procedures and calculations yield a sensitive and stable overall index of workload.

2.9.3 Subjective workload assessment technique

Similar to NASA-TLX, the subjective workload assessment technique (SWAT) is a multidimensional scale of workload. SWAT integrates three parameters: time load, mental effort load, and psychological stress load (Reid and Nygren, 1988).

1. *Time load* evaluates the time available and time overlap for a task. If the time required to perform a task exceeds the time available, the worker has a time load problem. In addition, time overlap may severely limit the worker's performance.

2. *Mental effort load* deals with the consumption of the worker's mental resources available for performing mental tasks such as calculations and decision making.
3. *Psychological stress load* deals with the general concept of psychological stress such as fatigue, health, and emotional state.

The SWAT method consists of two stages: scale development (Stage 1) and event scoring (Stage 2). During the first stage, a preliminary card sort is performed by each subject to sort 27 different combinations of three levels (low, medium, and high) of the SWAT parameters with respect to their importance. Subjects rank these cards by recalling work-related situations and subjectively assessing the time load, mental effort load, and psychological stress load associated with each situation, followed by conjoint analysis techniques to provide an interval scale of the overall workload tailored for individual differences. Then, in the second stage, subjects provide ratings of low, medium, or high for the three parameters following the performance of each experimental task. A single rating of the overall workload is obtained by referring to the position on the interval scale (developed in the first stage) identified by that combination of values (Reid and Nygren, 1988).

2.9.4   Position analysis questionnaire

The position analysis questionnaire (PAQ) is a job analysis questionnaire that includes job elements that are essentially worker oriented in nature. Most of the elements are to characterize human behaviors (or work behaviors) instead of activities that are described in technological or strictly job terms (McCormick et al., 1969).

PAQ classifies the domain of mental task demands into two subdomains: sensory and perceptual processes, and mental processes. The first subdomain (sensory/perceptual process) consists of several elements such as far visual differentiation, depth perception, color perception, sound pattern recognition and differentiation, body movement sensing and balance, estimating the speed of moving parts/moving objects/processes, judging condition/quality, inspecting, as well as estimating quantity/size/time. The second subdomain (mental processes) is made of elements such as decision making, level of reasoning in problem solving, amount of planning, information-processing activities, combining information, analyzing information or data, compiling, coding/decoding, transcribing, and the use of short-term memory (McCormick et al., 1969).

Each element of the sensory/perceptual and mental processes is evaluated in terms of its importance to task performance and the result can range from "Does not apply" to "Extreme." A major limitation is that the PAQ does not develop a procedure to integrate the different pieces of information into a global mental workload index such as the NASA-TLX.

2.10   Industrial psychology efforts

The basic premise of industrial psychology efforts is that opportunities should be created in the work environment to satisfy human needs. In other words, employees need motivation to maximize their work productivity and output quality.

Motivation is not behavior. It is a complex internal state that cannot be observed directly but affects behavior. Motivation is made up of all internal striving conditions described as wishes, desires, and drives. It is an internal state that activates or moves individuals. Motivation is generally considered to be rooted in human needs and an individual responds to needs by doing something about them. Behavior is an attempt to satisfy the needs that motivate the individual. In other words, behavior is the means by which the individual seeks to satisfy needs. The basis for understanding motivation in organizations lies in understanding the needs that motivate the behavior of the people in those organizations. Four major theories about human needs and motivation are

- Hierarchy of needs theory
- Motivation—hygiene theory
- Job characteristics theory
- Expectancy theory

2.10.1   Hierarchy of needs theory

In this theory, human needs are categorized into five different hierarchies. When the lowest order of needs in the hierarchy is satisfied, a higher-order need appears, and since it has a greater potency at the time, this higher-order need causes the individual to attempt to satisfy it. Abraham Maslow's theory is a general theory of motivation that is applicable not only to work life but also to all aspects of life (Maslow, 1958a,b). According to Maslow's theory, the five different orders of human needs are

1. *Basic physiological needs*: Includes food, water, shelter, and the like. In modern society, the basic drives of human existence cause individuals to become involved in organizational life. People become participants in the organization that employs them. Thus, at the simplest level of human needs, people are motivated to join organizations, remain in them, and contribute to their objectives.
2. *Security and safety*: Security means many things to different people in different circumstances. For some, it means earning a higher income to assure freedom from what might happen in case of sickness or during old age. Thus, many people are motivated to work harder to seek success that is measured in terms of income. It can also be interpreted as job security. To some people such as civil servants and teachers, the assurance of life tenure and a guaranteed pension may be strong motivators for their participation in employing organizations.
3. *Social affiliation*: An employee with a reasonable well-paying and secure job will begin to feel that belonging and approval are important motivators in his/her organizational behavior.
4. *Esteem*: The need to be recognized, to be respected, and to have prestige (self-image and the view that one holds of oneself). There is a dynamic interplay between one's own sense of satisfaction and self-confidence on one hand, and feedback from others in diverse forms such as being asked for advice on the other.
5. *Self-actualization*: The desire to become more and more of what one is, is to become everything that one is capable of becoming. The self-actualized person is strongly inner directed, seeks self-growth, and is highly motivated by loyalty to cherished values, ethics, and beliefs. Not everyone reaches the self-actualized state. It is estimated that these higher-level needs are met about 10% of the time.

Porter (1961) reformulated Maslow's original hierarchy of managers on the assumptions that few managers are motivated by basic needs such as (i) the need for security (lowest in the hierarchy), (ii) the need for affiliation, (iii) the need for self-esteem, (iv) the need for autonomy (the only addition to the hierarchy refers to the individual's need to participate in making decisions that affect him/her, to exert influence in controlling the work situation, to have a voice in setting job-related goals, and to have authority to make

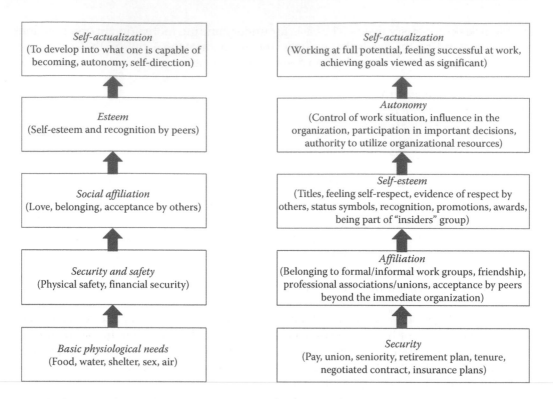

Figure 2.11 Hierarchy of needs in Maslow's theory of motivation (left) and Porter's concept of the hierarchy of needs for managers (right).

decisions and latitude to work independently), and (v) the need for self-actualization. Figure 2.11 represents both Maslow's theory of motivation and Porter's concept of the hierarchy of needs for managers side by side.

2.10.2   Motivation: Hygiene theory

Motivation hygiene theory suggests that the factors involved in producing job satisfaction (and motivation) are separate and distinct from the factors that lead to job dissatisfaction. In other words, the two feelings (job satisfaction and dissatisfaction) are not opposites of each other, that is, the opposite of "job satisfaction" is not "job dissatisfaction," but rather "no job satisfaction." Similarly, the opposite of "job dissatisfaction" is not "job satisfaction," but "no job dissatisfaction" (Herzberg, 1982, 1987).

In one of his studies, Herzberg asked people to recall the circumstances in which (a) they, at specific times in the past, felt satisfied with their jobs, and in which (b) they had similarly been dissatisfied with their jobs. Analysis of the responses indicated that there was one specific group of factors that was associated with motivation and satisfaction at work and another, equally specific, group of factors that was associated with dissatisfaction and apathy. An important concept in the theory is that people tend to see *job satisfaction* as being related to *intrinsic* factors such as success, the challenge of work, achievement, and recognition, while they tend to see *dissatisfaction* as being related to *extrinsic* factors such as salary, supervision, and work conditions (Herzberg, 1982, 1987). The extrinsic factors that are associated with job dissatisfaction are called hygiene factors such as

1. *Company policy and administration*: Ineffectiveness of the organization and personnel policies and disagreement with company goals
2. *Supervision*: Competency, delegation of work, supervisor consistently critical, and showing favoritism
3. *Relationship with the supervisor*: Friendly, learning, support with the management, willing to listen and honest, and give credit for the work done
4. *Work conditions*: Work in isolation, work in social surroundings, quality of physical surroundings and facilities, and amount of work
5. *Basic compensation or salary*: Compare favorably with others doing a similar or same job, receive wage increase, and amount of salary
6. *Relationship with peers*: Liking of people with whom you work, cooperation, and isolated from the group
7. *Personal life*: Family problems, community and other outside situations, and salary-wise family needs and aspirations
8. *Relationship with subordinates*: Quality of relationship
9. *Status*: Having a given status
10. *Security*: Objective signs of job security such as company stability

Similarly, the intrinsic factors that are associated with job satisfactions are called motivation and include

1. *Achievement*: Successful completion of job seeing results of work
2. *Recognition*: Work praise (with or without rewards), credit for work taken by the supervisor or others, and an idea accepted by the company
3. *Work itself*: Work variety, challenging or creative work, and opportunity to do a whole job
4. *Responsibility*: Responsible for own work
5. *Advancement*: Promotion
6. *Growth*: Growth in skills

As shown in Figure 2.12 (modeled after Herzberg, 1982, 1987), a factor such as *achievement* was selected as a reason for job dissatisfaction by 10% of the responders, while more than 40% of the participants chose *achievement* as an important factor for job satisfaction. Similarly, more than 30% of participants indicated *company policy and administration* as a reason for job dissatisfaction, while <10% responders saw it as a factor for job satisfaction.

2.10.3   Job characteristics theory

The job characteristics theory proposes that positive, personal, and work outcomes (i.e., high internal motivation, high work satisfaction, high-quality performance, and low absenteeism and turnover) are obtained when three "critical psychological states" are present for a given employee (Hackman and Oldham, 1976). The three critical psychological states are

1. *Experienced meaningfulness of the work*: The degree to which the individual experiences the job as one that is generally meaningful, valuable, and worthwhile.
2. *Experienced responsibility for work outcomes*: The degree to which the individual feels personally accountable and responsible for the results of the work he/she does.
3. *Knowledge of results*: The degree to which the individual knows and understands, on a continuous basis, how effectively he/she is performing the job.

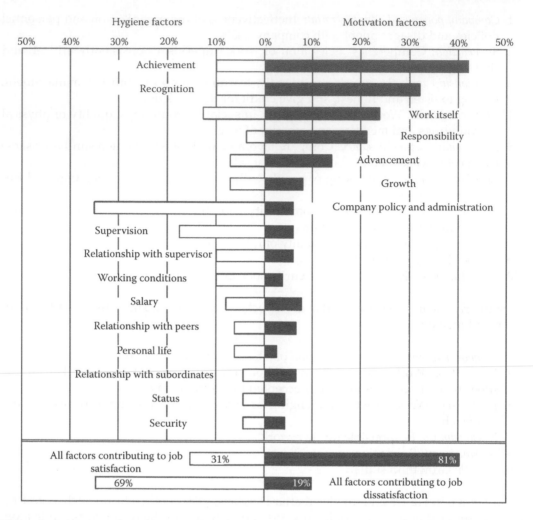

Figure 2.12 Motivation and hygiene factors.

Figure 2.13 (modeled after Hackman and Oldham, 1976) demonstrates how core job dimensions can affect critical psychological states of an individual that eventually lead to a specific level of personal and work outcome. To achieve positive, personal, and work outcomes, it is important to note that all three of the psychological states must be satisfied via the presence of five "core" job dimensions, which are

1. *Skill variety*: The degree to which a job requires a variety of different activities in carrying out the work that involves the use of a number of different skills and talents of the person.
2. *Task identity*: The degree to which the job requires completion of a "whole" and identifiable piece of work; that is, doing a job from the beginning to an end with a visible outcome.
3. *Task significance*: The degree to which the job has a substantial impact on the lives or work of other people, whether in the immediate organization or in the external environment.

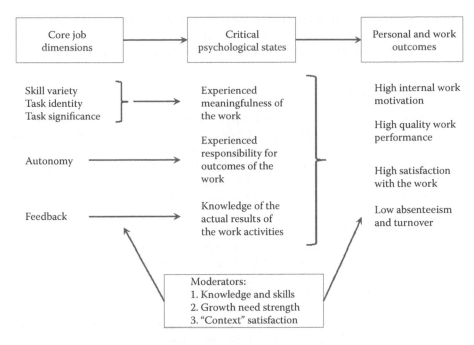

Figure 2.13 Job characteristics theory.

4. *Autonomy:* The degree to which the job provides substantial freedom, independence, and discretion to the individual in scheduling the work and in determining the procedures to be used in carrying it out.
5. *Feedback:* The degree to which carrying out the work activities required by the job results in the individual obtaining direct and clear information about the effectiveness of his/her performance.

The "experienced meaningfulness of the work" is primarily enhanced by three of the core dimensions: (1) skill variety, task identity, and task significance, (2) "experienced responsibility for work outcomes," which is increased when a job has high autonomy, and (3) "knowledge of results," which is increased when a job offers substantial feedback. The theory suggests that the strength of individual growth need has a moderating effect between the core job dimensions and the three psychological states, and between the critical psychological states and positive, personal, and work outcomes. These work outcomes include *general satisfaction* (an overall measured by the degree to which the employee is satisfied and happy with the job) and *internal work motivation* (the degree to which the employee is self-motivated to perform effectively on the job). In other words, the employee experiences positive internal feelings when working effectively on the job and negative internal feelings when doing poorly (Hackman and Oldham, 1976).

Individual growth need strength is a measure that is viewed as an individual difference characteristic predicted to influence how positively an employee will respond to a job with an objectively high-motivating potential. There are two possible links for its moderating effects (Hackman and Oldham, 1976):

1. The link between core job dimensions and psychological states—People with high growth need are more likely or better able to experience the psychological states when the job conditions are good than their counterparts with low growth need.

2. The link between psychological states and outcome variables—Nearly everybody may experience the psychological states when job conditions are right, but individuals with high growth needs respond more positively to that experience.

2.10.4   Expectancy theory

Expectancy theory is based on three assumptions: (i) people do not respond to events after they occur (they anticipate [or expect] that things will occur and certain behaviors in response to those events will probably produce predictable consequences), (ii) humans usually confront possible alternative behaviors (and their probable consequences) in rational ways, and (iii) through experience, individuals learn to anticipate the likely consequences of alternative ways of dealing with events and, through this learning, modify their responses (Porter and Lawler, 1968).

Expectancy theory starts with a given situation in which there are alternative possible courses of action. Each such alternative has one or more possible outcomes associated with it, and there are three concepts associated with each outcome:

1. *Valence*: The value to the individual of each of such outcome, such as the value of a raise.
2. *Expectancy*: The person's estimate of probabilities, or odds, of any given outcome taking place, such as the estimated probability of a raise.
3. *Force*: The result of adding up the combinations of valences and expectancies associated with the various possible outcomes.

In sum, expectancy theory of motivation focuses on rational expectations held by the worker that desirable rewards are likely to be the predictable outcome of certain behaviors. There are various models of expectancy theory. One of the most popular models is Porter–Lawler expectancy model that contains the following variables (Porter and Lawler, 1968):

1. *Value of reward*: How attractive or desirable a potential outcome of an individual's behavior is in the work situation.
2. *Effort–reward probability*: An individual's expectations concerning the likelihood that the given amounts of rewards depend on the given amounts of effort on his/her part.
3. *Effort*: The energy expended to perform some task, but does not necessarily correlate with how successfully the task is carried out in a given situation.
4. *Abilities and traits*: Relatively stable, long-term individual characteristics that represent the individual's currently developed power to perform.
5. *Role perceptions*: The direction of effort; that is, the kinds of activities and behaviors in which the individual believes he/she should engage to perform his/her job successfully. In other words, the way in which the individual defines his/her job—the types of effort the person believes are essential to effective job performance.
6. *Performance*: A person's accomplishment on tasks that comprise his/her jobs.
7. *Rewards*: Desirable outcomes or returns to a person that are provided by himself/herself or by others.
8. *Perceived equitable rewards*: The level or amount of rewards that an individual feels is fair, given his/her performance on the tasks the person has been asked to undertake by the organization.

9. *Satisfaction*: The extent to which the rewards actually received meet or exceed the perceived equitable level of rewards. The greater failure of actual rewards to meet or exceed perceived equitable rewards in a given situation, the more dissatisfied a person is considered to be.

The mechanism of the Porter–Lawler model is described in Figure 2.14 (modeled after Porter and Lawler, 1968). It starts with step 1 where a worker assigns some value to the possible reward for performing work [1], and he/she also has some subjective judgment as to the likelihood that the proposed effort on the job actually will result in receiving the reward [2]. Then, the combination of wanting the reward plus the perceived chances of actually getting the reward leads to a certain level of effort [3]. Of course, effort alone does not produce results as the model shows, as effort is mediated by a combination of the worker's ability and traits [4] and the perception that he/she has about the role on the job [5]. The interplay of effort, abilities, and role perception is seen as yielding certain results, that is, actual accomplishment from the performed behavior [6]. On the basis of a person's performance, the individual expects to receive a fair reward [8] and is, in fact, rewarded. Rewards can be intrinsic (e.g., satisfaction) [7A] or extrinsic (e.g., pay and recognition) [7B]. At this point, these rewards are perceived to be equitable to some degree or another, in terms of the individual's subjective values and expectations, and it is this complex interplay that leads to some level of satisfaction [9]. The model is completed with two feedback loops. The first loop shapes the individual's perception of the likelihood that the effort will actually yield the expected reward. The other loop shapes the individual's judgment as to the value of the rewards obtained. In short, the model suggests (i) that performance leads to satisfaction, and (ii) that the level of satisfaction obtained shapes future effort to perform (Porter and Lawler, 1968).

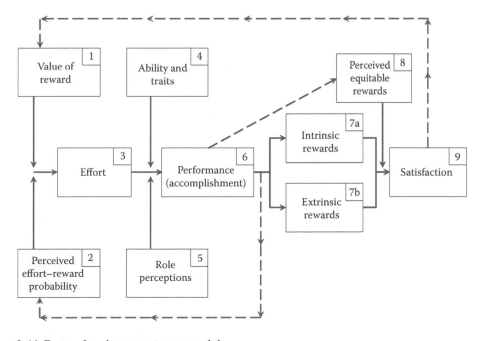

Figure 2.14 Porter–Lawler expectancy model.

2.11   Psychosocial and work stress efforts

The general view of psychosocial and work stress efforts is that exposure to psychological (mental), social, and organizational risk factors should be minimized to reduce the likelihood of injuries or illnesses. To achieve such a goal, perceived demands should be balanced with perceived capabilities to reduce the likelihood of work stress. There are numerous definitions of psychological factors, and one major definition is developed by a Joint Committee on Occupational Health formed by ILO (International Labour Organization) and WHO (World Health Organization): psychosocial factors refer to interactions between and among the work environment, job content, organizational conditions, and human factors (workers' capacities, needs, culture, and personal extra job considerations) that may, through perceptions and experience, influence health, work performance, and job satisfaction (The Joint ILO/WHO Committee on Occupational Health, 1984).

A negative interaction between occupational conditions and human factors may lead to emotional disturbance, behavioral problems, and biochemical and neurohormonal changes that present added risks of mental and physical illness. Adverse effects of work performance and satisfaction can also be expected. An optimum balance between human factors and occupational conditions would suggest a psychosocial situation at work having a positive influence, particularly as it relates to health (The Joint ILO/WHO Committee on Occupational Health, 1984).

There are three different approaches to define work stress. One approach treats stress as a response variable describing it in terms of the person's response in disturbing or harmful environments. The second approach describes stress in terms of the stimulus characteristics of those disturbing or harmful environments, and thus, treats it as an independent variable. The third approach views stress as a reflection of a "lack of fit" between the person and his/her environment. There are different assessment methods for work stress that are currently used, such as

- Transactional model of stress (TMS)
- Person–environment fit model
- Job stress model
- Demand–control and demand–control–support models

2.11.1   Transactional model of stress

The mechanism of TMS is made of five different stages. As presented in Figure 2.15 (modeled after Cox, 1978), the first stage is represented by the sources of demand related to the person and is part of the person's environment. Demand is usually regarded as a factor of the person's external environment; however, the model distinguishes between external and internal demands. A person has psychological and physiological needs, and fulfillment of these needs is important in determining his/her behavior. Thus, these needs constitute internally generated demands (Cox, 1978).

The second stage is about the person's perception of the demand and his or her own ability to cope with it. Stress arises when there is an imbalance between the perceived (not the actual or objective) demand and the perceived capability to meet that demand. The important factor at this stage is the person's cognitive appraisal of the potential stressful situation and of his or her ability to cope with it. If a situation demands too much of a person, but he or she has not realized his or her limitations, the person will continue without being stressed until it becomes obvious that he/she cannot cope. Then, the person must

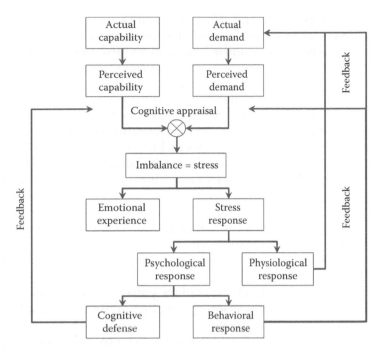

Figure 2.15 Transactional model of stress.

recognize his or her limitations and the imbalance between demand and capability, which follows from experiencing stress. The presence of this perceptual factor allows for the operation of a wide variety of organismic variables such as personality that contribute to interindividual differences. The critical imbalance is accompanied by the subjective (emotional) experience of stress (Cox, 1978).

The emotional experience of stress is accompanied by psychophysiological responses that can be regarded as the third stage of the model. The responses to stress are sometimes thought of as the end point of the stress process but should be regarded as methods of coping available to the person. The fourth stage is concerned with the consequences of coping responses, which are actual and perceived consequences (Cox, 1978).

The fifth stage is the feedback link that occurs at all other stages in the model and which is effective in shaping up the outcome at each of these stages. For example, feedback occurs when (i) a physiological response such as the release of adrenaline influences the organism's perception of the stressful situation, or (ii) a behavioral response alters the exact nature of the demand (Cox, 1978).

Another important example concerns the effectiveness of the stress response in coping. Inappropriate and ineffective response strategies will invariably prolong or even increase the experience of stress. Abnormal coping develops at this point, and this may accelerate the development of damage. It is suggested that functional and structural damage can occur as a result of a prolonged or severe experience of stress.

2.11.2   Person–environment fit model

The person–environment (P–E) fit model is conceptually similar to the TMS. The basic idea of P–E fit theory is that stress arises neither from the person nor environment separately, but rather by their fit or comparison with one another. Two types of P–E fit are considered in this model: needs–supplies fit and demands–abilities fit. Needs–supplies fit is the fit between an employee's needs and the job's supplies, resources, or opportunities for meeting those needs. The needs encompass innate biological and psychological requirements, values acquired through learning, socialization, and motives to achieve desired ends. The supplies refer to extrinsic and intrinsic rewards that may fulfill the person's needs such as food, shelter, money, social involvement, and the opportunity to achieve (Edwards et al., 1998).

The demands–abilities fit is the fit between the job's demands and the person's ability. The demands include quantitative and qualitative job requirements, role expectations, and group and organizational norms. The abilities include the aptitudes, skills, training, time, and energy the person may collect to meet demands (Edwards et al., 1998).

As shown in Figure 2.16 (modeled after Edwards et al., 1998), the P–E model is made of different components that are briefly described as

1. *Objective environment:* Includes physical and social situations and events as they exist independent of the person's perceptions
2. *Subjective environment:* Refers to situations and events as encountered and perceived by the person
3. *Objective person:* Refers to the attributes of the person as they actually exist
4. *Subjective person:* Signifies the person's perception of his/her own attributes (i.e., the person's self-identity or self-concept)
5. *Objective P–E fit:* Refers to the fit between the objective person and objective environment
6. *Subjective P–E fit:* Refers to the fit between the subjective person and subjective environment
7. *Contact with reality:* Means the degree to which the subjective environment corresponds to the objective environment
8. *Accuracy of self-assessment:* Represents the match between the objective person and subjective person

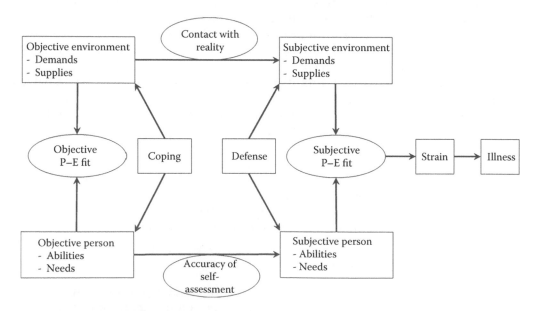

Figure 2.16 P–E fit model.

There are four different possible outcomes in case there is a misfit in P–E model. In other words, if a person does not fit into his or her environment, the person might react with strain, illness, coping, or defense (Edwards et al., 1998).

Strain includes deviations from normal functioning and can be in the form of (i) psychological (e.g., dissatisfaction or anxiety), (ii) physiological (e.g., elevated blood pressure, elevated serum cholesterol, and compromised immune system functioning), or (iii) behavioral (e.g., smoking, frequent utilization of health-care services, overeating, and absenteeism).

Illness is the cumulative experience of strain over time that can lead to mental and physical illnesses such as chronic depression, hypertension, coronary heart disease, and peptic ulcer.

Coping is the efforts to improve objective P–E fit either by changing the objective person (i.e., adaptation) or the objective environment (i.e., environmental mastery). For example, a person experiencing excessive demands at work may seek to enhance his/her ability or attempt to negotiate a decreased workload with his/her supervisor.

Finally, defense involves efforts to enhance the subjective P–E fit through a cognitive distortion of the subjective person or movement (e.g., denial and repression) without changing objective counterparts. For example, a person may respond to role overload by overestimating his/her abilities or by downplaying or ignoring excess demands. It may also include the denial of experienced strain. For example, the person acknowledges the subjective P–E misfit but discounts its resulting negative impacts on health. Another form of defense is when a person may respond to subjective misfit by reducing the perceived importance of the dimension on which misfit occurs as when a person disengages from unattainable goals.

According to the P–E fit model, stress arises when the environment does not provide adequate supplies to meet the person's needs or when the abilities fall short of demands that are prerequisite to receiving supplies. The theory maintains, as in the transactional model, that the subjective fit is the critical pathway from person and environment to strain.

The correlation between needs–supplies and strain is demonstrated in Figure 2.17 (modeled after Edwards et al., 1998). If supplies increase and get closer to needs, then, strain will decrease. However, if the supplies exceed the needs, three different scenarios are possible that are indicated by curves A, B, and C (Edwards et al., 1998).

- Curve "A"—When excess supplies do not influence need fulfillment on other dimensions, strain should remain constant (e.g., food and water reduce strain until hunger and thirst are satisfied, and additional consumption of food will not further reduce strain).
- Curve "B"—Strain may decrease when excess supplies for one dimension are used to satisfy needs on another dimension (e.g., once a person's need for control is satisfied, excess supplies for control may be used to bring about desired changes at work thereby attaining needs–supplies on other dimensions) or when excess supplies can be preserved for later use (e.g., when funds beyond one's current expenses are saved for later use).
- Curve "C"—Excess supplies may increase strain when they prevent the fulfillment of needs on other dimensions (e.g., interaction with coworkers may fulfill one's need for companionship as supplies increase toward needs but then interfere with one's need for privacy as supplies exceed needs).

Similarly, the correlation between demand–ability and strain is demonstrated in Figure 2.18 (modeled after Edwards et al., 1998). Strain should increase as demand exceeds

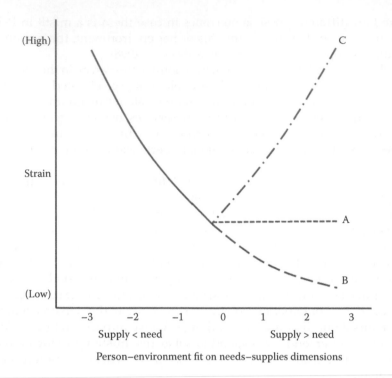

Figure 2.17 Needs–supplies and strain correlation.

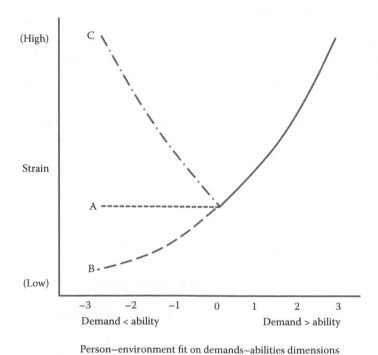

Figure 2.18 Demands–abilities and strain correlation.

abilities, assuming that excess demand inhibits the receipt of supplies required to fulfill needs. However, in case abilities exceed the demand, three different scenarios can occur as indicated by curves A, B, and C (Edwards et al., 1998).

- Curve A—Excess abilities will not influence strain when they cannot be used to acquire supplies (e.g., excess technical skills specific to a particular job demand may be of little use for meeting other demands or fulfilling other work needs or goals).
- Curve B—Excess abilities may decrease strain by providing supplies for needs, in the way that being able to complete one's work more quickly than required creates time for reading, socializing, or other pleasurable activities; conversely, excess abilities may decrease strain by allowing the person to conserve personal resources (e.g., time and energy to apply toward future demands).
- Curve C—Excess abilities may increase strain by creating insufficient supplies for motives (e.g., the inability to utilize valued skills results in boredom and lowered self-esteem) or by threatening the fulfillment of future demands (e.g., unused skills or knowledge may be forgotten, making the person susceptible to task overload if demands increase in the future).

There are different types of P–E fit models with their own implications for the well-being of employees (Caplan, 1987):

Demands–supplies fit (E–E, environment–environment): This model determines whether or not the environment has provided its members with the types of resources required to meet the demands it imposes (e.g., is there adequate time allotted by the system or organization to meet the demand?).

Demands–abilities fit (E–P, environment–person): Demands–abilities fit explains if the employee has the ability to satisfy the needs of the system or organization. In other words, the employee is able and capable to fulfill what is expected of him/her.

Demands–needs fit (E–P): Demands–needs fit deals with an environment measure relevant to the need to satisfy others ("How much do others expect of you?") and a person measure relevant to the need to satisfy oneself ("How much do you expect from them?").

Supplies–abilities fit (E–P): This fit determines if the resources available to an employee are enough and appropriate to enable him/her to fulfill what is expected of this person.

Supplies–needs fit (E–P): This model deals with an environment measure relevant to satisfy the need of oneself ("How much do they provide you?") and a person measure relevant to the need to satisfy others ("How much do you provide to them?").

Abilities–needs fit (P–P, person–person): Of consequence is determining whether the person has the capability and ability required to satisfy his/her needs.

Demands–abilities fit is defined in terms of the requirements of others while needs–supplies fit is defined in terms of the requirements of the person. It is assumed that demands–abilities fit will be more important than needs–supplies fit to persons with a strong need to satisfy others. Similarly, it is assumed that needs–supplies fit will be particularly important to persons with a strong need to satisfy themselves rather than others.

2.11.3 Job stress model

The job stress model uses a questionnaire survey examining psychosocial stress at work involving workers' self-reports of job characteristics and health-related complaints—the former achieves the status of "stressors" correlated with the latter. The model is made of different components as displayed in Figure 2.19 (modeled after Hurrell and McLaney,

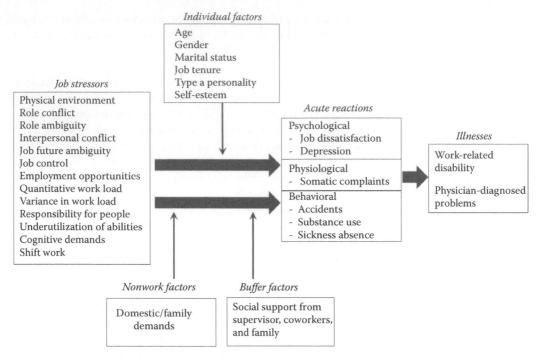

Figure 2.19 Job stress model.

1988). *Job stressors* are work conditions leading to *acute reactions* or *strains* in the worker. Strains represent more or less transient, psychological, physiological, and behavioral responses. The short-term strains lead to *long-term indicators* of mental and physical health. Three other components are *individual factors, nonwork factors,* and *buffer factors,* which are about personal attributes, domestic and familial conditions, as well as the level of support individuals receive from their society. NIOSH has developed the *"Generic Job Stress Questionnaire"* to assess the different model parameters (Hurrell and McLaney, 1988; NIOSH Generic Job Stress Questionnaire).

2.11.4   Demand–control model

The demand–control model suggests that psychological strain results not from a single aspect of the work environment, but instead from the joint effects of the demands of the job and the range of decision-making authority available to the worker facing such demands (Karasek, 1979).

As shown in Figure 2.20 (modeled after Karasek, 1979), when moving along line A, low strain (square 3) can be the result of low psychological/job demands and high decision latitude (controls). Inversely, high strain (square 1) occurs when there is high psychological/job demand with little decision latitude or control. On the other hand, moving along line B, a job is defined as active (square 2) when both job demands and job decision latitude are high. It is hypothesized that this condition leads to developing new behavior patterns both on and off the job. Inversely, passive job (square 4) happens when both job demands and job decision latitude (controls) are low that can lead to a decline in productivity and problem-solving activities (Karasek, 1979).

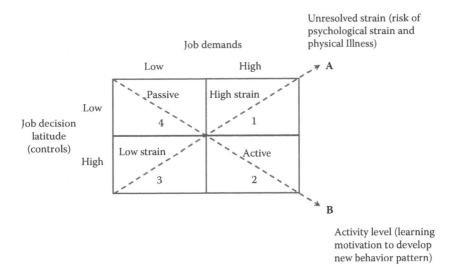

Figure 2.20 Psychological demand/decision latitude model.

The demand–control model was expanded to take into account social factors too. Thus, the third dimension added to the demand–control model is social support (Figure 2.21, modeled after Karasek and Theorell, 1990). Social support at work refers to overall levels of helpful social interaction available on the job from both coworkers and supervisors. Social supports can affect the worker's well-being in different mechanisms: (i) it can work as a buffering mechanism between stressors and adverse health effects, (ii) it can affect the basic psychological processes important to maintaining long-term health, (iii) it can also affect the coping patterns, and (iv) it can help develop active and positive

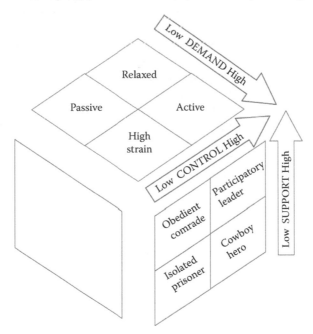

Figure 2.21 Demand–control–support model of the psychosocial work environment.

behavior patterns. There is, however, lack of evidence for some of the mechanisms mentioned above. Different surveys such as "Job Content Instrument" have been developed to measure the three dimensions of the demand–control–support model (Karasek and Theorell, 1990).

2.12   *Future of work system analysis*

There are limitations regarding all the models discussed in this chapter. Most psychosocial and work stress instruments concentrated on the measurement of subsets of psychological, social, and organizational factors. Physical task demands and physical environment conditions were not emphasized in some instruments. In the ergonomic models, the psychological and social aspects of the work environment are not considered. To date, there is no universal instrument that can be used to measure the work–individual balance or fit, and no instrument is available to comprehensively assess the demands and rewards in the workplace. Development of a comprehensive instrument that integrates all aspects of human factors in work system analysis could benefit the industrial engineering, management, occupational physicians or nurses, and safety expert communities.

There are several researchers who have attempted to develop the work compatibility theory in which a multidimensional comprehensive instrument called a work compatibility questionnaire is used to identify all the factors that can determine if an individual is fit for any given job. In other words, with the help of this questionnaire, it can be determined if the person with all his/her physical and mental traits and capabilities is compatible with the job and work environment (Genaidy et al., 2002; Genaidy and Karwowski, 2003; Abdallah et al., 2004; Salem et al., 2006). One potential disadvantage of this method is that the questions are supposed to be answered subjectively by workers. This can produce better results for some dimensions such as the social and psychological aspect of the analysis, but might lead to poor results in other dimensions such as physical or chemical aspects of the work environment that can be assessed better with direct and objective observation or measurements.

Just like any aspect of daily life, new advancements in technology have profound effects on workplaces, too. The recent innovations and trends in technology such as smart phones, information technology, smart systems, nanotechnology and nanomaterials, social media, etc., have created new sources of stress and hazards for workers, and, at the same time, new resources to improve the work environment. Some of these changes are happening at a relatively fast pace, and more studies are required to understand the compatibility between workers and their work system. This means that because of both workers' mental and physical traits and capabilities as well as the work environment, content, and demands that are constantly changing as a result of changes in technology, there remains the need for further research in work system analysis.

References

Abdallah, S., Genaidy, A., Shell, R., Salem, O., and Karwowski, W. 2004. The concept of work compatibility for improving workplace human performance in manufacturing systems. *Human Factors and Ergonomics in Manufacturing and Service Industries*, 14(4), 1–24.

Bailey, R. W. 1996. *Human Performance Engineering: Designing High Quality Professional User Interfaces for Computer Products, Applications and Systems* (3rd ed.). Upper Saddle River, NJ: Prentice-Hall Inc.

Borman, W. C. and Peterson, N. G. 1982. Selection and training of personnel. In G. Salvendy (Ed.), *Handbook of Industrial Engineering* (pp. 5.2.1–5.2.31). New York, NY: John Wiley and Sons, Inc.

Caplan, R. D. 1987. Person–environment fit in organizations: Theories, facts and values. In A. W. Riley and Zaccaro, S. J. (Ed.), *Occupational Stress and Organizational Effectiveness* (pp. 103–140). New York, NY: Praeger Publishers.

Chaffin, D. B., Andersson, G. B. J., and Martin, B. J. 2006. *Occupational Biomechanics* (4th ed.). Hoboken, NJ: John Wiley & Sons, Inc.

Cox, T. 1978. *Stress*. Baltimore, MD: University Park Press.

Donmez, B., Brzezinski, A. S., Graham, H., and Cummings, M. L. 2008. *Modified Cooper Harper Scales for Assessing Unmanned Vehicle Displays*. Cambridge, MA: MIT Department of Aeronautics and Astronautics.

Edwards, J. R., Caplan, R. D., and Harrison, R. V. 1998. Person–environment fit theory: Conceptual foundations, empirical evidence, and directions for future research. In C. Cooper (Ed.), *Theories of Organizational Stress* (pp. 28–67). New York, NY: Oxford University Press.

Fraser, T. M. 1992. *Fitness for Work*. London, UK: Taylor & Francis.

Genaidy, A. and Karwowski, W. 2003. Human performance in lean production environment: Critical assessment and research framework. *Human Factors and Ergonomics in Manufacturing and Service Industries*, 13(4), 317–330.

Genaidy, A., Karwowski, W., and Christensen, D. 1999. Principles of work system performance optimization: A business ergonomics approach. *Human Factors and Ergonomics in Manufacturing*, 9(1), 105–128.

Genaidy, A., Karwowski, W., and Shoaf, C. 2002. The fundamentals of work system compatibility theory: An integrated approach to optimization of human performance at work. *Theoretical Issues in Ergonomics Science*, 3(4), 346–368.

Hackman, J. R. and Oldham, G. R. 1976. Motivation through the design of work: Test of a theory. *Organizational Behavior and Human Performance*, 16(2), 259–279.

Hackman, J. R. and Oldham, G. R. 1980. *Work Redesign*. Reading, MA: Addison-Wesley Publishing Company.

Hart, S. G. and Staveland, L. E. 1988. Development of NASA-TLX (task load index): Results of empirical and theoretical research. In P. A. Hancock and Meshkati, N. (Ed.), *Human Mental Workload* (pp. 139–183). Amsterdam, the Netherlands: Elsevier Science Publishers (North-Holland).

Herzberg, F. 1982. *The Managerial Choice: To Be Efficient and to Be Human* (2nd ed.). Salt Lake City, UT: Olympus Publishing Co.

Herzberg, F. 1987. One more time: How do you motivate your employee? *Harvard Business Review*, 65(5), 109–120.

Hurrell, J. J. and McLaney, M. A. 1988. Exposure to job stress—A new psychometric instrument. *Scandinavian Journal of Work, Health, and Environment*, 14(1), 27–28.

Karasek, R. 1979. Job demands, job decision latitude, and mental strain: Implications for job redesign. *Administrative Sciences Quarterly*, 24, 285–307.

Karasek, R. and Theorell, T. 1990. *Healthy Work. Stress, Productivity, and the Reconstruction of Working Life*. New York, NY: Basic Books.

Karwowski, W. and Rodrick, D. 2001. Physical tasks: Analysis, design and operation. In G. Salvendy (Ed.), *Handbook of Industrial Engineering* (3rd ed., pp. 1041–1110). New York, NY: John Wiley and Sons, Inc.

Maslow, A. H. 1958a. A dynamic theory of motivation. In C. L. Stacey and DeMartino, M. F. (Ed.), *Understanding Human Motivation* (pp. 26–47). Cleveland, OH: Howard Allen, Inc. Publishers.

Maslow, A. H. 1958b. Higher and lower needs. In C. L. Stacey and DeMartino, M. F. (Ed.), *Understanding Human Motivation* (pp. 48–51). Cleveland, OH: Howard Allen, Inc. Publishers.

McAtamney, L. and Corlett, E. N. 1993. RULA: A survey method for the investigation of work-related upper limb disorders. *Applied Ergonomics*, 24(2), 91–99.

McCormick, E. J. 1979. *Job Analysis: Methods and Applications*. New York, NY: AMACOM, A Division of American Management Association.

McCormick, E. J. 1982. Job evaluation. In G. Salvendy (Ed.), *Handbook of Industrial Engineering* (pp. 5.3.1–5.3.17). New York, NY: John Wiley and Sons, Inc.

McCormick, E. J., Jeanneret, P. R., and Mecham, R. C. 1969. *The Development and Background of the Position Analysis Questionnaire*. Lafayette, IN: Occupational Research Center, Purdue University.

Moore, J. S. and Garg, A. 1995. The strain index: A proposed method to analyze jobs for risk of distal upper extremity disorders. *American Industrial Hygiene Association Journal*, 56(5), 443–458.

NIOSH Generic Job Stress Questionnaire (discussed in Hurrell and McLaney, 1988). Cincinnati, OH: National Institute for Occupational Safety and Health.

Porter, L. W. 1961. A study of perceived need satisfaction in bottom and middle-management jobs. *Journal of Applied Psychology*, 45(1), 1–10.

Porter, L. W. and Lawler, E. E., III. 1968. *Managerial Attitudes and Performance*. Homewood, IL: Richard D. Irwin, Inc., and The Dorsey-Press.

Rasmussen, J. 1983. Skills, rules, and knowledge: Signals, signs, and symbols, and other distinctions in human performance models. *IEEE Transactions on Systems, Man, and Cybernetics*, SMC-13(3), 257–266.

Reid, G. B. and Nygren, T. E. 1988. The subjective workload assessment technique: A scaling procedure for measuring mental workload. In P. A. Hancock and Meshkati, N. (Ed.), *Human Mental Workload* (pp. 185–218). Amsterdam, the Netherlands: Elsevier Science Publishers (North-Holland).

Rohmert, W. and Landau, K. 1983. *A New Technique for Job Analysis*. London, UK: Taylor & Francis.

Rummler, G. A. and Brache, A. P. 1995. *Improving Performance: How to Manage the White Space on the Organization Chart* (2nd ed.). San Francisco, CA: Jossey-Bass Publishers.

Salem, S., Paez, O., Holley, M., Tuncel, S., Genaidy, A., and Karwowski, W. 2006. Performance tracking through the work compatibility model. *Human Factors and Ergonomics in Manufacturing and Service Industries*, 16(2), 133–153.

Tayyari, F. and Smith, J. L. 1997. *Occupational Ergonomics—Principles and Applications*. London, UK: Chapman & Hall.

The Joint ILO/WHO Committee on Occupational Health. 1984. *Psychosocial Factors at Work: Recognition and Control*. Occupational Safety and Health Series, No. 56, Geneva.

Waters, T. R., Putz-Anderson, V., and Garg, A. 1994. *Applications Manual for the Revised NIOSH Lifting Equation*. Cincinnati, OH: U.S. Department of Health and Human Services.

Wickens, C. D., Gordon, S.E., and Liu, Y. 1998. *An Introduction to Human Factors Engineering*. Reading, MA: Addison Wesley Longman Inc.

Wierwille, W. W. and Casali, J. G. 1983. *A Validated Rating Scale for Global Mental Workload Measurement Application*. Paper presented at the *27th Annual Meeting of the Human Factors Society*, October 10–14, Norfolk, VA.

chapter three

Measurements of environmental health

LeeAnn Racz and Dirk Yamamoto

Contents

3.1 Introduction

For decades, the world has been becoming increasingly aware of and sensitive to the state of its environment, and how it impacts the health of humans and other species. Rachel Carson's book *Silent Spring*, 1962 (Carson 1962) highlighted the detrimental effects of artificial pesticides on birds. *Time* magazine's article in 1969 on the Cuyahoga River in Ohio catching fire as a result of pollution sparked outrage (*Time* 1969). These and other pivotal events led to the creation of the National Environmental Policy Act (NEPA) and the U.S. Environmental Protection Agency (USEPA) in 1970. In the following decades, there have been dozens of additional environmental protection legislations in the United States, all intended to protect human health and the environment.

Environmental health may be defined as, "the segment of public health that is concerned with assessing, understanding, and controlling the impacts of people on their environment and the impacts of the environment on them" (Moeller 2011). This field of study should not be confused with environmental protection that may be regarded as primarily protecting ecosystems. Nevertheless, the goals and consequences of each field of study are not mutually exclusive. Furthermore, in defining the "environment," we may include the occupational setting and consider the field of occupational health or industrial hygiene. Therefore, the study of environmental health casts a wide net and encompasses a great variety of environments and conditions ranging from the ambient environment to the personal and the occupational environment.

Moeller (1992, 2011) framed common environmental health evaluations by using a systems approach to assess issues in their entirety. He identified four major steps in this approach.

1. Determine the source and nature of each environmental contaminant or stressor.
2. Assess how and in what form it comes into contact with people.
3. Measure the physical effects.
4. Apply controls when and where appropriate.

3.2   Determine source and nature of environmental contaminant or stressor

There are many ways to determine the source of an environmental contaminant or stressor. Perhaps one of the most intuitive is to use experience and professional judgment to antici-pate when and where environmental contaminants may occur as a way to narrow down a search. For example, sites that have been used as dumping grounds for industrial waste prior to the days of environmental regulation are often susceptible to chemical contamina-tion in the groundwater. Physical evidence of stressed vegetation or even records of past activities may be clues to where to install groundwater-monitoring wells or to conduct further investigation as to the extent of contamination. In the case of current activities where an industrial process can be well characterized, it is even easier to identify specific chemicals or physical hazards that may reach the environment or to which a worker may be exposed.

An important method for determining the nature of an environmental contaminant or stressor is rooted in the field of toxicology. The Society of Toxicology (2009) defined toxicol-ogy as, "the study of adverse effects of chemical, physical, or biological agents on living organisms and the ecosystem, including the prevention and amelioration of such adverse effects." There are tens of thousands of chemicals in common use with hundreds more cre-ated each year. While they are created and marketed to benefit society, these pharmaceuti-cal, antimicrobial, personal care, and industrial compounds have the potential to reach the environment and potentially have negative effects. Ideally, we would know everything about a chemical that there is to know such as if it affects any particular organs and of which species, how it interacts with other compounds, how various species metabolize it, and how it breaks down under various environmental conditions. However, this is a daunting task for even a few chemicals, and impossible for all chemicals we use in the world today.

Toxicologists use as much science as is reasonable to collect data about specific com-pounds or other agents. Given ethical limitations of using human subjects to test chemi-cals, toxicologists instead rely on surrogate organisms such as animals or even bacteria to predict the effects on humans. However, there are several problems with these studies that make it difficult to make such predictions accurately. One such problem is that ani-mals do not necessarily metabolize compounds the same way as humans do. In addition, these studies often use high doses in order to both see effects rapidly and to introduce margins of safety, but extrapolations to lower doses or chronic uses may not be proper. Furthermore, these studies tend to focus on effects of a single compound rather than inter-actions of multiple compounds.

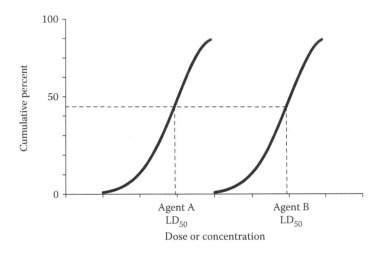

Figure 3.1 Example LD$_{50}$ for two different agents.

When toxicologists are not able to conduct costly in-depth studies using animals or other surrogate organisms, they may rely on structure–activity relationship (SAR) analyses. Where a compound has a structure similar to that of another that is well understood, we might be able to predict the nature of the new compound based on what we know about the first compound. While quite helpful in toxicological studies, these SAR computer models are only as good as the quality of data upon they are based, and the user must understand the boundaries surrounding their predictive capabilities (McKinney et al. 2000).

Regardless of how the toxicology of a compound or stressor is assessed, we normally express its toxic effects by common terminology. The observed endpoint in toxicity studies may be some definitive response such as death or cancer. The cumulative percent of exposed subjects may be plotted against the dose or concentration to give a dose–response curve (Figure 3.1). Dose is typically expressed as mass of agent per mass of subject (e.g., milligram of chemical per kilogram of animal), whereas concentration of an agent is expressed as mass of an agent per mass or volume of the media it is in (e.g., milligram of chemical per liter of water or parts per million [ppm]). Since the response among subjects typically follows a Gaussian (normal) distribution, we have the curve as shown. The dose at which 50% of the population dies is called LD$_{50}$ or the lethal dose for 50% of the population. For curves, which plot the response against concentrations, we have the lethal concentration for 50% of the population instead, LC$_{50}$. In this figure, the LD$_{50}$ for agent A was lower than agent B making agent A more toxic than agent B. Similar dose–response curves can be created to identify doses or concentrations at which there is no observable effect level (NOEL), lowest observable effect level (LOEL), and no observed adverse effect level (NOAEL).

In the case where animal studies are used to predict the effect an agent would have on humans, toxicologists employ safety factors to determine a reference dose (RfD) using Equation 3.1 (USEPA 1993). The RfD is intended to predict the dose at which a human could be exposed to for a lifetime, and suffer no harm

$$RfD = \frac{NOEL}{UF1 \times UF2 \times UF3 \times MF}$$ (3.1)

where

RfD = Reference dose

NOEL = Dose at which there is no observable effect level in the animal study. Sometimes the NOAEL is used instead

UF1 = Uncertainty factor 1 to account for extrapolation from animals to humans (typically 10)

UF2 = Uncertainty factor 2 to account for sensitive populations such as children, the elderly and immunocompromised (typically 10)

UF3 = Uncertainty factor 3 to account for extrapolation from short-term studies to long-term results (typically 10)

MF = Modifying factor to account for additional data uncertainty according to professional judgment (typically 1–10)

3.3   Assess how and in what form contaminants and stressors come into contact with people

In order for there to be a hazard from a contaminant or stressor, there must be a pathway for it to reach a receptor (Figure 3.1). The mere existence of a contaminant does not necessarily pose a risk that needs mitigation unless there is a population that is exposed to it and may experience some detrimental effect. Exposure occurs when a source (e.g., chemical or stressor) is present in a medium (e.g., air, water, soil, or food) that comes into contact with a receptor (human) via an exposure route (e.g., oral, dermal, or inhalation) (Figure 3.2) (Sexton et al. 1992).

Disrupting the transport of a contaminant between any one of these elements reduces the hazard. If, for example, contaminated groundwater is not used for drinking water and does not otherwise come in contact with people, there is not much risk of a health hazard. This situation would be a much lower threat than a situation if such a contaminated groundwater source was used for drinking water or led to a surface water body with aquatic species. However, if the groundwater contamination included a volatile chemical and passed underneath occupied buildings, then there could be a vapor intrusion problem.

Another important consideration is the form of a chemical contaminant. For instance, mercury comes in three forms: elemental, inorganic (primarily mercuric chloride), and organic (methyl mercury). All three forms can be toxic though in different ways. At room temperature, elemental mercury evaporates from a liquid to a vapor form, which can be transported long distances and inhaled. Elemental mercury affects the central nervous system in humans. In its vapor phase, mercury can be changed into other forms of mercury

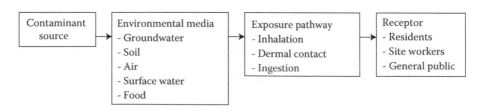

Figure 3.2 Common environmental contaminant exposure pathways.

and can be transported to water or soil in rain or snow. Exposure to inorganic mercury via the oral route (ingestion) can lead to nausea, vomiting, severe abdominal pain, and can damage kidneys. Microorganisms can convert inorganic mercury to organic mercury such as methyl mercury. Methyl mercury can be incorporated into tissues of living organisms and bioaccumulate as higher order organisms consume lower order organisms with methyl mercury. Because of the ease with which methyl mercury can enter living tissue, it is often regarded as the form of mercury with the greatest concern. High levels of methyl mercury can also result in effects to the central nervous system such as blindness, deafness, malaise, speech difficulties, and developmental effects (Agency for Toxic Substances and Disease Registry [ATSDR] 1999).

Exposure pathways are defined by the development of conceptual site models (CSMs). Either pictorially or via a list, a CSM may identify the source of a contaminant, the contaminant itself, the environmental media in which it exists, the route of exposure, whether there are any exposure controls, the affected population, the frequency or duration of exposure, the severity of potential health threats, and the probability of potential health threats. Note that a single contaminant source may have multiple exposure pathways with more than one environmental media, routes, or receptor populations. The goal of a CSM is to assist in evaluating the potential for health impacts. Use of a CSM provides a rational framework to organize what is known about a contaminant, identify what is not known, and helps project managers to make decisions about site remediation.

A CSM should assist in assessing the risk of a health threat contamination source. Consideration of both the severity and probability of a hazard can be used to generate a qualitative risk assessment as per Table 3.1. While there are other approaches to assessing risk, this approach helps decision makers prioritize resources to the sources and exposure pathways that present the greatest risk.

Figure 3.3 presents an example CSM for a landfill containing organic contaminants. In this example, the organic compounds may enter groundwater as landfill leachate, which could enter a drinking water source. The drinking water could then be consumed by the public presenting a critical health threat. However, since there is a landfill liner and leachate recovery system in place, there is little likelihood that the public would actually consume the contaminants. Using the risk assessment in Table 3.1, a critical severity with an unlikely probability indicates a low risk. The organics could also become volatilized into the air directly or from the contaminated groundwater. The volatilized compounds might then enter nearby buildings either on-site, exposing workers to the compounds or off-site, exposing the public. Although there may be a marginal health threat if people inhaled these compounds, a landfill gas recovery system mitigates the probability of the health threat to seldom and presents a low risk.

Table 3.1 Risk assessment matrix

Hazard severity	Hazard probability				
	Frequent	Likely	Occasional	Seldom	Unlikely
Catastrophic	Extremely high	Extremely high	High	High	Moderate
Critical	Extremely high	High	High	Moderate	Low
Marginal	High	Moderate	Moderate	Low	Low
Negligible	Moderate	Low	Low	Low	Low

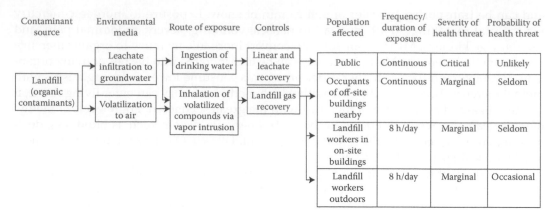

Figure 3.3 Example CSM for organic contaminants from a landfill.

3.4   Measure the physical effects

3.4.1   Sampling and analysis

Measuring the physical effects of a contaminant or stressor begins with measuring the extent of the contamination or condition. Field-deployable direct reading instruments (DRIs) are continuously improving and may be used extensively. Benefits of DRIs include providing real-time results and the ability to detect periods of high concentration for certain instruments such as during monitoring of airborne contamination. While they may provide reasonable qualitative data and reasonable estimates of quantitative data, they may not provide the required specificity, detection limit, or precision. Therefore, more definitive evaluation of contamination remains the result of collecting samples of contaminated media (e.g., air, water, or soil) and conducting analysis in a certified laboratory. It is critical that before a sampling event is conducted, we determine the specific compound to be analyzed (analyte) and the method of analysis. Several organizations have established standard methods for sampling analysis. The USEPA, the National Institute for Occupational Safety and Health (NIOSH), and the Occupational Safety and Health Administration (OSHA), to name a few, each has its own standard methods that specify how to collect samples and analyze for concentrations of certain analytes. These methods describe the required sampling media, collection method, pump flow rate (for air samples), sample holding times and temperatures, number and kinds of sample blanks and replicates, analytical instrument, and other key considerations. Almost always, a sampling event must be preceded by the consultation and coordination with the laboratory that will perform the analysis. Not all laboratories have the same capabilities, and it is critical that the laboratory can indeed perform the needed analysis before samples are collected.

Each sample must be accompanied by a chain of custody, the chronological documentation of the collection, handling, transfer, analysis, and disposition of a sample. If a sample's analytical results were ever called into question in a court of law, the chain of custody would likely be one of the first items to be scrutinized. A properly maintained chain of custody is key evidence that there is not a lapse in control over a sample.

3.4.2 Accounting for sampling and analytical error

It is important to recognize that even under well-controlled conditions there is considerable variability in field and laboratory methods. Carefully calibrated sample air pumps may not perform the same between samples. Highly sophisticated laboratory analytical instruments rarely give the same result for even the same sample. In order to account for this inherent sampling and analytical variability, the USEPA has reported the percent relative standard deviation (%RSD) in their standard analytical methods. The %RSD is the absolute value of the coefficient of variation, which indicates the repeatability and precision of a method. Similarly, OSHA and NIOSH have adopted the sampling and analytical error (SAE) concept, which is a measure of total error of a method and represents the variation from the true sample result. SAE includes sampling, analytical, and pump-related errors. However, it does not capture the inherent variability in concentration levels due to workplace exposure factors.

Terminology used across OSHA and NIOSH differs, as OSHA refers to this combined error as the total coefficient of variation, CV_T, while NIOSH refers to it as the overall precision, S_{rT}. The CV_T and S_{rT} values are specific to the sampling method and should be obtained from the analytical laboratory. For NIOSH sampling methods, the S_{rT} values listed in the method itself can be used if the laboratory does not deviate from the published method (Heline 2012).

For compliance inspections, OSHA uses the SAE to calculate the lower confidence limit (LCL) and upper confidence limit (UCL) with a 95% statistical confidence level, expressed as $LCL_{95\%}$ and $UCL_{95\%}$. Mathematically, the full-period air sampling result, X, is divided by the permissible exposure limit (PEL) to determine the exposure severity, Y, as shown in the following equation:

$$Y = \frac{X}{PEL} \tag{3.2}$$

Next, the SAE is calculated by multiplying the CV_T or S_{rT} by constant 1.645, which is the z-score (i.e., critical value) for a one-tailed test with a 95% level of confidence

$$SAE = CV_T \times 1.645 \tag{3.3}$$

$$SAE = S_{rT} \times 1.645 \tag{3.4}$$

The SAE is then subtracted or added to the exposure severity, in order to establish the $LCL_{95\%}$ and $UCL_{95\%}$, respectively.

$$LCL_{95\%} = Y - SAE$$
$$UCL_{95\%} = Y + SAE \tag{3.5}$$

As detailed in the OSHA Technical Manual (2014), the "OSHA compliance officer's test" (shown further down), utilizes the confidence limits to determine compliance status. For the "possible overexposure" scenario, OSHA would contend that insufficient data exist to document noncompliance. The employer would be encouraged to voluntarily reduce exposures and, ideally, conduct additional air monitoring (OSHA 2014).

- If $UCL_{95\%}$ < 1, then no violation
- If $LCL_{95\%}$ < 1 and $UCL_{95\%}$ > 1, then a possible overexposure exists
- If $LCL_{95\%}$ > 1, then a violation exists

3.4.3 Distribution of sample concentrations

In both environmental and industrial hygiene-related sampling, normal (Gaussian) and lognormal distributions are of particular interest. For example, random SAE, described earlier, is typically assumed to be normally distributed (Ignacio et al. 2006). In the industrial hygiene context, random fluctuations in exposures from shift to shift or within a shift tend to be lognormally distributed. Likewise, concentrations of pollutants in the environment also tend to be lognormally distributed (Gilbert 1987).

In general, an assumption of a lognormal distribution for sampling data is reasonable when there is a physical lower limit (e.g., zero). Graphically, this distribution is typically presented with concentration units on the x-axis and frequency or probability of occurrence on the y-axis, with a skewed-right behavior (Figure 3.4).

The assumption of lognormality also implies that occasional large values are expected, and the environmental health practitioner should not mistakenly dismiss such samples as being invalid. Repeated occurrence of unusual high sample results, beyond an occasional large value, might indicate a data trend worth investigating. For example, an industrial hygienist noting repeatedly high air sample results likely would investigate possible failure of an engineering control (e.g., industrial ventilation system used to reduce occupational exposures) or a change in the industrial process (e.g., increased quantities of chemicals being used, more units being produced, etc.).

Lognormal distributions, however, should not be analyzed using parametric techniques. Parametric methods, such as those that use the arithmetic mean ($\bar{x}$ or μ) and variance (s^2 or σ^2), apply only to normal distributions (Figure 3.5). In order to use parametric approaches properly, there are certain assumptions about normality that must be made.

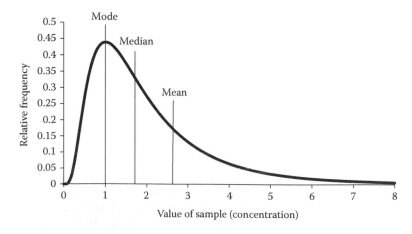

Figure 3.4 Example of a sampling data represented by a lognormal distribution.

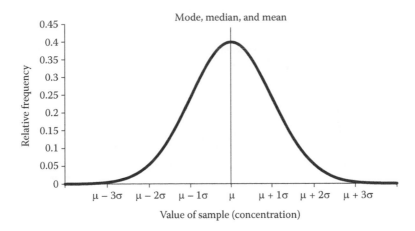

Figure 3.5 Example of a sampling data represented by a normal distribution.

1. Observations are independent of each other (uncorrelated data).
2. Observational error is independent of potential confounding effects (homoscedasticity).

However, environmental health studies rarely involve data sets that meet these criteria for normality. Correlated data is a common condition that prevents the data analyst from properly assuming normal distributions in these kinds of studies. In these cases, including lognormal distributions, nonparametric (distribution-free) methods may be used since they do not rely on the assumption of any particular distribution (Gilbert 1987).

Nonparametric statistical analyses often rank-order or categorize the data. An alternative to using nonparametric techniques is to transform the data to resemble a normal distribution. If the data is lognormally distributed, the random variable, X, may be transformed as $Y = \ln X$. The transformed variable, Y, then follows a normal distribution and can be analyzed using parametric techniques. It is important to use this approach with caution since estimated parameters in the transformed scale may lead to bias when they are transformed back to the original scale.

On an occasion, the environmental health practitioner may conclude that an assumption of lognormality does not apply. One way to test the assumption of lognormality is using the Shapiro and Wilk test, also known as the W-test. A description of how to calculate the W-test statistic can be seen in the literature (Ignacio et al. 2006).

Patterns in the data such as clustering might indicate that samples were taken from dissimilar sources or exposure groups. As an example, environmental samples taken over a period from two separate sources might display more of a bimodal distribution as there are two separate underlying distributions being observed simultaneously.

3.4.4 Decision statistic

The goal of statistical analyses in environmental health data is to decide whether exposures or pollutant concentrations exceed some threshold, usually a regulatory standard. The choice of which decision statistic to use depends on the level of confidence the sample data represent the actual exposures or pollutant concentrations.

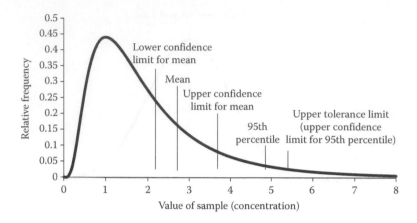

Figure 3.6 Arithmetic mean, 95th percentile, and confidence limits in a lognormal distribution.

In normal distributions, the mean, median, and mode should equal each other (Figure 3.5). The sample mean and standard deviation are also reasonable approximations of the population mean and standard deviation. Parametric techniques can then be easily used for hypothesis testing under these circumstances. For normally distributed data, we can use the sample mean with a confidence interval of a certain number of standard deviations.

Lognormal distributions, however, are not quite as straightforward. Figure 3.4 illustrates bias inherent in attempts to use the arithmetic mean, median, or mode on a lognormally distributed data set.

One approach for lognormal distributions is to use the geometric mean and geometric standard deviation. However, these values are generated from log-transformations, which, as noted earlier, have an inherent bias. Using the geometric mean as an estimate of the true median has a positive bias. However, use of the geometric mean tends to underestimate the true mean. These biases can be reduced if the sample size is sufficiently large (Gilbert 1987).

Many environmental health professionals choose a considerably conservative approach in order to have greater confidence that an exposure will not exceed some standard. Such an approach, especially for lognormally distributed data, uses the 95th percentile as the decision statistic. In order to account for additional uncertainty, we may use an upper tolerance limit of the 95% confidence interval about the 95th percentile (Ignacio et al. 2006). Figure 3.6 illustrates how this more conservative approach compares with using a confidence interval about the arithmetic mean.

3.5   *Apply controls when and where appropriate*

We have heretofore referenced "standards" as promulgated by various regulations. If exposure to a chemical or stressor exceeds such a standard of what may be considered safe, then controls may be required to reduce the exposure to acceptable levels. However, consideration of standards must include examination of which population is to be protected.

Figure 3.7 illustrates the health criteria continuum in which exposure or concentration standards adopted by various agencies tend to focus relative to each other. The lower (left hand) end of the spectrum is in the microgram per cubic meter ($\mu g/m^3$) range, and the upper (right hand) end is the milligram per cubic meter (mg/m^3) range. The criteria

Figure 3.7 Health criteria continuum.

ranges on the bottom show standards from agencies charged with protecting workers and the public. For comparison purposes, the ranges shown on the top of the figure apply to military personnel engaged in military activities.

3.5.1   Protecting the public

One of the primary purposes of the USEPA is to, "ensure that all Americans are protected from significant risks to human health and the environment where they live, learn and work" (USEPA 2014a,b). Because this agency's charge includes protecting populations that may be more vulnerable to the effects of pollution such as the young, elderly, and ill, its exposure standards tend to be quite strict with orders of magnitude safety factors built into the standard. Another purpose of the USEPA is to ensure the diversity, sustainability, and economic productivity of ecosystems (USEPA 2014a,b). In addition, the USEPA has two categories of standards: primary and secondary. Primary standards are intended to protect human health. Secondary standards, which are no less important to the USEPA, are intended to protect the environment. It may be that secondary standards are even more stringent than primary standards. Air pollution secondary standards aim to protect agricultural crops and property. Water pollution secondary standards are intended to ensure that rivers and streams are suitable for fishing and swimming (Moeller 2011).

3.5.2   Protecting workers

The overarching mission of the OSHA is to, "assure safe and healthful working conditions for working men and women by setting and enforcing standards" (OSHA 2014). As part of the U.S. Department of Labor, the OSHA administrator reports to the Secretary of Labor. Occupational health and safety is generally safeguarded by a standards–compliance–enforcement model, in which OSHA standards are promulgated to protect workers, employers are expected to demonstrate compliance with such standards, and OSHA enforces compliance with those standards.

    In general, occupational exposure limits are not meant to be applied against exposures to the general public. Exposure standards for workers are higher (i.e., less stringent) than for the general public because the industrial workforce is typically deemed "healthier" than the general public (Klonne 2011). The industrial workforce usually includes those who can do at least some physical exertion and excludes the very young and very old, along with those with physical ailments limiting their ability to perform work. Combined, these phenomena are referred to as the "healthy worker effect" (Klonne 2011).

    In the United States, OSHA standards normally apply to occupational workplaces, except under certain exclusions. OSHA standards apply to most private sector employers and their workers, along with some public sector employers and their workers in the 50 states and certain territories/jurisdictions under U.S. federal authority. Self-employed workers and workers otherwise regulated by another federal agency (e.g., Mine Safety and Health Administration, Federal Aviation Administration, and U.S. Coast Guard) are not covered by the Occupational Safety and Health (OSH) Act of 1970 and resulting standards.

The OSH Act, which created OSHA, has verbiage encouraging states to develop and operate their own safety and health programs. Employers in states without a plan fall under federal OSHA jurisdiction. Regardless of whether under a state or the federal plan, employers must follow all applicable OSHA safety and health standards. Examples of key requirements include identifying and correcting safety and health hazards; informing workers about chemical hazards through training, labeling, and safety data sheets; notifying OSHA within 8 h of a workplace fatality; providing required personal protective equipment at no cost to workers; and keeping accurate records of work-related injuries and illnesses (OSHA 2014).

OSHA standards are only issued after an extensive process, which includes public notification and comment periods. Further, OSHA must show that a significant occupational risk to workers exists and that there are feasible measures employers can take to protect their workers (OSHA 2014). OSHA standards, under Title 29 of the Code of Federal Regulations (29 CFR), are arranged by industry (e.g., construction, general industry, maritime, and agriculture).

An overarching clause of the OSH Act is what is commonly referred to as the "General Duty Clause." This clause, under Section 5(a)(1) of the OSH Act, requires employers to provide a place of employment, "free from recognized hazards that are causing or likely to cause death or serious physical harm." OSHA, through its inspection process, can issue citations under the General Duty Clause in the absence of specific OSHA standards applying to the hazard(s).

For enforcement, OSHA may perform inspections based on the following priority scheme:

1. Imminent danger
2. Catastrophies—fatalities or hospitalizations
3. Worker complaints and referrals
4. Targeted inspections: for example, particular hazards or industries with high injury rates
5. Follow-up inspections

Violations of OSHA standards are categorized as other-than-serious, serious, willful, repeated, and failure to abate. Penalties, if assigned, may range up to $7,000 for each serious violation and up to $70,000 for each willful or repeated violation. Reductions in penalties may occur based on the employer's "good faith," inspection history, and size of business (OSHA 2014).

Employers may contest any part of the citation including the overall validity of the citation. These employers are offered an opportunity for an informal conference with the OSHA area director, after which a settlement agreement is established to resolve the matter. As an alternative, employers can formally contest the alleged violations and/or penalties by sending written notice to the area director. This request is then forwarded to the Occupational Safety and Health Review Commission (OSHRC), for independent review.

3.5.3   Protection during emergencies

During emergencies, response personnel will likely accept a higher level of risk of adverse health effects in order to mitigate the emergency and protect the health and safety of others. There are several lists of standards for which short-term exposures may be acceptable during emergencies. For example, the National Oceanic and Atmospheric Administration (NOAA) has published *Temporary Emergency Exposure Limits* (TEELs), which are guidelines designed to predict how the general public may respond to different concentrations of chemicals during an emergency response incident (NOAA 2014). *Acute Exposure Guideline Levels* (AEGLs) (USEPA 2014b) and *Emergency Response Planning Guidelines* (ERPGs) (American Industrial Hygiene Association [AIHA] 2013) similarly indicate the airborne chemical concentrations above which different types of health effects may be expected in the unprotected general public. At the extreme end of this spectrum are concentrations which are considered immediately dangerous to life or health (IDLH) as defined by the U.S. National Institute for Occupational Safety and Health (Centers for Disease Control and Prevention [CDC] 2014).

3.5.4   Protection during military operations

Members of the military work in environments that are inherently hazardous and often accept higher levels of risk in order to accomplish the mission. However, this population is predominantly young, healthy, and fairly resilient. In addition, the durations of exposures in deployed environments rarely exceed 1 year at a time for any given individual. In order to estimate exposure standards for these populations, the U.S. Army Public Health Command (USAPHC) has developed *Military Exposure Guidelines* (MEGs) (USAPHC 2010). MEGs have been established for chemicals in air, water, and soil at exposure durations ranging from 1 h to 1 year.

3.6   Conclusions

In order to achieve environmental health conditions with which we would be comfortable living and which protect people and ecosystems, we must be able to assess the extent of exposure to pollutants and determine whether they pose a hazard. Here, we presented a four-step systems approach to making these determinations. These steps incorporate consideration of the source and nature of the contaminant, the exposure route, measurement of the contaminant (taking into consideration important statistical tests), and implementing controls where required depending on the composition of the exposed population. This framework assesses the different dimensions of these issues in their entirety and helps to ensure healthy environments at where we work and live.

References

Agency for Toxic Substances and Disease Registry (ATSDR), *Toxicological Profile for Mercury*. Public Health Service, US Department of Health and Human Services, Atlanta, GA, 1999.

American Industrial Hygiene Association (AIHA), *ERPG/WEEL Handbook*, 2013, https://www.aiha.org/get-involved/AIHAGuidelineFoundation/EmergencyResponsePlanningGuidelines/Documents/ERPGIntroText.pdf, accessed May 30, 2014.

Carson, R., *Silent Spring*, Houghton Mifflin, Boston, MA, 1962.

Centers for Disease Control and Prevention (CDC), http://www.cdc.gov/niosh/idlh/intridl4.html, accessed May 30, 2014.

Gilbert, R.O., *Statistical Methods for Environmental Pollution Monitoring*, John Wiley & Sons, Inc., New York, NY, 1987.

Heline, T.R., *US Air Force School of Aerospace Medicine, Laboratory Sampling Guide*, US Air Force School of Aerospace Medicine, Wright-Patterson Air Force Base, Ohio. AFRL-SA-WP-SR-2012-0008, 2012.

Ignacio, J.S., Bullock, W.H., *A Strategy for Assessing and Managing Occupational Exposures*, Third Edition, American Industrial Hygiene Association (AIHA) Press, Fairfax, VA, 2006.

Klonne, D., Occupational exposure limits. In *The Occupational Environment: Its Evaluation, Control, and Management*, Third Edition. American Industrial Hygiene Association (AIHA) Press, Fairfax, VA, 2011, pp. 57–81.

McKinney, J.D., Richard, A., Waller, C., Newman, M.C., Gerberick, F., The practice of structure activity relationships (SAR) in toxicology, *Toxicological Sciences*. 5(1), 8–17, 2000.

Moeller, D.W., *Environmental Health*, Harvard University Press, Cambridge, MA, 1992.

Moeller, D.W., *Environmental Health*, Fourth Edition, Harvard University Press, Cambridge, MA, 2011.

National Oceanic and Atmospheric Administration (NOAA), http://response.restoration.noaa.gov/oil-and-chemical-spills/chemical-spills/resources/temporary-emergency-exposure-limits-teels.html, accessed May 30, 2014.

Occupational Safety and Health Administration (OSHA), https://www.osha.gov/dts/osta/otm/otm_toc.html, accessed June 6, 2014.

Sexton, K., Selevan, S.G., Wagener, D.K., Lybarger, J.A., Estimating human exposures to environmental pollutants: Availability and utility of existing databases. *Archives of Environmental Health*, 47(6), 398–407, 1992.

Time, The cities: The price of optimism. 94(5), 51–52, 1969.

US Army Public Health Command (USAPHC), *Technical Guide 230, Environmental Health Risk Assessment and Chemical Exposure Guidelines for Deployed Military Personnel*, US Army Public Health Command, Aberdeen Proving Ground, Maryland. June 2010.

US Environmental Protection Agency (USEPA), *Reference Dose (RfD): Description and Use in Health Risk Assessments*, Background Document 1A, Integrated Risk Information, US Environmental Protection Agency, Washington, DC, March 15, 1993.

US Environmental Protection Agency (USEPA), http://www2.epa.gov/aboutepa/our-mission-and-what-we-do, accessed May 30, 2014a.

US Environmental Protection Agency (USEPA), http://www.epa.gov/oppt/aegl/, accessed May 30, 2014b.

chapter four

Measurement of environmental contamination

Stuart A. Willison, Matthew L. Magnuson,
Adrian S. Hanley, and David W. Nehrkorn

Contents

4.1   Introduction

The intentional or unintentional chemical contamination of our nation's infrastructure, including urban dwellings and drinking water distribution systems, continues to be a significant concern due to threat vulnerabilities and our reliance on the aforementioned infrastructure as part of our day-to-day lives. Such contamination can occur as a result of a deliberate act, accidental acts (e.g., oil spills), or natural disasters. An important step for investigating intentional and accidental contamination incidents includes screening samples of water, air, and soil. Tools have been created to assist in such screening, including a U.S. Environmental Protection Agency (USEPA) developed "Response Protocol Toolbox: Planning for and Responding to Drinking Water Contamination Threats and Incidents" (RPTB) (Magnuson and Allgeier 2003, Magnuson et al. 2005) and more recently the Water Security Initiative (WSI) (USEPA 2011a). The RPTB analytical module (USEPA 2003) provides guidance for screening a sample for suspected unknown contaminants. This process relies on a range of analytical technologies and methods for certain target compounds. For instance, EPA Method 525.3 (USEPA 2011b) can be used for a collection of semivolatile organic compounds analyzed by a gas chromatograph mass spectrometer (GC/MS). For compounds of interest not in the target list, the toolbox recommends some generalized screening procedures. Metals are screened by inductively coupled plasma mass spectrometry (ICP-MS) or inductively coupled plasma atomic emission spectrometry (ICP-AES), and organic compounds are screened by mass spectrometry. For those compounds amenable to gas chromatography, the procedure generally follows common GC/MS protocols to identify unknown compounds. For compounds not amenable to GC/MS, one of the approaches suggested is the use of liquid chromatography coupled to a mass spectrometer (LC/MS).

At the time of the release of the toolbox in 2003, GC/MS was used as a powerful tool to assist in identifying unknowns, in part due to decades of development of GC/MS technology, mass spectral libraries, and necessary software. Further, unlike GC/MS that had extensive, transferrable mass spectral libraries, LC/MS did not have reliable libraries available to assist the analyst in identifying unknown compounds. Factors favoring the reliability of GC/MS libraries included a robust, comparatively simple (relative to LC/MS) interface design, high ionization energies, the use of a gas as the chromatographic mobile phase, and more than 50 years of history, scientific advances, and consensus about how to create GC/MS libraries. However, not all potential contaminants are amenable to GC/MS, whereas LC/MS has the potential to identify such compounds quickly, although caution in the application of these LC/MS techniques is recommended as they are subjected to a number of interferences.

LC/MS hardware has matured quickly over the last two decades. Nevertheless, LC/MS libraries continue to be a complex subject. There has been a considerable interest in creating advanced libraries for these instruments to rival those of GC/MS libraries, particularly in areas where some form of sample screening is desired, such as (1) rapid field toxicological screening of patient samples for exposure to drugs; (2) food analysis to investigate pesticide residues, adulterants, and counterfeit products; (3) identification of pharmacologically or toxicologically relevant compounds in drug development; and any other hazardous chemicals of concern with nonvolatile, polar properties that are more amenable to LC/MS analysis than GC/MS.

Screening of environmental samples has received much less attention, perhaps because environmental regulations are typically directed at specific compounds; so, analytical method development involving LC/MS has focused on the reliable determination of those specific compounds in lieu of general screening (USEPA 2011b,c). For instance, EPA Method 538 can be used for the analysis of organic compounds in drinking water by direct aqueous LC/MS.

However, in cases where liquid or aqueous environmental samples need to be screened for contamination, such as drinking water samples, LC/MS is one of a limited number of technical means to achieve this goal, especially for contaminants not amenable to GC/MS. Proper identification is extremely important when screening for contaminants, and the use of any tools made available to assist in the identification process would prove advantageous during an incident. For this reason, the need for libraries for identification of unknowns using LC/MS is just as important as the need for libraries for use with GC/MS analysis. Although there are many types of interfaces that can be combined with a mass spectrometer, the discussion will focus on the LC/MS analysis since it will most likely be the preferred analysis method for nonvolatile, water-soluble compounds during a screening process.

Described herein are the advantages and complications associated with LC/MS analysis, from the attempt to simplify complex instrumentation for an enhanced user base to the development of libraries for use as they are intended, similar to GC/MS. An overview is presented focusing on the potential for LC/MS to identify unknown compounds during screening of suspect samples via library matching. Diverse study designs are presented within the literature examples. Suggestions are provided in three steps for optimizing the usefulness of future studies. Regardless, the best utilization of LC/MS for screening of questioned samples in any discipline is not well understood. Consequently, eight recommendations are suggested for the use of LC/MS for identification of unknown contaminants during screening of suspect samples. These recommendations span considerations ranging from instrument hardware to library software, with particular emphasis on quality control (QC) and analyst expertise. Despite advances in instrument control and data-analysis software, LC/MS library matching is still an area where advanced skill in the analysis of LC/MS data, which is far harder to acquire than skill in the actual mechanical operation of the instrument itself, is required. Nevertheless, all the suggestions provided below indicate the potential to identify an unknown compound quickly, and, in the hands of a skilled analyst, an LC/MS library can be a valuable tool. It should be noted that even with a skilled analyst and proper QC, the library is not the exclusive means of identification of an unknown, but instead is intended to provide additional information to assist the trained analyst in an investigation.

4.2   LC/MS instrumentation and libraries

Chemical signature databases are important to modern analytical sciences for structural determination and identification of unknowns, and such databases were recently reviewed for several types of spectroscopic and spectrometric techniques (Borland et al. 2010). A major goal of the use of chemical signature databases is to enable the use of sophisticated analytical instruments by operators of varying skill level, which increases the number of operators while controlling costs and limiting reliance on analyst expertise. The chapter summarized three characteristics that define their applicability and utility, regardless of an analytical technique: (1) database file formats, (2) spectrometer operating conditions, and (3) data-matching tools. Accordingly, it is instructive to briefly discuss the latter two characteristics as they relate to LC/MS.

4.2.1   LC/MS operation

Two current reviews of LC/MS operation are particularly relevant because they outline the recent developments and future directions for LC/MS in the field of analytical toxicology (Grebe and Singh 2011, Himmelsbach 2012). These reviews are meaningful to sample screening because many LC/MS approaches have been investigated in the field of analytical toxicology. This technique has the potential to quickly and accurately screen a large number of samples with minimal sample preparation, in comparison to other techniques, and reduce the need for tedious sample processing prior to analysis. Limited sample preparation steps will also lower potential human and/or sample error associated with additional handling and processing. Complex matrices may not always allow for limited sample preparation as sample preparation steps may be inevitable. Automated sample preparation procedures can be useful by reducing sample preparation times and minimizing matrix effects and false negatives prior to sample analysis. Automated online procedures may also limit the need for an analyst to handle samples, thus reducing human error and potential cross-contamination.

The modern LC/MS can be divided into essentially three components: the chromatograph, an interface between the chromatograph and the mass spectrometer, and the mass spectrometer itself (Figure 4.1). While various chromatographs can be used, the interface and the spectrometer are perhaps the two components that have the most variability in design and the resulting operation. Multiple varieties of interfaces and mass spectrometers are commercially available to impart the resulting LC/MS system with a range of unique and powerful applications, ranging from identification of unknowns to quantification of regulated compounds. While multiple interfaces are intended to once again assist the analyst with identification and quantification, care with respect to analytical interpretation is needed to ensure that the data are properly managed.

Among interfaces, the development of atmospheric pressure ionization technology such as electrospray ionization (ESI) revolutionized the LC/MS field. ESI has found enormous application in many fields, mainly due to its ability to ionize low volatility and nonvolatile substances, which has also led to its use in applications involving the identification of unknowns. In addition to ESI, a variety of ionization techniques are available for a wide variety of analytes in the mass spectrometer because ESI is not applicable to all analytes and sample types. Examples of additional ionization techniques commonly used include matrix-assisted laser desorption ionization (MALDI) and atmospheric pressure chemical ionization (APCI). MALDI is an ionization technique that uses a laser to bombard a sample mixed with a matrix that will absorb the laser radiation and transfer a proton. APCI is very similar to ESI, but instead of applying voltage to the spray, it is placed on a needle that creates a corona discharge at atmospheric pressures. Furthermore, APCI has overcome some of the pitfalls associated with ESI and is viewed as complementary to ESI because this technique is capable of analyzing low-polarity species. The opportunity to identify unknowns with such a wide array of techniques can complicate the issue and become a limitation if not properly used. For example, data comparability complications will result if multiple techniques are associated with identification of an unknown during screening for a contaminant. The use of

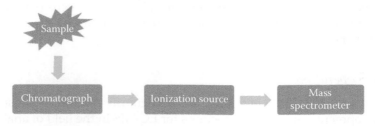

Figure 4.1 Schematic representation of the three components of an LC/MS instrument.

multiple techniques can be an inherent limitation in the identification of unknowns, but it is straightforward to investigate experimentally whether ESI or other ionization techniques will ionize particular compounds of interest.

In addition, there is no standard design among commercial instruments for the ESI interface because there are many ways of employing the physical phenomenon known as electrospray. Unfortunately, the physics underlying the operation of the ESI process is very complex and sensitive to small differences between hardware designs and the physical condition of the interface such as cleanliness and wear. The result is that the appearance of the mass spectra can be greatly influenced across and even within the same design(s).

Among mass spectrometers applied to the identification of unknown compounds, tandem mass spectrometers are popular because they offer potentially high sample throughput, ease of sample preparation, potentially high specificity, and low limits of detection. There are many variants on tandem mass spectrometers, equipped with a variety of mass analyzers ranging from ion traps to quadrupoles to time of flight (TOF). High-resolution mass analyzers, such as TOF or several ion traps, can provide empirical formulas, an additional type of information. Regardless of the analyzer, when combined with an LC, it performs what is nominally known as liquid chromatography–tandem mass spectrometry (LC/MS/MS).

Many of the recent advances in LC/MS/MS technology are designed to co-optimize throughput, specificity (Champarnaud and Hopley 2011), and sensitivity (Grebe and Singh 2011, Himmelsbach 2012) while minimizing some pitfalls inherently associated with LC/MS/MS, recently summarized by Vogeser and Seger (2010). These pitfalls arise due to the ionization process, physics/chemistry occurring within the ion source, and the intricate nature of simultaneously optimizing chromatography conditions and mass spectrometer settings. Background contaminants from sample processing or sample extraction, such as alkali metals or basic compounds are also known to create signal suppression effects, resulting in specificity and sensitivity complications, which will be described later. There are also important considerations regarding the risk of human-related errors from the lower level of automation capabilities available for LC/MS/MS data acquisition and analysis compared to typical GC/MS analysis. The summary of pitfalls (Vogeser and Seger 2010) concludes, "There is a preconceived notion that LC/MS/MS analysis are always highly reliable, but this innovative technology is subjected to gross handling errors as well as the general pitfalls of quantitative chromatographic analyses." Heavy reliance on an instrument's software to perform chromatographic analyses can also generate data with misleading implications. For example, manufacturers have attempted to create more user-friendly and automated approaches for instrumentation usage in an effort to eliminate the need for interpretation of complex data analysis and human error; yet, the results produced from the analysis are not always correct.

4.2.2 *LC/MS data matching through libraries*

Library development for mass spectrometry began in the 1950s for electron ionization (EI), and the proliferation of GC/MS instrumentation among the analytical community in multiple fields has contributed to the development of libraries containing hundreds of thousands of compounds. Much effort has been devoted to understanding how to match mass spectral data to acquired spectra, and many approaches were investigated to help increase the quality of the library match (McLafferty et al. 1991a,b, 1998, 1999), thereby increasing confidence by the user that whatever compound was tentatively identified by the library match was indeed the compound that was present in the unknown sample.

As LC/MS was developed, it was an initial goal for that instrument to produce EI or EI-like spectra (typical of GC/MS), which would then be amenable to matching to existing libraries. Progress in that area was limited, and mass spectrometer designs that sought to achieve that goal resulted in limited applications for the use of LC/MS libraries. Efforts have largely been relegated to history in favor of the modern designs discussed above. There has been a corresponding interest in developing libraries of LC/MS data for these instruments, recognizing that the spectra they produce are not EI spectra, but that perhaps many of the approaches used for data/library matching may be applicable. This quest has led to a deluge of scientific information to understand these libraries and also of commercial and proprietary products to help meet the hardware and software requirements of these libraries.

4.3   General LC/MS screening of unknowns

Screening of unknown samples by LC/MS/MS through the use of library matching has been the subject of many papers spanning a range of scientific disciplines. The purpose of the libraries in these papers falls into two broad categories: (1) small libraries of specific compounds created in a specific laboratory for specific purposes (e.g., to evaluate the presence of a particular compound in a reaction mixture) and (2) large libraries to be used by multiple laboratories in more general screening procedures to determine if a questioned sample contains the compounds represented in the library. The latter purpose is more representative of the goals of screening for contamination in various samples (e.g., water), and will comprise the majority of the following section.

Many of these large libraries have been investigated in the field of analytical toxicology because of the potential of this technique to quickly and accurately screen a large number of samples with less sample preparation than other techniques. The goals related to screening of samples are similar to those when investigating water contamination. Table 4.1 summarizes recent studies, mostly in the area of analytical toxicology, although others are included from a variety of fields with the similar goal of screening samples for unknown contaminants.

4.3.1   General discussion of LC/MS libraries

The column headings in Table 4.1 are meant to allow comparison of the cited literature with regard to several topics that emerge as cross-cutting issues for the application of LC/MS libraries in screening questioned water samples. These topics are divided into three broad categories: library design, reported performance, and implementation issues. The following section attempts to objectively standardize the various representations of the information that appear in the literature. While there is often a plethora of information within these literature articles, each broad category is divided into several subcategories relevant to the goal of this chapter. Below, each of these categories and subcategories will be discussed in general. Important lessons, which often involve a combination of these factors, that emerge from Table 4.1 will be summarized.

4.3.1.1   Study design

Study design refers to a general approach reported in the specific literature article for studying an aspect of the use of LC/MS for library screening. The study designs usually reflect the particular purposes of the studies, which range from general screening of particular types of matrices to fundamental studies of the science of LC/MS libraries.

Table 4.1 Selected representative literature relevant to understanding the role of LC/MS libraries for water contamination screening

	Design						Performance			Implementation
Purpose of study	Number of platforms	Number of compounds in library	Library-matching approach	Type of data in library	Number of compounds used to verify library[a]	Verification approach	% of successful identification	Measure of success	Approach to interferences[b] and QC measures
Fundamental study of library comparability, followed up by casework samples	4	402	Sophisticated in-house algorithm, including up to 5 step filtering Use ROC curves to balance sensitivity and specificity in setting up algorithm	MS/MS spectra acquired with up to 10 different collision energies	22 (acquired) 98 (from literature)	Overall successful ID across platforms, comparison to public database of spectra, comparison to casework [27], and to another library [18]	98.1% >95% 98.1, 97.3, 91.9% for 9 collision spectra, then 3, then 1 For 21/22 casework samples, results were same as GC/MS when used with a high resolution analyzer	Top hit on library match	1. Visual check for "noticeable discrepancies" 2. Match probability of known compounds had to be above a threshold 3. For casework samples, sample preparation was dilution followed by centrifugation
Cross-platform comparability between quadrupole and linear ion traps	2	3200	Algorithm in [99]	MS/MS and MS/MS/MS spectra	Not applicable	Comparison of number of matches in 100 clinical sample	477/533 (89%) compounds gave similar IDs on both platforms	Algorithm match score	Not applicable
Environmental water	1	22	Vendor supplied software	MS spectra	16	Prepared samples	60	2/3 CID spectra match	
Fundamental study of MS/MS library search capability	Several—review of 65 articles	1743	NIST software	Various	Not applicable	Searches of different combinations of spectra	60%–77% depending on library size and number of times the compound appears in the library	Top rank on library match	Not applicable
Fundamental study of MS/MS pesticide libraries	Severalreviews of 168 articles	490	NIST software	Various	Not applicable	Searches of different combinations of sample spectra and reference spectra	80%–100% depending on search approach and number of sample and reference spectra used	Top rank	Not applicable

(*Continued*)

Table 4.1 (Continued) Selected representative literature relevant to understanding the role of LC/MS libraries for water contamination screening

	Design					Performance			Implementation
Purpose of study	Number of platforms	Number of compounds in library	Library-matching approach	Type of data in library	Number of compounds used to verify library[a]	Verification approach	% of successful identification	Measure of success	Approach to interferences[b] and QC measures
Multi-target screening for analytical toxicology	3	1253	Vendor supplied software	MS/MS date acquired with three collision energies Retention times	25	Comparison with GC/MS for casework sample and proficiency samples Comparison across platforms and to another library [18]	18/31 (casework) 6/8, 5/7, and 4/7 (different proficiency samples) 15/25 or 18/25, depending on platform [28]	Top rank on library match—papers rely on software but notes match can be done manually	Internal deuterated standards for QC Known samples for every 10 injections Samples prepared by dilution or extraction. Results vary with preparation conditions
Analytical toxicology, including extensive utilization of urinary metabolites	1	3200	Two sets of different commercial software	MS/MS and MS/MS/MS spectra	87	Spiked samples, proficiency samples	54/87 studied compounds were detected, in accordance with GC/MS results	Library-matching rank	Protein precipitation Confirmation by a skilled analyst
Transformation products in environmental water	1	52	Vendor supplied software	Exact mass, RT (predicted if no standard), isotopic ratios, ionization mode (+/−), examination of fragment	29	Fortification of samples (showing 19) with reference material	19/29 (identified 19 in the sample, and identified reasons for missing the others)	Top rank on library match	

(Continued)

Table 4.1 (Continued) Selected representative literature relevant to understanding the role of LC/MS libraries for water contamination screening

	Design				Performance				Implementation
Purpose of study	Number of platforms	Number of compounds in library	Library-matching approach	Type of data in library	Number of compounds used to verify library[a]	Verification approach	% of successful identification	Measure of success	Approach to interferences[b] and QC measures
Analytical toxicology—part of screening involving GC and immunoassay	1	800	Vendor (NIST)	MS/MS spectra with ramped collision energies Retention time	104 but unclear if these are unique. Only 15 drugs max were in the test sample	Proficiency test	83%	Top rank on library match	Liquid–liquid extraction and solid-phase extraction
Targeted analytes in drinking water	3	129	Vendor (NIST search engine)	MS/MS spectra for several cone and collision voltages	129	70% NIST probability" or operator judgment	107/129	Identified in two of three laboratories	Drinking water samples were spiked with standards to avoid coelution
Rapid targeted analysis of equine drugs of abuse	1	302	Vendor	MS/MS data Retention time	30 evaluated for long term stability	Fortified samples	Tabulates match performance for various concentrations	Rank in library match and retention time match	Liquid–liquid extraction
Pharma/toxicologically relevant compounds	1	50,500	Vendor	Exact mass	86		Ave # hits with same formula 1.82 ± 2.27 (median 1, range 1–39)		
Pesticide analysis in foods	2 High res MS compared to ESI/MS/MS	297	Vendor, supplemented by authors' expertise	Exact mass (TOF), LC RT, isotope ratios of parents and fragments, exact mass	150		50/60 gave same result by both methods. 94% of compounds were identified by both techniques	Comparison with LC/MS/MS library	Sample pretreatment with QuEChErS, ensuring adequate chromatography (not too short of run)

[a] This number usually represents a subset of the total number of compounds in the library.
[b] Interferences include coelution, matrix effects, etc.

The first column presented in Table 4.1 lists a general description for the purpose of the study. Some of the studies investigated an important additional purpose that was how well the library performed on multiple LC/MS "platforms," referring to different instrument designs. Investigating the performance across different platforms can be particularly meaningful because, as discussed above, LC/MS libraries are inherently not as transferable between instruments as other libraries (e.g., EI GC/MS). The literature reported in Table 4.1 ranges from 1 to 4 platforms (instrument designs). Certainly, the ability of a particular library methodology to perform on several types of LC/MS instruments suggests value to the approach; however, even experimental designs focusing on single platforms serve to illustrate interesting and important practical aspects relevant to the goal of this chapter.

The number of compounds in the library varies considerably among the presented literature in Table 4.1, reflecting the varied purposes for each study. One experimental approach lists 50,500 entries for a library of pharmacologically relevant compounds, as would be desirable for drug discovery. The desire to build libraries encompassing such a vast quantity of compounds is not always necessary for each field. For a field such as analytical toxicology, the goal of general screening is frequently directed at a targeted number of prescription and illicit drugs. As expected, and depending on the library's intended use, it is a matter of some controversy about the number of compounds that need to be in the library. This controversy is related to the type of mass spectral data that are present in the library. MS/MS transitions are frequently used to increase sensitivity to the levels corresponding to targeted compound levels of interest. To increase the quantity of spectral data, these MS/MS spectra are generated under a variety of instrumental conditions.

While MS/MS spectra can lead to better sensitivity compared to full-scan MS, the performance of the library-matching algorithm can ironically be degraded because only those transitions, akin to the application of selected ion monitoring in GC/MS, are monitored. Thus, the possibility of generating false negatives is related to both the library size and the sensitivity of the instrument for the compounds in the library. The trade-off between sensitivity and specificity is actually a long-standing, complex issue that has not yet systematically been addressed in the literature. Nonetheless, even without detailed consideration, there is much that can be learned from the literature presented in Table 4.1 with regard to library size, relative to the goal of general screening of water samples.

4.3.1.2 Library performance

A confounding factor in Section 4.3.1 is the software's ability to be "successful" toward identifying an unknown compound. Success is often related to the ability to generate a high match number using the software supplied by the LC/MS manufacturer. However, it is rarely clear or even known to the user how the software actually generates this number, or how this number can be compared to the software from other manufacturers due to proprietary rights associated with commercial software. The use of standardized programs such as those provided by NIST helps compare results, but only with other studies using the same version of the matching software. Further, some studies report the top match from the software as the identified species, whereas other studies do not. A top match is dependent on identical softwares matching the unknown among the top hits as the measure that the library was successful in identifying the compound. Caution is necessary when such heavy reliance is placed on the software's ability to identify a successful match. An analyst's expertise may also aid in determining whether a successful match may be plausible or incorrect. However, the analyst's experience with a specific compound or specific software may also limit the success rate if libraries are not accurate in identifying unknowns.

The definition of success is related to the verification process used for the library performance. Approaches range from comparing the results of identical library software across several LC/MS platforms to comparing different analytical techniques such as GC/MS. Other verification approaches include proficiency samples or studying a subset of the compound standards in detail within the library software. A verification approach appears to be primarily related to the purpose of the study. For instance, studies that explore fundamentals of LC/MS library science tend to use verification approaches aimed at the fundamental issue. Other more practical studies utilize more realistic samples.

Further complicating issues dealing with the validity of library software stems from various organizations and entities that have proposed different definitions of what constitutes identification (Magnuson et al. 2012a). Using the example of water contamination investigations, the goal of a library match is to provide the possible identity of compounds present in a questioned water sample, while minimizing false negatives. With this in mind, the success of the library is presented as fulfilling this goal, gleaned with utmost respect from the chapter and data presented. Because the intention is to relate the data presented to identifying compounds in water samples, the metric of success summarized in Table 4.1 attempts to present the information in uniform terms across the literature and may differ from the description in the literature that is based on the intended application.

The column labeled "% successful identification" represents an attempt to present the optimal percentage of successful identifications in the respective experiment in terms of the goal of practical aspects of screening water samples. Table 4.1 may therefore present slightly different results than the original literature intended. For instance, some of the investigations presented in Table 4.1 will note the reason why the LC/MS library did not produce a match (e.g., the compound was not in the library) and do not count those results that can be explained against their description of library performance because presumably the library could be modified accordingly. However, the approach in Table 4.1 does count those explainable results in the calculation of the success rate using the philosophy that if a user were to rely on the library as developed, the success rate provides an indication of what can be expected. For instance, there is some probability that any given compound will not be in the library, even for a carefully selected list of target analytes, especially for water contamination where the list of *possible* contaminants is very large (essentially all water-soluble compounds or compounds that can be made soluble).

4.3.1.3   Implementation considerations

The fundamental studies of LC/MS libraries presented in Table 4.1 help define the bounds of LC/MS implementation. The studies that seek to practically implement LC/MS libraries are the ones of most interest in describing practical solutions to problems with LC/MS libraries, such as matrix effects related to particular types of samples. Some authors present solutions to these challenges, which are informative for purposes of this chapter. The solutions range from QC measures to sample preparation techniques to data review philosophies. Most approaches resulting in practical applications incorporate solutions into their overall analytical methodology, beginning from sample collection and ending at data presentation, rather than focusing on improving selected steps within the overall analytical process.

4.3.2   Summary of topics with regard to drinking water screening

4.3.2.1   Full-scan screening versus MS/MS transition monitoring

Sauvage and Marquet (2010) who are proponents of full-scan screening for analytical toxicology, elegantly sum up a lesson learned as, "…we must admit that, owing to the use of the total scan mode as survey scan, our method is much less selective than the target screening method [using MS/MS transition monitoring]…" but argue that it is "…much more suited to clinical or forensic toxicology—where the question asked by clinicians or the judiciary is whether an individual had been intoxicated at all, and not whether they had taken a molecule in a necessarily limited pre-defined list…" The trade-off between sensitivity and selectivity is a timeless debate and will not be resolved here or in the near future. However, the lesson is that for the most effective utilization of any library, it is necessary to understand the trade-off. Morrison et al. (2003) and Swets (1988) illustrate a very effective approach to accomplishing and understanding this trade-off, namely through the use of receiver–operator curves (ROCs) that have gained popularity in comparing the accuracy of diagnostic tests. They can also help the user optimize the library-matching software (Oberacher et al. 2009a,b, 2011, Pavlic et al. 2010). ROCs, because they represent the trade-off between sensitivity and selectivity for every combination, can provide increased understanding over simple tabulation of sensitivity at a few concentrations, which is the approach utilized in some of the examples presented in Table 4.1.

4.3.2.2   Low resolution versus high resolution

High-resolution instruments can achieve better performance than lower-resolution instruments. However, higher-resolution instruments tend to be more expensive, less common, and frequently less sustainable in laboratories that tend to have numerous analytical responsibilities, such as water analysis laboratories. The data provided by high-resolution instruments are reported to be more often complemented by other information, such as prediction of isotopic patterns (for applicable molecules), retention times, and careful examination of MS fragmentation behavior and intensity. Thus, the seeming simplicity of the exact mass can be offset by the complexity required for spectral interpretation, not necessarily lending itself to high-throughput, rapid screening of some types of samples. High-resolution results are typically presented as a range of possibilities, and selection criteria must be applied to choose the "best match" from the set of possibilities. Indeed, retention time and fragmentation can be difficult to predict and/or measure if standards are not available. The literature in Table 4.1 that utilizes high-resolution platforms tends to: (1) spend a lot of effort in either tweaking the data interpretation to account for the effects of retention time and unknown fragmentation behavior or (2) simply note these factors as the largest contributing factors to lower library-matching success.

4.3.2.3   Matrix effects

Matrix effects result from constituents of the sample being studied interfering with the intended analysis and can occur for all types of analytical instrumentation. However, the nature of the electrospray interface makes it vulnerable to some unique matrix effects (Dams et al. 2003, Taylor et al. 2005, Cappiello et al. 2008, Kruve et al. 2008, Gosetti et al. 2010). These include suppression *and* enhancement of the mass spectral intensity, along with formation of adducts between the target analyte and common substances in samples, such as sodium, potassium, and ammonium ions. These matrix effects contribute to the pitfalls of LC/MS/MS summarized earlier (Vogeser and Seger 2010).

The ESI interface has been deftly described as akin to a chemical reactor since many chemical reactions with complex physics can take place in the interface. Consequently, the interface is subject to the effects of analyte composition, design of the interface, matrix effects, and a number of aspects related to the physics of the ionization process. Some consider matrix effects as the "Achilles Heel" of ESI (Taylor et al. 2005). Sauvage et al. (2008) dramatically demonstrated how unexpected compounds in the sample matrix can confound even a well-constructed LC/MS/MS library. Such effects will most likely occur in other fields of interest as well. A recent description of method validation specifically includes matrix effects in ESI mass spectrometry as a validation parameter especially important to ESI MS, alongside the usual metrological considerations that must be validated for all methods such as precision, bias, ruggedness, calibration, etc. (Peters et al. 2007). Many analysts are conscious of these considerations and carefully develop QC procedures to manage these effects in the method of operation or interpretation. Maurer (2006, 2007) noted with surprise that some papers do not address ESI matrix interferences, even though ways to deal with them have been discussed frequently over the past 25 years. Perhaps, part of this is that the popularity of LC/MS instrumentation has caused faster proliferation of instrumentation compared to education about the complex physics and chemistry of the technique. Another part may be reliance on chromatography to resolve analytes from interferences; yet, for complex samples, complete chromatographic resolution is not always possible. Further, matrix effects for ESI can arise after chromatography, for example, by lack of cleanliness of the ESI hardware. While there has been rapid advancement in LC/MS instrumentation, further analyst expertise is still needed to keep pace with the continuously growing field. Such a challenge may be difficult to overcome considering that manufacturers are designing instruments that will require less interaction between the user and instrument. However, it will be extremely important for analysts to acquire such skills to aid in the interpretation of spectra, resolve matrix effects, and interpret library software for proper identification of unknown compounds in samples.

Table 4.1 indicates which studies consider matrix effects in detail. The results from the studies that do not consider matrix effects are not necessarily invalid, but the reader should use caution if he chooses to apply the library approach used in the respective paper. Indeed, some studies that do not consider matrix effects provide excellent insight into fundamental aspects of the science of library matching for LC/MS.

4.3.2.4   Approaches to optimizing library-matching performance

The studies in Table 4.1 allude to approaches used for dealing with interferences and matrix effects. The main approaches include sample pretreatment, optimized library-matching algorithms, and various QC measures. Unfortunately, the same study frequently utilizes these techniques simultaneously so it is difficult to discuss them separately. However, as a general trend, studies that do not report using a cleanup procedure or other approaches for dealing with interferences tend to report lower success rates in library matching. Conscious application of one or a combination of the approaches used to minimize interferences appears to increase the success from approximately 40% to 80% to approximately 95%, which appears to be the upper bound on performance of the most sophisticated mass spectrometers and realistic, appropriate preprocessed matrices. Milman and Zhurkovich (2011) reported a 90%–100% match rate for libraries of selected spectra. The QuEChErS approach, which represents "Quick, Easy, Cheap, Effective, Rugged, and Safe" (Anastassiades et al. 2003), is touted as an easy, streamlined, and inexpensive cleanup procedure for pesticide residue analysis allowing for less matrix effects within sample extracts. The QuEChErS cleanup, when combined with a data-rich, high mass resolution library (Mezcua et al.

2009), leads to about the same performance as a simple dilution, but with a more sophisticated matching algorithm and a high-resolution MS (Pavlic et al. 2010). Note that the fundamental studies involving this algorithm reported a 98% upper bound (Oberacher et al. 2009a,b).

The general observation from Table 4.1 is that commercially available library-matching approaches will not produce performance above the 40%–80% success rate. However, this seems to be a fundamental limitation in mass spectral searching in general. Indeed, one of the fundamental studies in Table 4.1 compares EI GC/MS and LC/MS/MS and finds that under similar library search considerations, the results of both library-matching approaches are approximately 70% (Milman 2005a,b). For those studies in Table 4.1 that provide comparison between EI GC/MS and ESI LC/MS library matches for various realistic samples, the match percentage is similar, although a careful examination of the data shows that the compounds "missed" may vary between GC/MS and LC/MS (Milman 2005a,b, Dresen 2009, 2010). It is important not to conclude that LC/MS libraries are just as "good" as GC/MS (EI) libraries. However, LC/MS libraries can serve the same purpose as GC/MS libraries: to provide the skilled analyst with additional information to help them deduce the composition or identify the unknown compounds. Accordingly, a reference library should therefore not be used as the exclusive means of identification of an unknown and analyst expertise is necessary.

4.3.2.5 *Careful creation of libraries*

The libraries associated with the most successful approaches are carefully created, including careful selection of instrument conditions to overcome the inherent limitations of ESI as a soft ionization technique. Creation of a "successful" library also includes selecting compounds for the library that are expected to interfere with the target compounds of interest. Finally, careful creation of libraries means that the reference spectra and unknown spectra may need to be manipulated and examined to achieve optimal spectral matching, which has been referred to as "filtering." One example of filtering involves eliminating masses outside a specified range around a parent ion to reduce the chance of interfering fragments (Oberacher et al. 2009a,b). Note that in EI GC/MS, filtering involves careful evaluation of the mass spectrum for anomalies not related to fragmentation mechanisms (McLafferty et al. 1991b, 1998). To achieve "filtering" for the LC/MS/MS algorithm with the 95% (Pavlic et al. 2010) to 98.1% match rate (Oberacher et al. 2009a,b) for the fundamental study, many steps are systematically used, reflecting how a skilled analyst would evaluate mass spectra. Another version of "filtering" described for LC/MS/MS libraries is evaluation of the library search result. Some studies will analyze the library search to determine why certain compounds were not identified by other laboratories and suggest/demonstrate that if the problem was corrected, a higher success rate would occur (Cahill et al. 2004, Rosal et al. 2009). However, in practical application, this approach leads to some complex requirements: a very skilled analyst, knowing the identity of all potential contaminants, and optimizing the instrumental conditions for all potential contaminants.

4.3.2.6 *Effect of variations in mass spectral features on library matching*

The way a particular manufacturer implements ESI on its LC/MS instrument can affect both the presence and intensity of mass spectral peaks, which in turn affects the ability to identify the compound using the library-matching software. However, the importance of the presence of peaks and their intensity can vary as a function of the particular matching algorithm. The sophisticated custom algorithm utilizing the "filtering" discussed above attempts to minimize the effects of variations in both peak presence and intensity (Oberacher et al. 2009a,b). A study of water samples across several LC/MS platforms demonstrated that library matches appeared to be acceptable, even though the visual appearance of the spectra in terms of peak intensities was different (Rosal et al. 2009). A small molecule identification approach called "X rank," dealing with variations in peak intensity in a unique way (Mylonas et al. 2009), has also been presented on a single platform to increase match confidence. The current NIST (2011) algorithm allows "weighting" to deemphasize peak intensities in favor of the presence of certain masses. However, this approach is mainly related to the EI process, in which peaks with higher masses tend to have more experimental variation in their intensity. While the weighting factors for EI GC/MS have been explored extensively, different physical mechanisms for the formation of ions and the occurrence of ionic reactions are at work in ESI LC/MS/MS so the optimal weighting factor can be very different from what is useful for EI. Thus, the required tolerance is a function of the software, and how the software works is not necessarily well described by the manufacturer since it is often proprietary. Once again, these challenges require increased skill on the part of the analyst to obtain reliable results.

4.3.2.7   Automation of screening

In terms of the ability of analysts to practically apply library matching for screening purposes, library matching should ideally operate with little or no user intervention; so, the required skill level of the analyst is as low as possible. Manufacturers are advertising instrumentation to be user friendly and with minimal analyst interaction through automated processes, including method development, instrument calibration and tuning, and library matching for LC/MS analysis. With attention to automated library matching, algorithms have been reported that seemingly increase the library match success rate. The seemingly dramatic improvement is not really related to fundamental analytical technology, but instead to the thoughtful application of mass spectral science toward solving an analytical problem. Indeed, an examination of many of the studies in Table 4.1 illustrates that the nominal success rate of the library can be improved by careful consideration and interpretation of the fundamental science involved. However, a common theme associates the increased success rate of libraries with an increased skill of the analyst. An analyst may be required to perform preprocessing steps, implement a sophisticated and unfamiliar matching algorithm, or examine the matching spectra to decide if they make sense. The resulting requirements inherently decrease the degree to which screening can be automated. Moreover, the idea of incorporating completely automated processes sound ideal in theory, but in practice, a skilled analyst's ability in conjunction with the advanced algorithms used for library software programs are necessary for a greater likelihood that the success rates of such libraries are high.

4.4   Recommendations

Oberacher et al. (2009a,b) efficiently sum up what makes library matching with LC/MS successful. To paraphrase, it is a carefully selected and applied combination of: (1) a platform that enables accurate and reproducible measurement of fragment ion masses; (2) compound-specific reference spectra collected at several different collision energies; (3) proper instrument calibration and appropriate QC measures to ensure an LC/MS/MS instrument is functioning appropriately for optimal utilization of library searching; (4) filtering of reference spectra; and (5) the use of an appropriate library search algorithm that is comparatively tolerant to mass spectral effects resulting from differing instrument design and maintenance upkeep requirements. The chapter also recommends specific QC measures.

With this summation, and the lessons learned from Table 4.1, the following aspects are recommended for libraries for screening samples suspected of being contaminated.

4.4.1   *Ensure that the user analyzing the data understands the limitations in library-matching performance*

Recognize that even with proper library formation, custom algorithms, peak identification should always be considered tentative, even if a 95% success rate is achieved. The required success rate and the "tentativeness" of the library match is related to the use of the LC/MS data in the overall contamination incident investigation (i.e., the consequence management plan implemented by the user as a whole when investigating and responding to contamination incidents). Recall that the laboratory itself often provides only a piece of the overall puzzle, and, in some instances, even a 60% success rate may be helpful for the investigation. Thus, it is vital to discuss and explain the limitations and meaning of the data to incident managers. If the management is relying on an overestimation of the performance of the LC/MS for library matching, tragic consequences may result from false-negative identifications.

4.4.2   *Understand the impact of variable sample characteristics on screening and resulting limitations for types of samples screened*

Recognize the importance of matrix effects in LC/MS performance and library matching. Matrix effects will affect all sample types. In the case of water samples, seasonal and even daily variation in water quality may be important, particularly if the source of the drinking water is subject to rapid fluctuations. The variation in water quality will be a much more significant problem for some water utilities than others. Typical sources of matrix effects and interferences should be documented. These sources may include components within the water samples themselves such as chromatographically unresolved substances. For example, "natural organic matter" (NOM) refers to a nebulous collection of substances that occur in water via the breakdown of plant materials. The NOM varies among waters and may elute as a very broad chromatographic peak, forming a background on top of which some or all analytes are superimposed. In addition to chromatographically unresolved substances in the water, components in all types of matrices may tend to build up within the hardware of the ESI interface, creating maintenance challenges. Sample preparation procedures prior to analysis may aid in preventing sample and instrument challenges; however, complications may be likely to continue to occur over time. Furthermore, some samples may interact with sample preparation equipment and supplies, releasing unexpected and puzzling interferents.

4.4.3   *Build libraries carefully*

Build and maintain libraries carefully, maintaining the quality of spectra and uniformity of the approach. The libraries should be constructed in the context of the intended use of the library, namely, screening of a particular sample type, such as drinking water samples. As noted above, drinking water from different sources may be subject to specific and unique matrix effects. Therefore, extreme care and consideration should be given when sharing libraries with other users. Library entries should be excluded if it is not understood how they were generated, especially by other users who may have different skill levels, instrument maintenance practices, water types, etc. Currently, there is no accepted standard for library creation for LC/MS/MS, although similar concerns have prompted the recent creation of workgroups to develop standards for other instrumental techniques for which libraries are important, such as Raman spectroscopy (Rizzo 2011).

4.4.4 Size and completeness of the library

Ultimately, the success of being able to provide a good match is related to database size and quality. While there may be a comparatively small number of acutely toxic compounds that could be observed as water contaminants, even contamination with less toxic compounds can be significantly disruptive and expensive to address.

In choosing compounds for a library, recognize that even among similar compounds, the mass spectra are subject to ambiguity due to subtle changes in the structure. Fate and transport of the contaminant in a particular matrix produces degradation by-products from hydrolysis, reaction from environmental conditions, or a residual disinfectant. This situation is similar to problems with drug metabolites discussed in the literature in Table 4.1. McLafferty et al. (1998) summarize the effects of structural ambiguities on EI GC/MS library matching, "This basic weakness of mass spectrometry cannot be overcome even by a perfect matching algorithm database." This is an eerily accurate statement for many aspects of LC/MS library performance.

The total number of contaminants that should be in any library should be very large, or false negative and positive matches will occur merely because of the limited number of compounds in the library. As McLafferty et al. (1991b, 1998) put it, "The only mass spectrometry user who does not need a larger database is the one who is sure that every new unknown spectrum will match satisfactorily with the user's present database." If possible, it may be helpful to add as much information, such as relative retention times and responses from other simultaneous LC detectors, to the library as possible.

4.4.5 Quality assurance/QC

In light of the recommendations above, ensure that the mass spectrometer can provide reproducible library matches. Be aware that this aspect of MS performance is heavily influenced by the design of a particular instrument and the library-matching software. Some instrument designs can be more affected by maintenance issues. Regardless of the design, if key areas of the instrument become slightly unclean, the mass spectra can be impacted, resulting in lower quality for library matches. Instrument maintenance can be corrected with a program for ongoing QC.

The QC program should address the performance criteria important for the purpose of the analysis, namely, screening water samples. While there are many examples from which to draw, it is useful to consider two sources that summarize important aspects of a QC program. The first example pertains to methods involving LC/MS/MS designed for compliance monitoring under the regulation of drinking water contaminants. Although these methods are designed for targeted analysis by multiple-reaction monitoring and not screening of unknown samples via full-scan MS/MS, the QC procedures involved are designed to deal with operational considerations and matrix effects specific to drinking water samples. These QC practices aid not only in monitoring method performance and data quality, but also in assessing instrument cleanliness that could impact the screening integrity of MS/MS libraries.

The reader is recommended to review and thoughtfully consider the initial and ongoing procedures, along with acceptance criteria, described in these methods (USEPA 2011b,c), including

1. Demonstration of low system background
2. Demonstration of precision and accuracy
3. Confirmation of minimum reporting level
4. Calibration confirmation
5. Detection limit determination
6. Laboratory reagent blanks
7. Continuing calibration check
8. Laboratory-fortified blank
9. Internal standards and surrogates
10. Laboratory-fortified sample matrix and duplicates
11. Field duplicates
12. QC materials from alternate sources

QC steps are specifically listed in these compliance-monitoring methods, but all procedural steps in these methods are also designed to contribute to the quality of the data. One example is that EPA Method 538 utilizes filtration of samples and notes that when other filter lots of the same brand (or other brands) are used, it is necessary to ensure that the filters do not introduce interferences, presumably reflecting in part how ESI is subject to matrix effects. Thus, the reader is encouraged to consider implications upon the quality of each step in the procedure. For example, from the observation that a specific filter may influence the result, the reader might infer that any changes in common supplies may have dramatic, unexpected effects on performance. Indeed, some sources of glassware have been reported to contribute to the presence of adducts (e.g., sodium), which can confound library matching.

The QC procedures listed above apply to compliance-monitoring methods targeted toward quantitation of specific analytes through the use of multiple-reaction-monitoring MS/MS, whereas most library-matching approaches utilize full-scan MS/MS. Thus, while encompassing important QC considerations for water samples, additional QC steps are necessary to incorporate the needs of qualitative identification. For this purpose, it is useful to consider the "Validation Guidelines for Laboratories Performing Forensic Analysis of Chemical Terrorism" (LeBeau et al. 2005) and "Guidelines for the Identification of Unknown Samples for Laboratories Performing Forensic Analyses for Chemical Terrorism," developed by Federal Bureau of Investigation's (FBI's) Scientific Working Group on Forensic Analysis of Chemical Terrorism (SWGFACT). The U.S. EPA's *Agency Policy Directive Number FEM-2010-01* describes application of SWGFACT guidelines related to the validity of methods of analysis developed for emergency response situations (USEPA 2010).

Screening water samples is related to forensic analysis, and the results of the screening could ultimately become forensic evidence so it seems prudent to adopt similar measures. Specifically, the SWGFACT document directs the establishment of the presence and/or absence of specified analyte(s) or classes. In this case, the performance characteristics are referred to as selectivity and limit of detection. SWGFACT offers recommendations for these performance characteristics that the accompanying quality assurance (QA)/QC program should address (LeBeau et al. 2004, 2005). Note that, as defined by SWGFACT, the term "analytical procedure," as applied to water screening, would involve the entire procedure, ranging from sample preparation to running samples to library matching to data reporting.

These recommended performance parameters include the limit of detection, which is the lowest concentration or amount of analyte that can be detected. The instrumental limit of detection is a measure of instrument performance and even more important is the analytical procedure limit of detection that incorporates the instrumental sensitivity and noise as well as the variability from the sample matrix. Another parameter is selectivity that indicates how well an analytical procedure is free from interferences, including matrix components. While we may determine the existence of interference, it is difficult to claim that no interferences exist. Matrix interferences should be addressed on a matrix-by-matrix basis (LeBeau et al. 2005).

4.4.6   Instrument considerations

The LC/MS instrument used by the analyst must be sustainable in the laboratory, particularly with regard to the availability of skilled analysts to interpret data or implement the sophisticated search algorithm. All but a very few laboratories will be unable to afford and sustain anything but a basic low-resolution LC/MS/MS and the manufacturer's software.

Each type of LC/MS/MS analyzer has its own strengths and weaknesses. Selectivity cannot be assumed simply because an MS/MS analyzer was used. While the MS/MS technique can provide a high degree of selectivity, the true selectivity depends on the number of monitored ions and transitions. The appropriate transitions must be selected. High-resolution MS/MS analyzers can provide empirical formulas, but the high-resolution spectrum does not provide unequivocal identification and must be supported by other information specific to the compound such as retention time, physical properties, expected fragmentation, etc. (Pavlic et al. 2006, Polettini et al. 2008, Mezcua et al. 2009, Thurman et al. 2006). A suitable search algorithm must be available for all these data.

4.4.7   Confirmatory analysis

The focus of this chapter is on screening potentially contaminated water samples. However, should the screening reveal the potential presence of a contaminant, confirmatory analysis may be necessary depending on the seriousness of the response actions suggested by the presence of particular contaminants. Deciding which contaminants may need to be confirmed is part of planning for implementation of the screening program and should be discussed with emergency response managers. For this purpose, confirmation may be a lengthy process and requires a different technique than LC/MS/MS, involves reference materials, and/or perhaps isotopically labeled standards (if available). Laboratories should be aware of resulting challenges.

4.4.8   Verify, validate, and document performance of library matching

The performance of library matching represents a trade-off between sensitivity and specificity. Documenting this could include ROC curves as discussed above or could include other presentations such as tabulations of statistically meaningful recovery as a function of concentration. The recovery will likely be subject to matrix effects so recovery is dependent on the characteristics of the sample being analyzed.

4.5   Summary

There is a considerable amount of valuable technical information available on the subject of performance of LC/MS library matching, but synthesis of all the information is complicated by study design and presentation. Future work might be most useful to the community via the following steps:

1. The use of standardized approaches to creating LC/MS libraries and characterizing the performance of library-matching approaches would allow an easier comparison of various reports. Some examples are alluded to in the recommendations regarding screening of samples, although approaches to other activities may vary.
2. Conscious concern for matrix effects and interferences is essential because these parameters are particularly problematic for ESI LC/MS, yet continue to be overlooked.
3. The literature generally does not deal with the subject of trade-offs between sensitivity and specificity, particularly as it relates to the number of compounds represented in the library. The field could benefit from greater emphasis and consistent, comparable approaches to describing this trade-off, such as ROC curves.

These three steps, along with the eight recommended aspects to consider when utilizing LC/MS/MS for analysis of questioned samples, are aimed mainly at avoiding pitfalls experienced over the years by LC/MS/MS practitioners in related fields. However, the literature review also suggests an additional way that LC/MS/MS can be used to enhance analysis. Recent LC/MS/MS work has sought to identify compounds more definitively by combining data from the parent compound with data from any metabolites, with emphasis on automation of this process. While "metabolites" are strictly present in clinical samples, samples often contain chemical transformation products of analytes. For example, transformation products from water samples result from drinking water disinfection practices, hydrolysis, or other types of reactions with naturally occurring substances in the water. Perhaps, water analysis can benefit from those techniques being developed to assist in metabolite analysis.

In summary, by using commercially available software and careful QC practices, many examples in the literature suggest that the "success" rates of unknown identification by conventional library matching alone are in the 40%–80% range, even for libraries exclusively consisting of target compounds. Improvements upward of around 95% result from better algorithms to interpret the spectra, along with the inclusion of additional data in the library besides mass spectral data. The best of these algorithms do take *some* of the skill out of the interpretation, but ultimately, the ability to identify many unknown peaks accurately *requires* a skilled analyst. Skilled manual review of the spectra of compounds that fail to match can avoid false negatives; conversely, finding false matches will prevent false positives. Without skill and training, it is difficult to know where in the 40%–95% range of success any particular identification falls, regardless of sophistication of the instrument and data-analysis software.

A final note of caution seems appropriate when utilizing LC/MS/MS for screening samples during an emergency situation. Because of the consequences of false negatives during an emergency and the inherent tendencies of existing LC/MS/MS libraries to produce them, it seems inadvisable, scientifically indefensible, and even irresponsible to conclude that a particular compound is absent on the basis of a library search. Such a conclusion should be supported by additional information from the investigation. Thus, the library is not the exclusive means of identification of an unknown but instead provides additional information for a skilled analyst to assist in the investigation.

Acknowledgment

The authors wish to thank Dr. Joan Bursey for helpful discussions. The USEPA, through its Office of Research and Development and Office of Water, collaborated with San Francisco Water, Power, and Sewer in preparing this chapter. This content has been peer and administratively reviewed and has been approved for publication. Note that approval does not signify that the contents necessarily reflect the views of the USEPA or San Francisco Water, Power, and Sewer. Reference herein to any specific commercial product, process, or service by trade name, trademark, manufacturer, or otherwise does not necessarily constitute or imply its endorsement, recommendation, or favoring by the U.S. government. The views and opinions expressed herein do not necessarily state or reflect those of the U.S. government or San Francisco Water, Power, and Sewer and shall not be used for advertising or product endorsement purposes.

References

Anastassiades, M., Lehotay, S.J., Štajnbaher, D., Schenck, F.J. 2003. Fast and easy multiresidue method employing acetonitrile extraction/partitioning and dispersive solidphase extraction for the determination of pesticide residues in produce. *Journal of AOAC International*. 86: 412–431.

Borland, L., Brickhouse, M., Thomas, T., Fountain III, A.W. 2010. Review of chemical signature databases. *Analytical and Bioanalytical Chemistry*. 397(3): 1019–1028.

Cahill, J.D., Furlong, E.T., Burkhardt, M.R., Kolpin, D., Anderson, L.G. 2004. Determination of pharmaceutical compounds in surface- and ground-water samples by solid-phase extraction and high-performance liquid chromatography–electrospray ionization mass spectrometry. *Journal of Chromatography A*. 1041(1–2): 171–180.

Cappiello, A., Famiglini, G., Palma, P., Pierini, E., Termopoli, V., Trufelli, H. 2008. Overcoming matrix effects in liquid chromatography–mass spectrometry. *Analytical Chemistry*. 80(23): 9343–9348.

Champarnaud, E., Hopley, C. 2011. Evaluation of the comparability of spectra generated using a tuning point protocol on twelve electrospray ionisation tandem-in-space mass spectrometers. *Rapid Communications in Mass Spectrometry*. 25(8): 1001–1007.

Dams, R., Huestis, M.A., Lambert, W.E., Murphy, C.M. 2003. Matrix effect in bio-analysis of illicit drugs with LC–MS/MS: Influence of ionization type, sample preparation, and biofluid. *Journal of the American Society for Mass Spectrometry*. 14(11): 1290–1294.

Dresen, S., Ferreirós, N., Gnann, H., Zimmermann, R., Weinmann, W. 2010. Detection and identification of 700 drugs by multi-target screening with a 3200 Q TRAP(A (R)) LC–MS/MS system and library searching. *Analytical and Bioanalytical Chemistry*. 396(7): 2425–2434.

Dresen, S., Gergov, M., Politi, L., Halter, C., Weinmann, W. 2009. ESI–MS/MS library of 1,253 compounds for application in forensic and clinical toxicology. *Analytical and Bioanalytical Chemistry*. 395(8): 2521–2526.

Gosetti, F., Mazzucco, E., Zampieri, D., Gennaro, M.C. 2010. Signal suppression/enhancement in high-performance liquid chromatography tandem mass spectrometry. *Journal of Chromatography A*. 1217(25): 3929–3937.

Grebe, S.K., Singh, R.J. 2011. LC–MS/MS in the clinical laboratory—Where to from here? *Clinical Biochemistry Review*. 32(21451775): 5–31.

Himmelsbach, M. 2012. 10 years of MS instrumental developments—Impact on LC–MS/MS in clinical chemistry. *Journal of Chromatography B*. 883–884(0): 3–17.

Kruve, A., Kunnapas, A., Herodes, K., Leito, I. 2008. Matrix effects in pesticide multi-residue analysis by liquid chromatography–mass spectrometry. *Journal of Chromatography A*. 1187(1–2): 58–66.

LeBeau, M. et al. 2004. Quality assurance guidelines for laboratories performing forensic analysis of chemical terrorism. *Forensic Science Communication*. 6, http://www.fbi.gov/about-us/lab/forensic-science-communications/fsc/april2004/index.htm/standards/2004_02_standards01.htm, accessed Dec. 2012.

LeBeau, M. et al. 2005. Validation guidelines for laboratories performing forensic analysis of chemical terrorism. *Forensic Science Communication.* 7, http://www.fbi.gov/about-us/lab/forensic-science-communications/fsc/april2005/index.htm/standards/2005_04_standards01.htm, accessed Dec. 2012.

Magnuson, M.L., Allgeier, S.C. 2003. *Overview of the Response Protocol Toolbox* [cited EPA Document No. EPA-817-D-03-007; U.S. Environmental Protection Agency, Washington DC, http://www.epa.gov/safewater/security/pdfs/guide_response_overview.pdf, accessed Dec. 2012].

Magnuson, M.L., Allgeier, S.C., Koch, B., De Leon, R., Hunsinger, H. 2005. Responding to water contamination threats—Planning ahead is the key to dealing with potential terrorism. *Environmental Science and Technology.* 39(7): 153A–159A.

Magnuson, M.L., Satzger, R.D., Alcaraz, A., Brewer, J., Fetterolf, D., Harper, M., Hrynchuk, R. et al. 2012a. Guidelines for the identification of unknown samples for laboratories performing forensic analyses for chemical terrorism. *Journal of Forensic Sciences.* 57(3): 636–642.

Maurer, H.H. 2006. Hyphenated mass spectrometric techniques—Indispensable tools in clinical and forensic toxicology and in doping control. *Journal of Mass Spectrometry.* 41(11): 1399–1413.

Maurer, H.H. 2007. Current role of liquid chromatography–mass spectrometry in clinical and forensic toxicology. *Analytical and Bioanalytical Chemistry.* 388(7): 1315–1325.

McLafferty, F.W., Stauffer, D.A., Twiss-Brooks, A.B., Loh, S.Y. 1991a. An enlarged data base of electron-ionization mass spectra. *Journal of the American Society for Mass Spectrometry.* 2(5): 432–437.

McLafferty, F.W., Stauffer, D.B., Loh, S.Y. 1991b. Comparative evaluations of mass spectral data bases. *Journal of the American Society for Mass Spectrometry.* 2(5): 438–440.

McLafferty, F.W., Stauffer, D.A., Loh, S.Y., Wesdemiotis, C. 1999. Unknown identification using reference mass spectra. Quality evaluation of databases. *Journal of the American Society for Mass Spectrometry.* 10(12): 1229–1240.

McLafferty, F.W., Zhang, M.-Y., Stauffer, D.A., Loh, S.Y. 1998. Comparison of algorithms and databases for matching unknown mass spectra. *Journal of the American Society for Mass Spectrometry.* 9(1): 92–95.

Mezcua, M., Malato, O., Garcia-Reyes, J.F., Molina-Diaz, A., Fernandez-Alba, A.R. 2009. Accurate-mass databases for comprehensive screening of pesticide residues in food by fast liquid chromatography time-of-flight mass spectrometry. *Analytical Chemistry.* 81(3): 913–929.

Milman, B.L. 2005a. Identification of chemical compounds. *TRAC Trends in Analytical Chemistry.* 24(6): 493–508.

Milman, B.L. 2005b. Towards a full reference library of MSn spectra. Testing of a library containing 3126 MS2 spectra of 1743 compounds. *Rapid Communications in Mass Spectrometry.* 19(19): 2833–2839.

Milman, B.L., Zhurkovich, I.K. 2011. Towards a full reference library of MSn spectra. II: A perspective from the library of pesticide spectra extracted from the literature/Internet. *Rapid Communications in Mass Spectrometry.* 25(24): 3697–3705.

Morrison, A.M., Coughlin, K., Shine, J.P., Coull, B.A., Rex, A.C. 2003. Receiver operating characteristic curve analysis of beach water quality indicator variables. *Applied Environmental Microbiology.* 69(11): 6405–6411.

Mylonas, R., Yann, M., Alexandre, M., Binz, P.-A., Budin, N., Fathi, M., Viette, V., Hochstrasser, D., Lisacek, F. 2009. X-Rank: A robust algorithm for small molecule identification using tandem mass spectrometry. *Analytical Chemistry.* 81(18): 7604–7610.

NIST. 2011. *Mass Spectrometry Tools,* http://chemdata.nist.gov/, accessed Dec. 2012.

Oberacher, H., Pavlic, M., Libiseller, K., Schubert, B., Sulyok, M., Schuhmacher, R., Csaszar, E., Kofeler, H.C. 2009a. On the inter-instrument and inter-laboratory transferability of a tandem mass spectral reference library: 1. Results of an Austrian multicenter study. *Journal of Mass Spectrometry.* 44(4): 485–493.

Oberacher, H., Pavlic, M., Libiseller, K., Schubert, B., Sulyok, M., Schuhmacher, R., Csaszar, E., Kofeler, H.C. 2009b. On the inter-instrument and the inter-laboratory transferability of a tandem mass spectral reference library: 2. Optimization and characterization of the search algorithm. *Journal of Mass Spectrometry.* 44(4): 494–502.

Oberacher, H., Weinmann, W., Dresen, S. 2011. Quality evaluation of tandem mass spectral libraries. *Analytical and Bioanalytical Chemistry.* 400(8): 2641–2648.

Pavlic, M., Libiseller, K., Oberacher, H. 2006. Combined use of ESI-QqTOF-MS and ESI-QqTOF-MS/MS with mass-spectral library search for qualitative analysis of drugs. *Analytical and Bioanalytical Chemistry.* 386(1): 69–82.

Pavlic, M., Schubert, B., Libiseller, K., Oberarcher, H. 2010. Comprehensive identification of active compounds in tablets by flow-injection data-dependent tandem mass spectrometry combined with library search. *Forensic Science International.* 197(1–3): 40–47.

Peters, F.T., Drummer, O.H., Musshoff, F. 2007. Validation of new methods. *Forensic Science International.* 165(2–3): 216–224.

Polettini, A., Gottardo, R., Pascali, J.P., Tagliaro, F. 2008. Implementation and performance evaluation of a database of chemical formulas for the screening of pharmaco/toxicologically relevant compounds in biological samples using electrospray ionization–time-of-flight mass spectrometry. *Analytical Chemistry.* 80(8): 3050–3057.

Rizzo, L. 2011. Bioassays as a tool for evaluating advanced oxidation processes in water and wastewater treatment. *Water Research.* 45(15): 4311–4340.

Rosal, C., Betowski, B., Romano, J., Neukom, J., Wesolowski, D., Zintek, L. 2009. The development and inter-laboratory verification of LC–MS libraries for organic chemicals of environmental concern. *Talanta.* 79(3): 810–817.

Sauvage, F.-L., Gaulier, J.-M., Lachatre, G., Marquet, 2008. Pitfalls and prevention strategies for liquid chromatography–tandem mass spectrometry in the selected reaction—Monitoring mode for drug analysis. *Clinical Chemistry.* 54(9): 1519–1527.

Sauvage, F.-L., Marquet, P. 2010. Letter to the editor: ESI–MS–MS library of 1,253 compounds for application in forensic and clinical toxicology. *Analytical and Bioanalytical Chemistry.* 396(5): 1947–1947.

Swets, J.A. 1988. Measuring the accuracy of diagnostic systems. *Science.* 240(3287615): 1285–1293.

Taylor, P.J. et al. 2005. Matrix effects: The Achilles heel of quantitative high-performance liquid chromatography–electrospray-tandem mass spectrometry. *Clinical Biochemistry.* 38(4): 328–334.

Thurman, E.M., Ferrer, I., Malato, O., Fernandez-Alba, A.R. 2006. Feasibility of LC/TOFMS and elemental database searching as a spectral library for pesticides in food. *Food Additives and Contaminants.* 23(11): 1169–1178.

USEPA. 2003. *Analytical Guide—Module 4* [cited EPA Document No. EPA-817-D-03-004; U.S. Environmental Protection Agency, Washington DC, http://www.epa.gov/safewater/watersecurity/pubs/guide_response_module4.pdf, accessed Dec. 2012].

USEPA. 2010. *Ensuring the Validity of Agency Methods Validation and Peer Review Guidelines: Methods of Analysis Developed for Emergency Response Situations,* http://www.epa.gov/fem/pdfs/FEM_Policy2010-01_Emergency_Response_Methods_Final_July_2010.pdf, accessed Dec. 2012.

USEPA. 2011a. *Water Security Initiative,* http://water.epa.gov/infrastructure/watersecurity/lawsregs/initiative.cfm, accessed Dec. 2012.

USEPA. 2011b. *Drinking Water Methods Developed by the National Exposure Research Laboratory (NERL),* http://www.epa.gov/nerlcwww/ordmeth.htm, accessed Dec. 2012.

USEPA. 2011c. *Analytical Methods Developed by the Office of Ground Water and Drinking Water,* http://water.epa.gov/scitech/drinkingwater/labcert/analyticalmethods_ogwdw.cfm, accessed Dec. 2012.

Vogeser, M., Seger, C. 2010. Pitfalls associated with the use of liquid chromatography–tandem mass spectrometry in the clinical laboratory. *Clinical Chemistry.* 56(8): 1234–1244.

chapter five

Measurement of land

Justin D. Delorit

Contents

5.1 Introduction

For thousands of years monarchs and elected people alike have raised, funded, and fielded armies to fight for a number of reasons. Chief among these was, and some might argue still is, control of land. The Crusades of the early part of the second millennium AD were fought to restore Christian access to Jerusalem. Soldiers, from as far away as England, marched across Europe in droves. The journey, treacherous and deadly unto itself, culminated in wars for control of land in modern-day Israel, which was considered holy by Jews, Muslims, and Christians. The "Holy Land" was not particularly arable. It was not a source of rare or valuable minerals, and it was hardly militarily strategic, at least for the Christians. It was a matter of principle that the Crusades were fought, driven by the intrinsic, as opposed to the extrinsic, value of the land. This battle over a swath of dusty desert, one of the biblical proportions, is fought to this day.

Fortunately, war is not always required to obtain land and protect interests; it can be bought. In 1803, during Thomas Jefferson's presidency, a purchase was negotiated that expanded the territory of the United States by nearly 830,000 square miles, at a cost of

42 cents per acre in $2012. The "Louisiana Purchase" gave the United States ownership of all, or part, of what would become 15 modern-day states, including the majority of the Great Plains and most of the western border of the Mississippi River. It is widely accepted that the land purchase was as much a strategic move, to remove French political and military influence from the newly formed United States, as it was an opportunity to increase potential for America's future land development and trade. If you could buy nearly 4 square miles for $1000 in today's money value, you would be foolish not to invest.

A final, albeit geopolitically complicated, historical reference of land grabbing is actually a set of events, which occurred between 1889 and 1895. As Native American tribes were displaced from their traditional lands and moved from reservation to reservation by the federal government, large tracts of virgin, fertile land were left abandoned and open for development by an agriculturally minded American public. In 1891, nearly 1 million acres of land, which had formerly been claimed by several tribes, including the Shawnee and Potawatomi, were divided into 160-acre parcels and sold to settlers looking to farm. In 1893, an 8-million-acre area known as the Cherokee Strip was purchased by the United States at a rate of little more than $1 per acre. These lands were part of what is known by historians as the "Land Rush." Photos from that time show families in covered Conestoga Wagons waiting for the shotgun start to access first-come-first-serve claims. Anyone willing to work the land while living an agrarian, and mostly Spartan, existence was welcome to claim a piece of dirt as long as they were not a Native American.

There is a common thread that binds the aforementioned historical anecdotes that span nearly 1000 years. The land is worth something, and that worth is a function of the sum of values associated with any number of attributes the land possesses. It should be understood that land *is* value, and aside from the pride and prestige associated with ownership of a clump of dirt of any size, we treat our land as investments. However, there is an issue with the way we assign a value to our land. The value is based largely on our subjective "feeling" for what the value of a piece of land ought to be. Even our methods of land measurement for valuation are based on feeling, or at best, some type of comparative, fuzzy math.

If you ever purchased or sold land or a home and used a realtor to help you negotiate a fair price, you are probably familiar with the techniques used to value land. One practice is the use of comparisons or "comps," in which realtors gather price data from the Multiple Listing Service (MLS) from properties of similar characteristics (Pagourtzi et al. 1999). If a buyer is looking for a 3000-square-foot home with four bedrooms and two bathrooms, a realtor will collect sale price data on homes with similar characteristics, which have been sold recently within the same relative vicinity, in order to determine the going rate for a comparable home. This is done in an effort to arm the homebuyer with a perspective on what he or she should expect to pay for that home. The same method is used to assign values to homes being listed by sellers. Typically, realtors give this measurement in price per square foot (ppsf). For example, if it is determined the average price of the aforementioned home is $375,000, based on the square footage the price would be $125 ppsf. However, there is typically a disparity between the asking price and the agreed purchase amount, where winning and losing and getting a "good" deal are often measured by how far above or below the ppsf mark the agreed purchase amount falls, and that is the issue. There is subjectivity involved in the measurement of land value, and aside from ensuring homes and land are roughly the same size and have some of the same major features, the pricing is largely left to a "gut feel."

Can land measurement as a function of value be made more scientific? Can a systems engineering approach be applied to land measurement that extends beyond a surveyor measuring the boundaries and calculating an area, or a realtor assigning a value based on comparably sized lands? This chapter investigates, through case study analysis, a way of categorizing and evaluating land value-altering components (VACs). Although it does not suggest a single model or intend to be all inclusive and exhaustive, it should at least shed light on the relative complexity of land valuation and describe ways in which value can be assigned quantitatively.

5.2   Value-altering component

With regard to measuring land and its value, that is to accurately describe a tract of land and break the subjective "comp" mold, whether it is a quarter-acre lot for a home or a 1000-acre tract of rich-glaciated soil capable of producing over 150,000 of bushels of corn, the individuals or firm performing the measurement must accurately identify the land's attributes. VACs are any measurable asset, tangible or intangible, which has the potential to add or subtract from the value of a tract of land. This chapter divides land VACs into three areas: environment, hydrogeology, and development (Figure 5.1). Each VAC has value-altering subcomponents (VASCs) which further define the specific area under which value can be added or subtracted depending on the health of that measurable attribute. While the list of VACs and VASCs provided in Figure 5.1 should not be considered exhaustive; as components vary based on geographic location and season among other factors, it should be understood that the value of land can be based on any number of significant, measurable attributes.

While each VAC and associated VASC might be considered separately as a function of its own value-altering potential (e.g., timber value as a component of flora, which is a subcomponent of environment), the synergistic effects of multiple components acting together could lead to exponential changes in the value of land. For example, consider a square, 40-acre parcel of forested land, or a quarter-mile-by-quarter mile. This land might have a measureable value under all VAC categories and most VASCs. If the timber is marketable as lumber or pulp, the value is altered. If the habitat is optimal for the presence of legally harvestable species of wildlife, the value is altered. If the parcel

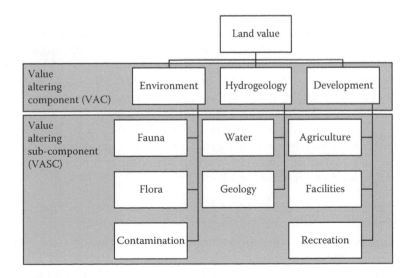

Figure 5.1 Land valuation hierarchy.

has direct access to off-road vehicle trails, government-owned forestland, and a river, the value is altered. If a spot for a home has been cleared, the value is altered. As you may have pictured this evolving parcel of land in your mind, you should have considered how adding VACs and associated VASCs could change the value of the land. In this case, the alterations to value are positive. However, what if the parcel is landlocked and an expensive easement would have to be obtained to ensure access, meaning a road does not adjoin the property, and currently access would have to be made overland? Certainly, the spot cleared for the house seems less valuable, as do the river and trail access. You might wonder how much an easement costs, and how much would it cost to have utilities brought to the site. If the land cannot be accessed, the timber, irrespective of the quality, may not be harvestable without great expense to move equipment on the parcel. The bottom line is that the land is currently inaccessible, and that truth alters the potential value of all the positive VASCs. As the number of VACs and VASCs considered when measuring land value increase, so should the complexity of assigning a value.

5.3   Complexity

The systems engineering discipline attempts to bring order and understanding to evolving, multivariate processes in order to optimize outcomes or outputs. As attributes or processes interact, possible outcomes and system predictability decrease (MITRE Systems Engineering Process Office 2005). Consider predicting the landfall time and location of a hurricane in a weather system. Models have been developed to predict intensity and location with respect to storm location (Kurihara et al. 1998). As time progresses from a forecasted prediction, the reliability and tolerance for intensity and location become poorer. This model is a function of the storm system's attributes and other variables such as surface water temperature, air temperature, prevailing wind speeds, and landmasses in the projected storm track. As variables are added to the model, complexity increases.

With regard to the complexity of land valuation, consider an acre of forested, but inaccessible and nearly undevelopable lowland. You would probably agree it is generally worth significantly less than an acre of choice, buildable beachfront property. The value of beachfront access is greater than the acre of marginally marketable timber on the lowland parcel. However, what if that beachfront property adjoins a harbor frequented by odiferous commercial fishing boats? What if the turquoise water, condo-lined, sugar-sand beaches you envisioned were actually the rocky, bluffed shores of Lake Superior? Do these changes to location change the value of the land? The answer should be, it depends. Maybe you had an eye to further develop the harbor with your acre of beachfront, or you feel drawn to the icy waves and basalt, stamp-sand beaches of Lake Superior. The bottom line, as mentioned earlier, is aside from comparable land association; that is, as lands are measured and therefore valued similarly, measurement is currently a marketing art rather than a science. However, no matter how we slice or compartmentalize land valuation techniques, we will be hard pressed to decrease complexity.

5.4   VAC: Environment

If you have ever read any U.S. environmental protection legislation, the term "environment" is vaguely defined, and intentionally so; it includes everything from subsurface soils to the air. With respect to the environment of any parcel of land, we are probably most concerned with natural VASCs (fauna and flora) and possibly contamination. Generally, we associate these VASCs with larger, recreational, agricultural, and timber producing lands, but they can apply to smaller parcels as well.

5.4.1   The Pike County effect

Pike County is located on the west central border of Illinois and Missouri. It is not unlike the rest of western Illinois, with large agricultural fields, fingers and hollows of hardwood timber, and rolling hills. Over the past 15 years, the land values have increased drastically, and it is not because the soil has miraculously become more productive. Whitetail deer hunters have flocked to Pike County to enjoy the large bucks and high deer density. The once sleepy farming county now boasts numerous guides and outfitters charging $4500 and upward for a chance to harvest a trophy buck during the peak of the whitetail deer breeding season. Annual reports on agricultural land sales are produced in Illinois and detail not only median land values, but also why land draws its value. If a hunter's pockets are deep enough, he or she can purchase a slice of land for $2250–$13,250 an acre depending on the quality of the land (Schnitkey 2012). In regions surrounding Pike County, most tracts of land sold are small (20–60 acres), wooded, and have irregularly shaped fields; these are trademark assets valued by recreational hunters and deer. Reports go as far as to state that the majority of sales reported by realtors are to deer hunters (Aupperle and Schnitkey 2013). Others might prefer to lease land on an annual basis to save on overall costs, paying $4–$60 per acre (Doye and Brorsen 2011).

Although the hunters are after the deer and they are ultimately the VASC causing land prices to increase, it is the rest of the environment that contributes to the exceptional quality of the hunting. The habitat is unparalleled for the deer with generally mild temperatures, abundant food including natural sources like acorns and other browse, and vast lands of soybeans and corn. Therefore, while it is the fauna that derives the value, it is the flora that supports the abundance and quality of the fauna. This same situation, the Pike County effect, has played out in other locations as well, including most of the Southern Plains, Southeast, and Delta Regions, where land value increases between 2000 and 2010 have been tied to hunting use (Nickerson et al. 2012). All have seen increased land values because of quality hunting opportunities.

5.4.2   Timber value in Michigan

Timber is big business in the Upper Peninsula of Michigan, and so is buying and selling timberland. Along with endless recreational uses, vast tracts covered with various types of pine are grown and harvested to produce lumber and paper in Michigan, Minnesota, and Wisconsin (Froese et al. 2007). Timber is generally selectively, or clear-cut harvested, based on the value goals of the landowner. Some owners, who plan to own their land in perpetuity, selectively cut portions of their timber to ensure they generate harvest income on a regular basis. Other owners buy a tract of land to clear the entire tract and then sell it at a reduced price.

In 2009, a nearly 300-acre tract of land was listed in Ewen, Michigan, for a price of nearly $348,000 or $1160 per acre. Although not specifically listed as a timber investment tract, further investigation revealed that over 90% was harvestable hemlock and white pine. At the time the property was listed, land real estate moved slowly, especially above $100,000. A lack of interested buyers and need for fiscal liquidity drove the owners to clear-cut harvest the marketable timber. The property was relisted in 2011 at $148,500 or $495 per acre, a nearly 57% decrease in the price of the land. We can safely assume the owners valued the timber VASC at almost $200,000, which is more than they valued the rest of the VACs combined! This simple story illustrates how the subtraction of a VASC can be quantified and used to assign value land.

5.4.3   Contamination and tax fraud

It is a nightmare situation for landowners when they discover the land they own, work, and live on is contaminated. Among the first questions they ask are, "What is my land contaminated with?" and "How did it get here?" The answers to these questions often result in serious litigation, which results in large settlements to pay for restitution and medical bills. At the forefront of our minds should be the story of Erin Brockovich, who fought against Pacific Gas and Electric Company, suspected by land owners of contaminating groundwater with hexavalent chromium (Brockovich 2014). The result was a settlement in the amount of $333 million for the cost of the health of several people.

We might think of contamination as a negative VASC, and in most cases, when considering sale of a parcel, we are correct. There are also others who seek to devalue their land based on the presence of contamination. In the case of *Schmidt versus Utah State Tax Commission*, the owner of a 2.7-acre, 7000-square-feet home sought a zero valuation of his property since part of it contained contaminated mine wastes (Wurtzler 1999). The tax commission argued that because the owner and his family lived on and grew vegetables on the property and the mine tailings were never identified as requiring testing, the ploy of Schmidt was to avoid paying property taxes. The property was of significant "value-in-use." The court ruled in favor of the tax commission. Rulings like *Schmidt versus Utah State Tax Commission* are not universal. Lawyers suggest there are no collective answers as to the valuation of contaminated land values. Sites in Aspen, Colorado and Midvale, Utah were both contaminated with the same classification of waste, and both were declared Superfund sites. When surveyed afterward, landowners reported the value of the land in Aspen was not affected, but the parcel in Utah was.

As you might suspect, there are many factors, facts, and layers of complexity that contribute to the valuation of contaminated lands. We should consider the nature of the contamination, whether the levels exceed established standards, whether cleanup has been mandated, the cleanup cost, the use of the property, and the relative strength of the surrounding real estate market. The literature suggests these factors be considered at a minimum before assessing the value of the land. For example, legal proceedings show decisions where buyers are awarded the cleanup cost of the land from the seller postsale or the same cost subtracted from the agreed selling price before the sale.

In general, buyers and sellers are probably most concerned with existing contamination, but latest research suggests that natural assets of land can actually clean the environment. Most people understand plants and trees help to convert polluted air, or at least carbon dioxide, into the air humans need to survive. A 2010 study by the National Park and Recreation Association highlighted the value of urban parks in the United States with respect to air pollutant removal, stating the estimated cost avoidance or savings is nearly $500 million per year (Nowak and Heisler 2010). The study also found that urban parks were, on average, 4°F cooler than surrounding areas. Buildings surrounding parks in urban settings have benefitted from the temperature difference with a lower cost of cooling during summer months, along with the aforementioned improved air quality. The study even suggests which tree species are best for temperature reduction and pollutant removal. Additionally, although not directly attributable to the environmental quality of lands, parks only comprise 6% of urban areas, and the estimated structural value of timber tops $300 billion.

Contamination, while typically a value-subtracting VASC, can be viewed from a cost avoidance perspective. The key to understanding any VASC is not to limit its bounds to conventional wisdom or sequester its potential based on preconceived notions. Contamination is the perfect example. We have shown how the presence of contamination can actually provide monetary incentives to land owners and users, both through lowered taxation via land devaluation, and through pollutant removal and temperature reduction.

5.5 VAC: Hydrogeology

5.5.1 Water and the Wetlands Bank

When people discuss land and water together, specifically when considering the value, they are probably talking about waterfront, water view, or accessible properties. Although these aspects are important and as a function of contributing to economic value are probably most impactful, consider the value of water from a hydrologic perspective.

Wetlands can contribute immensely to land value if managed properly. The benefits of wetlands include water storage, nutrient transformation and removal, and production of aquatic life (Novitzki et al. 1997). The degree to which a wetland performs the aforementioned processes is based on many factors including size and location. While it may be difficult to assign an economic value to the benefits of a wetland, federal and state environmental protection and natural resources agencies have established programs to protect wetlands, preserve and enhance existing wetlands, or establish new wetlands, all at the cost of a developer seeking to impact a wetland. It is probably easiest to think of the value created in this situation as a negative VASC or as a disincentive to those wishing to fill in or otherwise destroy productive wetlands.

The Oregon Department of State Lands–Wetlands Program has, like many other states, adopted compensatory mitigation actions or programs establishing requirements that must be followed by people applying for permits to fill in wetlands. At a macro level, the prospective permittee must first determine if the impact can be avoided, then look at options to minimize unavoidable impacts, and lastly consider compensatory mitigation (Oregon Department of State Lands 2004). Compensatory mitigation requires the permittee to perform actions, in an "eye for an eye" manner, that would "pay" for destroying an existing wetland. Such compensatory mitigation actions may include wetland replacement or enhancement actions elsewhere. In the state of Oregon, there are options. Wetlands may be replaced either on the same parcel as the wetland that is being destroyed or elsewhere (Table 5.1). The ratios of compensatory mitigation areas to lost wetlands areas are a function of the probability of successful establishment or construction and further survival of restored, created, or enhanced wetlands as determined (Oregon Department of State Lands 2004). Nevertheless, the emphasis is placed on protecting and fixing what already exists as nature intended, rather than producing an artificial system.

Table 5.1 Oregon Department of State Lands wetlands replacement ratios

Action to be performed by permittee	Ratio of action to be performed to acre lost
Restoration	1 ac. restored: 1 ac. lost
Creation	1.5 ac. created: 1 ac. lost
Enhancement	3 ac. enhanced: 1 ac. lost
Enhancement of cropped	2 ac. enhanced: 1 ac. lost

If the permittee is unable to or does not have the capability to replace the wetlands, individuals may buy credits from a mitigation bank. An individual may purchase a wetland preservation credit, usually at the rate of one credit per acre of impact. Of course, the cost of a credit is variable and based on the quality of the wetland being impacted by the permittee. The proceeds from credit purchases typically are used to create or preserve existing wetlands at the determination of the mitigation bank owners. Keep in mind, regulations of mitigation measures slightly differ from state to state.

While a developer may see wetlands as a roadblock to executing a development plan, and a negative VASC to his profit margins, wetlands mitigation banks are the enforcement of wetlands protection laws and exist to force developers to think about how and where they can develop or to dissuade development of wetlands altogether. When a developer is considering buying a parcel of land, he will generally order a wetlands delineation study in which an environmental firm will analyze the parcel to determine the presence, location, quality, and type of wetlands. The cost of delineation and developing areas not considered wetlands or buying a different parcel altogether is far less than the cost of impacting a wetland and later being forced to pay mitigation costs.

We think of wetlands as a landowner issue and not impactful on a greater or systematic scale, but consider floods that overtook much of the lowlands in the Upper Mississippi and Missouri River Basins in the summer of 1993 (Parrett et al. 1993). Damage to communities, property, and croplands in those fertile regions was substantially worsened by the once sponge-like local flood-plain wetlands having been altered and in many instances filled. The flood control value of wetlands can, therefore, be expressed in terms of excess damage to facilities and infrastructure.

When a parcel of land is listed for sale, seldom are wetlands listed as a positive VASC or at all because of their detriment to development. Understanding the value of water on a piece of land requires us to understand the connectedness of water systems. Recall the hierarchy of wetland replacement ratios from the Oregon Department of State Lands program, where fixing or restoring an existing wetland is the most preferred option by design (Oregon Department of State Lands 2004). Oregon, as well as many other states, recognizes that existing wetlands are part of larger systems of water bodies and are, therefore, more vital to protecting and restoring regional water quality. States impose a negative VASC, or at least the threat of a negative VASC, on those individuals wishing to assign a value to lands, which include water bodies.

5.5.2 Geology

A quick search of vacant land properties in eastern Ohio and West Virginia revealed two commonalities in short order. First, if you are interested in buying land cheaply, do not expect to own anything below the surface of the land, and second, if you want to own what is below, you are expected to pay much more per acre. Mineral rights have always been big money in oil-rich places like Texas and Oklahoma, but a new oil and gas boom has erupted in northern Appalachia in a practice known as fracking. This hot-button issue typically pulls at the heartstrings of those concerned with fracking's impact on environmental media, but a second conflict is quickly rising in the area of land values (Resource Media 2013).

In areas from the Catskills to Texas, fears surrounding the prospect of fracking and other drilling have caused significant impacts to property values. In 2010, a study conducted near Dallas, Texas revealed a 3%–14% decrease in property values for homes located within 1000 ft of drilling platform (Integra Realty Resources–DFW 2010). Another report conducted by Duke University added to the complexity by suggesting two opposing statistics. It found that homes within 1.25 miles of a fracking well actually saw a 11% increase in home values, while homes within the same range of a fracking well using groundwater saw a 13% decrease in home values (Muoio 2012). The disparity of home values found by the studies of Integra Realty Resources and Duke University can be attributed to the method of drilling. Fracking allows for horizontal drilling from a single wellhead. This means resources can be obtained from under the lands of many property owners. If a landowner also owns mineral rights, the drilling operator must obtain permission at great expense to drill through the landowner's property. In the Dallas area, however, drilling for crude oil is generally done vertically, and therefore, only provides positive impacts to the landowner under which the resource sits.

While not only unsightly, both vertical and horizontal fracking drilling platforms and associated infrastructure produce a consistent barrage of noise and noxious fumes. Real estate has been labeled speculative and highly volatile under normal circumstances, but fears over fracking and the associated drilling have undoubtedly impacted regional markets. Speculation and the market for properties in areas suitable for fracking have led to mortgage lenders refusing to assume liability for properties with wells or those subject to leasing (Morrison 2013). Not only have mortgage lenders declined financing, but they have also devalued properties abutting those where drilling platforms or other operations associated with fracking have been established.

As negatively as fracking has been portrayed in terms of environmental impact and property value, certainly we can imagine there is money to be made if a prospective buyer can purchase mineral rights to a parcel of land, or a seller can retain the same rights as part of a sale. Since most individual landowners do not have the ability to perform drilling and fracking operations themselves, rights to frack are sold or leased to drillers or those who own fracking operations. The value can easily be assigned to the ability to lease or sell mineral rights.

Depending on which side of the argument you fall, the VASC of the underlying geology of the parcel an individual owns or is considering selling can be positive or negative. If the individual is looking to retain the parcel for a home or to develop residentially, the VASC may be negative if neighboring landowners decide to install a drilling platform. However, if the owner buys the land to conduct fracking operations or for recreational use, the VASC may be positive. We should understand that the discovery of a new asset or VASC in an area generally leads to a dispute as to whether the discovered asset has a positive or negative effect on land value. The topic of fracking provides us with a vivid example of how the value of land is truly in the eye of the beholder.

5.6 VAC: Development

You have probably driven by countless plywood signs on the edge of a field or in front of an abandoned building that read, "Available," "Build to Suit," or "Zoned Commercial." You may have wondered what you would build there or whether you could afford it. In the age of house flipping and real estate investing, we are tempted to buy that vacant lot or ramshackle house and make our millions. We should consider that development, while a moneymaking and value-adding endeavor, requires considering many VASC facets and complexities, which, when used appropriately or altered, can significantly alter a parcel's value.

Vacant or raw land valuation can be very difficult because VACs and VASCs can be assessed differently by different buyers. Real estate agents suggest navigating the steps outlined in Figure 5.2 as a quantitative guide to satisfying qualitative objectives (Portesky 2014).

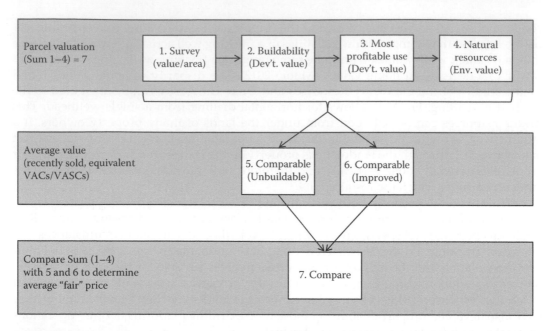

Figure 5.2 Quantitative value comparison of multifaceted lands. (Adapted from United States Department of Agriculture Economic Research Service. 2014. http://www.ers.usda.gov/, accessed June 4, 2014.)

The first step is to obtain an accurate survey of the land you either own or would like to own. Just as with any engineering problem, appropriately framing the question or, in this case, the land and its notable facets, is a necessity. Step two asks parties to define whether or not the land is suitable to build on. County or city development boards typically have this data as well as deeded access records and utility connection information. The third step introduces a novel concept, which can be useful if you intend to use the parcel for a specific purpose beyond what it is advertised and priced as. Determine what the best, in other words, highest value use of the land could be regardless of the buyer's intent. Just because land is not buildable, does not mean it is worthless. (The owners of the 300-acre Michigan parcel proved that when they harvested the timber!) Step four of the process suggests assessing the value of the natural resources. We may include the timber, minerals, and other harvestable resources. The penultimate step five suggests pulling comparative data from recent sales of land with similar assets that were deemed to be unbuildable. It is suggested to convert estimates to a per-square foot or acre price as a parcel of the same size may not have been sold recently. Beware when converting parcels of too big and too small a size to compare reasonably. For example, do not try to compare the per-acre prices for a 40-acre-goal parcel to a 1-acre or 500-acre parcel. Smaller parcels can command higher or lower prices depending on many factors. The same is true with larger parcels. Step six suggests finding recent sales of similar land that has been improved, and subtracting the value of the improvements. For example, if the land you are buying is a great hunting property and others you have researched have cabins and improved game feeding fields or food plots, subtract the value of the dwelling and improvements and derive the per acre price. Steps five and six allow you to formulate a true comparative analysis of surrounding lands. The larger the number of properties you can find and analyze, the better the average you will get, and the keener you will be at identifying overpriced or overvalued properties. The last step requires the summation of the raw, unbuildable land value and any harvestable resources to assign a true value to a property.

5.6.1 Agriculture and the USDA report

The production of grain, produce, and stock for consumption have been a cornerstone of the American economy and way of life since harvest-centric Native Americans and later Europeans first settled on the continent. To this day, "amber waves of grain" still embody and exemplify parts of the United States, and as it is one of the largest agricultural producers in the world, much of its arable land is devoted to this type of production. As with any development industry, agricultural engineers, farmers, and landowners seek to use the natural growth potential of the earth to add value to the economy and to their own pockets.

The United States Department of Agriculture (USDA) publishes an annual report that describes the agricultural land value in the United States with respect to prior year values (United States Department of Agriculture 2013). Major classification divisions are provided for farms, defined as establishments producing at least $1000 of agricultural products. Farm real estate is defined as the value of buildings and dwellings used for agricultural production. Cropland is land used to grow produce only, and not lands rotated between cropland and pasture.

In the month of June, using a "probability-based land-area sampling frame," the USDA selects 9900 land segments distributed throughout 10 economic regions of the contiguous United States (Figure 5.3). Each land segment is approximately 1 square mile or 640 acres. USDA agents contact producers within the boundaries of the preselected land segments and collect value data of land and facilities.

The 2-week effort results in the publishing of average land values, focusing on the percentile increase or decrease in the value of the lands by economic region and state. According to the 2013 USDA report, farm real estate values, which again include the

Figure 5.3 USDA-defined agricultural economic regions. (Adapted from United States Department of Agriculture Economic Research Service. 2014. http://www.ers.usda.gov/, accessed June 4, 2014.)

land and buildings, averaged $2900 per acre, which was a 9.4% increase over 2012; value has doubled since 2000. Value differences by economic region varied from +23.1% in the Northern Plains to no change in the Southeast. The average increase in cropland value, which only covers land, for the Northern Plains and Corn Belt increased by 25.0% and 16.1%, respectively. The report breaks fluctuation of land values from national, to economic region, and to state, over the last decade. Take, for example, state cropland values. By the acre, Illinois average cropland value is $7900. Compare that to a more arid and mountainous Colorado at $1780. It makes sense that Illinois, a state within the Corn Belt, would have higher cropland values, as corn production is a major industry.

According to the Economic Research Service and the Rural Policy Research Institute, in 2008, agriculture generated $130 billion into the U.S. economy, which accounted for 5% of the gross domestic product. It makes sense that there is plenty of literature with regard to agricultural land values, because of its influence on the U.S. economy, and because it employs 14% of the U.S. workforce. For that reason, agriculture is most likely the single most deterministic VASC discussed in this chapter.

5.6.2 Facilities

In Davidson County, Tennessee, the Assessor of Property values homes on a 4-year appraisal cycle. The assessor will visit residences and assign "market value" prices to homes. According to Davidson County, market value is what an informed seller and buyer would expect to receive and pay for the property in a competitive market, respectively (Assessor of Property: Davidson County, TN 2013). It is important to the State of Tennessee that homes and business are frequently valued to ensure that the appropriate amounts of taxes are collected. For example, imagine you bought a foreclosed home in Davidson County in 2013, after it was recently appraised. You plan to do major renovations to the kitchen, bathrooms, and basement with the money you saved by buying the foreclosure. In 2017, the Assessor of Property will revalue your renovated property, at what most likely will be a higher price. You will pay higher taxes, and likely insurance, on your home because of the improvements you have made. Of course, you can file appeals for exemptions, and the Davidson County Assessor of Property provides a litany of information for special cases. However, in the majority of instances, it is in the interest of the state, and ultimately you as a homeowner, to have your home accurately valued.

Facilities are a key piece of land development, and for residential, commercial, and industrial applications, facility values outweigh the value of the land itself. Assessors in Greenwich, Connecticut use a "land factor" which attempts to describe, for a given area or land use type, a method of valuation of an entire market population (Table 5.2) (Gwartney 2000). Assume the assessor in Greenwich inspects properties mentioned in Table 5.2, assigning appropriate values. The land factor is the quotient of the land value and total value of the property. Based on an adequate sample size of properties, an average land portion can be calculated, notwithstanding statistical abnormalities.

The advantage of determining the average land portion value is that it can be used to calculate land values for properties where only the total value of the property is known, as in the case of a recent sale of a home or a business, to assign property taxes or for insurance purposes (Table 5.3).

Note that while potentially useful in valuing facilities and land, this method of average valuation is strictly applicable to a particular category of real estates, such as single-family homes or duplexes. This method becomes particularly difficult when determining

Table 5.2 Example of land value factors

Unique land identification number	Total value of property	Facilities value	= Land value	Land factor (land/total)
17,113	$130,000	$85,000	$45,000	35%
314	$160,000	$111,000	$49,000	31%
116	$125,000	$83,000	$42,000	34%
4036	$157,000	$108,000	$49,000	31%
		Average land portion		33%

Table 5.3 Land values based on calculated land factor

Unique land identification number	Total value of property	×Land factor	= Land value
772	$147,000	0.33	$45,800
1332	$162,000	0.33	$53,500
612	$129,000	0.33	$42,600
420	$138,000	0.33	$45,500

the upper and lower bounds of the category of real estate. Major divisions might include residential, commercial, and industrial properties, as well as classifications by size. A land factor for a two-bedroom condominium, which is a part of a four-condominium facility, might not have the same value as a three-bedroom-detached home in the same neighborhood. Adequately defining the limits of a particular category of real estate is the key to success with regard to using any average value model.

5.6.3   Recreation

While lands valued for recreation derive their value from naturally occurring landforms, fauna, and flora, meaning the environment, it is the development, manipulation, and shaping of these lands, which give them value. So, as recreation straddles the bounds of the environment VAC and development VAC, here, we discuss it specifically with regard to its contribution to the economy through land purchases.

A study conducted by Mississippi State University in 2006, evaluated sales of 100 privately held rural lands, the majority of which (70%) were within in the Mississippi River Delta region from 2002 to 2005 (Jones et al. 2006). The study indicated that of the 13,559 ha surveyed, all were nearly split between forested and agricultural croplands. Buyers surveyed indicated they purchased the land for various reasons, including hunting various wildlife, off-road vehicle access, horseback riding, wildlife watching, ecotourism, and fishing. Research indicated several property characteristics influenced sale prices increasing nearly 36% ($808.73/ha) during the aforementioned period of time. Properties containing lowland hardwood forest, pine-hardwood forest, and wildlife supplemental feeding were preferred to open croplands. Not unexpectedly, the results suggested that wildlife recreation, conservation, and management have produced increases in land values. Of special note was that outdoor recreation is a "cost-effective approach to sustainable economic development" for the state of Mississippi. Additionally the report recommended formal training be developed and delivered to rural land appraisers and lenders to compensate for the recreational potential value being assigned to lands. Whether or not formal training is adopted for appraisers and lenders, it is clear researchers are noticing the recreational value of the land. Not unlike the Pike County effect discussed earlier, this study puts into perspective the potential value that developmentally unimproved or unimprovable land can hold.

5.7   Case study: Michigan revisited

While two people may agree on VACs and VASCs of a parcel of land, their views may be diametrically opposed as to the positivity and negativity assigned to particular components. As hard as we may try to quantify the cost of contamination, the value of trees, or the quality of a wetland, they may never agree to have adequately captured every component and its true value. The work of land measurement and valuation must be a mixture of art and science. Consider a final example of land valuation and measurement, which involves many VACs and VASCs presented in this chapter.

Many have relished ownership of a large tract of land on which to build a cabin, recreate, and retire. The 300-acre parcel discussed earlier in this chapter is an example of such a property before all the timber was cut. However, the lack of timber may not be the only facet that draws a potential buyer away from the property. Most real estate moguls talk about location as though it has more to do with selling a property or a home than the parcel or dwelling itself. However, the discussion of location is in many cases multifaceted and encompasses much more than a lot that backs up to a duck pond or the spot next to the superretailer.

Again, consider the 300-acre parcel in Michigan. It was advertised as surrounded by corporate forestland, meaning the adjoining lands were owned privately but open for recreation to provide the owner a tax break. With no neighbors except the wild animals that had lived there since the glaciers, the parcel may be quite attractive. However, the deeded legal access to the parcel was a two-track road which was not paved, required constant maintenance, and was over a half mile long. Since the area of Michigan in which this property sits gets well over 150 in. of snow annually, its value may not be what it could have been, if it had easier access. Some people may have seen this VASC as an addition through the lens of complete and rugged isolation, but others may find it less appealing.

There was yet another wrinkle to the story with the 300-acre parcel. While vast and relatively flat, it was too low to build. The majority of accessible areas was either behind or surrounded by wetlands. As noted earlier, wetlands are considered a very valuable resource with regard to water quality and are expensive to disturb. Although federal and state regulatory agencies would prefer avoiding development in and around wetlands, a premium can be paid to preserve a wetland elsewhere.

The last geographically challenging facet of the 300-acre tract was utilities. As stated previously, a half-mile, two-track trail had been established to connect the parcel to the main road. Utilities had not been run from the road to the parcel, and this provision was not covered in the possible sale of the property. Running the necessary utilities to the edge of the property, while expensive, would also require disturbing wetlands and purchasing an additional easement.

The cost of developing this property for a residence mounted to a point where the property was far less valuable than it had at first seemed. In other words, the cumulative VACs and VASCs were not worth the advertised price of the property, at least in the opinion of the prospective buyer.

5.8   Conclusions

This chapter has focused on the scientific measurement of land components for valuation purposes. Many of the reference cases presented were selected because the VASC within which they fit could be viewed from many different perspectives, and that is the point. While the list of VACs and VASCs is not all-inclusive, it does provide major subdivisions under which value altering conditions and assets can be assigned. Surely, many more VACs and VASCs can be added, and as you read, you probably thought of several or maybe even a different methodology altogether. As cliché as it may seem, like beauty, land value is ultimately in the eye of the beholder. While facts, figures, measurements, and comparisons can be made and used to derive a value or suggest a notional range of likely values for a given parcel, a landowner, buyer, or seller will, for themselves, decide what a piece of dirt is worth. While land is value, the debate over that value remains.

References

Assessor of Property: Davidson County, TN. 2013. *Assessor of Property*. Retrieved February 27, 2014, from Frequently Asked Questions: http://www.padctn.com/questions.htm#3.

Aupperle, D. E. and Schnitkey, G. 2013. *Illinois Farmland Values and Lease Trends*. Illinois Society of Professional Farm Managers and Rural Appraisers.

Brockovich, E. 2014. Unscripted: Erin Brokovich (E. Brokovich, Interviewer).

Doye, D. and Brorsen, B. W. 2011. *Pasture Land Values: A Green Acres Effect?* Champaign, IL: University of Illinois.

Froese, R., Hyslop, M., Chris, M., Garmon, B., Mc Diarmid, H., Shaw, A. et al. 2007. *Large-Tract Forestland Ownership Change: Land Use, Conservation, and Prosperity in Michigan's Upper Peninsula*. National Wildlife Federation, Merrifield, VA.

Gwartney, T. 2000. *Estimating Land Values*. Greenwich: Understanding Economics.

Integra Realty Resources–DFW. 2010. *Flower Mount Well Site Impact Study*. Dallas/Fort Worth: Integra Realty Sources.

Jones, W. D., Ring, J. K., Jones, J. C., Watson, K., Parvin, D. W., and Munn, I. 2006. Land valuation increases from recreational opportunity: A study of Mississippi rural land sales. *Proceedings of the Annual Conference of SEAFWA*, 60, 49–53. Mississippi State: Southeast Association of Fish and Wildlife Agencies.

Kurihara, Y., Tuleya, R. E., and Bender, M. A. 1998. The GFDL hurricane prediction system and its performance in the 1995 hurricane season. *Monthly Weather Review*, 126 (5), 1306, 17.

MITRE Systems Engineering Process Office. 2005. *Perspectives on Complex-System Engineering*. Bedford, MA: The MITRE Corporation.

Morrison, D. 2013. *SECU Pulls Back from Financing Fracking Property*. Retrieved February 27, 2014, from *Credit Union Times*: http://www.cutimes.com/2013/11/14/secu-pulls-back-from-financing-fracking-property.

Muoio, D. 2012. *The Chronicle: Research*. Retrieved February 27, 2014, from *The Chronicle*: http://www.dukechronicle.com/articles/2012/11/16/duke-researchers-show-dip-home-value-caused-nearby-fracking.

Nickerson, C., Morehart, M., Kuether, T., Beckman, J., Ifft, J., and Williams, R. 2012. *Trends in U.S. Farmland Values and Ownership*. Washington, DC: Economic Research Service/United States Department of Agriculture.

Novitzki, R. P., Smith, R. D., and Fretwell, J. D. 1997. *Restoration, Creation, and Recovery of Wetlands: Wetland Functions, Values, and Assessment*. Sandusky, OH: United States Geological Survey.

Nowak, D. J. and Heisler, G. M. 2010. *Air Quality Effects of Urban Trees and Parks*. Ashburn, VA: National Recreation and Park Association.

Oregon Department of State Lands. 2004. *Just the Facts ... about Compensatory Mitigation for Wetland Impacts*. Salem, OR: Oregon Department of State Lands.

Pagourtzi, E., Assimakopoulos, V., Hatzichristos, T., and French, N. 1999. Real estate appraisal: A review of valuation methods. *Journal of Property Investment and Finance*, 54(4), 18–21.

Parrett, C., Melcher, N. B., and James, R. W. 1993. *Flood Discharges in the Upper Mississippi River Basin—1993*. Sandusky, OH: U.S. Geological Survey Circular 1120-A, 14p.

Portesky, S. 2014. *How to Appraise Vacant Unbuildable Land*. Retrieved February 27, 2014, from Global Post: http://everydaylife.globalpost.com/appraise-vacant-unbuildable-land-16143.html.

Resource Media. 2013. *Resource Media*. Retrieved February 27, 2014, from Resource Media web site: http://www.resource-media.org/drilling-vs-the-american-dream-fracking-impacts-on-property-rights-and-home-values/.

Schnitkey, G. D. 2012. *Land Values Report*. Champaign, IL: University of Illinois, Department of Agriculture and Consumer Economics, Urbana, IL.

United States Department of Agriculture. 2013. *Land Values: 2013 Summary*. Washington, DC: USDA, National Agricultural Statistics Service.

United States Department of Agriculture Economic Research Service. 2014. http://www.ers.usda.gov/, accessed June 4, 2014.

Wurtzler, G. L. 1999. Environmental law: Even contaminated land invariably has some value. *The Colorado Journal*. September 1999, p. 7.

chapter six

Measuring building performance

Steven J. Schuldt

Contents

6.1   Introduction

The construction industry is as old as civilization itself. From skyscrapers in Dubai to single-family homes in rural Iowa, the goal is always the same: to construct a facility that meets the end users' needs. The needs and priorities of facilities are what make the construction industry, quite possibly, the most diverse in the world. There is an endless list of priorities and constraints that the architect, construction manager, and the user must consider before breaking ground. Furthermore, the priorities considered and importance placed on each will likely vary from project to project. For example, accounting for snow load when designing an office building will be of vital importance in St. Paul, Minnesota, but can be completely ignored in San Diego, California. Ultimately, the success of a construction project is typically determined by how well metrics are met at the culmination of construction. For example, was the project completed on time? Was the facility within its allotted budget? Was it constructed according to the design, meeting all legal, environmental, and engineering requirements? Answering yes to these questions speaks of the quality of the construction process but does not truly determine if the project can be called a success. The success of a facility cannot be determined at the point construction is completed, but rather, can only be determined after the facility is in use and the occupants can confirm their needs have been successfully met. If the goal of the new office building in St. Paul was to create a space that would increase worker efficiency by 20%, and if after 5 years occupancy efficiency stayed the same, it would not matter if the project was completed on time and within the budget. Ultimately, the user requirement was not met and the project might not be considered a success.

There has been a transformation in the construction industry over the last few decades. Topics such as climate change and responsible use of natural resources have been brought to the forefront by such events as the Kyoto Protocol and have caused a global movement toward building more sustainable facilities, commonly called "Green Buildings," with decreased energy and resource usage. A number of green building rating and certification methods have emerged, such as Green Globes, the United Kingdom's Building Research Establishment's Environmental Assessment Method, German Sustainable Building Certificate, and the U.S. Green Building Council's Leadership in Energy and Environmental Design (LEED). LEED is the most popular and widely accepted certification in the United States, having been adopted by the Department of Defense (DoD), which requires all new construction (NC) to be LEED certifiable.

Under LEED, buildings are rated and receive credits according to a number of criteria: sustainable sites, water efficiency, energy and atmosphere, materials and resources, and indoor environmental quality (U.S. Green Building Council 2014). Certification level (silver, gold, or platinum) is then awarded based on the total number of credits earned. While LEED and other similar programs are highly accepted, there are potential downfalls. Earning a certain level of certification does not necessarily equate to a sustainable or successful facility. More importantly, being certified as a sustainable facility does not necessarily mean the facility users' requirements identified at the onset of construction have been met.

Sustainability is just one piece of the successful construction puzzle. This chapter will address five areas that, when determined holistically, help measure the true performance of a facility: cost, facility and personnel security, occupant satisfaction, energy consumption, and environmental performance.

6.2   Cost

When calculating the cost of owning and operating a facility, it is imperative to include costs from cradle to grave, that is, from land acquisition and building construction to operation and maintenance, sustainment, and potential demolition. The recipe for creating accurate estimates for these numbers is two-parts art, one part science, and perhaps, a pinch of luck.

Cost estimating involves a degree of prediction and potential forecasting of future trends. For example, let us consider the potential impact of changes in diesel fuel prices as we plan a construction project. Diesel prices impact nearly every aspect of a project, including contractor fuel prices and the cost of all materials due to changes in transportation fees. In March 2009, the average cost of diesel fuel in the United States was $2.09/gallon. Twenty-four months later, the cost skyrocketed to $3.91/gallon (U.S. Energy Information Administration 2014). This 87% increase in fuel costs would likely have had a significant impact on the total construction cost. This example also underscores a certain degree of unpredictability that goes into the cost estimation process.

Before construction can begin, there may be numerous sources of cost on a project to include land acquisition, design fees, environmental studies, fees, and permits. Once construction has begun, there are five basic areas of cost: materials, labor, equipment, subcontracts, and overhead and profit (OH&P).

The material estimation must be done in a manner that makes sense, from a unit of measure used to the amount of waste accounted for. Items are most commonly measured by area, volume, length, or unit (Table 6.1).

Accounting for material waste varies by item and should not exceed 3%–10% of the total material needed. Two examples of accounting for item waste are estimating the number of 2 × 4 studs and length of electrical wiring needed to construct a new home. When framing the home's walls using 10-ft 2 × 4s for a ceiling that is only 9 ft tall, each board will have

Table 6.1 Example units of measure for construction materials

Item	Unit of measure
Carpet, wallpaper, tile	Area, typically square feet (ft^2)
Concrete	Volume, typically cubic yard (yd^3)
Wire	Length, typically linear feet (ft)
Toilets, sinks, showers	Unit, each

about a foot of waste. However, the builder will not stub together nine wasted 1-ft pieces to make a whole 2 × 4. Therefore, there will be 1 ft of waste, or 10%, on each 10-ft 2 × 4. When wiring the electrical outlets and lights, it may be determined that a total of 2345 linear feet of wire is needed; however, wiring is typically sold in increments of 1000 and 500 linear feet. The contractor would be best suited to purchase 2500 linear feet of wire, leaving 155 linear feet of waste (6.2%). Items that are purchased on a unit basis (e.g., toilets and sinks) may not have any waste.

Labor estimates assess how long it will take a workforce to complete a project. Labor time and cost may vary dramatically based on the complexity of the project and the experience of the workers. Labor is typically measured in man-hours per unit produced (0.1 man-hours/ft^2) or units produced per time (50 ft^2/h). Other factors that may affect labor costs are location, presence of labor unions, and associated costs. Direct costs are the wages paid directly to a worker. Associated costs include insurance, taxes, and retirement contributions. It is not atypical for associated costs to be nearly as high as, or even higher than, the salary paid directly to the worker. When using a price book to estimate labor costs, it is important to determine what is included in labor rates and account for associated costs as appropriate.

Equipment costs are estimated at a time-needed basis: hourly, daily, or longer-term rates. The cost of equipment usage is influenced by a number of factors such as fuel costs, maintenance and repair requirements, taxes, insurance and registration, owner depreciation, and availability. If the equipment is common, contractor owned, and already depreciated, the rates may be very low. If the equipment is specialized and mobilized for a short duration, the equipment cost may be significantly higher.

Subcontractor costs can be included in the discussion of OH&P as the presence of subcontractors simply adds another layer of OH&P. *Overhead* is defined by the Dictionary of Construction as *the cost to conduct business other than direct job costs* (Dictionary of Construction 2014a). Examples of overhead costs include both general overhead costs (indirect expenses) like the cost to run the home office and other company expenses such as accountant and lawyer fees, the cost of business insurance, bonds, fees, and licensures, and job overhead costs (direct expenses) such as temporary office facilities, sanitation facilities, temporary enclosures, and construction site drinking water. *Gross profit* is defined as *earnings from an ongoing business after direct and project indirect costs have been deducted from sales revenues for a given period* (Dictionary of Construction 2014b). Subcontractors are typically hired by a general contractor, or prime, to perform specialized labor that the prime does not have organically. Primes most commonly subcontract mechanical, plumbing, and electrical work. Both the prime and subcontractor charge OH&P for their work: the subcontractor for performing the work and the prime for supervising and overseeing that work.

Now that we know the main components of developing a cost estimate, we must discuss where to find sources of the required cost data. It is possible to produce locally developed estimating guides based on a company's years of experience and knowledge of a geographical region; however, the most common cost estimating tools are commercially developed systems. R.S. Means, Dodge, and Craftsman, among others, have created commercial guides. R.S. Means is perhaps the best-known commercial cost-estimating system, producing several manuals (construction, renovation, etc.) and a number of automated systems such as Costworks and E4clicks. Other agencies may require the use of specific cost-estimating tools. For example, the U.S. DoD has developed *The DoD Facilities Pricing Guide*, Unified Facility Criteria (UFC) 3-701-01 (Whole Building Design Guide 2013a).

UFC 3-740-05, *Handbook: Construction Cost Estimating* (Whole Building Design Guide 2011) outlines the four general methods used to perform construction cost estimates. Each sequential method provides a greater likelihood of accuracy but requires additional information about project specifications and local conditions. Table 6.2 gives a summary of the estimation methods.

Even with a detailed project cost estimate, numerous factors can influence the final cost. These factors can be broken into four categories: economy, location, size, and time. The earlier example of increased diesel fuel costs is an example of an economic factor. Other economic factors include labor, equipment, or tool availability. The location certainly

Table 6.2 Methods of project cost estimating

Method of estimating	Level of accuracy	When to use	Information required
Project comparison	−25% to +40%	Early planning stages	Uses historical data of similar, previously completed projects in the same geographic area
Square foot	−14% to +25%	Planning phase	Use of R.S. Means or similar database to determine project SF cost by facility type and location
Parametric cost	−10% to +15%	Design 10%–35% complete	Group several trades into a single item for estimating purposes (e.g. a foundation usually requires excavation, formwork, reinforcing, concrete, and backfill)
Quantity take off	−7.5% to +10%	>35% Design	Divide all work into smallest possible increments, "unit price." Unit prices are then multiplied by total quantity of each item and summed to obtain project cost

Source: Adapted from Whole Building Design Guide. 2011. Unified Facilities Criteria 3-740-05, *Handbook: Construction Cost Estimating*, http://www.wbdg.org/ccb/DOD/UFC/ufc_3_740_05.pdf, accessed Jun 1, 2011.

impacts the project cost, for example, construction in Minnesota is 1.13 times the national average while the office building in San Diego may cost 1.04 times the national average. Constructing in remote locations like Northern Alaska can make delivery of materials and availability of a skilled labor pool incredibly challenging. The size of a project also affects a project's total cost, particularly if additional floors are being added to a facility. The fourth major consideration is time. Perhaps the greatest of all sins in a construction project is not staying on schedule. Delaying certain steps in the construction process can cause significant increases to the total project cost.

6.3   Facility and personnel security

At 9:02 a.m. on April 19, 1995, Timothy McVeigh, an ex-Army soldier and security guard committed what was the most deadly instance of homegrown terrorism in U.S. history. A total of 168 people were killed and several hundred more were injured as a result of the car bomb, which was made out of a mixture of fertilizer, diesel fuel, and other chemicals, that he set off in front of the Alfred P. Murrah Federal Building in Oklahoma City, Oklahoma (Federal Bureau of Investigation 2014). Research of the Oklahoma City bombing showed that the damage went far beyond what was originally caused by the bomb due to a progressive collapse of the building's structural members (Corley et al. 1998). The research of Corley, Sozen, and Thornton, using information developed for the Federal Emergency Management Agency and the Department of Housing and Urban Development, identified numerous structural modifications, including compartmentalized construction, that could be made to help a building survive the effects of catastrophic events such as earthquakes and explosions. These types of changes in the construction industry are instrumental in protecting buildings, although many other simpler and less costly building systems and architectural countermeasures can be made to increase building performance in the areas of facility and personnel security (Smart Buildings 2009).

While very few individuals have faced a devastating situation like the Oklahoma City bombing, most individuals have likely felt a moment of fear as they walk through an unlit parking lot at night. Perhaps in this moment of fear, they look to the corner of the building to check for a surveillance camera. Simple technologies such as exterior lighting and surveillance systems are examples of building systems that can add to the sense of security amongst building occupants. Other examples include robust security and fire alarm systems, facility access controls, and mirrored corners which allow individuals to see what or who may be around the corner of an upcoming turn, whether in a hallway or the building exterior. Another example is shelter-in-place rooms within the facility. The National Terror Alert Response Center describes shelter-in-place rooms as small, interior rooms with no windows in which people can take refuge during an emergency in which hazardous materials may have been released into the atmosphere (The National Terror Alert Response Center 2014). Thus, shelter-in-place rooms can be as simple as an interior bathroom, but they can also be very elaborate, often having their own mechanical systems, allowing cutoff of exterior air in the event of an emergency.

In addition to building systems, architectural countermeasures assist in protecting buildings and its occupants. The most common use of architectural countermeasures is preventing vehicular access to buildings, an action that may have saved the lives of those involved in the Oklahoma City bombing. Vehicular access can be denied through numerous means, employed independently or in unison from the use of landscaping around the facility perimeter to more permanent devices like bollards. Additionally, parking lot locations and incorporating standoff distances can prove vital in protecting a facility. The

DoD has an inherent need to protect its facilities and has created standards for establishing minimum standoff distances (UFC 4-010-01, *DoD Minimum Antiterrorism Standards for Buildings*) based on such criteria as building material, the presence of load bearing versus non-load bearing walls, applicable level of protection (relative threat), applicable explosive weight, and building category (e.g., inhabited building versus family housing) (Whole Building Design Guide 2013b). Additional architectural countermeasures include walls, fences, trenches, trees, ponds, and offsetting of vehicle entrances to reduce vehicle speed.

While not a physical aspect of the building or its surroundings, one cannot complete a discussion on facility and personnel security without considering training and individual preparedness. A safe facility must incorporate emergency preparedness training and policies in order to ensure its occupants can take maximum advantage of the building systems and architectural countermeasures mentioned above. Overall, the security performance of a building can be evaluated by conducting occupant surveys, tracking the number and type of security incidents in the building, analyzing trends, and comparing the data collected to crime statistics in the area (Smart Buildings 2009).

6.4   Occupant satisfaction

The success of a building project typically depends on whether it was completed on time and within budget, as well as whether it met all legal, environmental, and engineering requirements. Occupant satisfaction or its impact on a business's bottom line is rarely considered. However, there is evidence that providing a workplace that focuses on meeting occupant needs, emphasizing worker comfort, can impact the health of the workforce and increase the productivity of employees. An American Society of Interior Designers survey of 200 business decision makers found that 90% of respondents believed an improved office design would increase worker productivity. Furthermore, respondents identified five components of the building environment which they believed would affect worker productivity: comfort and aesthetics, privacy, distractions, flexibility of space and customization, and access to people and resources (Wheeler 1998). To simplify things, we can consolidate these five components into two categories: indoor building environment and facility layout. Let us first look at the effect of an indoor environment.

In 2000, the study of health and productivity gains from better indoor environments reported by William Fisk using theoretical and limited empirical data concluded that there is strong evidence that the number of instances of communicable respiratory illness, allergy and asthma symptoms, sick building syndrome, and worker performance are significantly influenced by the characteristics of the building and the quality of its indoor environment (Fisk 2000). Furthermore, he determined that improving building indoor environments has the potential to produce tremendous annual savings and an increase in workplace productivity to the tune of \$9–\$24 billion from reduced respiratory disease, \$1.5–\$6 billion from reduced allergies and asthma, \$15–\$45 billion from reduced sick building syndrome symptoms, and \$30–\$239 billion from direct improvements in worker performance that are unrelated to health (all values adjusted from 1996 dollars to 2014 dollars using U.S. Inflation Calculator 2014). The estimated savings are a combination of reduced healthcare costs, reduced amount of worker sick leave, and accounting for reduced worker performance during times of illness (Fisk 2000).

Supporting Fisk's claim, a review of 20 years of occupant surveys conducted in the United Kingdom by Leaman (1999) reported a direct relationship between increased control over indoor environment in facilities with natural ventilation and air conditioning and increased worker comfort and perceived productivity. Results from multiple studies and surveys support these findings. Preller et al. (1990) conducted a study of 11,000 workers in 107 buildings in Europe, finding that workers with control over their temperature and ventilation conditions had increased perceived productivity, fewer illness symptoms, and fewer missed days of work. An intervention study in Canada conducted by Menzies et al. (1997) showed a productivity increase of 11% for workers who were given control of air ventilation in their work areas versus a 4% decrease in productivity for the control group. Additionally, reported sick building syndrome symptoms decreased significantly in the intervention group, but not in the control group. In an attempt to quantify the relationship between control of workplace temperature and its effects on worker productivity, Wyon (1996) estimated that a temperature control range of just 3°, plus or minus, would result in worker productivity increases of 7% for typical clerical tasks, 2.7% for logical thinking tasks, 3% for skilled manual work, and 8.6% for very rapid manual work. Brager and deDear (1998) concluded that there is a strong correlation between control over one's environmental conditions, especially temperature and ventilation, and worker comfort, resulting in enhanced work performance.

The second major category affecting workplace productivity is facility layout, specifically the type and amount of space allotted to each worker, and the effect it has on controlling distractions and providing privacy. Sundstrom et al.'s (1980) study of workplace privacy showed a strong link between increased privacy, both psychological and architectural, and increased job satisfaction and job performance. Sundstrom defines privacy in two ways: a psychological state and a physical feature of the environment, he describes psychological privacy as having a sense of control over access to oneself or one's group and architectural privacy as the visual and acoustic isolation provided by an environment. Despite the fact that this study, completed more than 30 years ago, emphasizes the importance of privacy and personal space in the workplace, space and privacy provided to modern-day workers is decreasing precipitously. Gensler's 2013 U.S. Workplace Survey reported that from 2010 to 2012, the average space per worker dropped from 225 to 176 ft². Furthermore, this number is predicted to drop to as low as 100 ft²/person by 2017 (Gensler 2013). Let us take a closer look at Gensler's study to see the impact decreased worker space and privacy is having on workplace productivity.

Gensler's 2013 U.S. Workplace Survey was comprised of a random sample of 2035 respondents representing a broad cross section of demographics, including age, gender, and location, with responses obtained from 10 industry segments. This study concluded that only one in four U.S. workers is in optimal workplace environments, citing a lack of privacy, choice, and autonomy as the biggest hindrances to productivity and job satisfaction. Beyond the loss of space per worker, the numbers are telling: 53% reported being disturbed by others when trying to focus; 42% use makeshift solutions to block out workplace distractions; 77% of employees prefer quiet when trying to focus; and 69% are dissatisfied with noise levels at their primary workspace. These statistics have contributed to a reported 6% reduction in workplace performance when compared to 2008 survey results (Gensler 2013).

Seeing the results of Gensler's survey, one might conclude that the emphasis on workplace collaboration is decreasing. In fact, the numbers support this conclusion as reported time spent collaborating has decreased by 20%; however, that does not mean that collaboration is no longer valued or necessary. In fact, the desire of the modern worker seems to be for more flexibility. Workers expressed the desire to collaborate without sacrificing their ability to focus; they want the best of both worlds. Those employers that provide a spectrum of when and where to work report the highest-performing employees. From a facility and construction standpoint, this underscores the need to construct flexible floor spaces, perhaps even incorporating modular capabilities to allow the space to adapt to changing workplace and personnel requirements. This concept of modular facilities is especially popular in academic settings from partition walls that can change the size and shape of classrooms to incredibly complex modularized laboratories with flexible building systems including the ability to completely move and reconfigure electrical and mechanical components.

In summary, the health, morale, and productivity of employees is influenced tremendously by the quality of his or her workplace and the level of control he has in manipulating the characteristics of his surroundings. Constructing facilities that allow for employee control of building systems, such as temperature and airflow, and workspace orientation creates a winning combination that has been shown to increase worker satisfaction and output and decrease instances of communicable respiratory illness, allergy and asthma symptoms, and sick building syndrome. Improving worker efficiency and decreasing healthcare costs can contribute to the success of a facility for the company owner as much as, if not more than, reducing operation and maintenance costs, extending the life of the facility, or reducing the cost of NC.

6.5 Energy

Energy conservation is an area of primary concern, and it has sparked research initiatives across all industries. Discussions of global warming and terms such as peak oil have driven numerous technological innovations such as hybrid vehicles and solar and wind energy. The market is being driven toward alternative methods of energy production that reduce the reliance on fossil fuels and reverse deleterious effects to our planet. The building industry may be the most important area on which to focus reducing energy use as buildings consume 70% of the total energy in the United States (U.S. Department of Energy 2011). The benefits of reducing energy usage are far reaching, affecting not only the building owner and building occupants but also society as a whole. This section will discuss the assessment of several of said benefits, such as how to measure energy usage from operating an appliance to an overview of building energy baseline codes. Many tools are in place to promote and certify building energy conservation, and issues with these programs should be addressed.

The benefit of energy reduction in buildings positively affects what is commonly known as the "Triple Bottom Line"—profit, people, and planet. First, building energy-efficient facilities reduces energy consumption. According to the U.S. Department of Energy (2011), energy improvements in residential and commercial buildings will reduce primary energy use by an estimated 0.5-quadrillion Btu/year by 2015 and 3.5-quadrillion Btu/year by 2030, the equivalent of eliminating 260 medium (450 MW) power plants. This energy reduction will yield tremendous savings to building owners and occupants. Estimates show that building owners and occupants will save over $30 billion dollars by 2030 (U.S. Department of Energy 2011). The energy improvements will also significantly reduce carbon dioxide (CO_2) emissions. In 2007, building operations (heating, cooling, lighting, water heating, etc.) accounted for 2.5 billion metric tons of CO_2 emissions, 40% of the total U.S. CO_2 emissions (U.S. Energy Information Administration 2007). Estimates show that the energy reductions will reduce CO_2 emissions by roughly 3% by 2030 (U.S. Department of Energy). The reduction of CO_2 emissions rounds out the triple-bottom-line impacts, positively affecting both people and planet by reducing the emission of harmful greenhouse gases into the environment.

How does one measure energy? For the purposes of this discussion, power is defined as the rate at which work is done, or energy is transmitted, and is measured in watts (W) or joules per second (J/s). Energy is defined as power integrated over time and has generic units of power multiplied by time, typically kilowatt hour (kWh) (Diffen.com 2014). Therefore, energy measures the amount of power used over a given amount of time according to Equation 6.1:

$$E = P \times T \tag{6.1}$$

where
   E = Energy (kilowatt hour, kWh)
   P = Power (kilowatts, kW)
   T = Time (hours, h)

We can calculate the cost of using energy in a building if we know the electricity rate charged by the service provider, using Equation 6.2:

$$C = r \times E \tag{6.2}$$

where
   C = Cost (dollars, $)
   r = Electricity rate (dollars per kilowatt hour, $/kWh)

On a smaller scale, we can use these equations to measure the energy use of an appliance. For example, we can calculate the total annual cost of watching a 60-in. flat-screen television for 3 h/day given that the television requires 120 W to operate, and the city's electricity rate is $0.11/kWh

$$E = P \times T = 120\ \text{W} \times \frac{1\ \text{kW}}{1000\ \text{W}} \times \frac{3\ \text{h}}{\text{day}} \times \frac{365\ \text{days}}{\text{year}} = 131.4\ \text{kWh}$$

$$C = r \times E = \frac{\$0.11}{\text{kWh}} \times 131.4\ \text{kWh} = \$14.45$$

There are two primary baseline codes, which are used to regulate the design and construction of new facilities. One of these codes is the International Energy Conservation Code (IECC), which addresses all residential and commercial buildings. The other code is the American Society of Heating, Refrigeration and Air Conditioning Engineers (ASHRAE) Standard 90.1, which addresses commercial buildings that are defined as buildings other than single-family dwellings and multifamily buildings three stories or less above grade. States are not required to adopt these codes, but they are the industry standard for addressing energy-efficiency requirements for the design, materials, and equipment used in construction and renovation projects. The codes address almost every facet of construction, including heating, ventilation, and air-conditioning (HVAC) systems, water heating, lighting, doors and windows, and walls, floor, and ceilings.

Of course, when determining the energy efficiency desired in a NC or renovation project, the building owner must balance desires with cost. A building owner could decide that he or she wants the most energy-efficient home possible, opting to frame the walls with 2 × 6s instead of 2 × 4s, allowing the use of R-21 insulation instead of R-11. He may decide to purchase the most energy-efficient systems possible by purchasing an air conditioner with a seasonal energy efficiency ratio (SEER) rating of 26 and opting to buy a tankless water heater versus a traditional gas or electric unit. However, at some point, the long-term savings resulting from energy efficiency may not outweigh the increases in the initial cost of the facility. ASHRAE 90.1 and IECC, coupled with intelligent designs and capable architects have the ability to take advantage of a space, optimize its orientation and layout, and incorporate appropriate energy-efficient systems and materials to tip the scales toward the benefit of the owner while pursuing desired energy efficiency.

In addition to the baseline codes mentioned above, numerous beyond-code programs exist which put additional emphasis on energy efficiency and sustainability. Well-known examples include the U.S. Environmental Protection Agency's ENERGY STAR certification, the Home Energy Rating System (HERS), the National Association of Home Builders (NAHB) Model Green Home Building Guidelines, and LEED for NC and renovation. While these programs have tremendous value in certifying energy-efficient initiatives in a building, it is imperative to discuss the accuracy of predictive models versus the actual energy usage of facilities. To put it in other words, will an energy-efficient certified building under any program perform as predicted?

Numerous studies have been conducted on the predicted versus actual energy usage of energy-efficient certified buildings. Results typically show that energy-certified buildings, when examined in groups, tend to perform at a higher energy efficiency compared to baseline facilities; however, there is a wide variability when analyzing buildings individually. Torcellini et al. (2004) analyzed six high-performance buildings and concluded that all the buildings performed worse than predicted by models, but that each performed better than comparable code-equivalent buildings. Diamond et al. (2006) analyzed 21 LEED-certified buildings concluding that, as a group, the buildings performed remarkably close to the simulated whole building design data with large variability in individual performance data.

The most extensive investigation was conducted by Turner and Frankel (2008) in which they analyzed 121 LEED NC buildings, representing 22% of the available LEED-certified building stock. In their study, measured performance showed that on average LEED buildings were saving energy, but again individual performance varied wildly, even for LEED platinum facilities, the highest possible certification. They concluded that program-wide, energy modeling was a good predictor of average building energy performance; however, energy modeling was not accurate on a case-by-case basis. Figure 6.1, adapted from their study, shows design versus measured energy use intensity (EUI). National EUI data is derived from the Commercial Building Energy Consumption Survey (CBECS), which is completed every 4 years by the federal Energy Information Administration. Figure 6.2 shows that even LEED gold and platinum buildings often performed well below predicted energy savings levels.

The data and numerous studies are clear; while energy-efficient certified buildings use less energy on average, having an energy-certified building does not automatically translate into energy savings. The fact is modeled energy use does not equal measured energy use. This begs the question, where is the disconnect between these models and actual energy usage in facilities? The solution is not clear, but there are likely two main culprits contributing to the disconnect. First, models and simulations have much room for improvement, particularly when it comes to accurately predicting the performance of

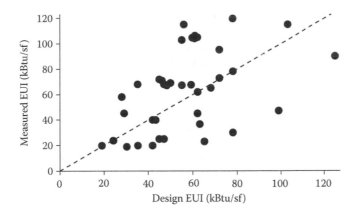

Figure 6.1 Design versus measured energy use intensity (EUI). (Adapted from Turner, C. and Frankel, M. 2008. Energy performance of LEED for new construction buildings. *Final Report to U.S. Green Building Council.*)

individual facilities. Second, no model can accurately predict the behavior and choices of individuals occupying a facility. For example, installing the most efficient air-conditioning unit in the market will save energy when compared to a less efficient air conditioner, but the facility may use more energy overall if the high-efficiency air conditioner is kept 5° colder than expected. The bottom line is builders should not rely on models and simulations to predict their facilities' energy use. The best possible outcome is for builders to construct facilities with the most energy-efficient components financially acceptable, and building occupants should make wise decisions when using energy-consuming systems. When used in tandem, this approach will result in the lowest possible energy use in a facility, regardless of how the actual numbers compare to predicted data.

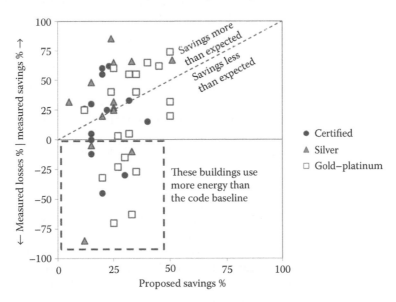

Figure 6.2 LEED gold and platinum buildings. (Adapted from Turner, C. and Frankel, M. 2008. Energy performance of LEED for new construction buildings. *Final Report to U.S. Green Building Council.*)

6.6   Environment

The U.N. World Commission on Environment and Development coined the phrase "sustainable development" in its 1987 Bruntland Report, defining it as development that "meets the needs of the present without compromising the ability of future generations to meet their own needs." From this definition stems the idea of the "Three Ps" or triple bottom line, mentioned in the previous section. A balanced and integrated approach considers price, people, and the planet to create win–win–win solutions in construction. To put it another way, the Whole Building Design Guide describes a sustainable approach as one that meets the intended function of the facility or infrastructure while dedicating a greater commitment to environmental stewardship and conservation, resulting in an optimal balance of cost, environmental, and social benefits (Whole Building Design Guide 2014). Are these concepts mutually exclusive? Can a building that reduces cost and has positive effects on the environment really be constructed? No one can argue the nobility of these statements; however, in order to be widely accepted and successful, this approach needs to make business sense.

Programs such as Green Globes, German Sustainable Building Certificate, and LEED attempt to create a standardized and consistent approach by which buildings can achieve varying levels of green or sustainable certification. Such programs have soared in popularity over the last several decades, even being adopted by the U.S. DoD. Unless these programs become legislated as standard requirements, the movement toward constructing green facilities will only survive the test of the time if it works as advertised. The private sector will only utilize it if it saves builders and facility owners money. As sustainability certification programs have matured, studies to analyze their claims and effectiveness have increased. Specifically, many parties are interested in determining the costs of constructing green facilities versus traditional facilities. The remainder of this section will discuss the major qualitative advantages of constructing green facilities, analyze the research on green building cost data, and investigate two green building success stories.

The Whole Building Design Guide lists six strategies, closely resembling LEED credit categories, which should be incorporated into a sustainable design project:

- Maximize site potential
- Minimize resource and energy consumption
- Conserve water
- Use environmentally preferable products and materials
- Enhance indoor environmental quality
- Optimize operational and maintenance practices (Whole Building Design Guide 2014)

Each of these areas can be utilized to have triple-bottom-line positive impacts. For example, consider maximizing site potential. In a partially wooded area, it may be possible to experience economic benefits by siting a building in such a way that it would not require clear-cutting and extra site preparation. The building may also be optimally oriented to reduce energy costs. A major societal benefit would be increased aesthetics as it is much more appealing to see a building integrated into its environment than surrounded by a massive parking lot and no vegetation. Finally, environmental benefits might include preservation of the land and protection of ecological resources.

Let us consider an example of water efficiency: an office building constructed with interior graywater reuse systems, a porous pavement parking lot, and low-impact development mechanisms designed to manage all on-site stormwater. The economic benefits would include reduced annual water costs and lower municipal costs for wastewater treatment. The societal benefits include water preservation for future generations and a reduced need for wastewater treatment plants. Environmental benefits include reduced potable water use, reduced pollutant discharges, and less strain on aquatic ecosystems in water-scarce areas.

Finally, let us analyze, using environmentally preferable products and materials, a project that reuses concrete and wood from a nearby demolishing project. Economically, there is a decreased capital cost due to the use of reused and recycled materials. There will also be a lower cost of disposal from the demolition project. Societal impacts include the benefit of having fewer or smaller landfills and expanding the market for environmentally preferable products. Environmental benefits include decreased use of virgin materials and less strain on landfills.

6.7   Life-cycle cost analysis

No one can deny the fact that building environmental friendly, energy-efficient facilities is a great idea, but is it worth it? Will the builder, owner, or occupant benefit in the long run? Before jumping into the conversation of cost, it is important to define the life-cycle cost (LCC) of a facility. Life-cycle cost analysis (LCCA) is a method for assessing the total cost of a facility over its lifetime, incorporating the costs of acquisition, owning, operating, and disposing of a facility (Whole Building Design Guide 2010). The Whole Building Design Guide uses the following LCC calculation (Equation 6.3):

$$LCC = I + Repl - Res + E + W + OM\&R + O \qquad (6.3)$$

where

LCC = Total LCC in present-value (PV) dollars
I = PV investment costs
Repl = PV capital replacement costs
Res = PV residual value (resale value and salvage value) less disposal costs
E = PV of energy costs
W = PV of water costs
OM&R = PV of non-fuel operating, maintenance, and repair costs
O = PV of other costs (e.g., contract costs)

In order to fairly and correctly decide between building alternatives, one must consider the LCC and not just the upfront costs. In fact, comparisons should be made, and alternatives should be considered by comparing LCC to the upfront cost. Furthermore, the building owner must decide how long of payback period he or she is going to consider when making design choices. Let us consider a novel example of assessment, choosing between two different air conditioners by analyzing upfront and LCCs. Air conditioner A costs $32,000 and energy bill estimates total $1000/month. Air conditioner B costs $20,800 with a $1250 monthly energy bill estimate. Purchasing an air conditioner based solely on upfront cost is simple; the builder would chose air conditioner B as it costs $11,200 less than air conditioner A. Let us now look at the payback period to determine how long the builder must own the facility to have the energy savings of air conditioner A surpass its additional cost when compared to air conditioner B.

Let *m* equal the number of months required for the cost of air conditioner A to equal air conditioner B:

Initial cost A + monthly energy cost A × *m* = Initial cost B + monthly energy cost B × *m*

$$\$32{,}000 + \$1000\ m = \$20{,}800 + \$1250\ m$$
$$\$11{,}200 = \$250\ m$$
$$m = 44.8$$

m = 44.8 months, or 3 years, 8.8 months.

Therefore, if the builder anticipates occupying the facility for 3 years, 9 months or longer, it is less expensive over the building lifespan to purchase the more expensive, more efficient air conditioner. Now that we understand LCC versus upfront cost and the impact of determining the payback period, we can analyze existing research results.

Davis Langdon (2007, now known as AECOM) analyzed 221 buildings, 83 of which were designed with the goal of achieving LEED certification, and the other 138 were projects of similar program types that did not have the goal of sustainable design. This study had two significant findings. First, many projects achieved LEED certification within their budget and in the same cost range as non-LEED projects. Second, many builders must overcome a mental obstacle that adding green features to facilities is often seen as extra facility requirements. While there are fees associated with obtaining LEED certification, building a facility to LEED standards often does not cost more than a similar, non-LEED facility. These researchers concluded that there was much variation in building costs, even amongst the same category of buildings. For example, a builder can construct an inexpensive, LEED-certified office building, an expensive LEED-certified building, an inexpensive non-green office building, or an expensive non-green office building.

In the cost and financial benefits of green buildings, Kats et al. (2003) reported to the California Sustainable Building Task Force that a 2% increase in upfront costs would, on average, result in a life-cycle savings of 20% of total construction costs. For example, an upfront cost of $20,000 on a $1 million project to incorporate green building features would result in a life-cycle savings of $200,000—a 10-fold return on investment. The U.S. Green Building Council has similarly stated that the upfront cost increase to construct a LEED-certified building is typically 2%–7% (sited w/in Fed Facilities). Table 6.3, derived from this study's analysis shows that the average total financial benefits of green buildings are over 10 times the cost to design and construct the facility. Interestingly, energy savings alone typically cover the entire cost of constructing a green facility.

To conclude this section, let us consider two examples of incredibly successful green buildings in the United States: the Environmental Protection Agency Campus at Research Triangle Park, North Carolina (a federal example) and Herman Miller Marketplace, Zeeland, Michigan (a private sector example).

Research Triangle Park started with a sustainable site that was constructed in a wooded area, emphasizing maximum preservation of old-growth trees, minimal disruption to habitats and wetlands, and an orientation to provide sunshading and daylighting. Many green technologies were incorporated including high-efficiency chillers and boilers, and water-conserving toilets, urinals, and faucets. Additionally, recycled building materials were used. To incorporate societal and workplace buy in, 700 parking spots were eliminated, a multistory parking garage was constructed instead of a large parking lot, and incentives were provided to employees to encourage carpooling. The project cost

Table 6.3 Financial benefits of green buildings summary of findings (per ft²)

Category	20-year NPV
Energy value	$5.79
Emissions value	$1.18
Water value	$0.51
Water value (construction only)—1 year	$0.03
Commissioning O&M value	$8.47
Productivity and health value (certified and silver)	$36.89
Productivity and health value (gold and platinum)	$55.33
Less green cost premium	($4.00)
Total 20-year NPV (certified and silver)	$48.87
Total 20-year NPV (gold and platinum)	$67.31

Source: Adapted from Kats, G. et al., 2003. The costs and financial benefits of green buildings, *A Report to California's Sustainable Building Task Force, October 2003,* http://www.usgbc.org/Docs/News/News477.pdf, accessed Oct 24, 2014.

was lowered by $30 million when compared to original estimates. Additionally, annual energy costs were 40% lower than similar facilities, resulting in savings of approximately $1 million/year.

Herman Miller's office space in Zeeland, Michigan used facility layout and orientation, high-efficiency technologies, recycled and locally procured building materials, indigenous, drought-resistant landscaping and societal encouragement to create a high-performance, environmentally conscious office space. Significant sustainable design features used include high ceilings, operable windows, occupant thermal controls, computerized building controls, and low-water use fixtures. Again, facility owners encouraged sound environmental practices by placing special parking spaces close to the facility designated for carpoolers. The results were incredible; Herman Miller constructed the facility at just $89/ft², compared to $135/ft² they spent on traditional offices, resulting in an upfront savings of over $4 million. In total, the facility saved 33% on construction cost and 44% in utility costs.

These success stories highlight the fact that when done right, green buildings can save a tremendous amount of money upfront and can result in even more significant LCC savings and reductions in energy and water use.

6.8 Conclusions

Despite the fact that construction has existed since the dawn of civilization, the field has changed dramatically over the last three decades, embracing a holistic approach to facility construction and analyzing the impact of a facility on its workers, surrounding environment, and the planet. Gone are the days of measuring a successful project by the final cost, adherence to schedule, and meeting minimum standards. While sustainable construction programs have taken center stage, creating a green-certified building does not guarantee success. A successful project considers and appropriately incorporates cost, security and life safety, energy consumption, environmental performance, and occupant satisfaction.

References

Brager, G.S. and deDear, R.J. 1998. Thermal adaptation in the built environment: A literature review. *Energy and Buildings,* 27: 83–96.

Corley, W., Sr, P., Sozen, M. and Thorton, C. 1998. The Oklahoma City bombing: Summary and recommendations for multihazard mitigation. *Journal of Performance of Constructed Facilities,* 12(3): 100–112.

Davis, L. 2007. Cost of green revisitied: Reexamining the feasibility and cost impact of sustainable design in the light of increased market adoption. http://www.usgbc.org/sites/default/files/leed-cost-of-green.pdf, accessed Oct 24, 2014.

Diamond, R., Opitz, M., Hicks, T., Von Neida, B. and Herrara, S. 2006. Evaluating the energy performance of the first generation of LEED-certified commercial buildings. *Proceedings of the 2006 ACEEE Summer Study on Energy Efficiency in Buildings,* American Council for an Energy-Efficient Economy, Washington, DC, 3:41–52.

Dictionary of Construction. 2014a. http://www.dictionaryofconstruction.com/definition/overhead. html, accessed Jan 10, 2014.

Dictionary of Construction. 2014b. http://www.dictionaryofconstruction.com/definition/profit. html, accessed Jan 10, 2014.

Diffen.com. 2014. http://www.diffen.com/difference/Energy_vs_Power, accessed Feb 20, 2014.

Federal Bureau of Investigation. 2014. http://www.fbi.gov/about-us/history/famous-cases/oklahoma-city-bombing, accessed Feb 5, 2014.

Fisk, W.J. 2000. Health and productivity gains from better indoor environments and their relationship with building energy efficiency. *Annual Review of Energy and the Environment,* 25: 537–566.

Gensler, M.A. 2013. http://www.gensler.com/uploads/documents/2013_US_Workplace_Survey_07_15_2013.pdf, accessed Feb 11, 2014.

Kats, G., Alevantix, L., Berman, A., Mills, E. and Perlman, J. 2003. The Costs and Financial Benefits of Green Buildings: A Report to California's Sustainable Building Task Force, October 2003, http://www.usgbc.org/Docs/News/News477.pdf, accessed Oct 24, 2014.

Leaman, A. 1999. UK study links productivity to ventilation systems. *HPAC Magazine,* 71(11): 14.

Menzies, D., Pasztor, J., Nunes, F., Leduc, J. and Chan, C.-H. 1997. Effect of a new ventilation system on health and well-being of office workers. *Archives of Environmental Health,* 52(5): 360–368.

Preller, L., Sweers, T., Brunekreef, B. and Bolej, J.S.M. 1990. Sick leave due to work related health complaints among office workers in the Netherlands. *Indoor Air '90, Toronto,* 1: 227–230.

Smart Buildings. 2009. How do we measure the performance of a building? http://www.smart-buildings.com/uploads/1/1/4/3/11439474/howdowemeasure0809.pdf, accessed Dec 12, 2013.

Sundstrom, E., Burt, R.E. and Kamp, D. 1980. Privacy at work: Architectural correlates of job satisfaction and job performance. *Academy of Management Journal,* 23: 101–117.

The National Terror Alert Response Center. 2014. http://www.nationalterroralert.com/shelterin place, accessed Feb 9, 2014.

Torcellini, P.A., Deru, M., Griffith, B., Long, N., Pless, S., Judkoff, R. and Crawley, D. 2004. Lessons learned from the field evaluation of six high-performance buildings. *ACEEE Summer Study on Energy Efficiency of Buildings,* Pacific Grove, CA.

Turner, C. and Frankel, M. 2008. Energy performance of LEED for new construction buildings. Final Report to U.S. Green Building Council.

U.S. Department of Energy. 2011. http://www.eia.doe.gov/emeu/aer/consump.html, accessed Feb 22, 2014.

U.S. Energy Information Administration. 2007. *Electric Power Annual 2007,* United States.

U.S. Energy Information Administration. 2014. http://www.eia.gov/dnav/pet/hist/LeafHandler.ashx?n=pet&s=emd_epd2dxl0_pte_nus_dpg&f=m, accessed Jan 15, 2014.

U.S. Green Building Council. 2014. http://www.usgbc.org/leed#credits, accessed Jan 15, 2014.

U.S. Inflation Calculator. 2014. http://www.usinflationcalculator.com, accessed Jan 18, 2014.

Wheeler, G. 1998. Presentation of ASID survey findings. *AIA Conference on Highly Effective Buildings,* Cincinnati, OH, March 12–14.

Whole Building Design Guide. 2010. Life-cycle cost analysis (LCCA), http://www.wbdg.org/resources/lcca.php, accessed Jun 28, 2010.

Whole Building Design Guide. 2011. Unified Facilities Criteria 3-740-05, *Handbook: Construction Cost Estimating*, http://www.wbdg.org/ccb/DOD/UFC/ufc_3_740_05.pdf, accessed Jun 1, 2011.

Whole Building Design Guide. 2013a. Unified Facilities Criteria 3-701-01, *DoD Facilities Pricing Guide*, http://www.wbdg.org/ccb/DOD/UFC/ufc_3_701_01.pdf, accessed Aug 5, 2013.

Whole Building Design Guide. 2013b. Unified Facilities Criteria 4-010-01, *DoD Minimum Antiterrorism Standards for Buildings*, http://www.wbdg.org/ccb/DOD/UFC/ufc_4_010_01.pdf, accessed Oct 1, 2013.

Whole Building Design Guide. 2014. http://wbdg.org/design, accessed Apr 26, 2014.

Wyon, D.P. 1996. Indoor environmental effects on productivity. In *Proceedings of IAQ '96, Paths to Better Building Environments*, Baltimore, MD, October 6–8.

chapter seven

Energy systems measurements

Olufemi A. Omitaomu

Contents

7.1 Introduction

The low-cost nature of various automatic measurement systems has created significant advantages over the traditional measurement and communication technologies used today in the electric power systems. These automatic measurement systems have been widely recognized as promising technologies for achieving the so-called smart grid. They provide tremendous opportunities for collecting timely valuable process and operational data that could help realize advanced monitorability of the power grid assets and operations to achieve an unprecedented level of situational awareness and controllability over its services and infrastructure. They can also provide fast and accurate diagnosis/prognosis as well as operation resiliency upon contingencies and malicious attacks. In this chapter, we describe some of these measurement devices and their applications. While many utilities understand the business values in the data collected by these measuring devices, the real challenge lies in how to develop processes and systems to continuously convert data into actionable information. Consequently, we also present some computational approaches for eliminating some of the inherent measurement errors in the collected data and extracting insights and knowledge from data collected to enhance situational awareness and controllability of the power grids.

7.1.1   Measuring devices used in power grids

Traditionally, the management of the power system is carried out using remote central stations that incur delays and cause ripple effects. Therefore, the existing power grid lacks pervasive and effective communications, monitoring, fault diagnosis, and automation. These deficiencies increase the possibility of wide-area system breakdown due to the cascading effect of a single fault (Kora et al., 2012). Consequently, the concept of *smart grid* has emerged as a vision for the next generation of the electric power system. At the core of this concept is the integration of information (and communication) technologies to the traditional power grid networks to achieve an unprecedented level of real-time data collection (and communication) about the conditions of power grid assets and operations. The integration will cut across the entire power grid supply chain—from generation to consumption. Here, we describe some of the existing measurement devices that are being deployed for operations.

7.1.2   Phasor measurement units

A phasor measurement unit (PMU) is one of the measurement units being deployed at substations. It measures the electrical waves on the electric grid, using a common time source for synchronization. The PMUs are usually called synchrophasors and are considered one of the most important measuring devices in the next generation of power systems. Synchrophasors measure frequencies, voltages, and currents at critical substations and can output time-stamped values in real time. Depending on the design and application, PMUs could stream between 10 and 60 synchronous reports per second. In the United States, more than 200 PMUs have been deployed in the last 5 years. The plan is to deploy thousands of PMUs in the coming decades.

Some of the proposed applications of PMUs data include

- Visualization and situational awareness tools—Deployment of PMUs provides a wealth of new information about the power system that could enhance how the entire network is managed.
- High-resolution state estimation/observation—The high-resolution data from PMUs provide opportunities for more datasets for state estimation. As the number of deployed PMUs reaches a critical penetration, the problem moves from state estimation to state observation with less emphasis on computing unknown values and more emphasis on identifying data errors.
- Model improvement and validation—With high-resolution measurement data from PMUs, existing system models and simulation results can be compared to the measurements. Differences between measurement data and simulation results can indicate erroneous portions of the system that need improvement.

- Real-time dynamic modeling—With an accurate dynamic model of the power system, measurement data can be processed in real time to determine the system operating point. As this processing is performed over time, the trajectory of the system can be estimated. When compared to the dynamic model, this estimate can be used to predict upcoming dynamic conditions and potential stability problems.
- Advanced relaying and other protective schemes—Current relaying and protection schemes are focused on local measurements. Improvements to these schemes may be implemented through the integration of remote measurement data. With a wider view of the power system, a relay can make better decisions about when and when not to trip.
- Advanced closed-loop control systems—An increase in the wealth and resolution of measurement data allows for more advanced control operations in the power system both in a local and wide-area sense.

7.1.3   Smart meters

The smart meter is an example of measurement devices being used on the consumption side of the electric grid supply chain. This is an electronic device that records energy consumption at a granular scale of an hour or less and communicates that information back to the utility for monitoring and billing purposes. This device is unlike a traditional energy meter that only measures total consumption and no information about when the energy was consumed. Thus, the smart meter technology enables two-way communication between the meter and the utility central station. Smart meter is often used when referring to an electricity meter, but it also may mean a device measuring natural gas or water consumption. The smart meter enables real-time or near-real-time notification of power outage and monitoring of power quality. The deployment of smart meters has been growing all over the world in the last 5 years; this trend will continue into the next decade.

Event information available from smart meters includes meter status information, power quality, and real-time device status, which provide powerful insights for improving utilities, operational efficiencies, and delivering enhanced and customized services to customers. Some of the potential areas where information from smart meters can be used to derive useful operational insights include

- Outage management—Detects real-time outage events and creates proactive restoration tickets as well as identifies power restoration progress after large-scale outage events.
- Customer experience—Events data can be used to develop customized services as well as alerts and notifications to customers regarding power outages and abnormal consumption patterns, which could signal the emergence of a problem.
- Power quality—Events such as "voltage sag" and "voltage swell" in correlation with other device status information will help to proactively identify open neutrals and flickering lights.
- Revenue assurance—Events such as meter inversion and reverse energy flow, along with meter reads to identify power theft and abnormal usage/demand patterns.
- Smart meter network operations and monitoring—Events to identify damaged/defective meters, access relays, and other devices for proactive maintenance/replacement strategies.
- Load assessment and forecasting (renewable and distributed generation integration)—As high-resolution smart meters are deployed, their information can be

aggregated to develop better load models. This information can provide insight into both the size and dynamic characteristics of loads throughout the system. Better forecasting of system loads can easily be achieved through analysis of these data. These meters can also provide information about how renewables and other distributed generation are interacting with the system.

7.1.4 Other measurement devices

Besides these two measurement devices, there are several other devices that are being used or deployed on the power grids all over the world. Kora et al. (2012) presented a summary of some of these devices. We summarized their presentation in this section. Interested readers should see Kora et al. (2012) for details.

7.1.4.1 Measurement devices on electric transformers

According to Kora et al. (2012), there are at least six sensor types being considered for use on transformers. They include

- Metal-insulated semiconducting gas in oil sensor: A low-cost hydrogen sensor on a chip.
- Acoustic fiber optic: Enables the measurement of internal partial discharges in a transformer.
- Gas fiber optic: Measures the presence of different gases at the tip of a fiber-optic cable.
- Online frequency response analysis: Developed by measuring the response of a transformer to normally occurring transients on the power system. Changes in frequency can identify changes in the internal geometry of the transformer.
- Load tap changer (LTC)— Monitoring gas ratios: Developed to monitor gas ratios in LTCs without measuring each gas individually.
- Three-dimensional (3D) acoustic emissions: Enables the detection and location of gassing sources in power transformers and LTCs.

7.1.4.2 Measurement devices for power transmission lines

Several measurement devices are being developed to monitor the health of transmission lines, especially in remote locations. Some of the devices discussed by Kora et al. (2012) include the following

- Circuit breaker—RF SF6 density: For measuring SF6 density and transmitting the results wirelessly.
- Current and potential transformers of RF acoustic emissions: Measure acoustic emissions due to internal discharge activity and transmit results wirelessly.
- Disconnect RF temperature sensors: Measure the disconnected jaw temperature to identify high-loading units.
- Shield wire RF-lightning sensors: Measure peak magnitude and time of lightning currents in the shield wires. These sensors could help in understanding the distribution of lightning currents on transmission lines and to validate lightning location systems.
- Overhead transmission structure sensor system: Used to solve specific issues such as unknown outages on the transmission lines.
- Wireless mesh sensors: For cost-effective substation online monitoring.

7.2   Denoising digital data from measuring devices

The ability to process data recorded by measuring devices for real-time system monitoring is a challenge if the recorded data are noisy. The noise may be due to weather conditions such as rain, snow, and smog; distance of the measuring device from the targeted object or system; and/or the presence of some forms of shield that may obstruct the sensing capability of the device. Despite these conditions, the data must be processed within a short allowable time frame (say, few seconds in power grid applications) during the precascading phase when operators may have time to take the appropriate control actions and avoid a cascading failure of the entire system. Thus, signal denoising is an important preprocessing step in measurement data analysis.

Signal denoising is an age-old problem that has been addressed using various techniques in different applications. Some of the widely used techniques include discrete Fourier transform (DFT), discrete wavelet transform (DWT), and empirical mode decomposition (EMD). A review of these and other methods of denoising measurement data can be found in Cohen (1995) and Kantz and Schreiber (1997). Omitaomu (2013) presents an application of DWT technique for denoising measurement data from manufacturing operations. Omitaomu et al. (2011b) present an application of EMD technique to measurement data in the surface transportation problem. In this chapter, a review of the EMD technique and its application to measurement data from power grid operations are presented.

7.2.1   The EMD technique

Even though the DFT technique uses linear filters that are easy to design and implement, these filters are not effective for denoising nonstationary data with sharp edges and impulses of short duration (Jha and Yadava, 2010). Such nonstationary data could be analyzed using the DWT technique, which has become quite popular as an alternative to the DFT technique. However, the DWT technique, like the DFT technique, uses linear and fixed *a priori* basis functions, which make the technique unsuitable for nonlinear data. The EMD technique was recently introduced as a data-driven technique for analyzing measurement data from nonlinear, nonstationary processes (Huang et al., 1996, 1998, 1999). Given such one-dimensional data, the technique adaptively decomposes the original data into a number of intrinsic mode functions (IMFs). Each IMF is an oscillatory signal that consists of a subset of frequency components from the original data. In this chapter, the word data and signal are used interchangeably. The EMD procedure can be described as follows:

Given a one-dimensional signal, x_j sampled at times t_j, $j = 1, ..., N$, the EMD technique decomposes the signal into a finite and often small number of fundamental oscillatory modes. The modes (or IMFs) into which the original signal is decomposed are obtained from the signal itself, and they are defined in the same time domain as the original signal. The modes are nearly orthogonal with respect to each other, and are linear components of the given signal. The following two conditions must be satisfied for an extracted signal to be called an IMF:

1. The total number of extrema of the IMF should be equal to the number of zero crossings, or they should differ by one, at most.
2. The mean of the upper envelope and the lower envelope of the IMF should be zero.

The process to obtain the IMFs from the given signal is called *sifting* (Huang et al., 1998). The sifting process consists of the following steps:

1. Identification of the maxima and minima of x_j.
2. Interpolation of the set of maximal and minimal points (by using cubic splines) to obtain an upper envelope (x_{jup}) and a lower envelope (x_{jlow}), respectively.
3. Calculation of the point-by-point average of the upper and lower envelopes $m_j = (x_{jup} + x_{jlow})/2$.
4. Subtraction of the average from the original signal to yield $d_j = x_j - m_j$.
5. Testing whether d_j satisfies the two conditions for being an IMF or not. If d_j is not an IMF, steps 1–4 are repeated until d_j satisfies the two conditions.
6. Once an IMF is generated, the residual signal $r_j = x_j - d_j$ is regarded as the original signal, and steps 1–5 are repeated to generate the second IMF, and so on.

The sifting is complete when either the residual function becomes monotonic, or the amplitude of the residue falls below a predetermined small value so that further sifting would not yield any useful components. The features of the EMD technique guarantee the computation of a finite number of IMFs within a finite number of iterations. At the end of the process, the original signal, x_j, can then be represented as

$$x_j = \sum_{i=1}^{M-1} d_{j,i} + r_{j,M} \quad i = 1,\ldots,M, \tag{7.1}$$

where $r_{j,M}$ is the final residue that has near-zero amplitude and frequency, M is the number of IMFs, and $d_{j,i}$ are the IMFs. A complete mathematical description of the EMD technique is beyond the scope of this chapter but can be found in Huang et al. (1996, 1998, 1999); Jha and Yadava (2010).

As an illustration of the sifting process, we consider the *Doppler* function

$$X = \sin\left[\frac{(2\pi \times 1.05)}{t + 0.05}\right]\sqrt{(t(1 - t))}, \tag{7.2}$$

where $t = i/N$, $i = 1,2, \ldots, N$ and $N = 1024$. We add white noise to X and define the original signal-to-noise ratio (SNR) (*original-SNR*) as

$$original\text{-}SNR(dB) = 20\log_{10}\frac{norm_{signal}}{norm_{noise}}, \tag{7.3}$$

where *noise* is the difference between the original and noisy signal and *norm* is the largest singular value of the signal.

The plots of noise-free Doppler function and the noisy Doppler function with *original-SNR* of −5 dB are shown in Figure 7.1. Using the sifting process, the decomposition of the noisy Doppler function is shown in Figure 7.2. The signal is decomposed into eight IMFs and a residue. On the basis of plots in Figure 7.2, one can argue that most, if not all, of the noise in the original signal is decomposed as a part of the first four or five IMFs (Figure 7.2).

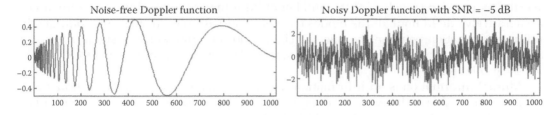

Figure 7.1 The plots of noise-free and noisy Doppler function.

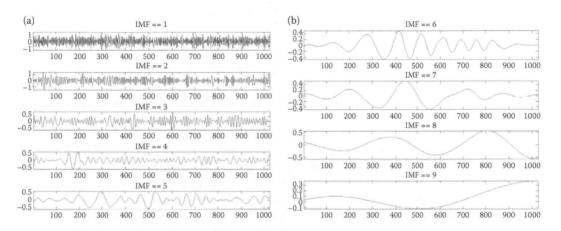

Figure 7.2 The decomposition of the noisy Doppler function in Figure 7.1 using EMD. (a) For IMF <= 5 and (b) for IMF >= 6.

7.2.2   An approach for denoising measurement data

In this section, we adopt a denoising approach original proposed by Omitaomu et al. (2011b) for power grid applications. The overall objective of the denoising approach is to reconstruct the original signal by selecting a set of the modes $\bar{x}_j$ that minimizes the noise contained in the original signal x_j as measured by mutual information $MI(\bar{x}_j, x_j)$. To achieve this objective, the approach consists of three steps:

1. The IMFs of the original noisy signal are partitioned into high-frequency (HF) and low-frequency (LF) groups using a modified forward selection strategy.
2. The HF group is denoised using a thresholding approach.
3. Finally, the LF group from the first step is linearly combined with the outputs from the second step to obtain the final denoised signal.

The following sections describe each of the three steps.

7.2.2.1   Separating the IMFs into HF and LF groups

For this approach, it is assumed that the given signal is corrupted by HF Gaussian noise; therefore, the noise should be captured by the first few IMFs as the case in the Doppler function shown in Figure 7.2. The question then is how to partition the IMFs into HF and LF groups. To answer this question, we employed a modified forward selection strategy. This strategy starts with a few modes, usually most of the second-half of the modes, selected by rounding one-half of the number of IMFs toward minus infinity. More modes are iteratively added until all the modes have been added. At each stage, we compute the mutual information between the selected signal and the original noisy signal $MI(\bar{x}_j, x_j)$. The idea of using mutual information for feature selection has been used in other applications (e.g., Fleuret, 2004).

Mutual information is an information-theoretic measure for quantifying how much information is shared between two or more random variables. Mutual information-based feature selection adds that variable $\hat{x}$ to the current set of chosen variables ϕ, which maximizes

$$MI(y, \hat{x}) - \zeta \sum_{x_j \in \phi} MI(\hat{x}, x_j), \tag{7.4}$$

where the first term attempts to maximize the relevancy of $\hat{x}$ to y, and the second term attempts to minimize the redundancy between $\hat{x}$ and the already-chosen set ϕ. The balance between maximizing relevancy and minimizing redundancy is determined by parameter ζ.

The main goal of feature selection is to select a small subset of features that carries as much information as possible to remove irrelevant and redundant variables. Such a task requires the estimation of 2^{K+1} probabilities, where K is the number of selected variables. This implementation could be computationally infeasible in real-time applications. As a result, we adapted an intermediate solution that considers the trade-off between individual power and independence by comparing each of the first-half of IMFs with the sum of the last-half of IMFs. We assume that an IMF mode, X_j, is good only if it carries more information about the unknown signal and less information about the noise in the original signal. Therefore, we will be minimizing the mutual information rather than maximizing it. By taking the modes that minimize the mutual information between them and the noisy signal, we ensure that the selected modes contain useful information about the unknown signal and are different from the known noisy signal. At the end of this task, we set the final partition index (k) to be equivalent to the position of IMF with the minimum MI.

The following summarizes our approach for partitioning the IMFs:

1. Decompose the given signal x_j into IMFs using the sifting process as explained in Section 7.1. The IMFs for a noisy Doppler signal with SNR of −5 dB are shown in Figure 7.2.
2. Determine the *initial partition index* (IPI) by rounding one-half of the number of IMFs toward minus infinity. The sum of all IMFs with index greater than IPI is called the initial LF group. The other IMFs are potential modes for the HF group. For the Doppler signal, the IPI is 4 since the number of IMFs is 9. Therefore, IMFs with index from 1 to 4 are potential modes for the HF group, and the sum of IMFs with modes from 5 to 9 is called the initial LF group.
3. Compute the mutual information between the initial LF group and the original noisy signal.
4. Add each of the remaining IMF modes iteratively bottom up starting with the IMF with index equal to the IPI, and compute the respective MI for each set of the new signal. Repeat this process until all the IMFs have been added.
5. The final partition index (k) is defined as the index of IMF with the minimum MI.

6. The sum of IMFs with index less than k gives the final HF group, and the sum of IMFs with index equal to and greater than k gives the final LF group.
7. Thus, the original signal can be represented as

$$x_j = d_{j,k} + s_{j,k}, \tag{7.5}$$

where $d_{j,k}$ is the HF group and $s_{j,k}$ is the LF group; they are defined as

$$d_{j,k} = \sum_{i=1}^{k-1} d_{j,i},$$

$$s_{j,k} = \sum_{i=k}^{M-1} d_{j,i} + r_{j,M}. \tag{7.6}$$

7.2.2.2 Denoising the HF group

Once the IMFs have been separated into HF and LF groups, the next step is to denoise the HF group. There are several approaches for handling the HF group. One simple approach would be to assume that the HF group is the noise in the original signal and the LF group is an approximation of the original signal. Then, the LF group becomes the denoised signal and the HF group is discarded. This approach is analogous to the smoothing technique. Another approach is to denoise the HF group using the shrinkage (thresholding) measure because this group may also contain some genuine information. The shrinkage measure has been widely used for denoising signals in the DWT literature (e.g., Stein's unbiased risk estimate [SURE] and VisuShrink shrinkage measures) as described in Donoho and Johnstone (1994, 1995). For this approach, the LF group would not be denoised at all. Denoising the LF mode could lead to loss of information. We note that the idea of thresholding only the HF data and keeping the LF data unmodified is also used in wavelet-thresholding procedures.

For EMD application, we adopt the SURE (Donoho and Johnstone, 1994) as our shrinkage measure for two reasons: it does not require an estimation of the noise in the original signal and is a good estimation of the true mean squared error (MSE) when the true function is unknown. The SURE measure is defined as follows:

$$SURE_{EMD} = N - 2 \times Q_{(j:\,|d_{j,k}|\leq\lambda)} + \sum_{j=1}^{N}\sum_{k=1}^{K-1}(|d_{j,k}| \wedge \lambda)^2, \tag{7.7}$$

where λ is the candidate threshold, $a \wedge b$ denotes min(a,b), $d_{j,k}$ is the HF group, N is the number of data points, and Q is the number of data less than λ. The selected threshold is defined as the value of λ that minimizes $SURE_{EMD}$. The shrinkage measure can be applied using both hard- and soft-thresholding methods.

The hard-thresholding method retains data whose absolute value is greater than the threshold and sets to zero data whose absolute value is less than or equal to the threshold. Consequently, the denoised signal is discontinuous around the threshold. More formally, the hard-thresholding method is defined as

$$\bar{d}_{j,k} = \begin{cases} d_{j,k} & \text{if } |d_{j,k}| > \lambda \\ 0 & \text{if } |d_{j,k}| \leq \lambda \end{cases}, \tag{7.8}$$

where $d_{j,k}$ and $\bar{d}_{j,k}$ represent the HF group before and after thresholding, respectively. On the other hand, the soft-thresholding method (Donoho, 1995) shrunk toward zero data whose absolute value is greater than the threshold and set to zero data whose absolute value is less than or equal to the threshold as defined by the following expressions:

$$\bar{d}_{j,k} = \begin{cases} sign(d_{j,k})(|d_{j,k}| - \lambda) & \text{if } |d_{j,k}| > \lambda \\ 0 & \text{if } |d_{j,k}| \leq \lambda \end{cases}. \tag{7.9}$$

The idea of using the shrinkage measure together with the EMD technique has also been investigated by Jha and Yadava (2010). However, the shrinkage measure was based on the VISU threshold measure.

7.2.2.3   Obtaining the denoised signal

Once the unmodified LF group is identified and the HF group is denoised, the overall denoised signal is obtained upon a linear combination of the LF group with the denoised HF group to obtain an approximation of the original signal defined as

$$\bar{x}_j = \bar{d}_{j,k} + s_{j,k}, \tag{7.10}$$

where $\bar{x}_j$ is an approximation of the original signal x_j, $\bar{d}_{j,k}$ is the denoised HF group, and $S_{j,k}$ is the LF group. Therefore, $\bar{x}_j$ contains most of the information in the original noisy signal, but much less (if any) of the noise in the original signal.

7.2.3   Applications of EMD to measurement data from the power grid

In this section, we present two applications of the EMD technique to sensor data from the power grid. In Section 7.3.1, the EMD-based denoising approach presented in Section 7.2 is applied to measurement data from distribution-level PMU data. In Section 7.3.2, the EMD-sifting process presented in Section 7.1 is applied to smart meters data.

7.2.3.1   PMU data preprocessing using EMD-based denoising approach

When a significant disturbance occurs in a power system, such as a generator trip, the voltage frequency varies in space and time (Bank et al., 2009). The PMU is used to monitor and record the changes in voltage frequency in real time at various locations. To provide a better understanding of any impending events on the grid using measurement data from PMUs, the data need to be properly conditioned first. In this section, we present an application of EMD-based denoising approach for preprocessing the PMUs data. For this implementation, we used real-world datasets recorded using single-phase PMUs (also called frequency disturbance recorders). The datasets used are frequency responses to a generator trip (see Figure 7.3).

The frequency of a power system provides a great deal of information about the health of the system and the prevailing operating conditions. The frequency trend indicates the power balance of the system. When the amount of power generated is equivalent to the amount of power consumed, the system is in steady state and the frequency remains constant. If there is an imbalance in generation/consumption, the system responds by converting some of its kinetic energy into electrical energy to make up for the power imbalance, causing acceleration toward a new quasi steady state. This acceleration can be seen as a change in frequency. An excess of generation causes the system to accelerate,

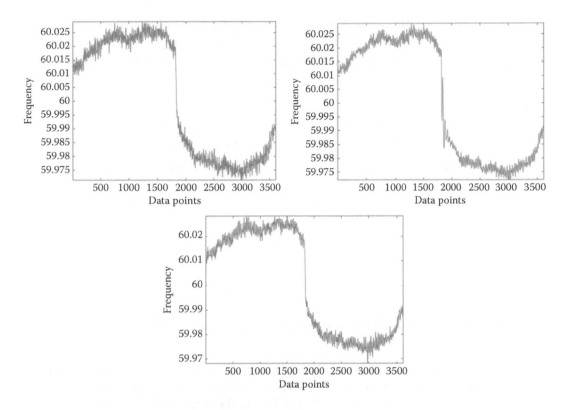

Figure 7.3 Frequency responses to a generator trip as recorded by three spatially distributed single-phase PMUs.

resulting in an increase in frequency. An excess of load causes the system to decelerate, depressing the frequency. This change in frequency is proportional to the magnitude of power imbalance.

We applied the EMD-denoising approach to these datasets to remove inherent system noise and enhance the data analysis process. The results for one PMU data are presented in Figure 7.4. According to these results (and results not shown in this chapter), we are able to remove the noise from the original without overfitting or underfitting the original signal. The MSE based on the denoised signal with respect to the original data is significantly small, which suggest that the original datasets have a small amount of noise. However, the removal of the noise is far more important for the next stage of data analysis, which may involve anomaly detection and prediction tasks.

7.2.3.2 Decomposing smart meter data using EMD

For this application, we consider smart meter data for a utility customer. The dataset is actual electricity consumption for 31 days in January. In this section, we are interested in decomposing the dataset for the customer to infer the portion of the total usage in each 15-min interval that may be attributed to various loads (e.g., general lighting, heating/cooling, appliance, etc.) in the house without any prior knowledge of the actual appliance landscape. The results are shown in Figure 7.5. In this figure, the horizontal axis shows the data points while the vertical axis is the amount of electricity in kilowatt-hour consumed in each of the 2976 15-min intervals in January. Looking at the figures, we can begin to

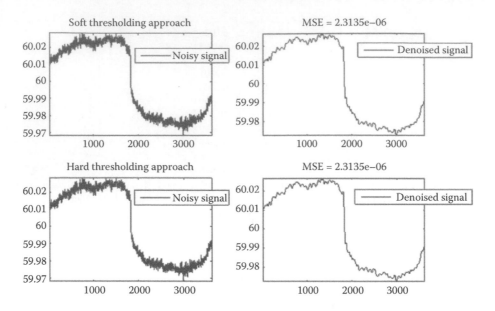

Figure 7.4 The results of the EMD-denoising approach applied to one PMU data.

infer what could be responsible for some of the periodic fluctuations and the different patterns at different times of the day as well as the correction in repeated patterns, which could suggest similar activities over different days of the week or the month.

We are currently validating some of the insights from these decompositions with field data. If EMD decomposition could provide deeper insights about temporal activities, this could help utilities in profiling their customers for customized services and programs.

7.3   Optimization-based data-mining approach for power grid applications

The support vector machines (SVMs) technique (Vapnik, 1995, 1998) is based on statistical learning theory, and it is used for learning classification and regression rules from data (Osuna et al., 1997). When used for classification problems, the algorithm is usually called support vector classification (SVC) and when used for regression problems, the algorithm is support vector regression (SVR). Unlike other predictive models, the SVM technique attempts to minimize the upper bound on the generalization error based on the principle of structural risk minimization (SRM) rather than minimizing the training error. This approach has been found to be superior to the empirical risk minimization (ERM) principle employed in artificial neural network (Vapnik et al., 1996; Gunn, 1998). In addition, the SRM principle incorporates capacity control that prevents overfitting of the input data (Bishop, 1995). The SVM technique has sound orientations toward real-world applications (Smola and Schölkopf, 2004); therefore, it is applicable to condition-monitoring problems (Omitaomu et al., 2006, 2007, 2011a).

The SVM technique continues to gain popularity for prediction because of its several outstanding properties (Muller et al., 1997; Fernandez, 1999; Cao and Tay, 2003). Some of these properties are the use of the kernel function that makes the technique applicable to both linear and nonlinear approximations, good generalization performance as a result

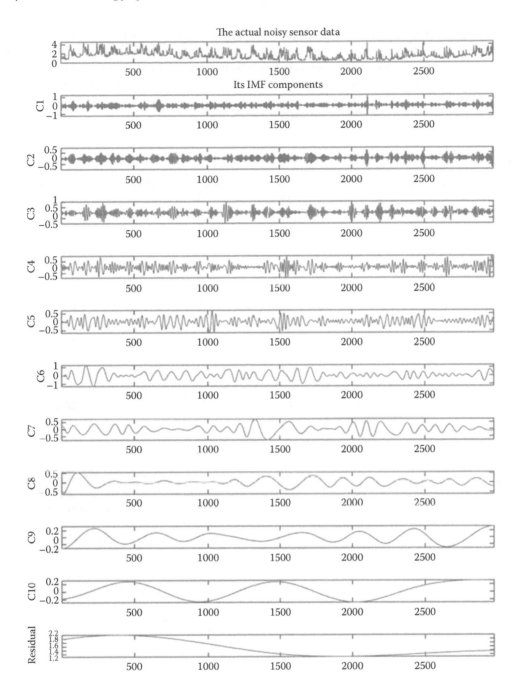

Figure 7.5 Decomposition of a 1-month 15-min interval smart meter data for a customer.

of the use of only the so-called support vectors (SVs) for prediction, the absence of local minima because of the convexity property of the objective function and its constraints, and the fact that it is based on SRM that seeks to minimize the upper bound of the generalization error rather than the training error. In this chapter, the focus is on SVR; but most of the discussions are also applicable to SVC. The SVR algorithm was developed after the successful implementation of the SVM algorithm for classification problems. The two key features in SVR implementation are mathematical programming and kernel functions. The model coefficients are obtained by solving a quadratic programming problem with linear equality and inequality constraints. The SVR technique has been successfully applied to a wide range of pattern recognition and prediction problems (e.g., Mattera and Haykin, 1999; Muller et al., 1999; Omitaomu et al., 2006, 2007, 2011a).

7.3.1 SVR formulation

A detailed formulation of SVR equations is provided by Vapnik (1995, 1998) and Omitaomu (2013). Given a set of training inputs (X) and their corresponding outputs (Y), they are defined respectively as

$$X = \{x_1, x_2, ..., x_m\} \subset \chi. \tag{7.11}$$

$$Y = \{y_1, y_2, ..., y_m\} \subset \gamma. \tag{7.12}$$

For a training set, T, it is represented by

$$T = \{(x_1, y_1), (x_2, y_2), ..., x_m, y_m)\}, \tag{7.13}$$

where $x \in X \subset \Re^n$ and $y \in Y \subset \Re$. Then, assume a nonlinear function, $f(x)$, given by

$$f(x) = \mathbf{w}^\mathrm{T}\Phi(\mathbf{x}_i) + b, \tag{7.14}$$

where $\mathbf{w}$ is the weight vector, b is the bias, and $\Phi(\mathbf{x}_i)$ is the high-dimensional feature space, which is linearly mapped from the input space x. Also, assume that the goal is to fit the data T by finding a function $f(x)$ that has a largest deviation ε from the actual targets y_i for all the training data T, and at the same time, is as small as possible. Hence, Equation 7.14 is transformed into a constrained convex optimization problem as follows:

$$\text{minimize} \quad \frac{1}{2}\mathbf{w}^\mathrm{T}\mathbf{w}$$
$$\text{subject to:} \quad \begin{cases} y_i - (\mathbf{w}^\mathrm{T}\Phi(\mathbf{x}_i) + b) \le \varepsilon \\ y_i - (\mathbf{w}^\mathrm{T}\Phi(\mathbf{x}_i) + b) \ge \varepsilon, \end{cases} \tag{7.15}$$

where $\varepsilon(\ge 0)$ is user defined and represents the largest deviation. Equation 7.15 can also be written as

$$\text{minimize} \quad \frac{1}{2}\mathbf{w}^\mathrm{T}\mathbf{w}$$
$$\text{subject to:} \quad \begin{cases} y_i - \mathbf{w}^\mathrm{T}\Phi(\mathbf{x}_i) - b \le \varepsilon \\ \mathbf{w}^\mathrm{T}\Phi(\mathbf{x}_i) + b - y_i \le \varepsilon. \end{cases} \tag{7.16}$$

The goal of the objective function in Equation 7.16 is to make the function as "flat" as possible; that is, to make **w** as "small" as possible while satisfying the constraints. To solve Equation 7.16, slack variables are introduced to cope with possible infeasible optimization problems. One underlining assumption here is that $f(x)$ actually exists; in other words, the convex optimization problem is *feasible*. However, this is not always the case; therefore, one might want to trade-off errors by flatness of the estimate. This idea leads to the following primal formulations as stated in Vapnik (1995):

$$\text{minimize} \quad \frac{1}{2}\mathbf{w}^T\mathbf{w} + C\sum_{i=1}^{m}(\xi_i^+ + \xi_i^-)$$

$$\text{subject to:} \quad \begin{cases} y_i - \mathbf{w}^T\Phi(\mathbf{x}_i) - b \leq \varepsilon + \xi_i^+ \\ \mathbf{w}^T\Phi(\mathbf{x}_i) + b - y_i \leq \varepsilon + \xi_i^- \\ \xi_i^+, \xi_i^- \qquad\qquad \geq 0, \end{cases} \tag{7.17}$$

where $C(>0)$ is a prespecified regularization constant and represents the penalty weight. The first term in the objective function $\mathbf{w}^T\mathbf{w}$ is the regularized term and makes the function as "flat" as possible whereas the second term $(C\sum_{i=1}^{m}(\xi_i^+ + \xi_i^-))$ is called the empirical term and measured the ε-insensitive loss function. According to Equation 7.17, all data points whose y-values differ from $f(x)$ by more than ε are penalized. The slack variables, ξ_i^+ and ξ_i^-, correspond to the size of this excess deviation for upper and lower deviations, respectively, as represented graphically in Figure 7.6. The ε-tube is the largest deviation and all the data points inside this tube do not contribute to the regression model since their coefficients are equal to zero. Data points outside this tube or lying on this tube are used in determining the decision function; they are called SVs and have nonzero coefficients. Equation 7.17 assumes ε-insensitive loss function (Vapnik, 1995) as shown in Figure 7.6 and is defined as

$$|\xi|_\varepsilon = \begin{cases} 0 & \text{if } |\xi| \leq \varepsilon \\ |\xi| - \varepsilon & \text{otherwise.} \end{cases} \tag{7.18}$$

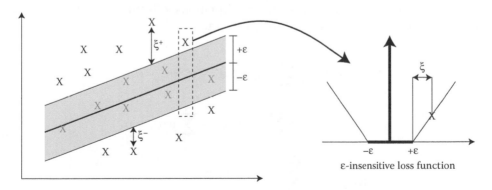

Figure 7.6 An illustration of ε-tube for SVR. (From Vapnik, V. *Statistical Learning Theory.* New York, NY: John Wiley & Sons, Inc.)

To solve Equation 7.17, some Lagrangian multipliers $(\alpha_i^+, \alpha_i^-, \eta_i^+, \eta_i^-)$ are introduced to eliminate some of the primal variables. Hence, the Lagrangian of Equation 7.17 is given as

$$L_p = \frac{1}{2}\mathbf{w}^T\mathbf{w} + C\sum_{i=1}^{m}(\xi_i^+ + \xi_i^-) - \sum_{i=1}^{m}(\eta_i^+\xi_i^+ + \eta_i^-\xi_i^-)$$

$$- \sum_{i=1}^{m}\alpha_i^+(\varepsilon + \xi_i^+ - y_i + \mathbf{w}^T\Phi(\mathbf{x}_i) + b)$$

$$- \sum_{i=1}^{m}\alpha_i^-(\varepsilon + \xi_i^- + y_i - \mathbf{w}^T\Phi(\mathbf{x}_i) - b) \tag{7.19}$$

$$s.t. \quad \alpha_i^+, \alpha_i^-, \eta_i^+, \eta_i^- \geq 0.$$

Two advantages of Equation 7.19 are that it provides the key for extending SVM to non-linear functions and it makes solving Equation 7.17 easier. It then follows from the saddle point condition (the point where the primal objective function is minimal and the dual objective function is maximal) that the partial derivatives of L_p with respect to the primal variables $(\mathbf{w}, b, \xi_i^+, \xi_i^-)$ have to vanish for optimality. Therefore

$$\partial_b L_P = \sum_{i=1}^{m}(\alpha_i^+ - \alpha_i^-) = 0, \tag{7.20}$$

$$\partial_{\mathbf{w}} L_P = \mathbf{w} - \sum_{i=1}^{m}(\alpha_i^+ - \alpha_i^-)x_i = 0, \tag{7.21}$$

$$\text{and} \quad \partial_{\xi_i^{(*)}} L_P = C - \alpha_i^{(*)} - \eta_i^{(*)} = 0, \tag{7.22}$$

where (*) denotes variables with + and − superscripts. Substituting Equations 7.20 and 7.22 into Equation 7.19 lets the terms in b and ξ vanish. In addition, Equation 7.22 can be transformed into $\alpha_i \in [0,C]$. Therefore, substituting Equations 7.20 through 7.22 into Equation 7.19 yields the following dual optimization problem:

$$\text{Maximize} \quad \frac{1}{2}\sum_{i,j=1}^{m}K(\mathbf{x}_i,\mathbf{x}_j)(\alpha_i^+ - \alpha_i^-)(\alpha_j^+ - \alpha_j^-)$$

$$+ \varepsilon\sum_{i=1}^{m}(\alpha_i^+ + \alpha_i^-) - \sum_{i=1}^{m}y_i(\alpha_i^+ - \alpha_i^-) \tag{7.23}$$

$$\text{Subject to:} \quad \begin{cases} \sum_{i=1}^{m}(\alpha_i^+ - \alpha_i^-) & = 0 \\ \alpha_i^+, \alpha_i^- & \in [0,C], \end{cases}$$

where $K(\mathbf{x}_i, \mathbf{x}_j)$ is called the kernel function. The flexibility of a kernel function allows the technique to search a wide range of solution space. The kernel function must be positive definite to guarantee a unique optimal solution to the quadratic optimization problem. It allows nonlinear function approximations with the SVM technique, while maintaining the simplicity and computational efficiency of linear SVM approximations. Some of the common kernel functions are polynomial kernel and Gaussian radial basis function (RBF) kernel. Descriptions of the common types of kernel used in SVM are discussed in Section 7.1.1.

The dual problem in Equation 7.23 has three advantages: the optimization problem is now a quadratic programming problem with linear constraints, which is easier to solve and ensures a unique global optimum. Second, the input vector only appears inside the dot product, which ensures that the dimensionality of the input space can be hidden from the remaining computations. That is, even though the input space is transformed into a high-dimensional space, the computation does not take place in that space but in the linear space (Gunn, 1998). Finally, the dual form does allow the replacement of the dot product of input vectors with a nonlinear transformation of the input vector. In deriving Equation 7.23, the dual variables η_i^+, η_i^- were already eliminated through the condition in Equation 7.22. Therefore, Equation 7.23 can be rewritten as

$$\mathbf{w} = \sum_{i=1}^{m} (\alpha_i^+ - \alpha_i^-) x_i. \tag{7.24}$$

Hence, Equation 7.14 becomes

$$f(x) = \sum_{i=1}^{m} (\alpha_i^+ - \alpha_i^-) K(\mathbf{x}_i, \mathbf{x}_j) + b. \tag{7.25}$$

This is the SVR expansion. That is, $\mathbf{w}$ can be completely described as a linear combination of the training patterns $\mathbf{x}_i$. Some outstanding advantages of Equation 7.25 are that it is independent of both the dimensionality of the input space χ and the sample size m.

7.3.1.1 Types of kernel functions used in SVM

$K(\mathbf{x}_i, \mathbf{x}_j)$ is defined in Equation 7.23 as the kernel function. Its value is equal to the inner product of two vectors $\mathbf{x}_i$ and $\mathbf{x}_j$ in the feature space $\Phi(\mathbf{x}_i)$ and $\Phi(\mathbf{x}_j)$. That is,

$$K(\mathbf{x}_i, \mathbf{x}_j) = \Phi(\mathbf{x}_i) \cdot \Phi(\mathbf{x}_j). \tag{7.26}$$

Therefore, the SVM techniques use a kernel function to map the input space into a high-dimensional feature space through some nonlinear mapping chosen *a priori* and are used to construct the optimal separating hyperplane in the feature space. This makes it possible to construct linear decision surfaces in the feature space instead of constructing nonlinear decision surfaces in the input space. There are several types of kernel function used in SVM. The type of SVM constructed is a function of the selected kernel function. This also affects the computation time of implementing SVM. According to Hilbert–Schmidt theory, $K(\mathbf{x}_i, \mathbf{x}_j)$ can be any symmetric function satisfying the following conditions (Courant and Hilbert, 1953):

Mercer conditions: To guarantee that the symmetric function $K(x_i, x_j)$ has the expansion

$$K(x_i, x_j) = \sum_{k=1}^{\infty} \alpha_k \psi_k(x_i) \psi_k(x_j),$$ (7.27)

with positive coefficients $\alpha_k > 0$ (i.e., $K(x_i, x_j)$ describes an inner product in some feature space), it is necessary and sufficient that the condition

$$\iint K(x_i, x_j) g(x_i) g(x_j) dx_i dx_j > 0,$$ (7.28)

be valid for all $g \neq 0$ for which

$$\int g^2(x_i) dx_i < \infty.$$ (7.29)

On the basis of this theorem, three of the popular kernel functions used in SVM are

- A polynomial kernel function can be constructed using

$$K(\mathbf{x}_i, \mathbf{x}_j) = (x_i, x_j)^d, \quad d = 1, 2, 3, \ldots.$$ (7.30)

An alternative form of Equation 7.30, which avoids the computation problem encountered in using Equation 7.30 is

$$K(\mathbf{x}_i, \mathbf{x}_j) = ((x_i, x_j) + 1)^d, \quad d = 1, 2, 3, \ldots.$$ (7.31)

- A Gaussian radial basis kernel function can be constructed using

$$K(\mathbf{x}_i, \mathbf{x}_j) = \exp\left(-\frac{(x_i - x_j)^2}{\sigma^2}\right),$$ (7.32)

where $\sigma(>0)$ is the kernel width.
- A sigmoid kernel function can be constructed using

$$K(x_i, x_j) = \tanh(b(x_i \cdot x_j) - c).$$ (7.33)

The Gaussian RBF kernels, usually called RBF kernels in SVM literature, are widely used in artificial neural networks (Haykin, 1999), SVMs (Vapnik, 1998) and approximation theory (Schölkopf and Smola, 2002). The RBF kernel is usually a reasonable first choice because of its outstanding features. For example, it can handle linear and nonlinear input–output mapping effectively and it requires fewer hyperparameters than the polynomial kernel, which reduces computation cost in terms of tuning for optimum hyperparameters. In addition, the kernel values for RBF range between 0 and 1; hence, there are fewer numerical difficulties since these values can range between 0 and infinity for the polynomial kernel. The sigmoid kernel is not always considered because it does not always fulfill the Mercer condition (Vapnik, 2000), which is a requirement for an SVR kernel. Furthermore, the sigmoid kernel is similar to the RBF kernel when the kernel width is a small value (Lin and Lin, 2003).

7.3.1.2 Methods of computing SVR parameters

The performance of the SVR technique depends on the setting of three training parameters (kernel, C, and ε) for ε-insensitive loss function. However, for any particular type of kernel, the values of C and ε affect the complexity of the final model. The value of ε affects the number of SVs used for predictions. Intuitively, a larger value of ε results in a smaller number of SVs, which leads to less-complex regression estimates. On the other hand, the value of C is the trade-off between model complexity and the degree of deviations allowed in the optimization formulation. Therefore, a larger value of C undermines model complexity (Cherkassky and Ma, 2004). The selection of optimum values for these training parameters (C and ε) that will guarantee less-complex models is an active area of research. There are several existing approaches for selecting the optimum value for these parameters.

The most common approach is based on users' prior knowledge or expertise in applying SVM techniques (Cherkassky and Mulier, 1998; Schölkopf et al., 1999). However, this approach could be subjective and is not appropriate for new users of SVR. It is also not applicable for online application since it requires manual intervention at each step of learning. This approach constitutes a source of uncertainty when used by nonexperts and experts of SVM not familiar with the characteristics of the dataset under consideration. Mattera and Haykin (1999) proposed that the value of C be equal to the range of output values; but this approach is not robust to outliers (Cherkassky and Ma, 2004), especially in condition-monitoring problems where data are prone to outliers due to faulty sensors or instruments. Another approach is the use of cross-validation techniques for parameter selection (Cherkassky and Mulier, 1998; Schölkopf et al., 1999). Even though this is a good approach for batch processing, it is data intensive; hence, it is very expensive to implement in terms of computation time, especially for larger datasets. Furthermore, resampling techniques are not applicable to online applications. One more approach is that ε values should be selected in proportion to the variance of the input noise (Smola et al., 1998; Kwok, 2001); this approach is independent of the sample size and is only suitable for batch processing where the entire dataset is available. Cherkassky and Ma (2004) presented another approach based on the training data. They proposed that C values should be based on the training data without resulting to resampling using the following estimation:

$$C = \max(|\bar{y} + 3\sigma_y|, |\bar{y} - 3\sigma_y|), \tag{7.34}$$

where $\bar{y}$ and σ_y are the mean and standard deviation of the y-values of the training data. One advantage of this approach is that it is robust to possible outliers. They also proposed that the value of ε should be proportional to the standard deviation of the input noise. Using the idea of central limit theorem, they proposed that ε must be given by

$$\varepsilon = 3\sigma\sqrt{\frac{\ln n}{n}}, \tag{7.35}$$

where σ is the standard deviation of the input noise and n is the number of training samples. Since the value of σ is not known *a priori*, the following equation can be used to estimate σ using the idea of k-nearest-neighbor's method

$$\hat{\sigma} = \sqrt{\frac{n^{1/5}k}{n^{1/5}k - 1} \cdot \frac{1}{n} \sum_{i=1}^{n} (y_i - \hat{y}_i)^2}, \quad 2 \le k \le 6, \tag{7.36}$$

where n is the number of training samples, k is the low-bias/high-variance estimators, and $\hat{y}$ is the predicted value of y by fitting a linear regression to the training data to estimate the noise variance. Again, this approach is only applicable to batch processing. Cao and Tay (2003) proposed the ascending regularization constant (C_i) and descending tube (ε_i) for batch SVR applications in financial time-series data. They adopt the following definitions:

$$C_i = C \frac{2}{1 + \exp(a - 2a(i/m))}, \tag{7.37}$$

and

$$\varepsilon_i = \varepsilon \frac{1 + \exp(b - 2b(i/m))}{2}, \tag{7.38}$$

where i represents the data sequence, C_i is the ascending regularization constant, ε_i is the descending tube, a is the parameter that controls the ascending rate, and b is the parameter that controls the descending rate.

7.3.1.3 Extension of the SVR technique for online prediction

Like the principal component regression (PCR) and partial least squares (PLS) techniques, the expansion of the SVR technique is not suitable for online prediction, because the addition of a data point requires the retraining of the entire training set. As a result, Ma et al. (2003) proposed accurate online support vector regression (AOSVR). The procedure involved in AOSVR is that whenever a new sample is added to the training set, the corresponding coefficient is updated in a finite number of steps until it meets the Karush–Kuhn–Tucker (KKT) conditions, while at the same time ensuring that the existing samples in the training set continue to satisfy the KKT conditions at each step. However, the same value is used for SVR parameters during the training and testing stages. Consequently, Omitaomu (2006, 2013) proposed adaptive online support vector regression (AOLSVR). Several approximate online SVM algorithms have previously been proposed (Li and Long, 1999; Cauwenberghs and Poggio, 2001; Csato and Opper, 2001; Gentile, 2001; Graepel et al., 2001; Herbster, 2001; Kivinen et al., 2002). A summarized presentation of the AOLSVR technique is presented in this section.

For AOLSVR, Equation 7.16 of SVR becomes

$$\text{Minimize} \quad \frac{1}{2} \mathbf{w}^T \mathbf{w}$$

$$\text{Subject to:} \quad \begin{cases} y_i - \mathbf{w}^T \Phi(\mathbf{x}_i) - b \le \varepsilon_i \\ \mathbf{w}^T \Phi(\mathbf{x}_i) + b - y_i \le \varepsilon_i, \end{cases} \tag{7.39}$$

where ε_i is the varying (adaptive) accuracy parameter. This idea leads to the following primal formulations:

$$\text{Minimize} \quad \frac{1}{2}\mathbf{w}^\mathsf{T}\mathbf{w} + \sum_{i=1}^{m} C_i(\xi_i^+ + \xi_i^-)$$

$$\text{Subject to:} \quad \begin{cases} y_i - \mathbf{w}^\mathsf{T}\Phi(\mathbf{x}_i) - b \le \varepsilon_i + \xi_i^+ \\ \mathbf{w}^\mathsf{T}\Phi(\mathbf{x}_i) + b - y_i \le \varepsilon_i + \xi_i^- \\ \xi_i^+, \xi_i^- \qquad\qquad \ge 0, \end{cases} \tag{7.40}$$

where C_i is the varying (adaptive) regularization constant. Therefore, the Lagrangian of Equation 7.40 is given as

$$\frac{1}{2}\mathbf{w}^\mathsf{T}\mathbf{w} + \sum_{i=1}^{m} C_i(\xi_i^+ + \xi_i^-) - \sum_{i=1}^{m}(\eta_i^+\xi_i^+ + \eta_i^-\xi_i^-) - \sum_{i=1}^{m}\alpha_i^+(\varepsilon_i + \xi_i^+ - y_i + \mathbf{w}^\mathsf{T}\Phi(\mathbf{x}_i) + b)$$

$$- \sum_{i=1}^{m}\alpha_i^-(\varepsilon_i + \xi_i^- + y_i - \mathbf{w}^\mathsf{T}\Phi(\mathbf{x}_i) - b) \tag{7.41}$$

s.t. $\alpha_i^+, \alpha_i^-, \eta_i^+, \eta_i^- \ge 0.$

Therefore, the necessary conditions for α to be a solution to the original optimization problem, Equation 7.40, are given by the following equation:

$$\partial_b = \sum_{i=1}^{m}(\alpha_i^+ - \alpha_i^-) = 0$$

$$\partial_\mathbf{w} = \mathbf{w} - \sum_{i=1}^{m}(\alpha_i^+ - \alpha_i^-)x_i = 0 \tag{7.42}$$

$$\partial_{\xi_i^+} = C_i - \eta_i^+ - \alpha_i^+ = 0$$

$$\partial_{\xi_i^-} = C_i - \eta_i^- - \alpha_i^- = 0.$$

We can rewrite Equation 7.41 as follows:

$$\frac{1}{2}\mathbf{w}^\mathsf{T}\mathbf{w} + \sum_{i=1}^{m}\xi_i^+(C_i - \eta_i^+ - \alpha_i^+) + \sum_{i=1}^{m}\xi_i^-(C_i - \eta_i^- - \alpha_i^-) + \sum_{i=1}^{m}\varepsilon_i(\alpha_i^+ + \alpha_i^-)$$

$$- \sum_{i=1}^{m} y_i(\alpha_i^+ - \alpha_i^-) - \sum_{i=1}^{m}\mathbf{w}^\mathsf{T}\Phi(\mathbf{x}_i)(\alpha_i^+ - \alpha_i^-) - b\sum_{i=1}^{m}(\alpha_i^+ - \alpha_i^-). \tag{7.43}$$

Substituting Equation 7.42 into Equation 7.43, to eliminate $\mathbf{w}, b, \xi_i^+,$ and ξ_i^-, results in the following dual optimization problem:

$$\text{Maximize } \frac{1}{2} \sum_{i,j=1}^{m} K(x_i, x_j)(\alpha_i^+ - \alpha_i^-)(\alpha_j^+ - \alpha_j^-)$$

$$+ \sum_{i=1}^{m} \varepsilon_i(\alpha_i^+ + \alpha_i^-) - \sum_{i=1}^{m} y_i(\alpha_i^+ - \alpha_i^-) \tag{7.44}$$

$$\text{Subject to: } \begin{cases} \displaystyle\sum_{i=1}^{m} (\alpha_i^+ - \alpha_i^-) = 0 \\ 0 \le \alpha_i^+, \alpha_i^- \le C_i. \end{cases}$$

Following the approach by Ma et al. (2003), the Lagrange of Equation 7.44 can be written as

$$\frac{1}{2} \sum_{i,j=1}^{m} K(\mathbf{x}_i, \mathbf{x}_j)(\alpha_i^+ - \alpha_i^-)(\alpha_j^+ - \alpha_j^-) + \varepsilon \sum_{i=1}^{m} (\alpha_i^+ + \alpha_i^-)$$

$$- \sum_{i=1}^{m} y_i(\alpha_i^+ - \alpha_i^-) - \sum_{i=1}^{m} (\delta_i^+ \alpha_i^+ + \delta_i^- \alpha_i^-) + \zeta \sum_{i=1}^{m} (\alpha_i^+ - \alpha_i^-)$$

$$+ \sum_{i=1}^{m} [u_i^+(\alpha_i^+ - C) + u_i^-(\alpha_i^- - C)], \tag{7.45}$$

where $\delta_i^+, \delta_i^-, u_i^+, u_i^-, \zeta$ are Lagrange multipliers. Optimizing Equation 7.45 leads to the following KKT conditions:

$$\frac{\partial}{\partial \alpha_i^+} = \sum_{j=1}^{m} K(\mathbf{x}_i, \mathbf{x}_j)(\alpha_j^+ - \alpha_j^-) + \varepsilon - y_i - \delta_i^+ + \zeta + u_i^+ = 0, \tag{7.46}$$

$$\frac{\partial}{\partial \alpha_i^-} = -\sum_{j=1}^{m} K(\mathbf{x}_i, \mathbf{x}_j)(\alpha_j^+ - \alpha_j^-) + \varepsilon + y_i - \delta_i^+ - \zeta + u_i^+ = 0, \tag{7.47}$$

$$\delta_i^{(*)} \ge 0, \quad \delta_i^{(*)} \alpha_i^{(*)} = 0, \tag{7.48}$$

$$\text{and} \quad u_i^{(*)} \ge 0, \quad u_i^{(*)}(\alpha_i^{(*)} - C) = 0. \tag{7.49}$$

Using the following definitions:

$$Q_{ij} = K(\mathbf{x}_i, \mathbf{x}_j), \tag{7.50}$$

$$\theta_i = \alpha_i^+ - \alpha_i^- \quad \text{and} \quad \theta_j = \alpha_j^+ - \alpha_j^-, \tag{7.51}$$

$$\text{and} \quad h(x_i) \equiv f(x_i) - y_i = \sum_{j=1}^{m} Q_{ij}\theta_j - y_i + b, \tag{7.52}$$

where $h(x_i)$ is the error of the target value for vector i. The KKT conditions in Equations 7.46 through 7.49 can be rewritten as

$$\frac{\partial L_D}{\partial \alpha_i^+} = h(x_i) + \varepsilon_i = \psi_i^+ = 0$$

$$\frac{\partial L_D}{\partial \alpha_i^-} = -h(x_i) + \varepsilon_i = \psi_i^- = -\psi_i^+ + 2\varepsilon_i = 0 \tag{7.53}$$

$$\frac{\partial L_D}{\partial b} = \sum_{i=1}^{m} \theta_i = 0,$$

where $\psi_i^{(*)}$ is the adaptive margin function and can be described as the threshold for error on both sides of the adaptive ε—tube. Modifying the approach by Ma et al. (2003), these KKT conditions lead to five new conditions for AOLSVR:

$$
\begin{aligned}
2\varepsilon_i < \psi_i^+ &\to \psi_i^- < 0, & \theta_i &= -C_i & i &\in E^- \\
\psi_i^+ = 2\varepsilon_i &\to \psi_i^- = 0, & -C_i < \theta_i &< 0 & i &\in S \\
0 < \psi_i^+ < 2\varepsilon_i, &\to 0 < \psi_i^- < 2\varepsilon_i, & \theta_i &= 0 & i &\in R \\
\psi_i^+ = 0 &\to \psi_i^- = 2\varepsilon_i, & 0 < \theta_i &< C_i & i &\in S \\
\psi_i^+ < 0 &\to \psi_i^- > 2\varepsilon_i, & \theta_i &= C_i, & i &\in E^+.
\end{aligned}
\tag{7.54}
$$

These conditions can be used to classify the training set into three subsets defined as follows:

$$\text{The } E \text{ set: Error support vectors: } E = \left\{i \,\middle|\, |\theta_i| = C_i\right\}$$

$$\text{The } S \text{ set: Margin support vectors: } S = \left\{i \,\middle|\, 0 < |\theta_i| < C_i\right\} \tag{7.55}$$

$$\text{The } R \text{ set: Remaining samples: } R = \left\{i \,\middle|\, \theta_i = 0\right\}.$$

On the basis of these conditions, we modify the AOSVR algorithm appropriately and incorporate the algorithms for computing adaptive SVR parameters for the online training as described in Section 7.1.4. For a detailed description of the AOLSVR approach, interested readers should be Omitaomu (2013).

7.3.1.4   Modified logistic weight function for computing adaptive SVR parameters

One of the most popular classical symmetric functions that use only one equation is the logistic function. It has wide applications in several areas including engineering, natural sciences, and statistics. It combines two characteristic exponential growth (exponential and bounded exponential). The logistic function has been widely used in the neural network as the preferred activation function, because it has good properties for updating training weights. However, the standard form of the logistic function is not flexible in setting lower and upper bounds on weights. For time-dependent predictions, it is reasonable to have a certain initial weight at the beginning of the training. Therefore, to extend the properties of the logistic function to estimate SVR parameters, Omitaomu (2006) proposed modified logistic weight function (MLWF) equations for adaptive SVR parameters. The adaptive regularization constant (C_i) is defined as

$$C_i = C_{min} + \left[\frac{C_{max}}{1 + \exp(-g \times (i - m_c))} \right] \tag{7.56}$$

and the MLWF equation for the adaptive accuracy parameter (ε_i) is defined as

$$\varepsilon_i = \varepsilon_{min} + \left[\frac{\varepsilon_{max}}{1 + \exp(g \times (i - m_c))} \right], \tag{7.57}$$

where $i = 1, \ldots, m$, where m is the number of training samples, m_c is the changing point, C_{min} and ε_{min} are the desired lower bound for the regularization constant and the accuracy parameter respectively, C_{max} and ε_{max} are the desired upper bound for the regularization constant and the accuracy parameter respectively, and g is an empirical constant that controls the curvature (slope) of the function; that is, it represents the factor for the relative importance of the samples. The essence of the lower bound is to avoid underestimation and the upper bound avoids overestimation of the parameters. The value of g could range from 0 to infinity and four special cases are given below. The behaviors of these four cases are summarized as follows:

1. *Constant weight*: This is the case with conventional AOSVR in which all data points are given the same weight. This is more suitable for data from the same process. This can be achieved in Equations 7.56 and 7.57 when $g = 0$; therefore, $C_i \cong C_{min} + C_{max}/2$ and $\varepsilon_i \cong \varepsilon_{min} + \varepsilon_{max}/2$.
2. *Linear weight*: This is applicable to cases in which the weight is linearly proportional to the size of the training set. This is the case when $g = 0.005$, then, the value of C_i is a linearly increasing relationship and the value of ε_i is a linearly decreasing relationship.
3. *Sigmoidal weight*: Different sigmoidal patterns can be achieved using different values of g in relation to the number of training sets. One possibility is when $g = 0.03$ the weight function follows a sigmoidal pattern. The value of g can also be set to achieve a pattern with a zero slope at the beginning and at the end of the training.
4. *Two distinct weights*: In this case, the first one-half of the training set is given one weight and the second one-half is given another weight. A possible application is the case of data from two different processes. This is possible when $g = 5$, then,

$$C_i \cong \begin{cases} C_{min}, & i < m_c \\ C_{min} + C_{max}, & i \geq m_c \end{cases} \quad \text{and} \quad \varepsilon_i \cong \begin{cases} \varepsilon_{min} + \varepsilon_{max}, & i < m_c \\ \varepsilon_{min}, & i \geq m_c. \end{cases}$$

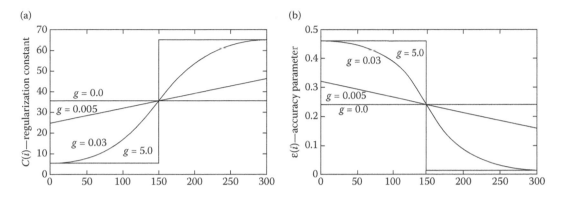

Figure 7.7 The pictorial representations of MLWF with different values of g: (a) regularization and (b) accuracy.

The pictorial representations of the different weights for these g values are shown in Figure 7.7a and b plots. Both plots show that the profile for MLWF is symmetric around the midpoint (m_c) of the total training set. As an illustration, for the plots in Figure 7.7, the C_{min} and C_{max} are set to 5.0 and 60.0 respectively and ε_{min} and ε_{max} are set to 0.01 and 0.45 respectively for m = 300 and m_c = 150.

Consequently, if the recent samples are more important than past samples, g must be greater than zero $g > 0$. Other values can also be used depending on the objectives of the problem. However, if past samples are more important than recent samples, then, g must be less than zero ($g < 0$) for all equations; note that this is a rare case.

Some of the advantages of the adaptive approach include:

- Selection of C and ε parameters directly in relation to the relative importance of data samples and/or their respective position in the data sequence without using resampling methods.
- Flexibility in setting lower and upper bounds for the parameters without using resampling techniques.
- The use of a curvature control parameter that has only fewer possible values that will enhance *a priori* settings.
- The use of parameters that can be set with the slightest knowledge of the characteristics of the incoming data.

7.3.2 Predicting power grid frequency using PMU data and AOLSVR

The ubiquitous PMUs are making real-time frequency information about the power grid readily available. Therefore, the accurate prediction of power grid frequency can increase reliability of the power systems. In this section, we explore the application of AOLSVR to predict frequency data from PMUs installed at the distribution level.

For this implementation, we used a typical online time-series prediction scenario as presented by Tashman (2000) and used a prediction horizon of one time step. Given a time series $\{x(t), t = 1,2,...\}$ and prediction origin O, the time from which the prediction is generated, we construct a set of training samples, $\mathbf{A}_{O,B}$, from the segment of time series

$$\{x(t), \ t = 1,...,O\} \quad \text{as} \quad \mathbf{A}_{O,B} = \{\mathbf{X}(t), y(t), t = B,...,O-1\},$$

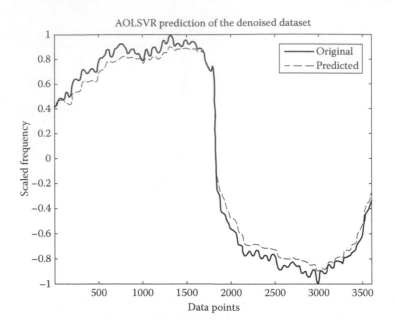

Figure 7.8 Plots of the denoised dataset 1 and its prediction.

where $\mathbf{X}(t) = [x(t), \ldots, x(t - B + 1)]^T$, $y(t) = x(t + 1)$, and B is the embedding dimension of the training set $\mathbf{A}_{O,B}$, which in this application is set to 10—being the number of samples collected every second. We train the predictor $P(\mathbf{A}_{O,B}; \mathbf{X})$ from the training set $\mathbf{A}_{O,B}$. Then, predict $x(O + 1)$ using $\hat{x}(O + 1) = P(\mathbf{A}_{O,B}; \mathbf{X}(O))$. When $x(O + 1)$ becomes available, we update the prediction origin; that is, $O = O + 1$ and repeat the procedure. As the origin increases, the training set keeps growing and this can become very computationally expensive. However, online prediction takes advantage of the fact that the training set is augmented one sample at a time and continues to update and improve the model as more data arrive.

The results of the implementation of this approach to the denoised data obtained from Section 7.3.1 (and presented in Figure 7.4) are shown in Figure 7.8. Looking at these plots (and others not shown in this chapter), we can conclude that the AOLSVR approach accurately predicts the original denoised data within an allowable operational margin for over- or underprediction. The fact that this approach uses varying parameters allows the technique to capture more of the properties of the original data than using fixed regression parameters.

7.4   Summary

Energy utility companies are making significant investments in rolling out new measurement devices that could help collect event information that will provide insights about how they can improve customer experience, grid reliability, operational efficiency, and revenue assurance. In this chapter, we described some of these measurement devices and the potential applications of the data they collected. Events data as recorded by various measurement devices are a raw data stream and are also associated with high volumes since hundreds of events are generated for normal operations, as well as for changed conditions. These events basically manifest the conditions of the power grid network and also some aspects of customer behavior. To help utilities analyze the collected data, we also presented an EMD-based approach for removing noise from the collected data. The EMD approach is applied to datasets from PMUs and smart meters. Furthermore, an AOLSVR approach that uses varying parameters is also presented for predicting signals from PMUs. The results obtained show that the predictions in all cases are within an allowable operational margin. As more of these measurement devices are deployed, utilities need to take a holistic approach toward analytics as the need to process massive amounts of operational data faster will become inevitable.

Acknowledgment

This chapter has been authored by UT-Battelle, LLC, under contract DE-AC05-00OR22725 with the U.S. Department of Energy. The U.S. Government retains, and the publisher, by accepting the article for publication, acknowledges that the U.S. Government retains a nonexclusive, paid-up, irrevocable, and worldwide license to publish or reproduce the published form of this chapter, or allow others to do so, for U.S. Government purposes.

References

Bank, J.N., Omitaomu, O.A., Fernandez, S.J., and Liu, Y. 2009. Visualization and classification of power system frequency data streams. *IEEE International Conference on Data Mining Workshop,* Miami, FL, 650–655, December 6.

Bishop, C.M. 1995. *Neural Networks for Pattern Recognition.* Oxford: Clarendon Press.

Cao, L. and Tay, F.E.H. 2003. Support vector machine with adaptive parameters in financial time series forecasting. *IEEE Transactions on Neural Networks,* 14 (6): 1506–1518.

Cauwenberghs, E. and Poggio, T. 2001. Incremental and decremental support vector machine learning. In T.K. Leen, T.G. Dietterich, and V. Tresp (Eds.), *Advances in Neural Information Processing Systems,* Vol. 13, 409–523. Cambridge, MA: MIT Press.

Cherkassky, V. and Ma, Y. 2004. Practical selection of SVM parameters and noise estimation for SVM regression. *Neural Networks,* 17 (1): 113–126.

Cherkassky, V. and Mulier, F. 1998. *Learning from Data: Concepts, Theory, and Methods.* New York: John Wiley & Sons, Inc.

Cohen, L. 1995. *Time-Frequency Analysis.* Englewood Cliffs, NJ: Prentice-Hall.

Courant, R. and Hilbert, D. 1953. *Methods of Mathematical Physics.* vol. 1, Berlin: Springer-Verlag.

Csato, L. and Opper, M. 2001. Sparse representation for Gaussian process models. In T.K. Leen, T.G. Dietterich, V. Tresp (Eds.), *Advances in Neural Information Processing Systems,* 13: 444–450. Cambridge, MA: MIT Press.

Donoho, D.L. 1995. De-noising by soft-thresholding. *IEEE Transactions on Information Theory,* 41: 613–627.

Donoho, D.L. and Johnstone, I.M. 1994. Ideal spatial adaptation by wavelet shrinkage. *Biometrika,* 81 (4): 425–455.

Donoho, D.L. and Johnstone, I.M. 1995. Adapting to unknown smoothness in wavelet shrinkage. *Biometrika,* 81 (4): 425–455.

Fernandez, R. 1999. Predicting time series with a local support vector regression machine. *Advanced Course on Artificial Intelligence* (ACAI '99). Available at: http://www.iit.demokritos.gr/skel/eetn/acai99/ (Downloaded on January 15, 2005).

Fleuret, F. 2004. Fast binary feature selection with conditional mutual information. *Journal of Machine Learning Research,* 5: 1531–1555.

Gentile, C. 2001. A new approximate maximal margin classification algorithm. *Journal of Machine Learning Research,* 2: 213–242.

Graepel, T., Herbrich, R., and Williamson, R.C. 2001. From margin to sparsity. In T.K. Leen, T.G. Dietterich, and V. Tresp (Eds.), *Advances in Neural Information Processing Systems*, Vol. 13, 210–216. Cambridge, MA: MIT Press.

Gunn, S.R. 1998. Support vector machines for classification and regression. *Technical Report*, Image Speech and Intelligent Systems Research Group, University of Southampton, UK. Available at: http://www.isis.ecs.soton.ac.uk/isystems/kernel/ (Downloaded on December 5, 2004).

Haykin, S. 1999. *Neural Networks: A Comprehensive Foundation*, 2nd ed. Upper Saddle River, NJ: Prentice-Hall.

Herbster, M. 2001. Learning additive models online with fast evaluating kernels. In D.P. Helmbold, and B. Williamson (Eds.), *Proceedings of the 14th Annual Conference on Computational Learning Theory*, pp. 444–460. New York: Springer-Verlag.

Huang, N.E., Long, S.R., and Shen, Z. 1996. Frequency downshift in non-linear water wave evolution. *Advances in Applied Mechanics*, 32: 59–117.

Huang, N.E., Shen, Z., and Long, S.R. 1999. A new view of non-linear water waves—The Hilbert spectrum. *Annual Review of Fluid Mechanics*, 31, 417–457.

Huang, N.E., Shen, Z., Long, S.R., Wu, M.C., Shih, S.H., Zheng, Q., Tung, C.C., and Liu, H.H. 1998. The empirical mode decomposition method and the Hilbert spectrum for non-stationary time series analysis. *Proceedings of the Royal Society of London*, A454: 903–995.

Jha, S.K. and Yadava, R.D.S. 2010. Denoising by singular value decomposition and its application to electronic nose data processing. *IEEE Sensors Journal*, 11 (1): 35–44.

Kantz, H. and Schreiber, T. 1997. *Non-Linear Time Series Analysis*. Cambridge: Cambridge University Press.

Kivinen, J., Smola, A.J., and Williamson, R.C. 2002. Online learning with kernels. In T.G. Dietterich, S. Becker, and R.C. Williamson (Eds.), *Advances in Neural Information Processing Systems*, Vol. 14, 785–792. Cambridge, MA: MIT Press.

Kora, S., Malu, P., and Wani, R. 2012. Role of smart sensors for making smart grid a reality. *IEEE PES Innovative Smart Grid Technologies Conference*, Washington, DC.

Kwok, J.T. 2001. Linear dependency between ε and the input noise in ε-support vector regression. In G. Dorffner, H. Bishof, and K. Hornik (Eds.), *ICANN 2001, LNCS 2130*, 405–410. New York: Springer Berlin Heidelberg.

Li, Y. and Long, P.M. 1999. The relaxed online maximum margin algorithm. In S.A. Solla, T.K. Leen, and K.-R. Müller (Eds.), *Advances in Neural Information Processing Systems*, Vol. 12, 498–504. Cambridge, MA: MIT Press.

Lin, H.T. and Lin, C.J. 2003. A study on sigmoid kernels for SVM and the training of non-psd kernels by smo-type methods. *Technical Report*. Available at: http://www.csie.ntu.edu.tw/~cjlin/papers/tanh.pdf (Downloaded on January 25, 2005).

Ma, J., James, T., and Simon, P. 2003. Accurate on-line support vector regression. *Neural Computation*, 15: 2683–2703.

Mattera, D. and Haykin, S. 1999. Support vector machines for dynamic reconstruction of a chaotic system. In B. Schölkopf, J. Burges, and A. Smola (Eds.), *Advances in Kernel Methods: Support Vector Machine*, pp. 211–142. Cambridge, MA: MIT Press.

Muller, K.-R., Smola, A.J., Ratsch, G., Scholkopf, B., Kohlmorgen, J., and Vapnik, V. 1997. Predicting time series with support vector machines. In W. Gerstner (Ed.), *Artificial Neural Networks—ICANN '97*, 999–1004. Berlin: Springer-Verlag.

Muller, K.-R., Smola, A., Ratsch, G., Schölkopf, B., Kohlmorgen, J., and Vapnik, V. 1999. Using support vector machines for time series prediction. In B. Schölkopf, J. Burges, and A. Smola (Eds.), *Advances in Kernel Methods: Support Vector Machine*, pp. 325–349. Cambridge, MA: MIT Press.

Omitaomu, O.A. 2006. On-line learning and wavelet-based feature extraction methodology for process monitoring using high-dimensional functional data. PhD dissertation, University of Tennessee, Knoxville, TN.

Omitaomu, O.A. 2013. *Intelligent Process Monitoring and Control Using Sensor Data*. Germany: Lap Lambert Academic Publishing.

Omitaomu, O.A., Jeong, M.K., Badiru, A.B., and Hines, J.W. 2006. On-line prediction of motor shaft misalignment using FFT generated spectra data and support vector regression. *ASME Transactions on Journal of Manufacturing Science and Engineering*, 128 (4): 1019–1024.

Omitaomu, O.A., Jeong, M.K., Badiru, A.B., and Hines, J.W. 2007. On-line support vector regression approach for the monitoring of motor shaft misalignment and feedwater flow rate. *IEEE Transactions on Systems, Man, and Cybernetics: Part C—Applications and Reviews*, 37 (5): 962–970.

Omitaomu, O.A., Jeong, M.K., and Badiru, A.B. 2011a. On-line support vector regression with varying parameters for time-dependent data. *IEEE Transactions on Systems, Man, and Cybernetics: Part A—Systems and Humans*, 41 (1): 191–197.

Omitaomu, O.A., Protopopescu, V.A., and Ganguly, A.R. 2011b. Empirical mode decomposition technique with conditional mutual information for denoising operational sensor data. *IEEE Sensors Journal*, 11 (10): 2565–2575.

Osuna, E., Freund, R., and Girosi, F. 1997. An improved training algorithm for support vector machines. In J. Principe, L. Gile, N. Morgan, and E. Wilson (Eds.), *Neural Networks for Signal Processing VII—Proceedings of the 1997 IEEE Workshop*, Amelia Island, FL, 276–285.

Schölkopf, B., Burges, J., and Smola, A. 1999. *Advances in Kernel Methods: Support Vector Machine.* Cambridge, MA: MIT Press.

Schölkopf, B. and Smola, A. 2002. *Learning with Kernels: Support Vector Machines, Regularization, Optimization, and beyond.* Cambridge, MA: MIT Press.

Smola, A.J., Murata, N., Schölkopf, B., and Muller, K. 1998. Asymptotically optimal choice of ε-loss for support vector machines. In L. Niklasson, M. Boden, and T. Ziemke (Eds.), *Proceedings of the International Conference on Artificial Neural Networks (ICANN 1998), Perspectives in Neural Computing*, Berlin: Springer, 105–110.

Smola, A.J. and Schölkopf, B. 2004. A tutorial on support vector regression. *Statistics and Computing*, 14: 199–222.

Tashman, L.J. 2000. Out-of-sample tests of forecasting accuracy: An analysis and review. *International Journal of Forecasting*, 16: 437–450.

Vapnik, V. 1995. *The Nature of Statistical Learning Theory.* New York, NY: Springer-Verlag.

Vapnik, V. 1998. *Statistical Learning Theory.* New York, NY: John Wiley & Sons, Inc.

Vapnik, V. 2000. Support-vector networks. *Machine Learning*, 20 (3): 273–297.

Vapnik, V., Golowich, S.E., and Smola, A. 1996. Support vector method for prediction, regression estimation, and signal processing. *Advances in Neural Information Processing Systems.* San Mateo, CA: Morgan Kaufmann Publishers.

chapter eight

Economic systems measurement

LeeAnn Racz

Contents

8.1 Introduction

Measuring economic performance or value is often accomplished by accounting for comparisons. We compare the output of an economic system from one time period to another. The input for production among industries is compared with each other. Different regions have varying costs of living. These comparisons of value are made among different states or alternatives. This approach also applies to evaluating alternative economic policies that would result in different economic conditions. However, reasonable comparisons are possible only if the methods used to compare the different accounts or conditions are comparable.

8.2 Consistent application of terms

There must be consistent understanding of an application of terms when measuring economic factors. Inconsistencies in definitions make it impossible to achieve a complete and comparable accounting system. Christensen and Jorgensen (1973) offer an important case in point where confusion in an understanding of depreciation and replacement leads to inconsistencies in an accounting system. Depreciation is the value of capital services lost in the course of producing output. Replacement, on the other hand, occurs because of a decline in the ability of a capital service to contribute to production. Where one may define the net product as "gross product less depreciation," another may define it as "the amount consumed plus the addition to its capital stock" or "the amount of output consumed without changing its stock of capital" (Denison 1961). Unless depreciation is equal to replacement, these definitions are not consistent with each other. Similarly, where the confusion between depreciation and replacement is carried over to the input side of production, the calculation of capital input is measured incorrectly. The term "depreciation" is also often confused with "deterioration." While depreciation refers to financial value, deterioration is a less-specialized concept relating to quantity.

Historically, there has also been confusion as to how to account for depreciation as well as efficiency. It is easy to establish the purchase price of a capital good. It is not so easy to account for how much the value of the capital good is reduced over time. For the sake of simplicity, there tends to be three approaches that are widely accepted. Under the constant efficiency pattern, or "one-hoss-shay" pattern, capital retains its full value until it is no longer usable. The straight-line pattern dictates that the value of capital is reduced linearly until it is no longer used. Using the geometric decay pattern, the productive capacity of a capital good decays at a constant rate. However, these patterns describing the change of efficiency over time should not be confused with depreciation. In fact, it is only the geometric pattern that can be used to describe the path of efficiency as well as depreciation (Hulten and Wykoff 1996).

Hulten and Wykoff (1996) point out that part of the problem in confusing these terms associated with capital is that there is some debate over the role of capital in economic growth. Capital provides service over a certain period of time and is typically used by its owner. Noncapital inputs, such as labor and materials, require imputational methods where values in the accounts are approximated or substituted. Furthermore, while labor is a purely input value, capital is both input and output of production. Therefore, analysis of production can be complicated, particularly if there is a lack of a clear understanding of definitions.

8.3   Accounting methods

Conventional accounting methods measure the real product and real assets. However, evaluation of comprehensive economic performance must also measure other factors such as output and factor input, expenditures, saving, accumulation, capital formation, and others. This is not a straightforward process, and there is not a universally accepted approach. Nevertheless, there have been attempts to create methods for comprehensive accounting systems that include all aspects of economics performance. Christensen and Jorgenson (1973) proposed an approach that accounts for income, wealth, production, income, expenditure, and accumulation in current and constant prices for the private domestic and private national economies. These researchers suggest that their approach could also be used for the U.S. economy that is largely governmental. Naturally, the success of this comprehensive approach is intrinsically rife with details, and the authors go to great lengths to clearly define the terms of their accounting method. However, it is important to recognize that the more comprehensive a metric is, the less meaningful it may be since more of the factors used in the metric are based on imputations or estimates (Stiglitz et al. 2009).

Even more recently, Stiglitz et al. (2009) outlined the problems with something as widely used as gross domestic product (GDP) being used as an indicator of economic performance. GDP is the sum of consumer spending, government spending, business spending on capital, and total net exports within a country during a specific time period:

$$GDP = C + I + G + (X - M) \tag{8.1}$$

where
   C = Private consumption
   I = Gross investment
   G = Government spending
   X = Exports
   M = Imports

GDP was originally devised as an indicator of market economic activity, although it is now commonly used as an indicator of economic performance of a country, and is even used to gauge a country's standard of living. However, GDP is actually only a measure of a nation's productivity, which is unrelated to income or social satisfaction. Although market activity is a poor indicator of material well-being, we continue to rely on such indicators in the absence of more appropriate measurements.

The problem of improper application of GDP and other measurements of national economic performance is not trivial. Comparisons of these measurements between countries are commonly used to design a policy. Since there are significant differences in how national accounts are computed, inferences from these accounts used to influence policy may be flawed. There are some individuals and organizations that use these metrics without understanding how they are constructed and their limitations. For example, it would be foolish for one country to adopt practices from another country with a higher-reported growth rate if the growth rates are computed differently. Where we may really desire measurements of performance (output), in reality we use measures of input (GDP).

A poignant example of the improper use of economic measurements can be found in the financial crisis of 2007–2008. There was a heavy reliance on GDP as an indicator of economic performance. However, this metric does not adequately account for sustainability, such as indebtedness. The financial sector reported impressive profits just before this time period, although they were often based on suspect valuations. Since financial deregulation led to the expansion of that industry as well as an increase in GDP, many regarded this policy as good. The United States and other advanced countries experienced unsustainable growth based on borrowing during the time leading to the 2007–2008 financial crisis. To make matters worse, this borrowing was not used for investment, but rather consumption. The excessive borrowing led to the so-called housing bubble and to people believing the country had more wealth than it actually had. Since the standard metrics did not account for this condition, borrowing and spending continued unabated. Perhaps, if the world had understood the limitations of the GDP measurement and not placed undue focus on it, we may have seen the warning signs more clearly and taken measures to mitigate the problems (Stiglitz et al. 2009).

Another example of improper application of metrics can be found in Argentina during the 1990s. During this time, the growth of Argentina's economy, which had been the third largest in Latin America, was based on consumption fueled by borrowing from other countries. Standard economic metrics indicated that their economy was functioning well. However, it did not account for indebtedness that contributed to the unsustainable nature of their situation. This economic growth was also disconnected from unemployment that rose to 12% in 1994. Other factors as well, including corruption, internal rigidity in monetary policy, and an inefficient tax system, contributed to the demise of Argentina's economic system and default on its US$155 billion public debt (The Economist 2002).

There exists a pervasive disparity between GDP and how well off individuals are. GDP does not necessarily indicate the level of prosperity among individuals or households. It could be possible to construct a metric more indicative of personal financial well-being such as median disposable income, though it would be complex when attempting to use data besides that used in GDP calculations (Stiglitz et al. 2009). Furthermore, GDP does not include household production. Goods and services produced for one's own consumption are omitted from this number. Those who plant their own gardens, make their own food, care for their own children, or do anything for themselves that they could, in principle, pay someone else to do, contribute to living standards not reflected in the GDP (Folbre 2009).

On a smaller scale, companies have traditionally relied on financial and accounting results to measure the performance of their business. However, we must recognize that accounting measures are lagging indicators. They do not cause business performance, but are rather the result of it (Eccles and Pyburn 1992). They can only reveal the consequences of decisions, and not necessarily predict future outcomes. A good analogy compares to the score using traditional financial measures in a basketball game. The score does tell who is winning and losing, but does not tell how the game should be played. Focusing on conventional accounting tools has the same limitations in business. Therefore, many firms have turned their attention to nonfinancial metrics such as quality, customer satisfaction, innovation, human resource development, and market share as well as accounting to indicate their level of performance. However, this approach can only be successful when the business performance model is agreed upon and understood by the management. The purpose of the metrics must be made clear; otherwise, the management is likely to revert back to traditional financial measures that are easier to understand (Eccles and Pyburn 1992).

There are a variety of sources of data and assessment methods that can be used to evaluate business performance. Data can be gathered directly from the subject organization or from outside the organization. Also, performance can be assessed using objective (based on an established or systematic method) or perceptual (based on judgments of individuals). Dess and Robinson (1984) concluded that objective measures are preferred over perceptual measures. Venkatraman and Ramanujam (1987), on the other hand, found results that were not so clear-cut. While many would dismiss the perceptions of managers as biased, this study found that managers were actually less biased in assessing their organizational performance than might have been assumed. This study also had mixed results that did not indicate a clear advantage to using one data source over another. Rather, it indicated that a combination of different approaches yields more reliable results less prone to methodological artifacts.

8.4   Measuring social well-being

Perhaps, one of the most insidious consequences of relying on GDP as a measure of prosperity is that it does not account for less-tangible measures of well-being. Consider the case of two households. One household consists of a married couple who enjoy each other's company while preparing a meal together using ingredients they grew in their garden and spend the evening playing games with their children. Another household consists of a single man who gets his dinner from a fast-food restaurant, goes to a bar, and, in his inebriated state, possibly gets into a car accident on the way home. While the first household is arguably happier than the second, the second household has contributed far more to the GDP. Increases in spending that result in an increased GDP do not necessarily correlate to increases in social or personal well-being.

GDP is also not necessarily tied to poverty. As the World Bank is the lead institution for reducing poverty across the globe, it has set a somewhat arbitrary "$1-a-day" poverty line at purchasing power parity. However, this method has received considerable criticism from some as underestimating poverty and from others as overestimating poverty. Both criticisms stem from divergent definitions of terms used to make this calculation as well as a variety of measurement errors. Some economists favor the use of national income accounts, while others insist that household surveys must be used to measure poverty since globalization may not be good for the poor. Factors such as adequate nutrition, good health, literacy, political freedoms, years of schooling, etc., determine well-being. Indeed, measurement of well-being can be considered more of an art than a science (Thorbecke 2011).

Health care expenditures also reveal problems with misuse of standard metrics. Spending on health care does not necessarily correlate to improved health or longevity. The United States spends more on health care than any other country (~US$2600 more per person than Norway, the second-highest spender) but is only ranked at the 27th highest in life expectancy. Japan, on the other hand, has the highest life expectancy but spends US$5000 less per person than the United States (24/7 Wall St 2012). Figure 8.1 shows the percent of GDP spent on health care between 2003 and 2011 by the United States and the European Union. Figure 8.2 shows the life expectancy for these countries and regions during the same time period. These figures further illustrate the disconnection between spending on health care and life expectancy.

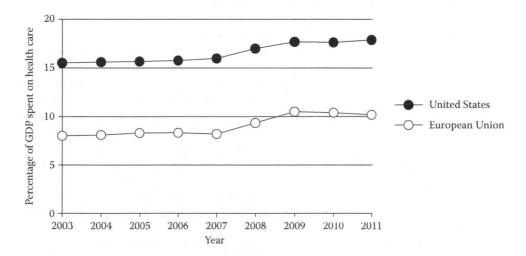

Figure 8.1 Percentage of GDP spent on health care in the United States and the European Union. (Adapted from World Bank, 2013, http://data.worldbank.org/, accessed Oct 17, 2013.)

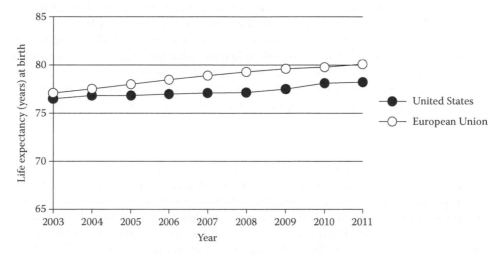

Figure 8.2 Life expectancy (years) at birth in the United States and the European Union. (Adapted from World Bank, 2013, http://data.worldbank.org/, accessed Oct 17, 2013.)

There is a growing recognition that well-being is closely tied to the quality of one's environment. The field of environmental economics aims to determine the value of ecosystems and the natural environment. However, this is an intrinsically complicated endeavor. Environmental economists attempt to ascribe value to ecosystems that is logically consistent with conventional cost–benefit analyses but quickly have difficulty understanding the interrelationships that connect effects of one part of an ecosystem to another part. Costanza et al. (1998) estimated the economic value of the entire planet's ecosystems at an average of US$33 trillion per year and the global gross national product at US$18 trillion per year. However, their approach has received considerable criticism. Bockstael et al. (2000), for example, took exception to how Costanza et al. (1998) regarded economic value. This fundamental question is not trivial and deserves substantial reflection when assigning value. Bockstael et al. (2000) remind us that value is assigned in comparative terms: value can be quantified when comparing trade-offs between different situations. Economic value is realized when measuring how much a person is willing to pay to have one condition over another. For example, consider the case where a coal mine is being proposed in a forested mountain. To mine the coal, the mountain would have to be deforested and have its topography drastically altered. Some people would value the beauty of the mountain and its ecosystem more than the coal and the jobs it would bring to the area. Other people who do not live near the mountain would value the electricity from the coal more than the quality of the mountain. Therefore, value is measured in the trade-off, and what each person is willing to pay. In the case of the value of the world's ecosystems, the willingness to pay to preserve the ecosystems must reflect the cost of what individuals would be willing to pay to give up the alternatives. Since Costanza et al.'s estimate of the value of the world's ecosystems (US$33 trillion per year) exceeds the world's ability to pay to avoid the loss of the ecosystems (US$18 trillion per year), the estimate of the value has little meaning (Bockstael et al. 2000).

Rather than attempting to measure an absolute value of the world's ecosystems, a more appropriate question would be what we would be willing to accept in place of the ecosystems. If the answer is that there would not be anything we would accept for losing the ecosystems and we would pay everything to avoid it, then, the ecosystems are, by definition, an essential good. A not-so-extreme example is of the value we place on environmental conditions depending on their abundance. If there is an abundance of, say, trees in a forest, the owner of the forest might be more willing to part with a few trees since many other trees remain. However, the more the trees are removed, the more the owner values the remaining trees. There is not a simple unit value for the trees. The applicability of the additive or multiplicative economic models further falls apart when we consider the intricate interactions between the elements of the ecosystem. By removing the trees, we are also removing habitats for numerous species and their subsequent interactions.

8.5   Index numbers

As we account for economic performance over time, we encounter the problem of measuring input and output in constant prices since price and quantity continuously change. One common approach is to rely on index numbers that are a statistical measure of changes in certain representative data points from a standard or base value. The use of index numbers is enticing as it offers a simplified framework by which we can analyze economic trends. However, by their very nature, they are aggregates of many data points. It is therefore critical that index numbers actually serve purposes for which they are intended without excessive loss of relevant information (Koopmans 1947).

Index numbers smooth out fluctuations in data. If long-term trends, such as annual averages, are desired, then, index numbers may be suitable. If, on the other hand, more fidelity is needed such as from quarterly or even shorter-duration time series, then index numbers should be used only with caution.

One form of index that has gained favor is the Divisia index. A Divisia index is a discrete approximation to continuous index numbers. It attempts to provide consistency between price and quantity indexes and has the benefit of applying weights to each component of the index according to its relative share. Divisia indexes are particularly useful when measuring welfare and productivity (Reinsdorf 1998). Christensen and Jorgensen (1973) have defined their Divisia index as

$$\frac{\dot{q}}{q} = \sum w_i \frac{\dot{q_i}}{q_i} \tag{8.2}$$

$$\frac{\dot{Y}}{Y} = \sum w_i \frac{\dot{Y_i}}{Y_i} \tag{8.3}$$

$$w_i = \frac{q_i Y_i}{\sum q_i Y_i} \tag{8.4}$$

where
  q is the price of output
  $\dot{q}$ is the rate of growth of price with respect to time
  w_i is the weight, or relative share, of the ith component of output
  Y is the quantity of output
  $\dot{Y}$ is the rate of growth of quantity with respect to time

Although often recognized as the preferred analytical measure, indexes are valid only under some unrealistic assumptions. Notably, the Divisia index is given by a path-dependent line integral under nonhomothetic preferences. However, Malaney (1996) and Reinsdorf (1998) make compelling arguments that despite this limitation, or even because of it, the Divisia index can be justified. So long as the paths avoid swerves or kinks, they are all likely to result in similar measures of change in welfare. If, however, there are drastic changes along the path, this index may not allow for reasonable interpretations (Reinsdorf 1998).

8.6   Conclusions

Economic value is measured in terms of comparison between conditions or alternatives, making absolute measurements problematic. It is therefore essential that consistent application of terms be used when attempting to assign value between scenarios. It is also critical to understand the true meaning behind measurements such as GDP. All computations have their limitations, and GDP is no exception. GDP had been, and continues to be, used for a variety of purposes beyond the original intent. While other measurements may be more meaningful, improperly used computations can lead to faulty conclusions and bad decisions.

References

Bockstael, N.E., Freeman, A.M., Kopp, R.J., Portney, P.R., Smith, V.K., 2000, On measuring economic values for nature, *Environmental Science and Technology*, 34, 1384–1389.

Christensen, L.R., Jorgensen, D.W., 1973, Measuring economic performance in the private sector, in *The Measurement of Economic and Social Performance*, Moss, M. Ed., pp. 233–352, UMI, http://www.nber.org/chapters/c3617, accessed Oct 28, 2013.

Costanza, R., d'Arge, R., de Groot, R., Farber, S., Grasso, M., Hannon, B., Limburg, K. et al., 1998, The value of the world's ecosystem services and natural capital, *Ecological Economics*, 25, 3–15.

Denison, E.F., 1961, Measurement of labor input: Some questions of definition and the adequacy of data, in *Output, Input, and Productivity Measurement, Studies in Income and Wealth*, Denison, E. F. Ed., pp. 347–372, Princeton University Press for NBER, Princeton, NJ.

Dess, G.G., Robinson, R.B., 1984, Measuring organizational performance in the absence of objective measures: The case of the privately-held firm and conglomerate business unit, *Strategic Management Journal*, 5, 265–273.

Eccles, R.G., Pyburn, P.J., 1992, Creating a comprehensive system to measure performance, *Management Accounting*, 74(4), 41–44.

Folbre, N., 2009, Including home production, G.D.P. might not look so bad, *New York Times*, http://economix.blogs.nytimes.com/2009/05/04/including-home-production-gdp-might-not-look-so-bad/?_r=0, accessed Oct 29, 2013.

Hulten, C.R., Wykoff, F.C., 1996, Issues in the measurement of economic depreciation: Introductory remarks, *Economic Inquiry*, 34, 10–23.

Koopmans, T.C., 1947, Measurement without theory, *The Review of Economics and Statistics*, 29(3), 161–172.

Malaney, P.N., 1996, The Index Number Problem: A Differential Geometric Approach, PhD thesis, Department of Economics, Harvard University, Cambridge, MA.

Reinsdorf, M.B., 1998, Divisia indexes and the representative consumer problem, http://www.ottawagroup.org/Ottawa/ottawagroup.nsf/home/Meeting+4/$file/1998%204th%20Meeting%20-%20Reinsdorf%20%20Marshall%20-%20Divisia%20Indexes%20and%20the%20Representative%20Consumer%20Problem.pdf, accessed Oct 28, 2013.

Stiglitz, J., Sen, A., Fitoussi, J.-P., 2009, The measurement of economic performance and social progress revisited: Reflections and overview, http://www.stiglitz-sen-fitoussi.fr, accessed Oct 28, 2013.

The Economist, 2002, Argentina's collapse: A decline without parallel, http://www.economist.com/node/1010911, accessed Oct 17, 2013.

Thorbecke, E., 2011, Debates on the measurement of global poverty (book review), *Journal of Economic Literature*, 49(3), 751–753.

Venkatraman, N., Ramanujam, V., 1987, Measurement of business economic performance: An examination of method convergence, *Journal of Management*, 13(1), 109–122.

24/7 Wall St, 2012, Countries that spend the most on health care, http://www.247wallst.com/special-report/2012/03/29/, accessed Oct 17, 2013.

World Bank, 2013, http://data.worldbank.org/, accessed Oct 17, 2013.

chapter nine

Measurement and quantum mechanics

David E. Weeks

Contents

9.1 Introduction

Physics stands apart from other philosophical thought by the requirement that the ideas and concepts of physics are testable through a comparison to the world around us. The validity of a physical concept, often expressed using mathematics, is determined through observation and measurement as part of an experiment. For example, the famous physicist James Prescott Joule developed the hypothesis that heat and work are equivalent forms of energy (Lemons 2009). Through a series of careful experiments, he was able to show that the temperature of a volume of water increased as it was stirred with a paddle wheel. The paddle wheel was caused to rotate by a weight connected to a rope that was wrapped around the wheel's rotational axis. With this experiment, Joule established the equivalence of the mechanical energy performed by the rotating paddle wheel to the change in heat energy of the water. He was able to quantify this equivalence by measuring the temperature of the water and by measuring how far the weight fell as it caused the paddle wheel to rotate. Additional measurements of the weight of the water and the falling mass allowed him to determine that a 1-lb weight falling at a distance of 778 ft will increase the temperature of 1 lb of water by 1°F (Richtmyer et al. 1969). Thus, through careful measurement, Joule was able to elevate his hypothesis regarding the mechanical equivalent of heat to a physical law.

It is interesting to note that to measure the temperature of the water, Joule observed the position of the top of a column of mercury relative to a scale on his thermometer. To measure the weight of the water and the weight of the falling mass, Joule observed the position of a mechanical pointer relative to a ruler on his scale. In both cases, careful calibration allowed the measured positions to be converted into temperature and weight. Nevertheless, these were fundamentally positions that were being measured, and when the measurements were being made, the column of mercury and the mechanical pointer were stationary and not moving. This tells us that in addition to knowing the position of the top of the column of mercury and the mechanical pointer, Joule also knew their speeds to be zero. This is a very natural part of our experience with nature where we intuitively expect to be able to simultaneously know both the position and speed of an object. For example, as we drive to work, we can read the speedometer while we pass under a bridge. We then simultaneously know the position of the car to be under the bridge, and the speed of the car as reported by the speedometer. While this is a very natural thing for us to expect to always be true, it turns out that it is not. As the scale of an object becomes much smaller than the scale, we can easily observe without specialized tools, a new and surprising picture emerges with regard to what we can measure. This new picture is a consequence of the laws of quantum mechanics as they are used to describe the behavior of particles at these smaller spatial dimensions.

9.2   Classical measurement

Before considering what is meant by measurement from the perspective of quantum mechanics, it is useful to review what is meant by measurement from the perspective of classical mechanics. For example, consider a ball rolling down a groove cut into an inclined ramp. Galileo Galilei used this arrangement to determine that near the surface of the earth, all things fall with the same acceleration independent of mass (Feynman et al. 1963). Our thought experiment will be somewhat different where we will focus on the position and velocity of the ball (Figure 9.1).

To measure the position of the ball, we can lay a ruler along the groove, and as the ball rolls down the ramp, use any number of methods to compare the ball with the ruler. We might even want to take a photograph at a particular time to refine our measurement. To measure the velocity of the ball, we could reflect electromagnetic radiation off the ball and measure the Doppler shift. An experiment to simultaneously measure both the position and the velocity of the ball would begin by releasing the ball at the top of the inclined plane at time $t = 0$. At some time $t = \tau$ later, we can bounce a pulse of electromagnetic radiation off the ball to determine its velocity v, and at the same time take a picture of the ball to determine its position x relative to the ruler. Using this experimental approach, we have simultaneously measured the position and velocity of the ball. It is important to note that when the electromagnetic radiation reflects off the ball, the momentum of the radiation is changed, and since momentum is conserved, the momentum of the ball must also change. As a result, the measurement of velocity will modify the measurement of the position of the ball to a very small degree. We have several approaches available to minimize the influence of our velocity measurement on the position of the ball. For example, we

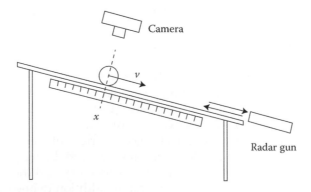

Figure 9.1 A schematic of our thought experiment to measure the position and velocity of a ball as it rolls down an inclined plane.

can lower the intensity of the electromagnetic radiation, or increase the mass of the ball. For a ball rolling down an inclined plane, the influence of the measurement of velocity on the position of the ball is negligible, and in classical mechanics, there is no built-in lower limit to the effect on position caused by a measurement of velocity. We are limited only by how creative we are in performing the experiment. In fact, the mathematical machinery of classical mechanics requires precise and simultaneous knowledge of both position and velocity or momentum. For example, the classical Hamiltonian for a ball rolling down the ramp is given by

$$H = \frac{p^2}{2m} - \alpha x \tag{9.1}$$

Here, H is the total energy of the system, the momentum $p = mv$ where m is the mass of the ball and v is the velocity along the ramp. The constant $\alpha = mg\sin\theta$ is the gravitational force along the ramp where g is the vertical acceleration at the surface of the earth and θ is the angle of inclination of the ramp relative to its horizontal position. The classical equations of motion that describe the ball as it rolls down the ramp are then given by

$$\frac{dx}{dt} = \{x, H\} = \frac{\partial x}{\partial x}\frac{\partial H}{\partial p} - \frac{\partial x}{\partial p}\frac{\partial H}{\partial x} = \frac{\partial H}{\partial p} = \frac{p}{m} \tag{9.2}$$

$$\frac{dp}{dt} = \{p, H\} = \frac{\partial p}{\partial x}\frac{\partial H}{\partial p} - \frac{\partial p}{\partial p}\frac{\partial H}{\partial x} = -\frac{\partial H}{\partial x} = \alpha \tag{9.3}$$

where a, b are the Poisson bracket (Goldstein et al. 2002). Equation 9.3 can be directly integrated to yield

$$p(t) = \alpha t + p_0 \tag{9.4}$$

and upon substituting Equation 9.4 into Equation 9.2 and integrating once again, we obtain

$$x(t) = \frac{\alpha t^2}{2m} + \frac{p_0 t}{m} + x_0 \tag{9.5}$$

Equations 9.4 and 9.5 summarize Galileo's observation that all bodies fall with constant acceleration α/m near the surface of the earth. These equations illustrate the requirement imposed by classical mechanics that you must have simultaneous knowledge of the initial position x_0 and initial momentum p_0 at time $t = 0$ to know the position and momentum at some later time t. In fact, if x_0 and $p_0 = mv_0$ are known and used together with Equations 9.4 and 9.5, then the velocity and the position of the ball will be simultaneously known for as long as the ball remains rolling down the ramp.

By any measure, classical mechanics is one of the great pillars of physics. However, as the nineteenth century came to a close, several experimental observations indicated that there were serious problems with some of the predictions made by classical mechanics. These problems were eventually resolved with the advent of quantum mechanics. While quantum mechanics eventually evolved into one of the most successful physical theories, it brought with it a number of counterintuitive ideas, the foremost of which is that nature is not deterministic as suggested by classical mechanics. Instead, quantum mechanics paints an entirely different picture where it replaces the classical determinism exhibited in Equations 9.4 and 9.5 with a probabilistic description. One consequence of this new way of looking at nature is the prohibition of simultaneously knowing both the position and velocity of a particle at the same time. A second consequence is the association of measurement with wave function collapse.

9.3 Quantum measurement

By the early 1900s, scientists were aware that matter consisted of negatively charged electrons and that most matter was neutral. This meant that along with the negatively charged electrons, matter had to also include some unknown arrangement of positive charge. Two pictures were proposed for the arrangement of positive charge in matter. The first was the "plum pudding" model where the positive charge was the pudding spread out evenly throughout the matter while the electrons, viewed as particles, were the raisins in the pudding. The second was the nuclear model where the nuclear charge was concentrated in highly localized nuclei about which the electrons orbited. We now know this second model is the correct way to view the positive charge in matter and this nuclear model was conclusively demonstrated by the Rutherford gold foil experiment performed by Hans Geiger and Ernest Marsden and overseen by Earnest Rutherford (Eisberg and Resnick 1974). In this experiment, a beam of helium nuclei, also known as α-particles, were aimed at a very thin sheet of gold foil. If the positive charge in the foil was spread out like pudding, there would be some small deflection of the positively charged α-particles as they passed through the foil, but otherwise the beam would not be greatly affected. On the other hand, if the positive charge in matter is localized in nuclei, then, occasionally, an α-particle would undergo a near-head-on collision and be backscattered. When the experiment was performed, backscattering was observed and the nuclear model for positive charge in matter won the day. This was a great success for understanding matter and essentially gave birth to nuclear physics. However, it posed a serious problem for classical mechanics. If the positive charge is concentrated in a highly localized nucleus, then, classically, the electrons must orbit the nucleus under the attractive electromagnetic force much like the planets orbit the sun under the attractive force of gravity. The difficulty with this picture is that the orbiting electrons are undergoing acceleration as they orbit and must therefore also radiate electromagnetic energy. This radiated electromagnetic energy is converted from the Coulomb potential energy between the nucleus and the electron. As a result, the electron must spiral in toward the nucleus, eventually coming into physical contact with the nucleus. The electrons would then "land" on a nucleus, giving rise to a neutral particle, and matter as we know would become unstable and simply fall apart.

Hints of how this paradox would ultimately be resolved were provided by Niels Bohr who proposed a model for the hydrogen atom where the electron orbited a proton without radiating in distinct orbits with radii r_n (Sears et al. 1977). These radii were determined by quantizing the angular momentum, of the electron, $\ell_n = mvr_n = n\hbar$ in units of Plank's constant $\hbar$. These discrete orbital radii were then used to compute a set of quantized energies

$$E_n = -\frac{e^2}{8\pi\varepsilon_0}\frac{1}{r_n} = -\frac{me^4}{32\pi^2\varepsilon_0^2\hbar^2}\frac{1}{n^2} \tag{9.6}$$

Bohr further hypothesized that as an electron goes from one orbit of radius r'_n to another of radius r_n, the atom will radiate a photon with frequency

$$\omega_{n'n} = \frac{E_{n'} - E_n}{\hbar} = -\frac{me^4}{32\pi^2\varepsilon_0^2\hbar^3}\left(\frac{1}{n'^2} - \frac{1}{n^2}\right) \tag{9.7}$$

The frequencies given by Equation 9.7 closely matched the observed atomic hydrogen spectrum and firmly established the need to replace classical mechanics with a new theory when describing very small systems. Several intermediate quantum theories were proposed including Bohr–Sommerfeld quantization where the classical action is quantized (Anderson 1971). When applied to the simple harmonic oscillator, Bohr–Sommerfeld quantization yields quantized energies $E_n = n\hbar\omega$ in agreement with Plank's postulate that energy of electromagnetic oscillators in a blackbody occur in integer multiples of $\hbar\omega$. Around the same time, Louis de Broglie proposed that particles have wave-like properties with a wavelength given by $\lambda = 2\pi\hbar/p$ (Blümel 2011). These ideas culminated in the current formulation of nonrelativistic quantum mechanics, essentially proposed simultaneously by Schrodinger in the form of his famous wave equation (Schiff 1949), and by Heisenberg and his matrix mechanics (Heisenberg 1949). We will start our discussion of quantum mechanics using Heisenberg's formulation because of its close similarity to the formulation of classical mechanics.

In the Heisenberg formulation of quantum mechanics, the classical observables of position x and momentum p are replaced with quantum operators $\hat{x}$ and $\hat{p}$, and Poisson bracket in Equations 9.2 and 9.3 is replaced with the commutator, $[\hat{a}, \hat{b}] = \hat{a}\hat{b} - \hat{b}\hat{a}$

$$i\hbar\frac{d\hat{x}}{dt} = [\hat{x}, \hat{H}] = \hat{x}\hat{H} - \hat{H}\hat{x} = i\hbar\frac{d\hat{H}}{d\hat{p}} \tag{9.8}$$

$$i\hbar\frac{d\hat{p}}{dt} = [\hat{p}, \hat{H}] = \hat{p}\hat{H} - \hat{H}\hat{p} = -i\hbar\frac{d\hat{H}}{d\hat{x}} \tag{9.9}$$

where $\hat{H}$ is the quantum Hamiltonian and is a function of the operators $\hat{x}$ and $\hat{p}$. If the quantum Hamiltonian is time independent, then, the solutions to Equations 9.8 and 9.9 are given by

$$\hat{x}(t) = \exp\left(\frac{i\hat{H}t}{\hbar}\right)\hat{x}(0)\exp\left(-\frac{i\hat{H}t}{\hbar}\right) \tag{9.10}$$

$$\hat{p}(t) = \exp\left(\frac{i\hat{H}t}{\hbar}\right)\hat{p}(0)\exp\left(-\frac{i\hat{H}t}{\hbar}\right) \tag{9.11}$$

as can be confirmed by substituting Equations 9.10 and 9.11 back into Equations 9.8 and 9.9. Using Equations 9.10 and 9.11, the quantum mechanics problem is reduced to the determination of the operators $\hat{x}$ and $\hat{p}$ at time $t = 0$. One clue to what these operators are is given by the observation that the operator $\hat{x}$ will also satisfy Equation 9.8 where, $[\hat{x}, \hat{x}] = \hat{x}\hat{x} - \hat{x}\hat{x} = i\hbar d\hat{x}/d\hat{p} = 0$. Now, if $\hat{x}$ and $\hat{H}$ satisfy Equation 9.8, then, so will any function $f = f(\hat{x}, \hat{H})$ of $\hat{x}$ and $\hat{H}$. Since the Hamilton $\hat{H}(\hat{x}, \hat{p})$ can be inverted to find $\hat{p} = \hat{p}(\hat{x}, \hat{H})$, it follows that the operator $\hat{p}$ also satisfies Equation 9.8 to yield (Razavy 2011)

$$[\hat{x}, \hat{p}] = \hat{x}\hat{p} - \hat{p}\hat{x} = i\hbar \tag{9.12}$$

Here, we see that the operators $\hat{x}$ and $\hat{p}$ do not commute and are therefore not simple numbers. Insight into the nature of $\hat{x}$ and $\hat{p}$ may be obtained by considering the effect of the following commutator on an arbitrary function $\psi(x)$:

$$\left[x, -i\hbar \frac{d}{dx}\right]\psi(x) = -i\hbar x \frac{d\psi(x)}{dx} + i\hbar \frac{d(x\psi(x))}{dx} = i\hbar\psi(x) \tag{9.13}$$

Since $\psi(x)$ is arbitrary, we can conclude that x and $-i\hbar d/dx$ satisfy Equation 9.12. An alternative choice for $\hat{x}$ and $\hat{p}$ is given by

$$\left[i\hbar \frac{d}{dp}, p\right]\phi(p) = i\hbar \frac{d(p\phi(p))}{dp} - i\hbar p \frac{d\phi(p)}{dp} = i\hbar\phi(p) \tag{9.14}$$

The first choice, $\hat{x} \rightarrow x$, and $\hat{p} \rightarrow -i\hbar d/dx$ is called the *coordinate representation* of $\hat{x}$ and $\hat{p}$, and the second choice $\hat{x} \rightarrow i\hbar d/dp$ and $\hat{p} \rightarrow p$ is called the *momentum representation* of $\hat{x}$ and $\hat{p}$. In the coordinate representation, eigenfunctions of the momentum operator are given by one-dimensional plane waves

$$-i\hbar \frac{d}{dx}\exp(ipx/\hbar) = p\exp(ipx/\hbar) \tag{9.15}$$

with eigenvalues given by p, and in the momentum representation, eigenfunctions of the coordinate operator are also given by plane waves

$$i\hbar \frac{d}{dp}\exp(-ipx/\hbar) = x\exp(-ipx/\hbar) \tag{9.16}$$

with eigenvalues given by x.

Now, if we identify $F(p)$ as the Fourier transformation of $f(x)$,

$$F(p) = \frac{1}{(2\pi\hbar)^{1/2}}\int dx f(x)\exp(-ipx/\hbar) \tag{9.17}$$

then, the derivative of $F(p)$ with respect to p is the Fourier transformation of the product $xf(x)$

$$i\hbar \frac{dF(p)}{dp} = \frac{i\hbar}{(2\pi\hbar)^{1/2}}\int dx f(x)\frac{d}{dp}\{\exp(-ipx/\hbar)\}$$

$$= \frac{1}{(2\pi\hbar)^{1/2}}\int dx x f(x)\exp(-ipx/\hbar) \tag{9.18}$$

Since the product of x with $f(x)$ is mapped by the Fourier transformation to the derivative of $F(p)$ with respect to p, we can identify the Fourier transformation as the bridge between the coordinate representation, where $\hat{x} \to x$, and the momentum representation, where $\hat{x} \to i\hbar d/dp$. As a result, if we compute the inverse Fourier transformation of the eigenstates of the coordinate operator in the momentum representation, we obtain eigenstates of the coordinate operator in the coordinate representation

$$\delta(x - x') = \frac{1}{2\pi\hbar} \int dp \exp(-ipx'/\hbar)\exp(ipx/\hbar) \tag{9.19}$$

where $\delta(x - x')$ is the Dirac delta function. In a similar manner, eigenstates of the momentum operator in the momentum representation are given by

$$\delta(p - p') = \frac{1}{2\pi\hbar} \int dx \exp(ip'x/\hbar)\exp(-ipx/\hbar) \tag{9.20}$$

To see how these ideas are related to quantum measurement, we need to introduce several postulates of quantum mechanics (Cohen-Tannoudji et al. 1977). To tie the postulates more firmly to the preceding discussion, we will introduce these postulates from the perspective of the coordinate representation. The first postulate that we will examine states that

Postulate 9.1

A quantum system is defined at time t by the specification of a wave function $\psi(x,t)$.

Specifying the wave function in quantum mechanics is analogous to specifying the phase space coordinates x, p in classical mechanics. As we shall see in a moment, the wave function carries information regarding the probability to make certain measurements. It is necessary when discussing the remaining postulates of interest to introduce Hermitian operators $\hat{O}$, that are equal to the complex conjugate of their own transpose, $\hat{O} = \hat{O}^{\dagger}$. For any Hermitian operator, $\hat{O}$, there exists a complete set of eigenfunctions φ_n and eigenvalues, o_n where, $\hat{O}\varphi_n = o_n\varphi_n$. For example, both $\hat{x}$ and $\hat{p}$ are Hermitian operators. With this in mind, the second postulate that we will introduce states that

Postulate 9.2

Measureable quantities correspond to Hermitian operators.

For example, position, momentum, and energy correspond to the Hermitian operators $\hat{x}$, $\hat{p}$, and $\hat{H}$, respectively. The next postulate states that

Postulate 9.3

The only possible result of a measurement is an eigenvalue of a Hermitian operator.

For example, the measureable quantity of momentum corresponds to the Hermitian operator $\hat{p} \to -i\hbar d/dx$ whose eigenvalues p constitute the full set of possible measurements of momentum. In a complete departure from the determinism of classical mechanics, the next postulate states that

Postulate 9.4

The probability of measuring an eigenvalue, o_n, is given by the square of the projection of the state of the system, $\psi(x)$, onto the corresponding eigenstate, $\phi_n(x)$.

$$P(o_n) = \left| \int dx \psi^*(x)\, \phi_n(x) \right|^2$$

For example, if the system is in the state $\psi(x)$, then, the probability density for measuring a particular value of x is given by

$$P(x) = \left| \int dx' \psi^*(x')\delta(x - x') \right|^2 = \psi^*(x)\psi(x) \tag{9.21}$$

The last postulate that we will introduce states that

Postulate 9.5

When a measurement is made and the result o_n is obtained, the system will immediately collapse into the corresponding eigenfunction $\phi_n(x)$.

For example, if the system is in the state $\psi(x)$ and a value of o_n is obtained during a measurement, then, immediately after the measurement, the state of the system instantaneously collapses from $\psi(x)$ to the corresponding eigenstate $\phi_n(x)$. As a consequence, any successful measurement of an eigenvalue of $\hat{O}$, no matter how subtly performed will make a profound and instantaneous change to the system as the wave function collapses from the state $\psi(x)$ into the state $\phi_n(x)$. Any subsequent measurement of the system will be governed by the state $\phi_n(x)$, and all information about the state $\psi(x)$ will have been erased as a result of its collapse.

Armed with these postulates, let us return to the measurement of position and momentum as applied to the quantum mechanical analog of Galileo's experiment where he used a ball rolling down an inclined plane. For this quantum mechanical analog, we will use an electron instead of a ball, and we will set the angle of the inclined plane to zero. In this scenario, the electron is a free particle moving along the x-axis. Using Postulate 9.1, the state of the system is determined by specifying a wave function $\psi(x)$. A complex Gaussian wave packet is a convenient choice

$$\psi(x) = \left(\frac{1}{2\pi\delta^2} \right)^{1/4} \exp\left\{ \frac{(x - x_0)^2}{4\delta^2} + i\frac{p_0}{\hbar}(x - x_0) \right\} \tag{9.22}$$

Using Postulate 9.4 and Equations 9.21 and 9.22, the probability density for measuring an eigenvalue of the $\hat{x}$ operator for a system prepared in this state is then given by

$$P(x) = \left(\frac{1}{2\pi\delta^2} \right)^{1/2} \exp\left\{ \frac{-(x - x_0)^2}{2\delta^2} \right\} \tag{9.23}$$

A series of measurements of the position of the electron can be performed, each of which will yield some value of x as determined by the probability distribution in Equation 9.23. Each time a measurement of position is made, the wave function will collapse into an eigenstate of the position operator in accord with Postulate 9.5. As a result of this collapse, any subsequent measure of position will not be governed by Equation 9.23. If the goal is to study the probability distribution given by Equation 9.23, then, the wave function must be reset back to the complex Gaussian given by Equation 9.21 in-between each measurement of position. After making a series of measurements of a system prepared in the state given by Equation 9.22, the average value of x is given by

$$\langle x \rangle = \int dx P(x)x = \int dx \psi^*(x)x\psi(x) = x_0 \tag{9.24}$$

and the variance is given by

$$(\Delta x)^2 = \langle x^2 \rangle - \langle x \rangle^2 = \int dx \psi^*(x)x^2\psi(x) - \left(\int dx \psi^*(x)x\psi(x) \right)^2 = \delta^2 \tag{9.25}$$

The Fourier transformation of Equation 9.22 yields the wave function in the momentum representation

$$\phi(p) = \left(\frac{2\delta^2}{\pi\hbar^2} \right)^{1/4} \exp\left\{ -\delta^2 \frac{(p - p_0)^2}{\hbar^2} - i\frac{p}{\hbar}x_0 \right\} \tag{9.26}$$

where the probability density of measuring an eigenvalue of the $\hat{p}$ operator is given by

$$P(p) = \left(\frac{2\delta^2}{\pi\hbar^2} \right)^{1/2} \exp\left\{ -2\delta^2 \frac{(p - p_0)^2}{\hbar^2} \right\} \tag{9.27}$$

with an average value and variance given by

$$\langle p \rangle = \int dp P(p)p = \int dp \phi^*(p)p\phi(p) = p_0 \tag{9.28}$$

$$(\Delta p)^2 = \langle p^2 \rangle - \langle p \rangle^2 = \int dx \phi^*(p)p^2\phi(p) - \left(\int dx \phi^*(p)p\phi(p) \right)^2 = \frac{\hbar^2}{4\delta^2} \tag{9.29}$$

By combining Equations 9.25 and 9.29, we arrive at the Heisenberg uncertainty relation for Gaussian wave packets

$$\Delta x \Delta p = \hbar/2 \tag{9.30}$$

Now, let us consider what happens if an attempt is made to measure both the position and momentum of an electron initially prepared in the state given by Equation 9.22. First, we measure the position and will observe an eigenvalue of the position operator, in accord with Postulate 9.3. The probability to measure a particular value $x = a$ is determined by the probability distribution given by Equation 9.23, in accord with Postulate 9.4. Upon measurement of the electron's position, the wave function instantly collapses into the corresponding eigenstate of the position operator, in accord with Postulate 9.5. Eigenstates of the position operator in the coordinate representation are given by Equation 9.19 where, for the measurement $x = a$, we obtain

$$\psi(x) = \delta(x - a) = \lim_{\eta \to 0} \left(\frac{1}{\pi \eta^2} \right)^{1/2} \exp \left\{ \frac{-(x - a)^2}{\eta^2} \right\} \tag{9.31}$$

In Equation 9.31, we use the definition of the delta function as the limit of a narrow Gaussian to emphasize that there is zero uncertainty in our measurement of the position of the electron. An appeal to the uncertainty relation in Equation 9.30 indicates that if $\Delta x \to 0$, then, $\Delta p \to \infty$ in such a way that that their product remains $\Delta x \, \Delta p = \hbar/2$. The Fourier transform of Equation 9.31 yields an eigenfunction of the coordinate operator in the momentum representation

$$\phi(p) = \left(\frac{1}{2\pi\hbar} \right)^{1/2} \exp \left\{ -i \frac{p}{\hbar} a \right\} \tag{9.32}$$

and is essentially an infinitely wide complex Gaussian with an associated probability density given by

$$P(p) = \left(\frac{1}{2\pi\hbar} \right) \tag{9.33}$$

This identifies that the probability for measuring the momentum of the electron is the same for all p and is consistent with the conclusion that $\Delta p \to \infty$ upon the collapse of the wave function into an eigenfunction of the coordinate operator. So, while we have perfect knowledge of the position of the electron, the probability distribution in Equation 9.33 tells us that any value of p is possible upon measurement of the momentum.

If we now measure the momentum, we will observe an eigenvalue of the momentum operator, in accord with Postulate 9.3. The probability of measuring a particular value $p = b$ is determined by the probability distribution given by Equation 9.33, in accord with Postulate 9.4. Upon measurement of the electron's momentum, the wave function instantly collapses into the corresponding eigenstate of the momentum operator, in accord with Postulate 9.5. Eigenstates of the momentum operator in the momentum representation are given by Equation 9.20 where, for the measurement $p = b$, we obtain

$$\phi(p) = \delta(p - b) = \lim_{\eta \to 0} \left(\frac{1}{\pi \eta^2} \right)^{1/2} \exp \left\{ \frac{-(p - b)^2}{\eta^2} \right\} \tag{9.34}$$

In Equation 9.34, we once again use the definition of the delta function as the limit of a narrow Gaussian to emphasize that there is zero uncertainty in our measurement of the momentum of the electron. An appeal to the uncertainty relation in Equation 9.30 indicates that if $\Delta p \to 0$, then, $\Delta x \to \infty$ in such a way that their product remains $\Delta x \, \Delta p = \hbar/2$. The Fourier transform of Equation 9.31 yields the momentum eigenfunction in the coordinate representation

$$\psi(x) = \left(\frac{1}{2\pi\hbar}\right)^{1/2} \exp\left\{i\frac{x}{\hbar}b\right\} \tag{9.35}$$

and is essentially an infinitely wide complex Gaussian with an associated probability density given by

$$P(x) = \left(\frac{1}{2\pi\hbar}\right) \tag{9.36}$$

This identifies that the probability for measuring the coordinate of the electron is the same for all x and is consistent with the conclusion that $\Delta x \to \infty$ upon the collapse of the wave function into an eigenfunction of the momentum operator. So, upon measurement of the electron's momentum, we have lost all knowledge of the electron's position.

It is important to emphasize that unlike a measurement of position or momentum in classical mechanics where we can reduce the effect of the measurement on a classical system to be as small as we want, the effect of a quantum mechanical measurement on a quantum system cannot be reduced by any refinement of the measurement. By Postulate 9.5, a successful quantum measurement will collapse the wave function of the system into an eigenstate corresponding to the measured eigenvalue. Here, we see that regardless of how careful we are not to disturb the system, a successful measurement will always coincide with the collapse of the systems wave function. So, while we can successfully measure the position x of the electron, this measurement eliminates all knowledge of the momentum of the electron. If we try to rectify the situation by measuring the momentum p of the electron, the wave function will collapse into an eigenfunction of the momentum operator and we lose all knowledge of the position x. This inability to simultaneously know the position and momentum of an electron occurs because the eigenfunctions of the position operator are not the same as the eigenfunctions of the momentum operator. On the other hand, if two operators share the same eigenfunctions, then it will be possible to simultaneously measure the corresponding eigenvalues. For example, the Hamiltonian of an electron moving freely along the x-axis is given by

$$\hat{H} = \frac{\hat{p}^2}{2m} \tag{9.37}$$

Eigenfunctions of $\hat{H}$ in Equation 9.37 are the same as those of $\hat{p}$ as identified in Equation 9.15. Using $\hat{p} \to -i\hbar d/dx$ to express $\hat{H}$ in the coordinate representation, we obtain

$$-\frac{\hbar^2}{2m}\frac{d^2}{dx^2}\exp(ipx/\hbar) = \frac{p^2}{2m}\exp(ipx/\hbar) \tag{9.38}$$

where the eigenvalues $E(p) = p^2/2m$ are the possible kinetic energies of the electron. If we prepare the system in the wave function given by Equation 9.22 and then measure the momentum of the electron to be $p = b$, the wave function will collapse into the corresponding eigenfunction of the momentum operator, given by Equation 9.34 in the momentum representation, and by Equation 9.35 in the coordinate representation. As we mentioned earlier, $\Delta p \to 0$ for an eigenfunction of the momentum operator and if we follow the measurement of momentum by a measurement of energy, we will obtain $E(p = b) = b^2/2m$ with complete certainty because we are already in an eigenfunction of $\hat{H}$. Since we are already in an eigenfunction of $\hat{H}$, the wave function collapses into itself and remains unchanged. As a result, we retain knowledge of the momentum of the electron and can simultaneously know this momentum and kinetic energy of the electron.

These considerations illustrate that if two operators share the same eigenfunctions, such as $\hat{H}$ and $\hat{p}$, it is possible to simultaneously measure their eigenvalues. On the other hand, if two operators do not share eigenfunctions, such as $\hat{x}$ and $\hat{p}$, then, it is not possible to simultaneously measure their eigenfunctions. A necessary and sufficient condition for two operators to share eigenfunctions is that they commute. The inability to know both the position and momentum of the electron in our example stems from the commutation relation in Equation 9.12, and ultimately from the replacement of the classical Poisson bracket with the quantum mechanical commutator.

9.4 Conclusion

Both classical mechanics and quantum mechanics provide an excellent description of the physical universe. Classical mechanics is most successfully applied to macroscopic objects and yields a deterministic picture of the universe where it is possible to measure position and momentum at the same time without significantly changing the classical state of the system. On the other hand, quantum mechanics is most successfully applied to microscopic objects and provides a nondeterministic or probabilistic description of the universe. This probabilistic description has a significant impact on what we mean by measurement in the quantum mechanical sense where we are limited to predicting only the probability for the outcome of some measurement and not the specific result. This probability is given by the square of the quantum mechanical wave function and upon making a successful measurement of position, for example, this wave function collapses into an eigenfunction of the position operator. This collapse cannot be avoided by some refinement of the measurement process. Indeed, in quantum mechanics, measurement and collapse of the wave function are so completely intertwined as to be synonymous. If two operators do not commute, they do not share a common set of eigenfunctions and it is not possible within the framework of quantum mechanics to simultaneously measure physical properties of the system associated with noncommuting operators. Since the position operator and momentum operator do not commute, a measurement of position collapses the wave function into an eigenfunction of the position operator and precludes any knowledge of momentum. If an attempt to measure momentum is performed, the system will collapse into an eigenfunction of the momentum operator and eliminate any knowledge of the position.

The postulates of quantum mechanics establish a link between measurement and the collapse of the wave function. However, the mechanism by which this collapse occurs is not predicted by Schrödinger's equation. The lack of a mechanism for the collapse of the wave function represents an important area of inquiry regarding the foundations of quantum theory (Greenstein and Zajonc 2006). Another foundational question is how the probabilistic quantum description of atoms and molecules on the microscopic scale relates to the deterministic classical description that works so well for macroscopic assemblies of these atoms and molecules (Bokulich 2008). Why for example, when I observe that my car is under a bridge, do I not become concerned that the wave function of the car will collapse making the momentum of my car completely uncertain?

Perhaps, the most remarkable difference between classical measurement and quantum measurement is our picture of what we are measuring before the measurement occurs.

Classically, an object has a preexisting position independent of our knowledge. The classical measurement of an object's position is just our determination of the object's preexisting condition. In quantum mechanics, an object does not have a preexisting position prior to a measurement. There is only a probability that it has a particular position. The act of making a quantum measurement creates the reality of an object's position. In this sense, a quantum mechanical measurement gives the observer an active role in generating reality.

References

Anderson, E. E., *Modern Physics and Quantum Mechanics* (W. B. Saunders Company, Philadelphia, 1971) p. 112.

Blümel, R., *Advanced Quantum Mechanics, the Classical Quantum Connection* (Jones and Bartlett, Boston, 2011) p. 106.

Bokulich, A., *Reexamining the Quantum-Classical Relation* (Cambridge University Press, New York, 2008).

Cohen-Tannoudji, C., Diu, B., and Laloë, F., *Quantum Mechanics*, Vol. I (Wiley, New York, 1977) p. 213.

Eisberg, R. and Resnick, R., *Quantum Physics of Atoms, Molecules, Solids, Nuclei, and Particles* (Wiley, New York, 1974) p. 99.

Feynman, R. P., Leighton, R. B., and Sands, M., *The Feynman Lectures on Physics*, Vol. I (Addison-Wesley, Reading, 1963) pp. 5–11.

Goldstein, H., Poole, C., and Safko, J., *Classical Mechanics*, 3rd Ed. (Addison-Wesley, New York, 2002) p. 388.

Greenstein, G. and Zajonc, A. G., *The Quantum Challenge, Modern Research on the Foundations of Quantum Mechanics*, 2nd Ed. (Jones and Bartlett, Boston, 2006) p. 219.

Heisenberg, W., *The Physical Principles of Quantum Theory*, Translated by Eckart, C. and Hoyt, F. C., (Dover, New York, 1949).

Lemons, D. S., *Mere Thermodynamics* (Johns Hopkins University Press, Baltimore, 2009) p. 21.

Razavy, M., *Heisenberg's Quantum Mechanics* (World Scientific, London, 2011) p. 43.

Richtmyer, F. K., Kennard, E. H., and Cooper, J. N., *Introduction to Modern Physics*, 6th Ed. (McGraw-Hill, New York, 1969) p. 28.

Schiff, L. I., *Quantum Mechanics* (McGraw-Hill, New York, 1949) p. 17.

Sears, F. W., Zemansky, M. W., and Young, H. D., *University Physics, Part II*, 5th Ed. (Addison-Wesley, Reading, MA, 1977) p. 759.

chapter ten

Social science measurement

John J. Elshaw

Contents

10.1 Introduction

Before we tread too deep into the waters of social science measurement, think about a person you have met who was a great leader and then come up with some words to describe that person. Would you describe that person based on certain traits they embodied? Are there specific behaviors displayed by people who are great leaders? Do these people behave differently depending on the situation or environment? Suppose you were asked to rate your immediate supervisor in regard to his/her leadership, how would you evaluate that person? Would your assessment of leadership agree with other coworkers' assessment of the same person? These are all questions that illustrate the challenge researchers have measuring social science constructs. The ability to precisely describe and define the phenomena of interest is at the heart of social science measurement.

10.1.1 Measurement defined

Measurement is the codifying of observations into data that can be analyzed, portrayed as information, and evaluated to support the decision maker (Badiru, 2013). In regard to traditional science, this might be a very straightforward task such as measuring production output per unit of time, determining the thrust-to-weight ratio of an airplane, or estimating the cost to develop a new communication system. Social science measurement on the other hand is not clear-cut. Human behavior and the attributions behind this behavior might be one of the most difficult *systems* to measure and predict. The underlying attributes that drive human behavior is at the core of social science measurement and is why the construct definition is so important.

10.2   Concepts and indicators

One of the assumptions in social science measurement is that there are real phenomena being measured that are referred to as a construct. A *construct* is a concept that is specifically defined for scientific study. For example, job performance is defined as the value of a set of employee behaviors that contribute to organizational goal accomplishment (Campbell, 1990). As a construct, however, job performance is additionally used to describe the relationship between other constructs in a theoretical representation and is defined and specified in a way that allows scientists to observe and measure it. Consider for a moment the job performance of an airline pilot. If you were the supervisor of a pilot working for a large commercial airline, you might judge the pilot's performance by assessing many of the behaviors relevant to the job. Does the pilot complete a proper preflight check? Are the aircraft logs complete following each flight? Does the pilot start the engines and operate the controls properly? Are flight plans accomplished according to government regulations using proper aeronautical charts? Has the cargo been checked to ensure baggage is loaded properly? You might also be interested in assessing pilot performance by examining behaviors that occur outside the task of flying such as whether the pilot adheres to training standards? Does the pilot represent the airline in a professional manner? Is the pilot up to date on required physical examinations, and does he adhere to proper crew rest procedures? (Colquitt et al., 2011).

All these questions may be irrelevant in regard to performance, and the real question an airline might be interested in is whether the pilot gets passengers to their destination on time in a safe manner. Passengers may not care at all about flight logs and training standards if the pilot cannot fly the plane without crashing. The safe and on-time arrival of flights has implications for the airline because it affects ticket sales, insurance rates, customer satisfaction, and so on. Surely, every late flight cannot be blamed on the pilot, however. There are weather delays that could affect on-time metrics, maintenance issues that may prevent a pilot from taking off on time, or delays in connecting flights that force subsequent flights to be delayed.

The example above highlights one of the difficulties in social science measurement. While evaluating job performance based on safety and flight arrival time seems like an easy thing to do, especially considering the objective nature of these assessments, it inadvertently simplifies the performance evaluation process. As previously stated, a pilot has no control over the weather or the maintenance performed on an aircraft. An airline may be very interested in measuring outcomes because they can be directly tied to passenger ticket sales and the bottom line of company earnings. The problem, however, is that assessing performance solely based on outcomes does not reveal anything that may help fix the issue. Simply tracking the number of on-time departures and arrivals tells you nothing about delays caused by maintenance issues. What if the maintenance team at a specific hub was continually late servicing an airplane because the training process did not meet standards? Assessing only performance outcomes would not help you address this problem. Similarly, if a pilot was guilty of skipping procedures or flying outside of accepted safety standards to get a delayed flight back on schedule, only examining performance outcomes would not help identify the problem. One of the keys to social science measurement is assessing behaviors that are related to relevant constructs, which then allow you to model and predict their impact on important organizational outcomes (Kerlinger, 1986).

10.3   Scales

Measurement scales are what makes it possible to collect and interpret data. Without a scale, there is no reference point to assess against a standard or any other level of performance. There are four scales that must be considered when collecting data. The type and nature of these scales will help determine what type of analysis is appropriate. The first is the *nominal* scale (also referred to as categorical). The nominal scale can be thought of as a grouping variable such as the type of aircraft, job location, or country of birth. Since this is simply a method of grouping data, there is no mathematical order or operations that can be applied to this scale. For example, one might be interested in organizing and describing different types of aircraft dependent on their mission. In doing so, you would collect data and then organize the information according to the specific airframe and mission (i.e., F-16, KC-10, HC-130, etc.). The groups you choose are nothing more than a way of organizing your data according to some predetermined criteria.

The next type of measurement scale is *ordinal*. This scale simply refers to the order in which measurements are placed. It is different from a nominal scale in that the order of the variables actually matters. For example, you can classify the experience level of workers as low, medium, or high. In this case, the order tells the researcher something about the experience level of each worker; however, it does not tell them anything about the magnitude of the difference between these workers (e.g., low and medium worker, medium and high worker). The relative order of individual work experience is known; however, the researcher does not know if one category of experience is greatly superior to another, or if the difference is only slightly superior. If the researcher was able to distinguish the relative amount of difference between groups, then, the scale would be interval.

An *interval* scale has equal distances between the scores obtained; however, it does not include a zero point. Temperature measured in either Fahrenheit or Celsius is a good example of an interval scale. To illustrate, the difference between 82° and 90° is the same as the difference between 28° and 36°. The interval provides information about the relative distance between measurements on the scale. Additionally, the zero point is arbitrary and does not mean anything as it is just another point on the scale. Another good example of an interval scale is the Likert-type scale used to measure behavioral constructs. A Likert scale is usually a five- or seven-point scale that allows an individual to select how much they agree or disagree with a particular statement. The distance between each of the points is assumed to be linear, and the level of agreement with each statement is also assumed to be equal and linear. Consider three individuals who score a two, four, and six respectively on a seven-point measure of transformational leadership. The difference in leadership ability between the two individuals who score a two and four is the same as the difference between the two who scored a four and six. The relative differences can be interpreted and explained whereas in the previous ordinal scale, only the order can be determined that does not provide any information about the level or amount of difference between the three leaders.

Finally, the *ratio* scale has all the characteristics of the previous scales, but the zero point is now relevant and means something specific that the researcher can interpret. It is a point where none of the quality being measured exists. Recalling our previous illustration of a pilot flying an airplane, ground speed is an example of a ratio scale because it has an absolute zero point. Consider the difference between an F-16 flying at 400 miles per hour (mph) and an HC-130 flying at 160 mph. The difference between these two is 240 mph, but more importantly, one is 400 mph away from being stopped while the other is 160 mph away. The zero point in this case actually means something to the researcher; the airplane is at rest. Another example of a ratio scale is the level of income measured in thousands of dollars per year. A person who earns $80,000 per year earns twice as much as a person who earns $40,000 per year; however, an income level of $0 per year indicates something totally different. Thus, *speed* and *income level* are both examples of a ratio scale because they contain a zero point that can be interpreted and means something.

10.4   Variable type

10.4.1   Latent and manifest variables

Oftentimes, when conducting research, there are variables involved that are not directly observable. Motivation, for example, is a variable that cannot be directly observed but is inferred from other more observable behaviors. Formally defined, motivation is a set of forces that originate both within and outside an employee and that determine direction, intensity, and persistence toward a specific behavior (Latham and Pinder, 2005). In this sense, motivation is a hypothetical construct as the *set of forces* are not observable; however, the result of these forces is. It is an aggregation of other more clearly defined behaviors. Manifest variables, on the other hand, are variables that can be directly observed and measured. Using the previous motivation example, persistence of effort (or time) spent working on a task can be measured. The intensity placed toward completing a task can also be measured. These observable, manifest variables are what is aggregated together to represent a latent variable. So, in a sense, the purpose of a latent variable is to provide a parsimonious description of observed data that combine multiple factors into a single, overall factor for analysis and interpretation (Harman, 1960, p. 5).

Sometimes, researchers will refer to latent variables as theoretical variables, and the way these variables are operationalized are called manifest variables. How a researcher defines a latent construct is very important because the operationalization of that construct will need to map onto specific behaviors that reflect that definition. Let us use another example to clarify this. If we were trying to assess leadership within an organization and we define leadership as the extent to which a supervisor facilitates our ability to get the job done, we might specifically ask about behaviors such as how often does he/she remove barriers that prevent me from completing a task? Does the supervisor offer ideas for solving job-related problems (Taylor, 1971)? However, if we define leadership as the extent to which a person facilitates interaction among and between employees, the behaviors we would measure are quite different. In this case, the manifest variables might be quantifying how often a supervisor holds group meetings where employees can discuss issues, or how often a supervisor encourages people to exchange opinions and ideas. The number of meetings or number of ideas a supervisor provides for solving work issues are operationalizations of the theoretical (latent) variable leadership.

10.4.1.1   Reflective constructs

The previous section illustrated the difference between latent and manifest variables. However, it assumed a more traditional structure where manifest variables are a reflection of the underlying construct. When developing and specifying variables, too little attention has been paid to the structural form of latent variables and the measurements used to create them. There are two primary models used to represent latent variables: the reflective

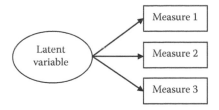

Figure 10.1 Reflective construct model.

model and the formative model. So far, most of our attention has been on the more common reflective variable. In a reflective variable, the "covariation among the measures is caused by, and therefore reflects, variation in the underlying latent factor" (Jarvis et al., 2003, p. 200). Notice in the model above (Figure 10.1) the arrows (which represent the direction of causality) are pointing from the latent variable to the manifest variables, or operationalized measures. Any changes in the latent variable should cause equal changes in all the measures. Any of the three measures above are assumed to be equally valid indicators of the latent variable, and therefore, any of the measures could be interchanged without affecting results assuming the reliability is the same (Bollen and Lennox, 1991). Any single measure or combination of measures should be equally valid in representing the latent variable.

Cognitive ability is a good example of a reflective construct from the social sciences. Cognitive ability refers to the capability associated with knowledge in problem solving and is typically assessed using measures that evaluate quantitative ability, verbal reasoning, spatial reasoning, and so on. Each of these measures *reflect* the underlying construct of cognitive ability. Consequently, cognitive ability is assumed to influence all these and any change in cognitive ability should be reflected in all these measures equally.

10.4.1.2 Formative constructs

A formative construct works with the arrows of causality pointing in the opposite direction (Figure 10.2). Any changes in the individual measures (manifest variables) are hypothesized to cause changes in the underlying latent variable. Therefore, the measures are *formative* in creating the latent variable of interest and the indicators jointly determine the conceptual and empirical meaning of the latent variable (Jarvis et al., 2003). In such a case, it is assumed that all the measures have an impact in contributing to the formation of the latent variable, but none of the individual measures themselves are actually caused by the latent construct. A change in one measure does not necessarily correlate to a change in another, and any of the measures may or may not be correlated; therefore, it would be inappropriate to calculate a reliability estimate (Cronbach alpha) on this type of measure.

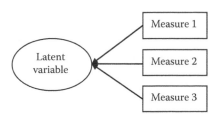

Figure 10.2 Formative construct model.

10.5　Reliability

If a researcher chose three individuals at random to evaluate the effectiveness of communication within their organization, it is unlikely that he or she would receive identical responses from all three people. In theory, there should only be one true score that represents the effectiveness of communication. However, the reality is the scores will vary among individuals. One reason for variability of scores among individuals is measurement error in the instrument. When dealing with social science measurement, it is important to keep error to a minimum to maximize reliability. Reliability addresses the consistency of a measure (a) within raters, (b) among raters, or (c) over time, and is the ability of a measure to produce consistent results when the same construct is measured under different circumstances (Field, 2009). In other words, it assesses the degree to which scores obtained are consistent across repetitions of measurement and free from random error (Schwab, 1999).

Internal consistency is a form of reliability that assesses similarity of scores *within each rater* (Nunnally, 1967). For example, if a researcher wanted to develop a three-item measure to assess perceived organizational support, he or she might ask to what degree an individual thinks the organization (1) contributes to their well-being, (2) is willing to help them, and (3) cares about their opinions (Eisenberger et al., 1986). If this is a reliable measure, one would expect similar responses to these questions. However, note that this does not assess whether these items are actually measuring the construct of interest (validity). It is only concerned with how consistent these items are *within* this measure of perceived organizational support.

Interrater reliability is the degree to which a group of observers agrees on the evaluation of a construct. In this case, the observers are repeated and not the items within a measure. Using our previous example, if five individuals are asked whether their organization cares about their opinions (assuming they all belong to the same organization), the extent to which these responses agree determines interrater reliability. The more agreement there is among observers, the more reliable the measure is. A more familiar example of interrater reliability is the Olympic judges we see rating an athletic performance. Each judge provides an individual rating, and the extent to which these ratings agree represent reliability of the measure.

Finally, stability reliability (test–retest) examines the extent to which a single measure is repeatable over time. In this case, the repetitions of measurement are time periods, and the same measure is administered at multiple points. A good example is a personality assessment used to collect data throughout the year. If we assume an individual's personality is stable, a measure with high test–retest reliability should yield similar results regardless of when the measure is used.

Presented below is a simple formulation of reliability as described in Schwab (1999). Since reliability is a gauge of the consistency of a measure, an observed score (within raters, between raters, or over time) can be thought of as

$$X_{oi} = X_{ti} + X_{ei} \tag{10.1}$$

where
　　X_{oi} = An observed score on the ith case
　　X_{ti} = Systematic score component of the ith case
　　X_{ei} = Random error score component of the ith case

The error score, X_{ei}, is assumed to vary equally above and below the true score and that the error score is unrelated to the systematic portion of the observed score. For example, let us assume an aircraft maintenance crew can change the engine on an airplane in 10 h. Ten hours represent the true score of that maintenance team. Sometimes, it will take longer than 10 h and sometimes it will take less, but over time, there will be an equal number of hours where it takes longer or shorter. Additionally, as the task is completed, the reason for completing ahead or behind schedule is not related to the systematic portion of the score. In other words, crews may be more tired on certain days or more alert and productive on others, and these reasons are tied to something such as lack of sleep, rushing to finish before the weekend, and so on, but they are not related to the systematic portion of the score (the true time it takes to complete the work). This can be represented by the following formula:

$$\rho X_t X_e = 0 \tag{10.2}$$

where $\rho X_t X_e =$ the correlation coefficient between the systematic and random error components.

If the systematic and error components are uncorrelated, then, the variance (σ^2) of the sum of these components is equal to the sum of the variances of each component. Thus

$$\sigma^2 X_o = \sigma^2 X_t + \sigma^2 X_e \tag{10.3}$$

where

$\sigma^2 X_o =$ variance of observed scores
$\sigma^2 X_t =$ variance of systematic scores
$\sigma^2 X_e =$ variance in random errors

Since the variance in the observed score includes variance from both the systematic and error components, variance of the observed scores will always be more than the variance of the systematic scores by themselves (Equation 10.4).

$$\sigma^2 X_o > \sigma^2 X_t \tag{10.4}$$

Therefore, reliability can be thought of as the systematic variance/total variance

$$P_{xx} = \frac{\sigma^2 X_t}{\sigma^2 X_o} \tag{10.5}$$

where P_{xx} is the reliability of scores for a given variable (Schwab, 1999). Given the above formula (Equation 10.5), we can see that a reliability score can range from 0.00 to 1.00. With a reliability score of 0.00, all the variance accounted for is random while a perfect score of 1.00 means all the variance accounted for is systematic. This formula applies regardless of whether we are calculating reliability based on within-person scores (internal consistency), between-person scores (interrater reliability), or scores over time (stability reliability). Coefficient alpha (also referred to as Cronbach alpha) is a generally accepted method to calculate reliability and is denoted as

$$\hat{P}_{XX} = \frac{k * \rho_{(\text{avg})ij}}{1 + (k - 1) * \rho_{(\text{avg})ij}} \tag{10.6}$$

where

$\hat{P}_{XX}$ = coefficient alpha estimated from observed scores

k = number of repetitions (items, observers, or time periods)

$\rho_{(avg)ij}$ = the average correlation among scores of repetitions (items, judges, time periods, etc.) (Schwab, 1999).

Let us look at our previous example of evaluating the job performance of an airline pilot and calculate the reliability of a hypothetical measure using Equation 10.6. Assume the job performance of an airline pilot is measured using a seven-point Likert scale to assess (1) on-time arrival, (2) performance in a flight simulator, and (3) accuracy of the pre-flight check. On the basis of these items, a supervisor could rate each of 50 pilots and score all of them individually on these three dimensions (Table 10.1).

If the average correlation among scores of measurement repetitions $\hat{P}_{XX}$ $\rho_{(avg)ij}$ were calculated as 0.40, the Cronbach alpha is

$$\hat{P}_{XX} = \frac{3 * 0.40}{1 + (3 - 1) * 0.40}$$

$$\hat{P}_{XX} = 0.67$$

The computed Cronbach alpha estimate of reliability is 0.67 based on our three-item measure of pilot performance. Current acceptance in the social science literature seems to be that any value of 0.70 or higher is adequate to demonstrate reliability (Cortina, 1993). However, there is no definitive cutoff and all scores should be interpreted with caution. Acceptance of 0.70 without consideration of the specific objective of the measure is risky. Let us look at our previous example of pilot job performance. If we were to use a 10-item measure of performance instead of a three-item measure (assume the average correlation among scores of repetitions is the same), Cronbach alpha would be calculated as

$$\hat{P}_{XX} = \frac{10 * 0.40}{1 + (10 - 1) * 0.40}$$

$$\hat{P}_{XX} = 0.87$$

Notice the sample size did not change, only the number of items included in the measure changed, thereby increasing the calculated Cronbach alpha score. The reliability of a measure is partially a function of the number of items used to calculate a score

Table 10.1 Example matrix for rating 50 pilots on three dimensions

Cases	Measurement repetitions		
	1	2	3
1			
2			
3			
.			
50			

and says nothing of the quality of those items in measuring the construct. Looking at our example from a different perspective, if we know the Cronbach alpha score for the three-item measure is 0.67, then, the average interitem correlation must be 0.40. However, if the 10-item measure had the same alpha of 0.67, the average interitem correlation is only 0.17. This example highlights the caution that must be exercised when interpreting scores and suggests Cronbach alpha measures are relatively meaningless when constructing measures with a large number of items. Similarly, if the number of items in a measure remain the same, a higher reliability can be obtained through the construction of higher-quality items, thus leading to less error and higher average correlation among scores of repetitions.

When attempting to assess the adequacy of a reliability score, one must ascertain the accuracy required for the decision that is to be made from the measure. For example, if a supervisor needed to determine the relative performance rankings of individuals to allocate bonus pay, a measure with high reliability may be desirable to accurately distinguish the order among top performers. If, however, he only needed to discriminate the top half from the bottom half, a measure with lower reliability might be appropriate.

10.6 Validity

From 1924 to 1932, there was a famous group of studies that are the basis for what we know today as the Hawthorne effect (Mayo, 1949; Franke and Kaul, 1978; Sonnenfeld, 1985). The studies were commissioned to determine if improved working conditions would increase worker productivity in a manufacturing environment. At first, industrial engineers increased the lighting and noted that efficiency of workers increased on the production line. However, when the lighting was decreased and the production line returned to its original condition, productivity of workers *increased* again. In the end, it was determined that the act of observing workers increased efficiency and that the changes to the actual working conditions of the production line itself had very little to do with increased performance. This example demonstrates the concept of validity, or lack thereof in assigning inference to the lighting conditions that had very little impact on worker efficiency. Validity is the approximate truth of an inference. It is the extent to which operationalizations of a construct actually measure what they are intended to. In our previous measure of pilot job performance, does a flight arriving on time really represent performance of the individual flying the plane, or is it more indicative of other factors that might influence arrival time? If we wanted to determine how fast an athlete is, we could observe him or her running 1 mile and use the time measured in minutes or seconds to calculate how fast he or she is. However, if we tried to use the weight of an athlete measured in pounds to infer how fast he or she can run, many would question the validity of that measure. The primary question to ask when determining validity is whether the measure is assessing the construct of interest. There are two additional types of validity this chapter will address: internal validity and external validity.

Internal validity refers to inferences about whether the covariation between variables reflects a causal relationship. To support a causal relationship between two variables (A and B), the researcher must establish that A preceded B in time, that there is a covariation between the two variables A and B, and that no other explanations for the relationship are plausible (Shadish et al., 2002). Internal validity is all about causal inferences and whether the researcher can really infer a change in one variable as being a direct result of a change in another. Threats to internal validity are those causes that could influence an outcome that is not attributed to the variable of interest.

External validity is the extent to which inferences about causal relationships hold true over variations in persons, settings, treatments, and outcomes (Campbell and Stanley, 1963; Campbell, 1986). In other words, can the inferred relationship among variables be generalized outside the specific research environment? Cronbach and Shapiro (1982) argue that most external validity questions are about applications that were not studied in the research experiment, and the threats to external validity are the reasons why a causal relationship might or might not hold true over changes in persons, settings, treatments, and outcomes.

10.6.1 *Reliability versus validity: Which is more important?*

When making an assessment of reliability and validity, it helps to think of reliability in terms of consistency, and validity in terms of covering the correct content. If we were to design a test that assesses cognitive ability and that test provided four different results when given to the same person at different times, that test would be considered unreliable assuming no other changes had occurred. In terms of validity, the test may very well assess areas that are relevant to the content of cognitive ability so the test could be valid, but not reliable. Conversely, the test might provide very similar results over time indicating a reliable measure. However, if the test items had nothing to do with assessing the content of cognitive ability, that test would have high reliability but low validity. This leads to an intuitive belief that validity is more important than reliability, and indeed, if you could only choose one, having a valid measure is more important than a reliable measure. However, to be useful as a tool for research, a measurement instrument should be both reasonably valid and reasonably reliable (Patten, 2009).

A common way to think about the relationship between reliability and validity is to imagine a target (construct) we are trying to assess using measurement items. In Figures 10.3 through 10.6, the measurement items are represented by arrows and the construct is represented by the target. If the arrows are properly aimed at the center of the target (valid) and all the arrows are straight so that they fly exactly as aimed toward the target (reliable), we have a representation of a measure that is both reliable and valid (Figure 10.3). However, if some of the arrows are warped so that they cannot fly true (low

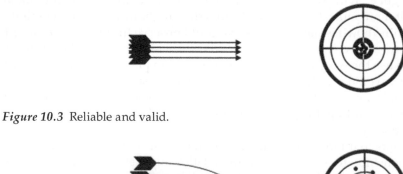

Figure 10.3 Reliable and valid.

Figure 10.4 Unreliable but valid.

Figure 10.5 Reliable but invalid.

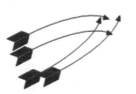

Figure 10.6 Unreliable and invalid.

reliability), you can aim them properly at the center target (high validity) but they will not hit the bulls-eye consistently. In this case, we have a measure that is valid, but unreliable (Figure 10.4). In Figure 10.5, we see an example where all the arrows are straight and fly exactly as predicted (high reliability); however, the arrows are not aimed at the center of the target (low validity) so we have a measure that is reliable, but invalid. Finally, if the arrows were not properly trained on the center of the target and they were also warped, you would end up with a measure similar to Figure 10.6 where the items are both unreliable and invalid.

10.7   Summary

Measurement in the social sciences is a very complex process. However, understanding all the related components that contribute to sound psychometric practices will help ensure more reliable, valid research results. All this starts with the construct definition. Without a clear understanding of the construct we are trying to measure, it would be impossible to develop an appropriate assessment tool. It would be the equivalent of trying to write a test before we know the topic of interest. Once we define the construct, an appropriate scale must be established to enable proper measurement. Scale types include a range of options such as nominal scales used to categorize data all the way up to ratio scales that provide information about the difference between measurements and an interpretable zero point. Reliability and validity must also be assessed, and while this is not always an easy prospect, a measure should be both reasonably valid and reasonably reliable if it is going to be used to draw inferential conclusions. In an effort to help future research in the social sciences, Table 10.2 provides a list of many common measures that have been established in the literature as being both reasonably valid and reasonably reliable. A final note of caution, this chapter is not intended to be used as a guide for creating, testing, or validating measures. Rather, this chapter should serve as an introduction and review of some of the main issues that must be considered when developing and choosing measures for social science.

Table 10.2 Sampling of measures from social science research

Variable name	Citation
Coworker exchange	Sherony and Green (2002)
Emotional intelligence	Wong and Law (2002)
Empowerment	Sprietzer (1995)
Goal orientation	Button et al. (1996)
	Hovarth et al. (2001)
	VandeWalle (1997)
Group characteristics	Campion et al. (1993)
	Denison et al. (1996)
Group interdependence	Wageman (1995)
Group process	Campion et al. (1993)
	Taylor and Bowers (1972)
Group psychological safety	Edmondson (1999)
Intrinsic motivation	Hackman and Lawler (1971)
	Hackman and Oldham (1980)
Job design	Edwards et al. (1999)
Job diagnostic survey (JDS)	Karasek (1979)
Task interdependence	Wall et al. (1995)
Job involvement	Hackman and Lawler (1971)
	Lodahl and Kejner (1965)
Leader appraisal of the member's performance	Liden et al. (1993)
Overall performance	Judge et al. (1999)
Overall performance	Motowidlo and Van Scotter (1994)
	Ostroff et al. (2004)
	Van Dyne and LePine (1998)
Job satisfaction survey	Spector (1997)
Leader–member exchange	LMX-7—Graen and Uhl-Bien (1995)
	LMX-7 (revised)—Liden et al. (1993)
	LMX-MDM—Liden and Maslyn (1998)
Member expectations of the leader	Liden et al. (1993)
Leader expectations of the member	Liden et al. (1993)
Leadership roles	Denison et al. (1995)
	Hart and Quinn (1993)
Likability	Schmitt et al. (1996)
Organizational citizenship behavior	Anderson and Williams (1996)
Organizational citizenship behavior	McNeely and Meglino (1994)
	Moorman and Blakely (1995)
	Motowidlo and Van Scotter (1994)
	Pearce and Gregersen (1991)
	Podsakoff et al. (1990)
	Puffer (1987)
	O'Reilly and Chatman (1986)
	Skarlicki and Latham (1996)
	Smith et al. (1983)
	Van Dyne et al. (1994)
	Van Dyne and LePine (1998)
	Van Scotter and Motowidlo (1996)
	Williams and Anderson (1991)

(Continued)

Table 10.2 (Continued) Sampling of measures from social science research

Variable name	Citation
Organizational commitment	Meyer and Allen (1997) Mowday and Steers (1979)
Organizational justice	Colquitt (2001) Niehoff and Moorman (1993)
Valence, instrumentality, and expectancy (VIE)	Lawler (1981)
Perceived organizational support	Eisenberger et al. (1986)
Person–organization fit	Bretz and Judge (1994) Cable and Judge (1996) Lovelace and Rosen (1996)
Personality—Big 5	Goldberg (1992)
Core self-evaluations	Judge et al. (2003)
Individualism–collectivism	Wagner (1995)
Manifest needs questionnaire	Steers and Braunstein (1976)
Positive and negative effect	Watson et al. (1988)
Proactive personality	Seibert et al. (2001)
Self-esteem	Robinson and Shaver (1973)
Uncertainty avoidance	Hofstede (1980)
Role ambiguity	Breaugh and Colihan (1994) House et al. (1983)
Safety climate	Burt et al. (1998)
Safety management/commitment and communication	Hofmann and Morgeson (1999) Cheyne et al. (1998)
Self-efficacy	Chen et al. (2001) Quinones (1995)
Self-monitoring	Gangestad and Snyder (1985)
Social skills	Ferris et al. (2001)
Team-member exchange (TMX)	Seers (1989)
Trust	Mayer et al. (1999) Clark and Payne (1997) McAllister (1995)
Turnover intentions	Cammann et al. (1986)
Work centrality	Paullay et al. (1994)

References

Badiru, A. B. (Ed.). 2013. *Handbook of Industrial and Systems Engineering*, Boca Raton: Taylor & Francis Group.

Bollen, K. A. and Lennox, R. 1991. Conventional wisdom on measurement: A structural equation perspective, *Psychological Bulletin*, 110(2), 305–314.

Campbell, D. T. 1986. Relabeling internal and external validity for applied social scientists. In Trochim, W. M. K. (Ed.). *Advances in Quasi-Experimental Design and Analysis*, San Francisco: Jossey-Bass, pp. 67–77.

Campbell, D. T. and Stanley, J. C. 1963. *Experimental and Quasi-Experimental Designs for Research*, Chicago: Rand McNally.

Campbell, J. P. 1990. Modeling the performance prediction problem in industrial and organizational psychology. In Dunnette, M. D. and Hough, L. M. (Eds). *Handbook of Industrial and Organizational Psychology*, Vol. 1, Palo Alto: Consulting Psychologists Press.

Colquitt, J. A., Lepine, J. A., and Wesson, M. J. 2011. *Organizational Behavior, Improving Performance and Commitment in the Workplace*, New York: McGraw-Hill Irwin.

Cortina, J. M. 1993. What is coefficient alpha? An examination of theory and applications. *Journal of Applied Psychology*, 78(1), 98–104.

Cronbach, L. and Shapiro, K. 1982. *Designing Evaluations of Educational and Social Programs*, San Francisco: Jossey-Bass.

Eisenberger, R., Huntington, R., Hutchison, S., and Sowa, D. 1986. Perceived organizational support. *Journal of Applied Psychology*, 71, 500–507.

Field, A. 2009. *Discovering Statistics Using SPSS*, London: Sage Publications Ltd.

Franke, R. H. and Kaul, J. D. 1978. The Hawthorne experiments: First statistical interpretation. *American Sociological Review*, 43, 623–643.

Harman, H. H. 1960. *Modern Factor Analysis*, Chicago: University Chicago Press.

Jarvis, C. B., MacKenzie, S. B., and Podsakoff, P. M. 2003. A critical review of construct indicators and measurement model misspecification in marketing and consumer research. *Journal of Consumer Research*, 30(2), 199–218.

Kerlinger, F. N. 1986. Constructs, variables, and definitions. In *Foundations of Behavioral Research*, Kerlinger, F. N. (Ed.), pp. 1–19, New York: Harcourt Brace & Company.

Latham, G. P. and Pinder, C. C. 2005. Work motivation theory and research at the dawn of the twenty-first century. *Annual Review of Psychology*, 56, 485–516.

Mayo, E. 1949. Hawthorne and the Western Electric Company. In *The Social Problems of an Industrial Civilisation*, Mayo, E. (Ed.), pp. 31–45, London: Routledge.

Nunnally, J. C. 1967. *Psychometric Theory*, New York: McGraw-Hill.

Patten, M. L. 2009. *Understanding Research Methods: An Overview of the Essentials*, Glendale: Pyrczak Publishing.

Schwab, D. P. 1999. *Research Methods for Organizational Studies*, Mahwah: Lawrence Erlbaum Associates, Inc.

Shadish, W. R., Cook, T. D., and Campbell, D. T. 2002. *Experimental and Quasi-Experimental Designs for Generalized Causal Inference*, Boston: Houghton Mifflin Company.

Sonnenfeld, J. A. 1985. Shedding light on the Hawthorne studies. *Journal of Occupational Behavior*, 6, 125.

Taylor, J. C. 1971. An empirical examination of a four-factor theory of leadership using smallest space analysis. *Organizational Behavior and Human Performance*, 6, 249–266.

chapter eleven

Systems interoperability measurement

Thomas C. Ford

Contents

11.1 Introduction

"Measuring, assessing, and reporting interoperability in a visible way is essential to setting the right priorities" (Kasunic and Anderson 2004). Due to an increasingly networked (technologically, culturally, commercially, etc.) world, the need for interoperability evaluation is becoming more important. The "need for more interoperability" is constantly heard throughout government, industry, and scientific communities. This chapter presents a general method of quantitatively measuring the interoperability of a heterogeneous set of systems (e.g., coalitions, technology, organizations, cultures, political philosophies, languages, people, and religions, among others) experiencing a wide variety of types of interoperations (e.g., enterprise, doctrine, force, joint, logistics, operational, semantic, and technical) in the context of an operational process (Ford et al. 2009). It improves upon extant interoperability assessment methods by accepting all system and interoperability types, describing collaborative and confrontational interoperability in the context of operations, and accommodating the ever-changing modes of interoperability rather than relying upon outdated interoperability attributes. The method is flexible and allows systems and their interoperations to be defined at any level of abstraction, resulting in precise interoperability measurements that are not limited to a small set of possible values.

Recognizing that interoperability is not an end unto itself, but that it facilitates operational advantage, the method mathematically assesses the impact of interoperability on operational effectiveness.

Interoperability occurs in both collaborative and noncollaborative (i.e., confrontational) ways. Collaborative interoperability is the most commonly understood type of interoperability today. When the term interoperability is used, collaborative interoperability, or the idea of systems, organizations, or services working together to mutual advantage, is implied. On the other hand, confrontational interoperability is also critically important, not just in military, but also in nonmilitary domains and occurs when sets of opposing systems attempt to control each other (i.e., a jammer's attempt to degrade an enemy communication system's effectiveness or a negotiation team's attempt to swing a deal in its favor). In general, if an operational process implies any form of opposition between systems (e.g., negotiation, attack, greediness, pushing, removing, limiting, and preventing, among others), then confrontational interoperability can be measured. This type of interoperability measurement is powerful because it can be directly related to operational effectiveness without discrete event or other types of simulation. Indeed, for these processes confrontational interoperability should receive focus as it describes the ability of friendly (blue) systems to control adversary (red) systems and prevent reciprocation. Confrontational interoperability implies and underpins effects-based activities in which all planning, preparation, execution, and assessment of an operation concentrates on the desired effects on opposing systems. In other words, the goal of confrontational interoperability is to ensure that blue systems are not only able to interoperate with each other but are also able to affect red systems as well.

11.2 Background

One of the earliest definitions of interoperability (first published in 1977) is still one of the most popular, "the ability of systems, units, or forces to provide services to and accept services from other systems, units, or forces and to use the services so exchanged to enable them to operate effectively together" (Department of Defense 2008). Although not perfect, this definition is adopted here because it (1) infers that interoperation occurs between many types of entities (e.g., systems, units, or forces), (2) describes interoperability as a relationship between entities, (3) implies that interoperation is an exchange between a "provider" and an "acceptor," and (4) explains that interoperation enables effective operation. These important concepts permeate and are foundational to a general method of measuring interoperability.

11.2.1 State-of-the-practice in interoperability measurement

Over two-dozen papers have been published specifically on interoperability measurement. Table 11.1 lists those researchers who have proposed a new interoperability measurement method or an extension/improvement to an existing one. Each published method generally applies to only one system and interoperability type and is classified as a *leveling* or *nonleveling* method. Several papers offer analyses of these methods (Sutton 1999; Brownsword et al. 2004; Kasunic and Anderson 2004; Morris et al. 2004; Ford et al. 2007a).

Leveling interoperability assessment methods are largely based upon the maturity model concept developed by the United States Air Force in 1987 (Humphrey and Sweet 1987) and represent maturity by thresholds of increasing interoperability capability (Department of Defense 1998). These methods exhibit numerous weaknesses including

Table 11.1 Main contributions and types of extant interoperability measurement methods

Method	Author	Type[a]	Main contribution
SoIM	LaVean (1980)	NL	Interoperability can be measured in levels
QoIM	Mensh et al. (1989)	NL	Interoperability is correlated to measures of effectiveness
MCISI	Amanowicz and Gajewski (1996)	NL	The distance between systems modeled as points in space indicates their interoperability
LISI	DoD (1998)	L	Systems possess interoperability attributes
IAM	Leite (1998)	NL	Same as LISI
OIM	Clark and Jones (1999)	L	Organizations interoperate but have different interoperability attributes than technical systems
Stoplight	Hamilton et al. (2002)	NL	Operations and acquisitions both have interoperability requirements
LCI	Tolk (2003)	L	Operational interoperability is an extension of technical interoperability
LCIM	Tolk and Muguira (2003)	L	Conceptual interoperability bridges system interoperability
NMI	NATO (2003)	L	Same as LISI
NCW	Alberts and Hayes (2003)	L	Interoperability occurs in the physical, information, cognitive, and social domains; lack of interoperability increases difficulty in accomplishing the mission
NTI	Stewart et al. (2004)	L	Social, personnel, and process interoperability are valid types of nontechnical interoperability
OIAM	Kingston et al. (2005)	L	There are levels of ability of organizations to be agile in their interoperations
NID	Schade (2005)	L	Levels of interoperability can be described in linguistic terms
i-Score	Ford et al. (2007b, 2008a)	NL	Interoperability measurements are operational process-specific and have a theoretical maximum value
GMIM	Ford et al. (2009)	NL	Generalized method of interoperability measurement

[a] NL = nonleveling and L = leveling.

(1) limited precision of measurement (usually an integer from 0–4), (2) fixed number of unchangeable interoperability attributes (usually 1–4) which can become outdated, (3) applicability to only one type of system, and (4) an interoperability measurement tied to a single system vice a pair of systems. These methods are special cases of the more general interoperability measurement method presented in this chapter.

Nonleveling interoperability assessment methods are a much more diverse group and, as a whole, generally predate the leveling methods. They also are specialized to a particular type of system or interoperability. Each leveling and nonleveling method made important contributions to the theory of interoperability measurement (Table 11.1), but could not claim to be a complete and general method of measuring the interoperability of a set of heterogeneous systems.

It should be pointed out that all these past models (both leveling and nonleveling) create a high level of abstraction as the foundational representation for the systems under study. They seem simplistic because they hide system complexity. These are not meant to be detailed simulations, but rather an attempt to understand critical design parameters affecting interoperability early in the design lifecycle of new systems and system modifications.

11.2.2   Impact on operational effectiveness

It is difficult to measure the impact of an entity or factor on an operation. In spite of this, many researchers have made great strides in doing so, albeit within the constraints of a specific type of entity or factor or even a single type of operation. Besides presenting a general method of measuring the interoperability of systems in the context of an operational process, the method in this chapter also gives sufficient conditions for relating that interoperability measurement to measures of operational effectiveness pertinent to the operational process.

11.3   Method preview

Friendly (blue) and adversary (red) systems implement the activities and decisions of an operational process and can be modeled as a sequence of states of system characteristics. If strictly interoperability-related characteristics are used to model the systems, then a fundamental result is obtained—that a measure of the similarity of a pair of system models is a measure of their associated systems' interoperability. Furthermore, given a measure of operational effectiveness (MoOE) for a confrontational operational process, another fundamental axiom states that if all systems and system interoperability features pertinent to the confrontational operational process are completely modeled, then if the blue systems are more directionally interoperable with red systems than vice versa, blue systems have operational advantage over red systems. This important result permits an interoperability measurement to be related to operational effectiveness. Each step in the general interoperability measurement process (Figure 11.1) is discussed in succeeding sections.

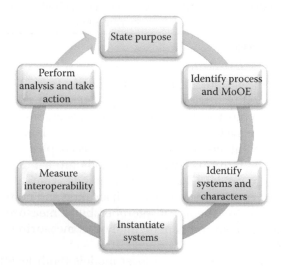

Figure 11.1 Interoperability measurement methods.

11.3.1 Purpose of interoperability measurement

Before an interoperability analysis is undertaken, the purpose of the analysis must be stated in order to keep the analysis focused and properly scoped. There are uncountable important reasons to measure interoperability, examples of which are included in Table 11.2.

11.3.2 Identify operational process

An operational process (see, e.g., in Figure 11.2) is defined as a set of tasks, logically sequenced, which when executed achieve a desired effect and, as will be shown later, is used to identify the set of blue and red systems whose interoperability is to be measured and to determine the MoOE to which the interoperability measurement will be correlated. Many methods of task, activity, process, and thread modeling have been devised over the years (Knutilla et al. 1998), which are graphical, mathematical, and linguistic in nature, any one of which can be used to describe an operational process in support of an interoperability measurement.

11.3.3 Identify measures of operational effectiveness

In order to relate interoperability to operational effectiveness, an MoOE must be chosen which ties the purpose of the interoperability measurement to the operational process. The MoOE can be a natural, constructed, or proxy measure originating from the physical, information, or cognitive domain and should be written at a level of abstraction equivalent to that of the operational process. Ideally the MoOE is relevant, measurable, responsive (sensitive), understandable, discriminatory, quantitative, realistic, objective, appropriate, inclusive, independent, valid, and reliable. Additionally, in order to apply

Table 11.2 Example purposes for measuring interoperability

Determine level of integration and synchronization of organizations
Measure effectiveness of coalition equipment
Troubleshoot logistics problems
Determine impact of lack of common procedures
Predict impact of liaison officers on coalition force mission success
Provide leaders with interoperability requirements
Determine impact of insertion of new communications technology
Assess shortfalls/deficiencies of communications on operational effectiveness
Determine ability to cooperate with other organizations
Facilitate CIO responsibility to enforce interoperability
Support business units in verifying operational interoperability procedures
Measure interoperability of communication systems
Determine product distribution bottlenecks
Validate key interoperability solutions prior to mission execution
Measure and predict future interagency or intergovernment interoperability
Establish requirements for interagency emergency response communications
Measure ability of government agencies to share information during crisis
Measure operational connection between agencies in a dynamic environment
Assess interoperability of supply chains

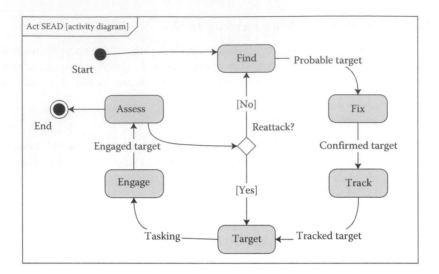

Figure 11.2 SEAD operational process.

the Interoperability Impact on Operational Effectiveness axiom described later, the MoOE must be written as a diametric pair $O = \{O_B, O_R\}$ which relates the effectiveness of the set of blue systems to the lack of effectiveness of the set of red systems. For example, $O_B + O_R = 1$ could describe the relationship of the diametric MoOE $O = \{O_B, O_R\}$, where O_B is the percent of market share gained and O_R is the percent of market share lost. All percentage-based MoOEs can be written as diametric MoOEs. Other MoOEs, such as those that represent a count (e.g., number of radio connections made), can be normalized in order to be written diametrically (e.g., the percent of expected radio connections made).

11.3.4   System identification

Everything can be expressed as a system (Guan et al. 2008) including processes, enterprises, organizations, coalitions, concepts, functions, objects, products, models, cultures, doctrines, forces, cities, public services, and applications, among others. A system can be defined as an entity comprising related interacting elements that act together to achieve a purpose. This definition is broad enough to include a wide variety of system types including, but not limited to, technical, biological, environmental, organizational, conceptual, physical, and philosophical systems, and ensures that the system interoperability measurement method presented in this chapter has extensive utility. The operational process is used to identify and constrain the set of systems. The operational process is chosen as the constraining mechanism because systems perform different interoperations in different scenarios, a limited number of systems implement any given operational process, effectiveness is often measured at the operational process level, and operational processes can be written at any level of abstraction enabling systems to be abstracted at the same level. The set of systems $S = \{s_1, s_2, \ldots, s_n\}$ is often diverse and can be small or large.

11.3.5   Character identification

A set of systems S can be modeled by a set of characters $X = \{x_1, x_2, \ldots, x_n\}$, which represent traits, attributes, or characteristics that describe the important features of the system.

These system characters can be morphological, physiological, interfacial, ecological, and distributional among others. Ideally, the set of characters chosen are natural (i.e., characters are not confounded with each other) and diagnostic (i.e., characters distinguish one system, or system type, from another). Additionally, the types of characters chosen should be related to the type of interoperability measurement that is to be undertaken. For example, interoperability measurement of devices on a network demands that functional characters be emphasized, whereas a spatial interoperability analysis, such as for a spacecraft docking assessment, requires that morphological characters are used.

Extending a definition from the phylogeneticists (Semple and Steel 2003), characters are functions which map systems in S to the states C of their characters X where the set of valid character states for a set of characters is $C = \{c_1, c_2 ..., c_n\}$ (see definition below). Character states are either qualitative (discrete) or quantitative (discrete or continuous) (Sneath and Sokal 1973) or a mixture of both. Generally, the set of character states is restricted to the binary numbers (absence/presence states) or the positive real numbers, although other states are certainly possible.

Definition (system characterization)

Given a set of systems S, then $X:S \rightarrow C$ is a function that maps systems to a set of character states C and X, called the characterization of S.

Although numerous types of system characters can be used to describe a system, in order to form a basis for an interoperability measurement, only system interoperability characters should be used to model a system. In their essence, system interoperability characters are pairs of actions that describe what systems do to each other (Table 11.3).

Interoperability characters can be extracted in a methodical fashion from sentences found in requirements, architecture, and a host of other acquisition, capabilities, and

Table 11.3 A selection of interoperability pairs and associated characters

Interoperability pairs	Interoperability type	Character
Provide ⇔ Accept	General	*Interoperate*
Transmit ⇔ Receive	Communication	*Communicate*
Publish ⇔ Subscribe	Net-centric	*Service*
Occupy ⇔ Accommodate	Spatial	*Accommodate*
Serve ⇔ Be served	Human	*Service*
Give ⇔ Take	Human	*Share*
Buy ⇔ Sell	Business	*Trade*
Pay ⇔ Get paid	Financial	*Transact*
Output ⇔ Input	Traditional system	*Output Input*
Lead ⇔ Follow	Organizational	*Dance*
Order ⇔ Obey	Human, organizational	*Command*
Produce ⇔ Consume	Business, human	*Economy*
Transport ⇔ Transported by	Business	*Transport*

Table 11.4 Guide for relating parts of speech to interoperability modeling

Part of speech	Relationship
Noun	System or refinement of an interoperability character
Verb	Interoperability character
Pronoun	System
Adverb	Refinement of an interoperability character
Adjective	Refinement of a system
Preposition	Refinement of an interoperability character
Conjunction	Not applicable
Interjection	Not applicable
Article	Not applicable

operations documents. A guide for relating the parts of speech found in sentences to system interoperability measurement is given in Table 11.4.

To illustrate, consider the following sentences taken from the United State's Department of Defense's Joint Publication 3-01 *Countering Air and Missile Threats*, "Suppression of enemy air defenses (SEAD) is an activity that neutralizes, destroys, or temporarily degrades surface-based enemy air defenses ... Each component has its own capabilities for suppression of enemy air defenses (SEAD) ... Some of these options include aircraft with special detectors, jammers and electronic countermeasures (ECM) equipment, antiradiation missiles (ARMs), and precision-guided munitions (PGM)" (Department of Defense 2007). Applying the guide in Table 11.4, the beginnings of one possible interoperability model results (Equation 11.1).

$$S = \left\{Enemy\ Air\ Defense\ System, Component\ Aircraft\right\}$$

$$X = \left\{ \begin{cases} SEAD(T), \\ SEAD.destroy(T), \\ SEAD.destroy.ARM(T), \\ SEAD.destroy.PGM(T), \\ SEAD.degrade(T), \\ SEAD.degrade.jam(T), \\ SEAD.detect(T) \end{cases}, \begin{cases} SEAD(R), \\ SEAD.destroy(R), \\ SEAD.destroy.ARM(R), \\ SEAD.destroy.PGM(R), \\ SEAD.degrade(R), \\ SEAD.degrade.jam(R), \\ SEAD.detect(R) \end{cases} \right\} \tag{11.1}$$

$$C = \left\{0,1\right\}$$

Several concepts can be observed from this example. First, interoperability characters should be modeled hierarchically to allow increased precision of the eventual interoperability measurement. And second, as interoperability characters are inherently directional, the direction of the interoperation (i.e., transmit or receive) must be captured by appending a suffix to the interoperability character as in $X = \{X_T, X_R\} = \{SEAD(T), SEAD(R)\}$. This characterization implies that a system can "act," "be acted upon," or both. If every interoperability character in X is bi-directional, then the (T) and (R) suffixes are not needed.

11.3.6 System instantiation

Once systems, their interoperability characters, and the states of those characters have been identified, then a specific system can be modeled, or instantiated, as a sequence (Amanowicz and Gajewski 1996) of states of system characters (Ford et al. 2008b).

Definition (system instantiation)

Given a specific $s \in S$ and a set $x \subseteq X$ of system characters descriptive of s, then $\sigma = x(s)$ is a sequence of system character states called the instantiation of s, which models s.

After all $s \in S$ have been instantiated, the system instantiations must be aligned with each other in order to support meaningful system comparisons and other mathematical operations. Hereafter, the term system instantiation implies an aligned system instantiation.

Definition (instantiation alignment)

Given a set $x' \subseteq X$ of system characters descriptive of s' and a set $x'' \subseteq X$ of system characters descriptive of s'', then two system instantiations σ' and σ'' are aligned if $\sigma' = X(s')$ and $\sigma'' = X(s'')$. The aligned instantiation of S is given by $\Sigma = X(S)$.

For example, let $S = \{s_1, s_2, s_3\}$ be a set of systems of interest, let $X = \{X_T, X_R\} = \{\{x_1, x_2\}, \{x_3, x_4\}\}$ be a set of directional interoperability characters used to characterize S, and let all character states be absence/presence states (i.e., $C = \{0,1\}$). Given individual system instantiations as in Equation 11.2, then an aligned instantiation of S is given by Equation 11.3.

$$\sigma_1 = \{x_1(s_1), x_2(s_1), x_3(s_1), x_4(s_1)\} = \{1,1,0,1\}$$
$$\sigma_2 = \{x_1(s_2)\} = \{1\} \tag{11.2}$$
$$\sigma_3 = \{x_2(s_3), x_4(s_3)\} = \{1,0\}$$

$$\Sigma = X(S) = \{\{1,1,0,1\}, \{1,0,0,0\}, \{0,1,0,0\}\} = \begin{bmatrix} 1 & 1 & 0 & 1 \\ 1 & 0 & 0 & 0 \\ 0 & 1 & 0 & 0 \end{bmatrix} \tag{11.3}$$

A system instantiation Σ in which interoperability characters are assigned binary character states is an underlying interoperability model upon which a performance-enhanced instantiation is based. The performance-enhanced instantiation can be used to facilitate data rate, cost, efficiency, or throughput analysis, among others. For example, given the interoperability model in Equation 11.1, a partial underlying SEAD system instantiation is given by Equation 11.4.

$$\Sigma^T =$$

	Air Defense System	SEAD Aircraft
SEAD(T)	0	1
SEAD.destroy(T)	0	1
SEAD(R)	1	0
SEAD.destroy(R)	1	0

$$\tag{11.4}$$

If a probability focused interoperability analysis is desired, a performance-enhanced system instantiation Σ_p (notional) can be defined as in Equation 11.5, which models the probability of the SEAD aircraft being able to implement the mission and the probability of the air defense system being destroyed.

$$\Sigma^T = \begin{bmatrix} & | & \textit{Air Defense System} & \textit{SEAD Aircraft} \\ \hline \textit{SEAD(T)} & | & 0 & 0.9 \\ \textit{SEAD.destroy(T)} & | & 0 & 0.8 \\ \textit{SEAD(R)} & | & 0.6 & 0 \\ \textit{SEAD.destroy(R)} & | & 0.6 & 0 \end{bmatrix} \tag{11.5}$$

11.3.7 Interoperability measurement

Metrology, or the science of measurement, defines measurement as, "the assignment of numbers to properties or events in the real world by means of an objective empirical operation, in such a way as to describe them" (Finkelstein and Leaning 1984). If interoperability is considered as a property of a set of systems, then an operation, called a system interoperability measurement, can be defined, which objectively and empirically assigns a number to systems interoperability. This operation, its derivation, and its varieties are defined next.

The foundation of an interoperability measurement is a system similarity measurement calculated using a similarity function that takes aligned system instantiations Σ as its arguments. A similarity function is the converse of a dissimilarity function (e.g., Euclidean distance) in that it gives a larger result if its arguments are more similar and a smaller result if they are more dissimilar. There are numerous types of similarity functions such as distance, association, correlation, and probabilistic measures (Sneath and Sokal 1973) and geometric, feature contrast, alignment-based, and transformational measures (Guan et al. 2008). These classifications are not mutually exclusive.

An interoperability measurement represents the ability of two systems to interoperate, or in other words, it represents the similarity between their capabilities to interoperate. Hence, the character (feature) contrast similarity function is appropriately chosen. The general form of a character contrast similarity function is given in Equation 11.6, where θ, α, β are weights (see definition below), f is a contrasting function, $\sigma' \cap \sigma''$ represents the characters that system instantiations σ', σ'' have in common, $\sigma' - \sigma''$ represents the characters that σ' possesses that σ'' does not, and $\sigma'' - \sigma'$ represents the opposite:

$$Sim(\sigma',\sigma'') = \theta f(\sigma' \cap \sigma'') - \alpha f(\sigma' - \sigma'') - \beta f(\sigma'' - \sigma') \tag{11.6}$$

An interoperability function is a similarity function that meets the criteria given in the definition below. Two interoperability functions, Sim_{Bin} for systems instantiated with binary absence/presence character states and Sim_{Real} for systems instantiated with real-valued character states, are also defined.

Definition (interoperability function)

An interoperability function I is a similarity function $Sim(\sigma',\sigma'')$ which (1) takes a pair of system instantiations as its arguments, (2) has a range of [0,1] (i.e., 0 indicates non-interoperable systems while 1 indicates perfectly interoperable systems), (3) rewards for shared characters and optionally penalizes for unshared characters (i.e., α, β), and (4) gives a greater reward (i.e., θ) to system pairs whose shared characters' states have a "better" value.

Definition (Sim_{Bin})

Given a pair of systems s', s'' instantiated as σ', $\sigma'' \in \{0, 1\}^n$, where $\{0, 1\}^n$ represents binary n-space and where $\wedge$ is the Boolean AND operator, then $I = Sim_{Bin} = Sim(\sigma',\sigma'')$ is an interoperability function that gives a weighted, normalized measure of the similarity of systems instantiated with the absence/presence character states where $f = \sum_{i=1}^{n}(\sigma' \wedge \sigma'')$ and $\theta = 1/n$, $\alpha = \beta = 0$.

Definition (Sim_{Real})

Given a pair of systems s', s'' instantiated as $\sigma',\sigma'' \in \mathbb{R}^n \cap [0,c_{max}]$, then $I = Sim_{Real} = Sim(\sigma',\sigma'')$ (Equation 11.7) is an interoperability function which gives a weighted, normalized measure of the similarity of systems instantiated with real-valued character states where f is the modified Minkowski similarity function (Equation 11.8), θ is the average character state value (Equation 11.9) of a pair of system instantiations, n is the number of characters used to instantiate σ', σ'', c_{max} is the maximum character state value, r is the Minkowski parameter (usually set to $r = 2$), and $\alpha = \beta = 0$.

$$I = Sim_{Real} = \left[\frac{\sum_{i=1}^{n} \sigma'(i) + \sum_{i=1}^{n} \sigma''(i)}{2nc_{max}} \right] \left[1 - \left(\frac{1}{\sqrt[r]{n}} \right) \left(\sum_{i=1}^{n} b_i \left(\frac{\sigma'(i) - \sigma''(i)}{c_{max}} \right)^r \right)^{1/r} \right] \qquad (11.7)$$

$$\text{Modified Minkowski similarity} = \left[1 - \left(\frac{1}{\sqrt[r]{n}} \right) \left(\sum_{i=1}^{n} b_i \left(\frac{\sigma'(i) - \sigma''(i)}{c_{max}} \right)^r \right)^{1/r} \right] \qquad (11.8)$$

$$b_i = \begin{cases} 0 & \sigma'(i) = 0 \quad or \; \sigma''(i) = 0 \\ 1 & else \end{cases}$$

$$\text{Average character state value} = \theta = \frac{\sum_{i=1}^{n} \sigma'(i) + \sum_{i=1}^{n} \sigma''(i)}{2nc_{max}} \qquad (11.9)$$

A fundamental axiomatic relationship between similarity of systems and interoperability of systems is given as follows. This axiom can then be used to formally define an interoperability measurement.

Axiom (system similarity and interoperability)

If a pair of systems is instantiated only with system interoperability characters, then a measure of the similarity of the instantiations is a measure of their associated systems' interoperability.

Definition (interoperability measurement)

Given two systems instantiated with bi-directional characters as σ', σ'' and an interoperability function I, then $m = I(\sigma',\sigma'')$ is a measure of the interoperability of s' and s'' and $m = 0 \rightarrow s'$, s'' are noninteroperable and $m = 1 \rightarrow s'$, s'' are perfectly interoperable. Similarly, $M = [m_{ij}]$, $i, j \leq |S|$ is a matrix of interoperability measurements for all system pairs in S.

If the characters used to instantiate systems are directional in nature (i.e., a system can provide an interoperation, but not accept it), then directional interoperability measurements (see the next definition) must be made.

Definition (directional interoperability measurement)

If two systems are instantiated as $\sigma' = \{\sigma'_T,\sigma'_R\}$, $\sigma'' = \{\sigma''_T,\sigma''_R\}$ with directional interoperability characters $X = \{X_T,X_R\}$, then $m = I(\sigma'_T,\sigma''_R)$ is a measure of the directional interoperability of s' to s''.

Interoperability is generally time variant. Atmospheric effects, phase of operations, random/un-modeled influences, or other effects may cause the variance. Time-variant interoperability measurements are useful in determining at which points in time interoperability lapses cause associated degradation in operational effectiveness.

It is not possible, or even desirable, for all systems to interoperate with each other, let alone interoperate perfectly. Indeed, there are often many limitations, specific to the operational process, which prevent systems from interoperating at their full potential. These limitations may be physical, operational, or reliability related, among many others. Therefore, in order to manage expectations on the final interoperability measurement, it is useful to define a realistic upper bound on the interoperability measurement, called the constrained upper bound which admits that the best possible interoperability measurement must be less than $m = 1$ due to these various degradations and limitations, both desired and undesired. It is calculated by building an interoperability model that includes all interoperability characters the set of systems could conceivably implement with their character states set to their best possible value in light of the predetermined limitations. The difference between the constrained upper bound on the interoperability measurement and the current interoperability measurement is called the interoperability gap and represents the trade space, design space, funding space, etc., in which process or system changes can occur, to the end of improving operational effectiveness.

11.3.8   Interoperability impact on operational effectiveness

In the context of an operational process, it is desirable for blue systems to maintain operational advantage over red systems through both improved blue-to-red directional interoperability and degraded red-to-blue directional interoperability. The following axiom gives a sufficient condition for relating directional confrontational interoperability to operational advantage.

Axiom (Operational Advantage)

Let the subscripts B and R refer to blue and red forces, respectively. Given a set of systems $S = \{S_B,S_R\}$ instantiated a $\Sigma = \{\Sigma_B,\Sigma_R\}$, then a sufficient condition for blue force operational advantage over red force is for all pairs (σ_B,σ_R), $I(\sigma_B,\sigma_R) > I(\sigma_R,\sigma_B)$ assuming Σ completely characterizes S.

As MoOEs quantify operational advantage, applying the Operational Advantage axiom, the impact of interoperability on operational effectiveness can be described. This important result is given below as the Interoperability Impact on Operational Effectiveness axiom, which, like the Operational Advantage axiom, describes a sufficient condition.

Axiom (Interoperability Impact on Operational Effectiveness)

Let the subscripts B and R refer to blue and red forces, respectively. Given a set of systems $S = \{S_B, S_R\}$ characterized by X, instantiated as $\Sigma = \{\Sigma_B, \Sigma_R\}$ and a diametric MoOE $O = \{O_B, O_R\}$, then if X completely characterizes all interoperations related to O_B and O_R then a sufficient condition for $O_B > O_R$ is that blue systems have Operational Advantage over red systems, or in other words blue-to-red directional interoperability exceeds red-to-blue directional interoperability for all pairs (σ_B,σ_R) (i.e., $I(\sigma_B,\sigma_R) > I(\sigma_R,\sigma_B) \leftrightarrow O_B > O_R$).

Two examples of the implementation of the interoperability measurement method are given in the following section. The first example demonstrates the use of interoperability measurement to assess the impact of interoperability on operational effectiveness in a military scenario. The second one uses an example of a United Nations peacekeeping effort to show how the precision of a legacy interoperability assessment can be improved by applying the method in this chapter.

11.4   Suppression of enemy air defenses example

The following SEAD example demonstrates the general interoperability measurement method, illustrates confrontational interoperability, and exemplifies the Interoperability Impact on Operational Effectiveness axiom which states that improved blue-to-red directional interoperability combined with degraded red-to-blue interoperability results in higher operational effectiveness.

Hypothesis: Applying the Interoperability Impact on Operational Effectiveness axiom, it can be shown that operational effectiveness of a notional SEAD operation is improved by (1) the addition of blue-force precision strike and electronic attack capability (i.e., increased blue-to-red confrontational interoperability) and (2) the addition of blue-force stealth (i.e., decreased red-to-blue confrontational interoperability).

SEAD is defined as, "activity that neutralizes, destroys, or temporarily degrades surface-based enemy air defenses by destructive and/or disruptive means" (Department of Defense 2008). In this example, the definition is further refined to include only activity that destroys enemy air defenses by destructive means. An operational process for this application is given in Figure 11.2.

An MoOE "percent of enemy air defenses destroyed" for the process can be written as the diametric pair given in Equation 11.10, which obeys the relationship $O_B + O_R = 1$

$$O = \{O_B, O_R\} = \begin{Bmatrix} \text{Percent of enemy air defenses destroyed} \\ \text{Percent of enemy air defenses protected} \end{Bmatrix} \qquad (11.10)$$

Typical SEAD systems are associated with the activities and decisions of the operational process (Figure 11.2) and are given in Equation 11.11. The *ISR* (intelligence, surveillance, and reconnaissance) system performs the Find, Fix, and Track activities, the *AOC* (air operations center) system performs the Target, Assess, and Reattack Activities and decision, and the *PSP* (precision strike package) system performs the Engage activity. Enemy *IADS* (integrated air defense) systems are targets. An operational view of the mission is given in Figure 11.3.

$$S = \{S_B, S_R\} = \{\{HB, ISR, AOC, PSP\}, \{IADS_1, IADS_2\}\} \qquad (11.11)$$

In order to apply the Interoperability Impact on Operational Effectiveness axiom, the characterization X of S must include all interoperability characters which describe every interoperation (collaborative and confrontational) between systems in S related to O. In other words, all interoperability characters related to the destruction and protection of the IADS systems must be included in X. In order to ensure the set of interoperability characters X chosen for the SEAD application in this section is complete and authoritative, they have been methodically identified and extracted from Joint Publications and Air Force Doctrine Documents (AFDDs) related not just to SEAD, but also to Joint and Air Force operations in general (Table 11.5). The top level of the interoperability character hierarchy is the set of joint operational functions given in Joint Publication (JP) 3-0 *Joint Operations* (Department of Defense 2006) and the second level is a pertinent subset of the operational functions of air and space power given in AFDD 1 Air Force Basic Doctrine (Department of the Air Force 2003). Lower levels of the interoperability character hierarchy have been extracted from Joint Publications and AFDDs.

Assign the set of states of X as absence/presence states $C = \{0, 1\}$ then the instantiation of S is given as Σ in Table 11.6. Although the joint operational functions *Movement & Maneuver*, *Protection*, and *Sustainment* are included in the instantiation for completeness, they have been assigned zero states, as their functionality is not critical to this example.

If the interoperability function $I = Sim_{Bin}$ is chosen, then assuming no self-interoperability, the resulting directional interoperability measurements are given as Equation 11.12.

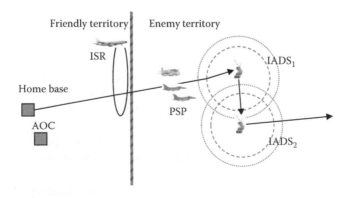

Figure 11.3 Operational view of SEAD mission.

Table 11.5 Explanation of SEAD characterization X

Interoperability character	Explanation
C2	Command and control/be commanded and controlled
C2.Comm	Communicate (T)/communicate (R)
C2.Comm.Blue	Communicate (T) on blue channels/communicate (R) on blue channels
C2.Comm.Blue.Target	Communicate (T) red targets on blue channels/communicate (R) red targets on blue channels
C2.Comm.Red	Communicate (T) on red channels/communicate (R) on red channels
Intel	Intel (T)/Intel (R)
Intel.ISR	ISR (T) = collect and ISR (R) = collected against
Intel.ISR.Detect	Detect = (collect)/be detected by (=collected against)
Intel.ISR.Detect.Blue	Detect blue systems (=identify blue)/be detected as a blue system (identified as blue)
Intel.ISR.Detect.Red	Detect red systems (identify red)/be detected as a red system (identified as red)
Fires	Attack/be attacked
Fires.CA	Attack threats to airborne systems/be attacked as a threat to airborne systems
Fires.CA.OCA	Attack threats to airborne systems prior to their employment/be attacked, prior to employment, as a threat to airborne systems
Fires.CA.OCA.Ground	Attack ground-based threats to airborne systems prior to their employment/be attacked, prior to employment, as a ground-based threat to airborne systems
Fires.CA.OCA.Ground. Cluster	Attack ground-based threats to airborne systems using cluster bombs/be attacked as a ground-based threat to airborne systems using cluster bombs
Fires.CA.OCA.Ground. Precision	Attack ground-based threats to airborne systems using precision-guided munitions/be attacked as a ground-based threat to airborne systems using precision-guided munitions
Fires.CA.DCA	Attack threats to airborne systems after weapon employment/be attacked as a threat to airborne systems after weapon employment
Fires.IO	Attack via info operations/be attacked via information operations
Fires.IO.EW	Attack via info operations using electronic warfare/be attacked via info operations by electronic warfare
Fires.IO.EW.EA	Jam/be jammed
Fires.IO.EW.EA.Barrage	Jam by barrage method/be jammed by a barrage method
Fires.IO.EW.EA.Reactive	Jam precisely by reactive method/be jammed precisely by a reactive method
Fires.IO.InflOps	Apply influence operations/receive influence operations
Fires.IO.InflOps.MILDEC	Apply military deception/receive military deception
Movement&Maneuver	Not used in this scenario
Protection	Not used in this scenario
Sustainment	Not used in this scenario

Table 11.6 SEAD instantiation Σ

	Transmit						Receive					
	HB	ISR	AOC	PSP	IADS1	IADS2	HB	ISR	AOC	PSP	IADS1	IADS2
C2	1	1	1	1	1	1	1	1	1	1	1	1
C2.Comm	1	1	1	1	1	1	1	1	1	1	1	1
C2.Comm.Blue	1	1	1	1	0	0	1	1	1	1	0	0
C2.Comm.Blue.Target	0	1	1	0	0	0	0	0	1	1	0	0
C2.Comm.Red	0	0	0	0	1	1	0	1	0	0	1	1
Intel	0	1	0	0	1	1	0	0	0	1	1	1
Intel.ISR	0	1	0	0	1	1	0	0	0	1	1	1
Intel.ISR.Detect	0	1	0	0	1	1	0	0	0	1	1	1
Intel.ISR.Detect.Blue	0	1	0	0	1	1	0	0	1	1	0	0
Intel.ISR.Detect.Red	0	1	0	0	0	0	0	0	0	0	1	1
Fires	0	0	0	1	1	1	0	0	0	1	1	1
Fires.CA	0	0	0	1	1	1	0	0	0	1	1	1
Fires.CA.OCA	0	0	0	1	0	0	0	0	0	1	1	1
Fires.CA.OCA.Ground	0	0	0	1	0	0	0	0	0	0	1	1
Fires.CA.OCA.Ground.Cluster	0	0	0	1	0	0	0	0	0	0	1	1
Fires.CA.OCA.Ground.Precision	0	0	0	0	0	0	0	0	0	0	1	1
Fires.CA.DCA	0	0	0	0	1	1	0	0	0	0	1	1
Fires.IO	0	0	0	0	0	0	0	0	0	0	1	1
Fires.IO.EW	0	0	0	0	0	0	0	0	0	0	1	1
Fires.IO.EW.EA	0	0	0	0	0	0	0	0	0	0	1	1
Fires.IO.EW.EA.Barrage	0	0	0	0	0	0	0	0	0	0	1	1
Fires.IO.EW.EA.Reactive	0	0	0	0	0	0	0	0	0	0	1	1
Fires.IO.InflOps	0	0	0	0	1	1	0	1	1	1	0	0
Fires.IO.InflOps.MILDEC	0	0	0	0	1	1	0	1	1	1	0	0
Movement&Maneuver	0	0	0	0	0	0	0	0	0	0	0	0
Protection	0	0	0	0	0	0	0	0	0	0	0	0
Sustainment	0	0	0	0	0	0	0	0	0	0	0	0

$$M = \begin{bmatrix} & HB & ISR & AOC & PSP & IADS_1 & IADS_2 \\ \hline HB & 0 & \frac{1}{9} & \frac{1}{9} & \frac{1}{9} & \frac{2}{27} & \frac{2}{27} \\ ISR & \frac{1}{9} & 0 & \frac{5}{27} & \frac{8}{27} & \frac{2}{9} & \frac{2}{9} \\ AOC & \frac{1}{9} & \frac{1}{9} & 0 & \frac{4}{27} & \frac{2}{27} & \frac{2}{27} \\ PSP & \frac{1}{9} & \frac{1}{9} & \frac{1}{9} & 0 & \frac{7}{27} & \frac{7}{27} \\ IADS_1 & \frac{2}{27} & \frac{5}{27} & \frac{5}{27} & \frac{10}{27} & 0 & \frac{1}{3} \\ IADS_2 & \frac{2}{27} & \frac{5}{27} & \frac{5}{27} & \frac{10}{27} & \frac{1}{3} & 0 \end{bmatrix} \qquad (11.12)$$

Table 11.7 SEAD interoperability impact on operational effectiveness analysis

Blue–red system pair	Analysis	Blue advantage?
$HB \leftrightarrow IADS$	$I(HB, IADS) = 2/27 = I(IADS, HB) \rightarrow O_B \leq O_R$	No
$ISR \leftrightarrow IADS$	$I(ISR, IADS) = 2/9 > I(IADS, ISR) = 5/27 \rightarrow O_B > O_R$	Yes
$AOC \leftrightarrow IADS$	$I(AOC, IADS) = 2/27 < I(IADS, AOC) = 5/27 \rightarrow O_B \leq O_R$	No
$PSP \leftrightarrow IADS$	$I(PSP, IADS) = 7/27 < I(IADS, PSP) = 10/27 \rightarrow O_B \leq O_R$	No

For this example, the appropriate analysis of M is the comparison of blue-to-red and red-to-blue interoperability (i.e., confrontational interoperability) to the end of applying the Interoperability Impact on Operational Effectiveness axiom to determine if the blue systems will enjoy operational effectiveness over the red systems. Four blue–red system pairs are possible and must be considered, $HB \leftrightarrow IADS$, $ISR \leftrightarrow IADS$, $AOC \leftrightarrow IADS$, and $PSP \leftrightarrow IADS$. The Interoperability Impact on Operational Effectiveness axiom (see Table 11.7 summary) shows that only one blue system (ISR) is operationally effective over the red $IADS$ systems. Two others (AOC, PSP) do not possess operational advantage over the $IADS$, and the third (HB) is at a standoff (i.e., equivalent directional interoperability measurements).

The directional interoperability measurements in Table 11.7 indicate that red targets can be detected but not effectively destroyed by blue systems. Additionally, the measurement of the directional confrontational interoperability from $IADS$ to PSP indicates that PSP is vulnerable to destruction by the $IADS$ systems. According to the Interoperability Impact on Operational Effectiveness axiom, in order to give blue systems operational effectiveness over red systems, blue-to-red directional interoperability must exceed red-to-blue interoperability. To this end, according to the hypothesis, blue-to-red interoperability will be increased by adding precision strike and electronic attack capability to the PSP system. Additionally, red-to-blue interoperability will be decreased by adding stealth capability to the PSP system (manifested in the model as an inability of the $IADS$ to detect the PSP). Assuming that red systems are also capable of being upgraded, the ability to resist all but reactive jamming will be given to the $IADS$ systems. Again using $I = Sim_{Bin}$ as the interoperability function, a set of interoperability measurements M_U for the upgraded systems is obtained and given in Equation 11.13. Changes from the original interoperability matrix are highlighted.

$$M_U = \begin{bmatrix} & HB & ISR & AOC & PSP & IADS_1 & IADS_2 \\ \hline HB & 0 & \tfrac{1}{9} & \tfrac{1}{9} & \tfrac{1}{9} & \tfrac{2}{27} & \tfrac{2}{27} \\ ISR & \tfrac{1}{9} & 0 & \tfrac{5}{27} & \boxed{\tfrac{7}{27}} & \tfrac{2}{9} & \tfrac{2}{9} \\ AOC & \tfrac{1}{9} & \tfrac{1}{9} & 0 & \tfrac{4}{27} & \tfrac{2}{27} & \tfrac{2}{27} \\ PSP & \tfrac{1}{9} & \tfrac{1}{9} & \tfrac{1}{9} & 0 & \boxed{\tfrac{4}{9}} & \boxed{\tfrac{4}{9}} \\ IADS_1 & \tfrac{2}{27} & \tfrac{5}{27} & \tfrac{5}{27} & \boxed{\tfrac{1}{3}} & 0 & \tfrac{1}{3} \\ IADS_2 & \tfrac{2}{27} & \tfrac{5}{27} & \tfrac{5}{27} & \boxed{\tfrac{1}{3}} & \tfrac{1}{3} & 0 \end{bmatrix} \qquad (11.13)$$

After upgrading the *PSP* system with precision strike, electronic attack, and stealth and countering with a red force upgrade of the *IADS* systems with resistance to all but reactive jamming, then $I(PSP, IADS) = 4/9 > I(IADS, PSP) = 1/3 \rightarrow O_B > O_R$. Hence, the blue force now has a slight edge over the red force, implying that the percentage of red targets destroyed will be greater than the percentage of red targets protected. Thus, the original hypothesis of this application is confirmed. Finally, it is interesting to note that one instance of blue-to-blue (i.e., collaborative) interoperability $I(ISR, PSP)$ decreased as a result of the system upgrades. The interpretation of this is that due to the addition of stealth capability the *PSP* system is also less detectable by the blue force *ISR* system.

This example demonstrated the general interoperability measurement method, selecting characters and instantiating systems, and then applying a straightforward, but simplistic, Sim_{Bin} interoperability function. The binary function is useful to capture sufficiency, but not proficiency.

11.5   Organizational interoperability example

Whereas the previous example assessed the interoperability of systems using binary character states, this last example demonstrates how Sim_{Real} can be used to provide a more precise interoperability measurement of systems using real-valued character states. Additionally, this final example shows that legacy maturity-model (leveling) interoperability assessment methods are a special case of the general interoperability measurement method described in this chapter.

Hypothesis: Maturity-model (leveling) interoperability assessment methods, such as the Organizational Interoperability Model (OIM), are a special case of the general interoperability method described in this chapter. To demonstrate, the method can be used to measure the interoperability of International Force East Timor (INTERFET) coalition forces and to arrive at more precise conclusions using Sim_{Real} than those made by Clark and Moon (2001) when they originally applied the more restrictive OIM model to the INTERFET operation.

In 1987, the U.S. Air Force developed the maturity model concept through a grant to the Carnegie Mellon Software Engineering Institute (CMU-SEI). Although the maturity model concept, which describes the stages through which a process progresses, was originally designed to assess contractor software engineering ability, it was adopted in 1998 by the MITRE Corporation as the basis of the first maturity model-based interoperability measurement method called Levels of Information System Interoperability (LISI) (Department of Defense 1998). In the LISI model, maturity was represented by five thresholds of interoperability capability. For nearly a decade, LISI was the template for numerous maturity model and maturity model-like (leveling) interoperability measurement approaches designed to measure both information and noninformation system interoperability. One such approach was the OIM (Clark and Jones 1999; Clark and Moon 2001; Fewell and Clark 2003). The usefulness of the OIM model was demonstrated when Clark and Moon used it to analyze the International Force East Timor (INTERFET) coalition sent to enforce peace in East Timor in 1999. The coalition consisted of forces from Australia (lead nation), the United States, New Zealand, Thailand, Philippines, and Republic of Korea, among others. Clark and Moon assessed the interoperability of INTERFET coalition forces with respect to the Australian Deployable Joint Force (ADJF) standard. Table 11.8 shows the framework they used, which consisted of Preparedness, Understanding, Command Style, and Ethos.

Table 11.8 Organizational Interoperability Maturity Model (OIM)

	Preparedness	Understanding	Command style	Ethos
Level 4 (Unified)	Complete normal day-to-day working	Shared	Homogeneous	Uniform
Level 3 (Combined)	Detailed doctrine and experience in using it	Shared communications and knowledge	One chain of command and interaction with home organization	Shared ethos but with influence from home organization
Level 2 (Collaborative)	General doctrine in place and some experience in using it	Shared communications and shared knowledge about specific topics	Separate reporting lines of responsibility overlaid with single command chain	Shared purpose; goals, value system significantly influenced by home organization
Level 1 (Cooperative)	General guidelines	Electronic communications and shared information	Separate reporting lines of responsibility	Shared purpose
Level 0 (Independent)	No preparedness	Voice communications via phone, etc.	No interaction	Limited shared purpose

Source: Adapted from Clark, T. and T. Moon. Interoperability for joint and coalition operations. *Australian Defence Force Journal* 151: 23–36, 2001.

Drawing from their paper, systems, interoperability characters, and character states are assigned as

$$S = \{AUS, US, NZ, Thai, Phil, ROK\} \tag{11.14}$$

$$X = \{Preparedness, Understanding, Command\ Style, Ethos\} \tag{11.15}$$

where these characters can take on one of the five (maturity) values as

$$C = \{0, 1, 2, 3, 4\} \tag{11.16}$$

Although their paper gives enough information to have used more precise real-valued character states, integer states were used to maintain consistency with their model.

$$\Sigma = \begin{array}{c|cccc} & Preparation & Understanding & Command\ Style & Ethos \\ \hline AUS & 2 & 3 & 3 & 1 \\ US & 2 & 3 & 3 & 1 \\ NZ & 2 & 3 & 3 & 1 \\ Thai & 1 & 1 & 1 & 1 \\ Phil & 1 & 1 & 1 & 1 \\ ROK & 0 & 1 & 1 & 1 \end{array} \tag{11.17}$$

Selecting $I = Sim_{Real}$ (Equation 11.7) as the interoperability function, with $c_{max} = 4$, $r = 2$ and $n = 4$, the following coalition interoperability measurement M results, scaled to OIM [0, 4]:

$$M_{SCALED} = \begin{bmatrix} & AUS & US & NZ & Thai & Phil & ROK \\ AUS & 2.3 & 2.3 & 2.3 & 1.0 & 1.0 & 0.9 \\ US & 2.3 & 2.3 & 2.3 & 1.0 & 1.0 & 0.9 \\ NZ & 2.3 & 2.3 & 2.3 & 1.0 & 1.0 & 0.9 \\ Thai & 1.0 & 1.0 & 1.0 & 1.0 & 0.9 & 0.8 \\ Phil & 1.0 & 1.0 & 1.0 & 0.9 & 1.0 & 0.8 \\ ROK & 0.9 & 0.9 & 0.9 & 0.8 & 0.8 & 0.8 \end{bmatrix} \qquad (11.18)$$

This assessment can be compared to the original results reported by Clark and Moon. Note, their assessment used a pair-wise minimum similarity measure

$$OIM = \begin{bmatrix} & AUS & US & NZ & Thai & Phil \\ US & 2 & & & & \\ NZ & 2 & 2 & & & \\ Thai & 1 & 0 & 0 & & \\ Phil & 1 & 0 & 0 & 0 & \\ ROK & 0 & 0 & 0 & 0 & 0 \end{bmatrix} \qquad (11.19)$$

It can be seen that among INTERFET member nations there were three clusters {AUS, US,NZ}, {Thai,Phil}, and {ROK}. Expectedly, the nations with Western-type philosophies and presumably more familiar with ADJF standards enjoyed a high degree of interoperability with each other, but less so with the Asian nations and vice versa. The lack of coalition interoperability between the Western and Eastern nations participating in INTERFET manifested itself in the fact that "the Thais, South Koreans, and Filipinos had their own areas of operation … and conducted their own operations" (Clark and Moon 2001). Similarly, the "divergent nature of the operational philosophies of the participating countries" was one of the "most difficult aspects of assembling and maintaining the coalition" [ibid]. Furthermore, some of the coalition officers "only understood half of what was said at briefings and conferences and … the Australians were unaware of this" [ibid]. Coalition interoperability could improve among these Western and Eastern nations if common philosophies on doctrine, training, information sharing, delegation, and cultural values and goals, acceptable to both East and West, are agreed upon, practiced, and implemented prior to future operations.

There are some notable differences between the general interoperability measurement and that resulting from the use of the OIM model. For example, whereas the OIM model scored the interoperability of the Republic of Korea with Australia as a zero, meaning the nations were operating completely independently of each other, the general method gives a more accurate result of 0.852. According to Clark and Moon, an OIM score of zero indicates that the two nations (1) had no level of preparedness to operate in a coalition together, (2) had no interaction among their commanders and forces, (3) were limited to telephone communication, and (4) shared a common purpose only in a limited fashion. But their paper indicates that the Koreans (1) did attend briefings and planning meetings, (2) understood at least half of the material presented in those briefings and meetings, (3) received task-ings from HQ INTERFET but operated in their own area of responsibility, (4) had personal contact between commanders, and (5) were not willing to participate in all aspects of the INTERFET operation but supported the humanitarian aspect of the operation. Then, con-sidering that an OIM score of one indicates (1) preparation was made by learning general guidelines (not met), (2) understanding is obtained through electronic communication and shared information (partially met), (3) command is implemented through separate lines of responsibility (met), and (4) the ethos of the operation is shared (partially met), it seems reasonable to assume that the Australian–Korean interoperability score should probably be somewhere between zero and one. Thus, the measurement of 0.852 is appropriate and reflects the true interoperability of the Republic of Korea with Australia more precisely and accurately than the assessment originally given by Clark and Moon.

11.6 Conclusion

Measuring interoperability has long been considered unquantifiable because of its com-plex nature (Kasunic and Anderson 2004). While interoperability has been defined and described, it is multifaceted and permeates many disciplines in many ways. It is reason-able to assume that interoperations occur in all human endeavors. Previous approaches to measuring or describing interoperability relied upon problem decomposition and limited methods of measuring specific types of interoperations of certain types of entities. The result was an eclectic set of somewhat related models useful within limited spheres. While the problem decomposition method was helpful in the short term for certain applications, it prevented the creation of a general interoperability measurement method because the answer to the problem could not be found by creating "a set of compatible mod-els that collectively address all the dimensions of interoperability" (Morris et al. 2004). Indeed, the set could never be complete, necessitating a general interoperability measure-ment method. The approach given in this chapter—a method which explicitly captures all types of systems and interoperations—produces a quantitative, operations-focused interoperability measurement which is as realistic, accurate, and precise as desired (Ford et al. 2009).

References

Alberts, D. and R. Hayes. *Power to the Edge*. Washington DC: DoD CCRP, 2003.

Amanowicz, M. and P. Gajewski. Military communications and information systems interoperabil-ity. *Proceedings of the 1996 IEEE Military Communications Conference*, 1996.

Brownsword, L. et al. *Current Perspectives on Interoperability*. CMU/SEI-2004-TR-009. Pittsburgh, PA: Carnegie Mellon University Software Engineering Institute, 2004.

Clark, T. and R. Jones. Organisational interoperability maturity model for C2. *Proceedings of the 9th International Command and Control Research and Technology Symposium*. Newport, RI, June 29–July 1, 1999.

Clark, T. and T. Moon. Interoperability for joint and coalition operations. *Australian Defence Force Journal* 151: 23–36, 2001.

Department of Defense. *C4ISR Architecture Working Group Final Report—Levels of Information System Interoperability (LISI)*. Washington DC: OSD (ASD(C3I)) C4ISR AWG, 1998.

Department of the Air Force. *AFDD 1 Air Force Basic Doctrine*. Washington DC: GPO, 2003.

Department of Defense. *JP 3-0 Joint Operations*. Washington DC: GPO, 2006.

Department of Defense. *JP 3-01 Countering Air and Missile Threats*. Washington DC: GPO, 2007.

Department of Defense. *JP 1-02 Department of Defense Dictionary of Military and Associated Terms*. Washington DC: GPO, 2008.

Fewell, S. and T. Clark. Organisational interoperability: Evaluation and further development of the OIM model. *Proceedings of the 8th International Command and Control Research and Technology Symposium*. Washington DC, June 17–19, 2003.

Finkelstein, L. and M. Leaning. A review of the fundamental concepts of measurement. *Measurement* 2.1: 25, 1984.

Ford, T. et al. A survey on interoperability measurement. *Proceedings of the 12th International Command and Control Research and Technology Symposium*. Newport, RI, June 19–21, 2007a.

Ford, T. et al. The interoperability score. *Proceedings of the 5th Annual Conference on Systems Engineering Research*. Hoboken, NJ, March 14–16, 2007b.

Ford, T. et al. Measuring system interoperability: An *i-Score* improvement. *Proceedings of the 6th Annual Conference on Systems Engineering Research*. Los Angeles, CA, April 4–5, 2008a.

Ford, T. et al. On the application of classification concepts to systems engineering design and evaluation. *Systems Engineering* Published online August 8, 2008b.

Ford, T. et al. A general method of measuring interoperability and describing its impact on operational effectiveness. *Journal of Defense Modeling and Simulation* 6.1: 17–32, 2009.

Guan, Y., X. Wang, and Q. Wang. A new measurement of systematic similarity. *IEEE Transactions on Systems, Man, and Cybernetics—Part A: Systems and Humans* 38.4: 743–758, 2008.

Hamilton, J., J. Rosen, and P. Summers. An interoperability roadmap for C4ISR legacy systems. *Acquisition Review Quarterly* 28: 17–31, 2002.

Humphrey, W. and W. Sweet. *A Method for Assessing the Software Engineering Capability of Contractors*. CMU/SEI-87-TR-23. Pittsburgh, PA: Carnegie Mellon University Software Engineering Institute, 1987.

Kasunic, M. and W. Anderson. *Measuring Systems Interoperability: Challenges and Opportunities*. CMU/SEI-2004-TN-003. Pittsburgh, PA: Carnegie-Mellon University Software Engineering Institute, 2004.

Kingston, G., S. Fewell, and W. Richer. An organisational interoperability agility model. *Proceedings of the 10th International Command and Control Research and Technology Symposium*. McLean, VA, June 13–16, 2005.

Knutilla, A. et al. *Process Specification Language: An Analysis of Existing Representations*. NISTIR 6133. Gaithersburg, MD: National Institute of Standards and Technology, 1998.

LaVean, G. Interoperability in defense communications. *Communications, IEEE Transactions on [legacy, pre-1988]* 28.9: 1445–1455, 1980.

Leite, M. Interoperability assessment. *Proceedings of the 66th Military Operations Research Society Symposium*. Monterey, CA, June 23–25, 1998.

Mensh, D., R. Kite, and P. Darby. A methodology for quantifying interoperability. *Naval Engineers Journal* 101.3: 251, 1989.

Morris, E. et al. System of Systems Interoperability (SOSI): Final Report. CMU/SEI-2004-TR-004. Pittsburgh, PA: Software Engineering Institute, Carnegie Mellon University, 2004.

NATO. NATO Allied Data Publication 34: NATO C3 Technical Architecture. Vol. 2. Brussels, Belgium: NATO, 2003.

Schade, U. Towards the edge and beyond: The role of interoperability. *Proceedings of the 10th International Command and Control, Research and Technology Symposium*. McLean, VA, June 13–16, 2005.

Semple, C. and M. Steel. *Phylogenetics*. New York: Oxford University Press, 2003.

Sneath, P. and R. Sokal. *Numerical Taxonomy*. San Francisco, CA: W. H. Freeman and Company, 1973.

Stewart, K. et al. Non-technical interoperability in multinational forces. *Proceedings of the 9th International Command and Control Research and Technology Symposium*. Copenhagen, Denmark, September 14–16, 2004.

Sutton, P. Interoperability: A new paradigm. *Proceedings of the 1999 International Conference on Computational Intelligence for Modeling, Control and Automation: Neural Networks and Advanced Control Strategies (CIMCA'99)*. Vienna, Austria, February 17–19, 1999.

Tolk, A. Beyond technical interoperability—Introducing a reference model for measures of merit for coalition interoperability. *Proceedings of the 8th Annual International Command and Control Research and Technology Symposium*. Washington DC, June 17–19, 2003.

Tolk, A. and J. A. Muguira. The levels of conceptual interoperability model. *Proceedings of the 2003 Fall Simulation Interoperability Workshop*. Orlando, FL, September 2003.

chapter twelve

A generalized measurement model to quantify health

The multiattribute preference response model

Paul F. M. Krabbe

Contents

12.1   Introduction

The measurement of health, which is defined as assigning meaningful numbers to an individual's health status, has proliferated ever since the World Health Organization (WHO, 1946) provided its definition of health. It was not until 1970 that Fanshel and Bush introduced the first instrument that was able to capture an individual's health state in a single metric value. Access to single metric values for health states is advantageous as these can be used in health outcomes research, disease modeling studies, economic evaluations, and monitoring the health status of patient groups in the general community. Often the values for health states are expanded by combining them with the duration of these states to obtain health summary measures. A well-known example of such a summary measure is the disability-adjusted life years (DALYs) approach that is being used by the WHO to compare different countries with one another on diverse aspects of health. Health economists often apply a rather comparable health summary measure, namely the quality-adjusted life year (QALY).

To quantify health states, these must be described and classified in terms of seriousness and assigned meaningful values (variously called utilities, strength of preference, index, or weights). The first step, thus, is to clarify the concept of health status. Essentially an umbrella concept, it covers independent health domains that together capture the rather loosely conceived notion of health-related quality of life (HRQoL). The second step is to assign a value to the health-state description by means of an appropriate measurement procedure. In the past, several measurement models have been developed to quantify subjective phenomena, and some of these models have found their way into the valuation of health states. Although the scientific enterprise of measuring health states has been going on for about 40 years, there are still concerns about its validity.

The aim of this chapter is to forge a linkage between two prominent measurement models to create a single general model that—at least in principle—resolves many of the problems posed by widely used but inferior valuation techniques. This new measurement framework for deriving health-state values is called the multiattribute preference response (MAPR) model. It combines the characteristics of hypothetical health states with a respondent's health-status characteristics to quantify both the hypothetical states as well as the location of the patient's state. In theory, this new model even allows individuals to choose the attributes (i.e., health domains) describing their health states. A health measurement model with such potential flexibility is unprecedented.

The first section of the chapter presents some concerns about the validity of current health-state valuation techniques followed by a section on the basic measurement principles of subjective phenomena (such as health). The next section explains the probabilistic discrete choice (DC) model and expands on its relationship to measurement models used in economics and psychology. The subsequent section sketches the history of the Rasch model and summarizes its underlying theory. Finally, the merits of integrating these two measurement models into the MAPR model are discussed. All examples and suggestions in this chapter apply to health-state valuation. It should be kept in mind that because the MAPR model is very general, it could also be applied in a number of other fields where the goal is to quantify other subjective phenomena.

12.2   Existing valuation techniques

The standard gamble (SG) and time trade-off (TTO) are frequently used to assign values to health states (Johannesson et al. 1994). The former emerged from the field of economics, the latter from the area of operations research (von Neumann and Morgenstern 1953, Torrance 1976). SG, considered the gold standard for years, was developed under the expected utility theory of von Neumann and Morgenstern (1953). However, as experience shows, assumptions underlying this theory were systematically violated by human behavior. In general, people have difficulty dealing with probabilities and may have an aversion to taking risk. As an alternative, Torrance and colleagues developed TTO, which is simpler to administer than SG. The main drawback of TTO is that the relationship between a health state, its duration, and its value is collapsed into a single measure. The problem is that this requires the values for health states to be independent of the duration of these states. Health-state values have also been derived by another technique, the visual analog scale (VAS), which stems from the field of psychology (Krabbe et al. 2006). Unfortunately, all these conventional measurement techniques (SG, TTO, and VAS) have theoretical and empirical drawbacks when used to value health states. With the possible exception of the VAS, they put a large cognitive burden on the respondents by demanding a relatively high degree of abstract reasoning (Green et al. 2000). The person trade-off (PTO) is another technique that has been used mainly in the area of policy making (Murray and Lopez 1996). This technique was named by Nord (1992), but the technique itself was applied earlier by Patrick et al. (1973). The PTO asks respondents to answer from the perspective of a social decision maker considering alternative policy choices.

The currently dominant valuation technique for quantifying health states, certainly in the field of health economics, is the TTO. It may be intuitively appealing for three reasons. First, it seems to reflect the actual medical situation. Second, it shows some correspondence to the general health-outcome framework (since the TTO is essentially a QALY equivalence statement). And third, it is grounded in economic thinking (the trade-off principle). Nevertheless, compelling arguments against the TTO have been raised by several authors (Gafni and Torrance 1984, Johannesson et al. 1994, Richardson 1994, Bleichrodt 2002, Drummond et al. 2005). In fact, TTO seems to be associated with many problems: practical (difficult for people to perform), theoretical (axiomatic violations and problems in dealing with states worse than dead), and biases (time preference). From a measurement perspective, the TTO technique has been criticized for its susceptibility to framing issues (e.g., duration of the time frame, indifference procedure, and states worse than dead). The same holds for the recently introduced technique known as lead-time TTO (Attema et al. 2013).

12.2.1 Patients versus general population

Conventionally, values for the health states used in economic evaluations are derived from a representative community sample (Gold et al. 1996), or in the case of the DALY approach, values for disease states were derived from medical experts (Murray 1996). Besides asserting that a sample of the general population is a reflection of the average taxpayer, which is considered fair grounds for arriving at resource allocation, other arguments are put forward. For example, it has been noted that patients may adapt to their health state over a period of time. As a result, they may assign higher values to their own poor health state. Patients may also strategically underrate the quality of their health state, knowing they will directly benefit from doing so. The proposition held in this chapter is that while adaptation is a real phenomenon, this effect can largely be reduced and eventually eliminated if the health-state values are derived in a fitting measurement framework. Moreover, it is reasonable to assume that healthy people may be inadequately informed or lack the imagination to make an appropriate judgment on the impact of severe health states; this is one of the reasons why researchers in the field of HRQoL are engaged in a debate about which values are more valid (Brazier et al. 2009, Krabbe et al. 2011). Many researchers assert that individuals are the best judges of their own health status (Gandjour 2010). Therefore, in a health-care context, it is sensible to defend the position that, from a validity perspective, it is the patient's judgment that should be elicited to arrive at health-state values, not that of a sample of unaffected members of the general population.This explains the rise of the so-called patient-reported outcome measurement (PROMs) movement (Devlin and Appleby 2010). Voices from another area have also stressed that such assessments from patients (experienced utility) should get more attention (Kahneman and Sugden 2005, Dolan and Kahneman 2008).

12.3   Measurement principles

12.3.1   Interval level

There are theoretical and methodological differences between the direct valuation techniques (SG, TTO, and VAS) and indirect (latent) measurement models such as probabilistic DC (see the next section). However, they all assume that individuals possess implicit preferences for health states that range from good to bad. Moreover, all the models maintain that it should be possible to reveal these preferences and express them quantitatively. Accordingly, differences between health states should reflect the increments of difference in severity of these states. For that reason, informative (i.e., metric) outcome measures should be at least at the interval level (cardinal data). This means that measures should lie on a continuous scale, whereby the differences between values would reflect true differences (e.g., if a patient's score increases from 40 to 60, this increase is the same as from 70 to 90). To arrive at health-state values with these qualities, two other basic measurement principles should be fulfilled, namely, unidimensionality and invariance.

12.3.2   Unidimensionality

The overall goal is to use health-state values for computational procedures (e.g., computing QALYs and Markov modeling). This implies positioning the values on an underlying unidimensional scale ranging from the worst health state to the best one. Specific analyses, for example, intraclass correlation statistics (inter-rater reliability) or specific mathematical routines closely related to factor analysis (e.g., singular value decomposition; SVD), can be applied to find empirical evidence that health-state values represent a unidimensional structure. An early application of the SVD technique compared TTO and VAS valuation data. The results showed a clear two-dimensional structure for the TTO (Krabbe 2006). Heterogeneous responses (or even distinct response structures) from individuals may indicate that a certain valuation technique is less appropriate to the task since it may not fulfill the need for unidimensional responses. Therefore, it is important to determine how similar individuals' judgments actually are. The unidimensionality condition is grounded in the implicit assumption that people tend to evaluate health states similarly. It is this presumed invariance that permits the aggregation of individual valuations to form group or societal values.

12.3.3   Invariance

Invariance is a critical prerequisite for fundamental measurement (see Section 12.5). It means that the outcome of judgments between two (or more) health states should not depend on which group of respondents performed the assessments. The resulting judgments among health states should also be independent of the set of health states being assessed (Engelhard 1992). Rasch models embody the invariance principle. Their formal structure permits algebraic separation of the person and health-state parameters. Specifically, the person parameter can be eliminated during the process of statistical estimation of the health-state parameters. Not surprisingly, the invariance principle is a key characteristic of measurement in physics (Engelhard 1992).

12.4   Discrete choice model

12.4.1   Background

Modern probabilistic DC models, which come from econometrics, build upon the work of McFadden, the 2000 Nobel Prize laureate in economics (McFadden 1974). DC models encompass a variety of experimental design techniques, data collection procedures, and statistical procedures that can be used to predict the choices that individuals will make between alternatives (e.g., health states). These techniques are applicable when individuals have the ability to choose between two or more distinct (discrete) alternatives.

In the mid-1960s, McFadden worked with a graduate student, Phoebe Cottingham, trying to analyze data on freeway routing decisions as a way to study economic decision-making behavior. He developed the first version of what he called the "conditional multinomial logistic model" (also known as the multinomial logistic model and conditional logistic model). McFadden proposed an econometric model in which the utilities of alternatives depend on utilities assigned to their attributes such as construction cost, route length, and areas of parkland and open space taken up (McFadden 2001). He developed a computer program that allowed him to estimate this model, which was based on an axiomatic theory of choice behavior developed by the mathematical psychologist Luce (Luce 1959).

The DC strategy was conceived in transport economics and later disseminated into other research fields, especially marketing. There, DC modeling was applied to analyze behavior that could be observed in real market contexts. Instead of modeling the choices people actually make in empirical settings, Louviere and others started modeling the choices made by individuals in carefully constructed experimental studies (Louviere and Woodworth 1983); this entailed presenting the participants with profiles containing features of hypothetical products. Originally, these profiles were known as simulated choice situations, but later they were called discrete choice experiments (DCEs). Therefore, instead of modeling actual choices, as McFadden had with the revealed preferences approach, Louviere modeled choices made in experimental studies with the stated preferences approach. This new approach also made it possible to predict values for alternatives that could not be judged in the real world. More recently, DC models have been used as an alternative way to derive people's values for health states (Hakim and Pathak 1999, Salomon 2003, Stolk et al. 2010).

12.4.2   Measurement model

The statistical literature classifies DC models among the probabilistic choice models that are grounded in modern measurement theory and consistent with economic theory (e.g., the random utility model) (Arons and Krabbe 2013). All DC models have in common that they can establish the relative merit of one phenomenon with respect to others. If the phenomena are characterized by specific attributes or domains with certain levels, extended DC models such as McFadden's model would permit estimating the relative importance of the attributes and their associated levels. DC modeling has good prospects for health-state valuation (McFadden 2001, McKenzie et al. 2001, McCabe et al. 2006, Ratcliffe et al. 2009, Bansback et al. 2012, Salomon et al. 2012). Moreover, DC models have a practical advantage: When conducting DCEs, health states may be evaluated in a self-completion format. The scope for valuation research is thereby widened. Most TTO protocols, for deriving values for preference-based health-state instruments, are interviewer-assisted, as studies have clearly shown that self-completion is not feasible or leads to inaccurate results (Oppe et al. 2014). The simplicity of DC tasks, however, facilitates web-based surveys (Arons and Krabbe 2013).

12.4.2.1 Discrimination mechanism

The modern measurement theory inherent in DC models builds upon the early work and basic principles of Thurstone's Law of Comparative Judgment (LCJ) (Thurstone 1927, Krabbe 2008). In fact, the class of choice- and rank-based models, with its lengthy history (1927 to the present), is one of the few areas in the social and behavioral sciences that has a strong underlying theory. It was Thurstone who introduced the well-known random utility model (RUM), although he used different notations and other terminology. The use of Thurstone's model based on paired comparisons to estimate health-state values was first proposed by Fanshel and Bush (1970) in one of the earliest examples of a composed QALY index model.

In Thurstone's terminology, choices are mediated by a "discriminal process," which he defined as the process by which an organism identifies, distinguishes, or reacts to stimuli. Consider the theoretical distributions of the discriminal process for any two objects (paired comparisons), such as two different health states s and t. In the LCJ model, the standard deviation of the distribution associated with a given health state is called the discriminal dispersion (or variance, in modern scientific language) of that health state. Discriminal dispersions may differ for different health states.

Let v_s and v_t correspond to the scale values of the two health states. The difference $(v_s - v_t)$ is measured in units of discriminal differences. The complete form of the LCJ is the following equation:

$$v_s - v_t = z_{st}\sqrt{\sigma_s^2 + \sigma_t^2 - 2\rho_{st}\sigma_s\sigma_t} \qquad (12.1)$$

where σ_s, σ_t denote the discriminal dispersions of the two health states s and t, ρ_{st} denotes the correlation between the pairs of discriminal processes s and t, and z_{st} is the unit normal deviate corresponding to the theoretical proportion of times health state s is judged greater than health state t. The difference is normally distributed with mean $v_s - v_t$ and variance σ_{st}^2 corresponding to $\sqrt{\sigma_s^2 + \sigma_t^2 - 2\rho_{st}\sigma_s\sigma_t}$, which reflects the standard deviation of the difference between two normal distributions. In its most basic form (Case V) the model can be represented as $v_s - v_t = z_{st}$, for which the probability that state s is judged to be better than state t is

$$P_{st} = \Phi\left(\frac{v_s - v_t}{\sigma_{st}}\right) \qquad (12.2)$$

where Φ is the cumulative normal distribution with mean zero and variance unity.

The discrimination mechanism underlying the LCJ is an extension of the "just noticeable difference" that played a major role in early psychophysical research, as initiated by Fechner (1801–1887) and Weber (1795–1878) in Germany. Later on, similar discrimination mechanisms were embedded in "signal detection theory," which was used by psychologists to measure the way people make decisions under conditions of uncertainty. Much of the early work in this research field was done by radar researchers (Marcum 1947).

12.4.2.2   Random utility model

Thurstone proposed that perceived physical phenomena or subjective concepts (e.g., health-states, treatment outcomes, and process characteristics) can be expressed as that a respondent r has a latent value (utility) for state s, U_{rs}, which includes a systematic component and an error term (Equation 12.3). This follows the fundamental idea of true score theory or classical test theory that includes an observed score with two components, namely the true score and an error term. It also summarizes different health domains by combining the scores on several items

$$U_{rs} = v_s + \varepsilon_{rs} \qquad (12.3)$$

Here, v is the measurable component and not determined by characteristics of the respondents. In other words, a given health state has the same expected value across all respondents. The assumption in the model proposed by Thurstone is that ε is normally distributed. This assumption yields the probit model. The choice probability is $P_{rs} = \Pr$ $(U_{rs} > U_{rt}$, all t not equal to $s)$, which depends on the difference in value, not on its absolute level. The fact that only differences in value matter has implications for the identification of this model and all its derivatives. In particular, it means that the only parameters that can be estimated are those that capture differences across alternatives.

Therefore, in Thurstone's LCJ, the perceived value of a health state equals its objective level plus a random error. The probability that one health state is judged better than another is the probability that this alternative has the higher perceived value. When the perceived values are interpreted as levels of satisfaction, HRQoL, or utility, this can be interpreted as a model for economic choice in which utility is modeled as a random variable. This assertion was made in 1960 by the economist Marschak, who thereby introduced Thurstone's work into economics. Marschak called his model the random utility maximization hypothesis or RUM (Marschak 1960, Strauss 1979). Like neoclassical economic theory, the RUM assumes that the decision maker has a perfect discrimination capability. However, it also assumes that the analyst has incomplete information, which implies that uncertainty (i.e., randomness) must be taken into account.

12.4.2.3   Multinomial model

Another way to analyze comparative data is with the Bradley–Terry–Luce (BTL) model, which was statistically formulated by Bradley and Terry (1955) and extended by Luce in 1959 (Luce 1959). It extends the Thurstone model by allowing a person to choose among more than two options. The BTL model postulates that measurement on a ratio scale level can be established if the data satisfy certain structural assumptions (Kind 1982). For mathematical reasons, the BTL model is based on the simple logistic function instead of the normal distribution of the Thurstone model. It is this mathematical model that McFadden used to develop and construct his own specific type of multinomial logit model. If only pairs of alternatives are judged, the BTL model is nearly identical to Thurstone's model. However, when more than two alternatives are judged, an important mathematical assumption must be made, namely the independence of irrelevant alternatives (see below).

Drawing upon the work of Thurstone, Luce, Marschak, and Lancaster (Lancaster 1966), McFadden was able to show how his model fits in with the economic theory of choice behavior. McFadden then investigated further the RUM foundations of the conditional multinomial logistic (MNL) model. He showed that the Luce model was consistent with the RUM model with IID (independent and identically distributed random variables) additive disturbances, if, and only if, these disturbances had a distribution called extreme value type I. More importantly, instead of one function, as in the classical Thurstone model (only values for health states can be estimated), the conditional MNL model comprises two functions. First, it contains a statistical model that describes the probability of ranking a particular health state higher than another, given the (unobserved) value associated with each health state. Secondly, it contains a valuation function that relates the value for a given health state to a set of explanatory variables (it will be shown that the same holds for the MAPR model).

12.4.2.4   *Assumptions*

MNL regression is based on three assumptions: (i) independence of irrelevant alternatives (IIA); (ii) error terms are independent and identically distributed across observations (IID); and (iii) no taste heterogeneity (i.e., homogeneous preferences across respondents). Luce's choice axiom states that the probability of selecting one item over another from a pool of many items is not affected by the presence or absence of other items in the pool (IIA assumption). The axiom states that if A is preferred to B out of the choice set {A, B}, then introducing a third, irrelevant, alternative X (thus expanding the choice set to {A, B, X}) should not make B preferred to A. In other words, whether A or B is better should not be changed by the availability of X. The IIA axiom simplifies the experimental collection of choice data by allowing multinomial choice probabilities to be inferred from binomial choice experiments. It is clear that assumptions (i) and (iii) bear some relation to the invariance principle from measurement theory.

12.4.3   *Mathematics*

In conditional logistic regression, none, some, or all of the observations in a choice set may be marked. McFadden's choice model (DC) is thus a special case of MNL regression. There is much confusion about the differences and similarities between conditional and multinomial logit models. MNL models are different from conditional (multinomial) logistic regression models. Conditional logit models unfortunately also include a wide array of submodels that depend on whether certain effects of interest are generic or whether these effects differ for one or more of the choice alternatives. In the conditional logit (CL) model, the explanatory variables assume different values for each alternative and the impact of level changes is assumed to be constant across alternatives. The model may be summarized in Equation 12.4:

$$v_{rs} = z_{rs}\gamma \tag{12.4}$$

whereby v are latent values or utilities of individuals choosing health state s, z_{rs} indicates a vector of *alternative-specific* explanatory variables for individual r, and γ represents a single vector of unknown regression coefficients. Under the assumptions described above, the probability that health state s is chosen is equal to

$$P_{rs} = \frac{e^{(v_{rs})}}{\sum\limits_{k=1}^{K} e^{(v_{rk})}} \tag{12.5}$$

or

$$P_{rs} = \frac{e^{(z_{rs}\,\gamma)}}{\sum\limits_{k=1}^{K} e^{(z_{rk}\,\gamma)}} \tag{12.6}$$

where K (one k has to be set as reference) is the number of alternatives (e.g., health states) in the choice set (e.g., two in most DC applications) and s is the chosen alternative.

The term MNL model refers to a model that generalizes logistic regression by allowing more than two discrete outcomes. It assumes that data are *case-specific*; that is, each independent variable has a single value for each case. Consider an individual choosing among K alternatives in a choice set. Let x_r represent the characteristics of individual r and β_s the regression parameters

$$\tilde{v}_{rs} = x_r\beta_s \tag{12.7}$$

The probability that individual r chooses health states s is

$$P_{rs} = \frac{e^{x_r\beta_s}}{\sum\limits_{k=1}^{K} e^{x_r\beta_k}} \tag{12.8}$$

Both models can be used to analyze an individual's choice among a set of K alternatives. The main difference between the two is that the conventional MNL model focuses on the individual as the unit of analysis and takes the individual's characteristics as explanatory variables. The CL model, in contrast, focuses on the set of alternatives for each individual, while the explanatory variables are characteristics of those alternatives.

It is possible to combine these two models. Doing so would simultaneously take into account the characteristics of both the alternatives and the individual characteristics, using them as explanatory variables. This combination is sometimes called a conditional MNL or mixed model.

$$\tilde{\tilde{v}}_{rs} = x_r\beta_s + z_{rs}\gamma \tag{12.9}$$

where $\tilde{\tilde{v}}_{rs}$ is the value of the alternative s assigned by the individual r. This value $(\tilde{\tilde{v}}_{rs})$ depends on both the alternative characteristics x and on the individuals' characteristics z. The probability that individual r chooses health states s is

$$P_{rs} = \frac{e^{(x_r\beta_s + z_{rs}\,\gamma)}}{\sum\limits_{k=1}^{K} e^{(x_r\beta_s + z_{rs}\,\gamma)}} \tag{12.10}$$

The most commonly applied types of DC models are presented above. A clear distinction is made between models that take an individual's characteristics as explanatory variables (MNL) and models with explanatory variables for characteristics of alternatives (i.e., health states). In the next section, the Rasch model will be explained. It will be shown that this model has a close similarity to the CL model (Equations 12.5 and 12.6). As the basic data structure underlying the Rasch measurement model should meet the invariance assumption (see Section 12.3), this rules out incorporating elements of the MNL model (Equations 12.7 through 12.10).

12.5   Rasch model

12.5.1   Background

The Rasch model—named after the Danish mathematician, statistician, and psychometrician Georg Rasch (1901–1980)—is a probabilistic measurement model. While primarily employed in attainment assessment, it is increasingly used in other areas (Rasch 1980). Its original setting was the field of reading skills, where it was intended for dichotomous response data (e.g., right or wrong). The field of health outcomes research has shown considerable interest in the topic of Rasch modeling. Recently, attempts have been made to apply the Rasch model to specific HRQoL domains (e.g., pain, depression, mobility) (Revicki and Cella 1997; Ten Klooster et al. 2008).

Rasch did not start from real data but rather from an axiomatic definition of measurement. He formulated a "model," that is, an equation, fixing the "ideal" relationship between the observation and the amount of the latent trait (i.e., variables that are not directly observed but are inferred, such as utility). At least three features of this relationship should be highlighted. First, the observed response (e.g., pass/yes/agree/right = 1, rather than fail/no/disagree/wrong = 0) depends on the difference between only two parameters, the "ability" of the individual and the "difficulty" of the item. No extraneous factors should bias this linear relationship. Second, "ability" and "difficulty" are independent of each other. As stated before, this invariance principle is also a theoretical requirement for measurement in the realm of physics. In his "separability theorem," Rasch demonstrated that his model is the only one that satisfies this requirement. Third, the model is probabilistic: uncertainty surrounds the expected response, which is consistent with the real-world situation.

A key element of the Rasch model is that the goal is to construct procedures or operations that provide data that meet the relevant criteria (Andrich 2004). It should be noted that the Rasch model makes relatively strong assumptions. Nonetheless, if the assumptions hold sufficiently, this measurement model can produce scales (i.e., health-state values) offering a number of advantages over those derived from standard measurement techniques or even contemporary DC models.

Rasch developed the model for dichotomous data. He applied it to response data derived from intelligence and attainment tests, including data collected by the Danish military (Rasch 1966). It does not confront the respondents with a paired comparison task or a ranking task. Instead, the responses are collected separately (monadic measurement) for a set of items. Versions of the Rasch model are particularly common in psychometrics; the field concerned with the theory and technique of psychological and educational measurement, where they are known as response models. The most important claim of the Rasch model is that due to the mode of collecting response data, in combination with the conditional estimation procedure, the derived measures comply with the three important principles: interval level, unidimensionality, and invariance. Because it uses a specific mechanism (see explanation below), the application of the Rasch model is sometimes referred to as fundamental or objective measurement.

12.5.2 Measurement model

The Rasch model is a mathematical function that relates the probability of a (correct) response on an item to characteristics of the person (e.g., ability) and to characteristics of the item (e.g., difficulty). For quantifying health states, this model would relate the probability of a response to a health state to characteristics of an individual (e.g., own health status) and to characteristics of given health states (e.g., severity).

The data structure required by the Rasch model is identical to that of another response model, namely Guttmann scaling, which had been developed independently at an earlier stage (Guttman 1950). While the Guttmann model is deterministic, the Rasch model is probabilistic (Figure 12.1). To obtain the specific structure of the data for Rasch analysis, the respondents (their health status) must be distributed over the whole unidimensional scale. Thus, a sample clustered at only one location on the scale (e.g., all healthy people) is not conducive to good estimations of the model. Moreover, the Rasch model can be seen as a practical realization of conjoint measurement (axiomatic theory for quantification of multiple rank-based attributes) with an underlying stochastic structure (Brogden 1977).

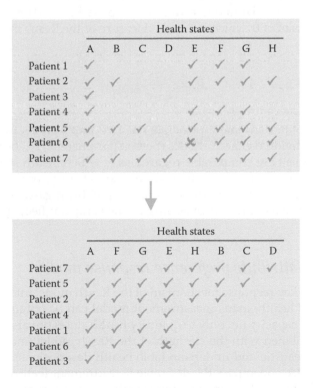

Figure 12.1 Schematic representation of the raw data and after sorting of the columns (health states) and the rows (patients) to arrive at the hierarchical Guttman/Rasch data structures (the check mark indicates that this health state is preferred over the next health state, the cross mark indicates a misfit).

For this and other reasons, many scientists consider the Rasch model as the pre-eminent means to measure subjective phenomena "objectively."

Extensions of the Rasch model have been developed independently and simultaneously; these are known as item response theory (IRT) models (Streiner and Norman 2008). The extensions differ from the original model in the sense that they have a parameter to express the discrimination of an item (the degree to which the item discriminates between persons in different regions on the latent continuum). These IRT models relax to some extent the strict requirements for responses (e.g., data) posed by the Rasch model. However, IRT models do not possess the specific fundamental measurement property of the Rasch model and, therefore, do not necessarily produce cardinal measures (Rasch 1980, Gandjour 2010).

12.5.3   Mathematics

In the Rasch model for dichotomous data, the probability that the outcome is correct (or that one health state is better than another) is given by

$$P_{rs} = \frac{e^{(\theta_r - v_s)}}{1 + e^{(\theta_r - v_s)}} \tag{12.11}$$

where θ_r identifies the health status θ of the person and v_s refers to the state s. By an interactive conditional maximum likelihood estimation approach, an estimate $v_s - v_t$ is obtained without the involvement of θ, which is a special feature of the Rasch model. Equation 12.11 can be rewritten as

$$P_{rs} = \frac{1}{1 + e^{-(\theta_r - v_s)}} \tag{12.12}$$

The invariance of measurement principle has two implications for the Rasch model. First, estimates of individual characteristics (person parameter θ; i.e., health status) as measured by the instrument are comparable regardless of which health states are included in the instrument. Second, estimates of the position (i.e., severity) of the health states (item parameter v) on the scale of the instrument are comparable regardless of the selection of the sample of respondents; this is true as long as the sample reflects the broad spectrum of the scale.

12.6   The multiattribute preference response model

By incorporating the key response mechanism of the Rasch model into the DC framework, a new and advanced health-status measurement model can be obtained. The strength of the DC models (their capacity not only to quantify health states but also to estimate a value function) can be combined with the strength of the Rasch model (individual patients are given responses to realistic and understandable health descriptions). In principle, such a new model should also encompass the desirable measurement features of the fundamental Rasch model (Figure 12.2). Moreover, the specific response mode of the Rasch model (patient's own health state versus other related health states) will largely prevent any adaptation effects. This combination of features from the DC and the Rasch models is referred to as the MAPR model.

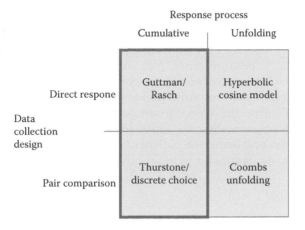

Figure 12.2 Schematic representation of the different data collection designs in combination with the specific response process of these designs and the appropriate measurement models for these four combinations (combination of DC model and Rasch model, block bounded by thick line is MAPR model).

In fact, the Rasch model is closely related to most DC models and their extensions. What makes the Rasch model unique is the person parameter (θ) (Andrich 1978, Jansen 1984). When dealing only with choice sets consisting of two health states (*s* and *t*), the left-hand side of Equation 12.5 can be expressed even more succinctly. Dividing the numerator and denominator by the numerator, and recognizing that $e^a/e^b = e^{(a-b)}$, Equation 12.5 becomes

$$P_{ts} = \frac{1}{1 + e^{-(v_s - v_t)}}.$$ (12.13)

In this equation, it is obvious that the basic formulas for the DC model and the Rasch model differ in only one parameter. The DC model requires two parameters, v_s and v_t. In the Rasch model, one of these parameters is "replaced" by a parameter θ that represents the location of the respondent (Equation 12.12). Equation 12.13 can, therefore, be rewritten as Equation 12.14. The latter is the basic formula for the MAPR model, created by adding a parameter, γ, for the attributes of the health states.

$$P_{rs} = \frac{1}{1 + e^{(\theta_r - z_{rs}\gamma)}}$$ (12.14)

In this model, a set of linear restrictions is imposed on the health-state parameters (*v*) of the Rasch model as $v_{rs} = z_{rs}\gamma$. In another setting and through a different approach, Fischer developed a model that has a close connection with the MAPR model (Fisher 1983, Boeck de and Wilson 2004).

12.6.1 Data collection

The Rasch model demands a specific data structure that is essentially different from that of the DC model. The implication is that data have to be derived from new and innovative response tasks. Judgments are required from a heterogeneous sample of people in various health conditions. This means that respondents should not be a representative sample of the general population. Instead, they should be patients who are currently experiencing one of the health states on the continuum from worst to best health status (Figure 12.3).

For the MAPR model, people respond to hypothetical health states by comparing these health states with their own health condition. For example: "Is this health state better than your own health state?" The conditional (multinomial) logit model then becomes similar to the Rasch model. This occurs when the following criteria are met: each comparison consists of two health states, one being the patient's own state. The patient's own health state is considered as a separate parameter in the conditional estimation procedure.

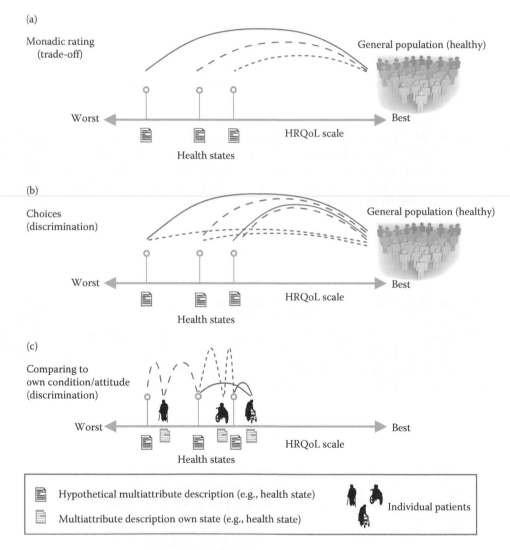

Figure 12.3 Schematic representation of the judgmental task for three health states by (a) conventional monadic measurement (SG and TTO) by a sample of the general population; (b) conventional DC task (paired comparison) by a sample of the general population; and (c) multiattribute preference response model for individual patients (three patients in this example).

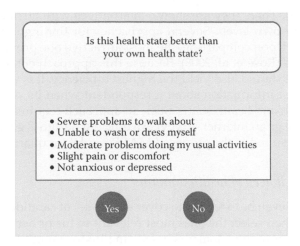

Figure 12.4 An example of a response task under the MAPR model (based on EQ-5D description).

The central mechanism in the MAPR model is that the response task is performed in two distinct stages. First, the classification of the individual's health status according to the set of health attributes generates a value for the description of that state (based on the underlying value function). In the next step (Figure 12.4)—to which not all respondents have to continue if the value function has reached a certain stage of predictive precision—individuals are confronted with a set of health states that cover the range from severe to mild or a set of other almost "equivalent" states (see Section 12.6.2) that are compared to their own state that is determined at the first stage (such a comparison may be easier than under DC). The information generated in this part is used to arrive at a more precise value function. Of course, this iterative mechanism only operates properly after a large number of individuals have gone through both stages. Therefore, there is clearly an initiation stage in which patients are performing the judgmental tasks to feed the statistical part of the MAPR model. At this stage, the value of the health condition of the patients themselves cannot be (precisely) estimated. It will take probably 1000 or more patients to conduct the inception of the MAPR model and arrive at a functional routine. A relatively large number of respondents is required as the MAPR model has to estimate health-state parameters and patient parameters, all based on binary data.

As a first exploration of the MAPR model, we may start with the most basic variant in which an existing health-status classification system is used (e.g., EQ-5D). First, patients (representing the whole continuum from bad to mild health conditions) classify their own health condition based on the EQ-5D classification. Then, they judge a fixed set (say 20) of EQ-5D health states (representing the whole continuum from bad to mild states).

12.6.2   Pivot designs

In standard DC studies, all respondents face the same choice situations (e.g., pair of health states) or a selection of them. While this is also possible for the MAPR model, it is not an efficient approach. Rather than presenting patients with a predetermined set of health states, it would be better to frame the choice task within an existing decision context, namely a situation pertaining to a person. This strategy makes use of pivot designs. First, the respondents classify their health status according to a standardized system (e.g., EQ-5D and HUI). Then, they are shown alternatives with attribute levels that deviate slightly from their own levels. Several approaches for finding efficient pivot designs have been developed. Upon comparison, the most attractive one proves to be based on the individual's responses (Rose et al. 2008). Because this approach entails a separate design for each respondent, it should also yield the highest efficiency. It is well known that a test item provides the most information about a respondent when its severity is roughly the same as that of the person's health status. Such an approach requires a computer adaptive testing environment (e.g., an internet survey). An efficient design is generated interactively, and is, therefore, well suited to the proposed MAPR model for quantifying health states.

12.6.3   Individual choices of health domains

The MAPR model can even be extended to cover a large set of candidate domains. An individual patient could then select the ones most relevant to his or her assessment (far more than the traditional four to nine domains in existing instruments). Recently, specific solutions (partial profile choice designs) for this situation have been introduced (Chrzan 2010). In such an extended variant of the MAPR, the precision of the underlying value function increases systematically, step by step. The reason is that individuals then use more and more attributes when comparing their own health status with other health states. Of course, the overall estimation and convergence of such a MAPR model requires substantial input from the patients. At present, many aspects of these partial profile designs remain to be investigated in more detail, in particular in the setting of health-state valuation—a challenging task. For example, one of the elementary assumptions in applying this type of model is that all the attributes (health domains) that are not part of individual judgmental tasks are to be set to "0" (no value). Studies in other areas have shown that this assumption does not always hold (Islam et al. 2007).

12.7   Discussion

This chapter presents a new measurement approach, the multiattribute preference response model (MAPR), to quantify subjective elements of health. The MAPR model can be considered as an adaptation of the standard DC model or as an extension of the standard Rasch model. This chapter also shows that seemingly different models are often very similar. Conventional valuation techniques are known to have many problems, such as adaptation, discounting, context, reference point, and other effects (Nord et al. 2010). With the MAPR model, unwanted mechanisms affecting valuations of health states may be largely eliminated.

From a theoretical and practical point of view, both the MAPR and the DC models are more attractive than TTO and other conventional valuation techniques. In the MAPR and DC models, the judgmental task and the analysis are executed within one unifying framework; this is different for the TTO, where the valuation technique and the statistical model to estimate a value function are distinct (e.g., using different regression techniques and different coding of the parameters). The strength of the DC and MAPR models is that the derived values relate to one particular aspect: the attractiveness of a health state. Such measures are not confounded by aspects such as time preference or risk aversion. Furthermore, discrimination is a basic operation of the judgment, and a means of generating knowledge. The core activity of measurement in probabilistic choice models is to compare two or more entities in such a way that the data yields compelling information on individuals' preferences, thereby imitating choices in daily life.

The Rasch model occupies a special position in the field of subjective measurement, although mathematically it is closely related to IRT. Broadly speaking there are two general schools of thought, each known for its particular approach. When the response data satisfy the conditions required by these mathematical models, the estimates derived from the models are considered robust. When the data do not fit the chosen model, two lines of enquiry are possible. In essence, when the data do not fit a given model, the IRT approach is to find a mathematical model that best fits the observed item-response data. By contrast, the Rasch measurement approach is to find data that better fit the Rasch model. Thus, it follows that proponents of IRT use a family of item-response models while proponents of Rasch measurement use only one model (Rasch model).

A major limitation of the MAPR model, as with any probabilistic choice model, is that it produces relative positions of all health states on the latent scale (Engelhard 1992, Louviere and Lancsar 2009). For the estimation of DALYs and QALYs, however, those values need to be on the "dead"–"full-health" scale. If the MAPR model is used to value health, a way must be found to link the position of "dead" with the derived values. A similar solution may also be relevant to locate the position of the best (dominant) health state of a multiattribute classification system to "full health." A strategy for rescaling DC values may be to anchor them on values obtained for the best health state and for "dead" using other valuation techniques, such as TTO or SG. However, the rationale for this approach is unclear. Part of the motivation for adopting probabilistic choice models as potential candidates to produce health-state values lays in the limitations of existing valuation techniques. Alternatively, the judgmental format may be set up in such a way that the derived health-state values can be related to the value of the state "dead." A simple manner to achieve this seems to involve making designs in which respondents are presented with at least one bad health state at a time and asked if they consider it better or worse than being dead. The procedure has been demonstrated by McCabe et al. (2006) and Salomon (2003). These authors mixed the state "dead" with the choice set as a health state so that a parameter for the state "dead" is estimated as part of the model. To investigate whether or not this strategy would produce health-state values with acceptably low bias, it will be necessary to draw comparisons with values obtained using alternative measurement methods. A problem associated with including "dead" as part of the choice set is that proper estimation of values seems only possible if almost all respondents consider some health states to be worse than dead. Otherwise, the estimated parameters of the model are likely to be biased (Flynn et al. 2008). Another approach could employ models that are suitable for dealing with dominant health states (the best health states in a multiattribute system) to calibrate the metric distances in this region (Craig et al. 2009). One example would be to apply multidimensional scaling (Krabbe et al. 2007). In sum, it is hard to say beforehand which approach to deriving DALY/QALY values that are anchored to "dead" would produce valuable and effective results for the MAPR model (DC models have the same problems). Theoretically, competent construction of MAPR models and subsequent experimentation with these models would, therefore, be required to see how these difficulties could be resolved in a particular situation.

One of the crucial elements that has to be decided before any measurement of health status can take place is the selection of the health domains to conceptualize health-state descriptions. An overwhelming number of health status or HRQoL instruments are available. Each has been developed with a particular concept in mind, resulting in instruments with a specific depth (basic or subordinate units of information; e.g., physical function, self-care, and bathing) and breadth (e.g., physical function, emotional function, pain, and cognitive ability). However, all current descriptive and preference-based health-status instruments are based on a predetermined restricted set of health domains. This common denominator has certain drawbacks. For example, for some patients or diseases, the predetermined set may miss some crucial health domains. Extending it to create a very broad set is not feasible for several reasons, both practical and analytical. In light of information theory, it is clear that a limited set of domains may deliver enough detailed information to adequately describe the health status of a person if these domains are well selected and reflect the domains most relevant to this specific person.

Therefore, an advantageous property of the MAPR model is that it can overcome the limitation of the restricted set of health domains, which is a drawback of the existing preference-based health-status systems (e.g., HUI and EQ-5D). In many instances, these instruments prove to be insensitive or invalid, due to their restricted and fixed set of health attributes. The MAPR model, in contrast, can be extended to include a large set (40–120) of candidate domains from which an individual patient can select a few (5–7), namely the most relevant ones. A large number of individual responses would be needed to collect enough data. As there is no shortage of patients, the estimation part of this model, from a practical point of view, is not too challenging (Chrzan 2010). Of course, the statistical routines for such an approach have to be developed, and ultimately, empirical research must prove the premise of such an extended MAPR model. An instrument that shows some resemblance to the measurement strategy of the MAPR model in which the content is derived from individual patients is the SEIQoL (Hickey et al. 1996). Although it is not embedded in a formal measurement model, this instrument permits individuals to select and value the health domains that are important to them in their HRQoL assessment. The EuroQol Group is planning to develop special "bolt-ons" (comprising additional health attributes) that will be added to the existing five health attributes of the EuroQol-5D system (Lamers et al. 2012). Yet, the analytical integration of these bolt-ons does not seem to be part of an overarching framework. The EuroQol Group has indicated that in the end the maximum number of predetermined attributes will be seven instead of five.

Existing multiattribute preference-based health-status systems are conceptualized as having two stages: one encompasses the valuation study to derive the value function; the second comprises an application stage, in which the health status of individuals is determined. One of the major strengths of the MAPR model is that it is a continuous process of valuation and application.

Patients may be regarded as the best judges of their own health status. Therefore, it is sensible to defend the position that it is the patient's judgment that should be elicited. However, it may be the case that values for health states worked out under the MAPR model both for patients and a sample of unaffected members of the general population are rather similar (Krabbe et al. 2011). Otherwise, empirical MAPR head-to-head studies may reveal that responses from the general population are evenly valid, except maybe for the worst health states.

The MAPR model may also be an avenue for developing a health-status measurement instrument that can be used in broader settings (e.g., medical interventions and medical care) and in distinctive patient populations (e.g., children, adults, and elderly). Another possible extension of the MAPR model would be to combine it with Monte Carlo simulation and then to use this technology in estimating response models within a full Bayesian framework (Albert 1992, Fox 2010). In principle, the MAPR model may be suitable to measure other unidimensional, subjective phenomena such as well-being, capabilities, and happiness; under certain conditions, it might be used in social value judgments (e.g., reimbursement decisions). For the overall quantification of quality, in general, the MAPR model may also be beneficial (Porter 2010).

12.8 Conclusions

Incorporating the basic elements of the Rasch model into the DC framework (or vice versa) produces an advanced model with fundamental measurement characteristics: the multiattribute preference response (MAPR) model. This new patient-reported outcome measurement model is more coherent than the conventional valuation methods and has a profound connection to measurement theories. The MAPR model can be applied to a wide range of research problems. Specifically, if extended with self-selection of relevant health attributes for the individual patient, this model will be more valid than existing valuation techniques.

Acknowledgments

I would like to thank David Andrich for his helpful comments during our discussions in Copenhagen (Rasch Conference, 13–16 June 2010) and London (International Society for Quality of Life Research, 27–30 October 2010). I would also like to express my thanks to Jeroen Weesie (University of Utrecht, The Netherlands) for his contribution to the Stata website (http://www.stata.com/support/faqs/stat/rasch.html), as it was his clear explanation of Rasch analysis that sparked many thoughts presented in this chapter. Karin Groothuis-Oudshoorn and Edwin van den Heuvel assisted me with the mathematical notations. This chapter was originally published in Public Library of Science One, 2013; 8(11): 1–12.

References

Albert J.H. Bayesian estimation of normal ogive item response curves using Gibbs sampling. *Journal of Educational Statistics* 1992; 17: 251–269.

Andrich D. Relationships between the Thurstone and Rasch approaches to item scaling. *Applied Psychological Measurement* 1978; 2(3): 449–460.

Andrich D. Controversy and the Rasch model: A characteristic of incompatible paradigms? *Medical Care* 2004; 42(1 Suppl): I7–I16.

Arons A.M.M., Krabbe P.F.M. Probabilistic choice models in health-state valuation research: Background, theory, assumptions and relationships. *Expert Review of Pharmacoeconomics and Outcomes Research* 2013; 13(1): 93–108.

Attema A.E., Versteegh M.M., Oppe M., Brouwer W.B.F., Stolk E.A. Lead time TTO: Leading to better health state valuations? *Health Economics* 2013; 22(4): 376–392.

Bansback N., Brazier J., Tsuchiyaa A., Anis A. Using a discrete choice experiment to estimate health state utility values. *Journal of Health Economics* 2012; 31(1): 306–318.

Bleichrodt H. A new explanation for the difference between time trade-off utilities and standard gamble utilities. *Health Economics* 2002; 11: 447–456.

Bradley R.A., Terry M.E. Rank analysis of incomplete block designs: I. The method of paired comparisons. *Biometrika* 1955; 39: 324–345.

Brazier J.E., Dixon S., Ratcliffe J. The role of patient preferences in cost-effectiveness analysis: A conflict of values? *Pharmacoeconomics* 2009; 27(9): 705–712.

Brogden H.E. The Rasch model, the law of comparative judgment and additive conjoint measurement. *Psychometrika* 1977; 42(4): 631–634.

Chrzan K. Using partial profile choice experiments to handle large numbers of attributes. *International Journal of Market Research* 2010; 52(6): 827–840.

Craig B.M., Busschbach J.J., Salomon J.A. Modeling ranking, time trade-off, and visual analog scale values for EQ-5D health states: A review and comparison of methods. *Medical Care* 2009; 47(6): 634–641.

Boeck de P., Wilson M. (Eds). *Explanatory Item Response Models: A Generalized Linear and Nonlinear Approach.* New York: Springer, 2004.

Devlin N., Appleby J. *Getting the Most Out of PROMS: Putting Health Outcomes at the Heart of NHS Decision-Making.* London: Kings Fund/Office of Health Economics, 2010.

Dolan P., Kahneman D. Interpretations of utility and their implications for the valuation of health. *The Economic Journal* 2008; 118: 215–234.

Drummond M.F., Sculpher M.J., Torrance G.W. et al. *Methods for the Economic Evaluation of Health Care Programmes* (3rd edn.) Oxford: Oxford University Press, 2005.

Engelhard G. Historical views of invariance: Evidence from the measurement theories of Thorndike, Thurstone, and Rasch. *Educational and Psychological Measurement* 1992; 2: 275–291.

Fanshel S., Bush J. A health-status index and its application to health-services outcomes. *Operations Research* 1970; 18(6): 1021–1066.

Fisher G.H. Logistic latent trait models with linear constraints. *Psychometrika* 1983; 48: 3–26.

Fox J-P. *Bayesian Item Response Modeling: Theory and Applications.* New York: Springer, 2010.

Flynn T.N., Louviere J.J., Marley A.A.J., Coast J., Peters T.J. Rescaling quality of life values from discrete choice experiments for use as QALYs: A cautionary tale. *Population Health Metrics* 2008; 6: 1–6.

Gafni A., Torrance GW. Risk attitude and time preference in health. *Management Science* 1984; 30: 440–451.

Gandjour A. Theoretical foundation of patient v. population preferences in calculating QALYs. *Medical Decision Making* 2010; 30: E57.

Gold M.R., Siegel J.E., Russel L.B., Weinstein M.C. *Cost-Effectiveness in Health and Medicine.* New York: Oxford University Press, 1996.

Green C., Brazier J., Deverill M. Valuing health-related quality of life. A review of health state valuation techniques. *Pharmacoeconomics* 2000; 17: 151–165.

Guttman L. The basis for scalogram analysis. In: Stouffer S.A., Guttman L., Suchman E.A., Lazarsfeld P.F., Star S.A., Clausen J.L. (Eds), *The American Soldier: Measurement and Prediction* (Vol. IV, pp. 60–90). New York: Princeton University Press, 1950.

Hakim Z., Pathak D.S. Modelling the Euroqol data: A comparison of discrete choice conjoint and conditional preference modelling. *Health Economics* 1999; 8: 103–116.

Hickey A.M., Bury G., O'Boyle C.A., Bradley F., O'Kelly F.D., Shannon W. A new short form individual quality of life measure (SEIQoL-DW): Application in a cohort of individuals with HIV/AIDS. *BMJ* 1996; 313: 29–33.

Islam T., Louviere J.J., Burke P. Modeling the effects of including/excluding attributes in choice experiments on systematic and random components. *International Journal of Research in Marketing* 2007; 24(4): 289–300.

Jansen P.G.W. Relationships between the Thurstone, Coombs, and Rasch approaches to item scaling. *Applied Psychological Measurement* 1984; 8: 373–383.

Johannesson M., Pliskin J.S., Weinstein M.C. A note on QALYs, time tradeoff, and discounting. *Medical Decision Making* 1994; 14: 188–193.

Kahneman D., Sugden R. Experienced utility as a standard of policy evaluation. *Environmental and Recourse Economics* 2005; 32: 161–181.

Kind P. A comparison of two models for scaling health indicators. *International Journal of Epidemiology* 1982; 11: 271–275.

Krabbe P.F.M. Valuation structures of health states revealed with singular value decomposition. *Medical Decision Making* 2006; 26: 30–37.

Krabbe P.F.M. Thurstone scaling as a measurement method to quantify subjective health outcomes. *Medical Care* 2008; 46: 357–365.

Krabbe P.F.M., Salomon J.A., Murray C.J.L. Quantification of health states with rank-based nonmetric multidimensional scaling. *Medical Decision Making* 2007; 27(4): 395–405.

Krabbe P.F.M., Stalmeier P.F.M., Lamers L.M., Busschbach van J.J. Testing the interval-level measurement property of multi-item visual analogue scales. *Quality of Life Research* 2006; 15(10): 1651–1661.

Krabbe P.F.M., Tromp N. Ruers T.J.M., Riel van P.L.C.M. Are patients' judgments of health status really different from the general population? *Health and Quality of Life Outcomes* 2011; 9: 31.

Lamers L.M., Gudex C., Pickard S., Rabin R. The 28th scientific plenary meeting of the EuroQol Group: Improving EQ-5D instruments and exploring new valuation techniques. *ISPOR Connections* 2012; Jan/Feb: 10–11.

Lancaster K.J. A new approach to consumer theory. *Journal of Political Economy* 1996; 74: 132–157.

Louviere J.J., Lancsar E. Choice experiments in health: The good, the bad, the ugly and toward a brighter future. *Health Economics Policy and Law* 2009; 4: 527–546.

Louviere L.L., Woodworth G. Design and analysis of simulated consumer choice or allocation experiments: An approach based on aggregate data. *Journal of Marketing Research* 1983; 20(4): 350–367.

Luce R.D. *Individual Choice Behavior: A Theoretical Analysis*. New York: Wiley, 1959.

Marcum J.L. *A Statistical Theory of Target Detection by Pulsed Radar*. Rand Corporation: RM-754, 1947.

Marschak J. Binary-choice constraints and random utility indicators. In: Arrow K.J., Karlin S, Suppes P. (Eds), *Mathematical Methods in the Social Sciences*, pp. 15–27. Stanford: Stanford University Press, 1960.

McCabe C., Brazier J., Gilks P., Tsuchiya A., Roberts J., O'Hagan A., Stevens K. Using rank data to estimate health state utility models. *Journal of Health Economics* 2006; 25: 418–431.

McFadden D. Conditional logit analysis of qualitative choice behavior. In: Zarembka P. (Ed.), *Frontiers in Econometrics*, pp. 52–73. New York: Academic Press, 1974.

McFadden D. Economic choices. *The American Economic Review* 2001; 91: 351–378.

McKenzie L., Cairns J., Osman L. Symptom-based outcome measures for asthma: The use of discrete choice methods to assess patient preferences. *Health Policy* 2001; 57: 193–204.

Murray C.J.L. Rethinking DALYs. In: Murray C.J.L., Lopez A.D. (Eds), *The Global Burden of Disease: A Comprehensive Assessment of Mortality and Disability from Diseases, Injuries, and Risk Factors in 1990 and Projected to 2020*, pp. 1–98. Cambridge: Harvard University Press, 1996.

Murray C.J.L., Lopez A.D. Evidence-based health policy–lessons from the Global Burden of Disease Study. *Science* 1996; 274(5288): 740–743.

Neumann von J., Morgenstern O. *Theory of Games and Economic Behavior*. Princeton: Princeton University Press, 1953.

Nord E. Methods for quality adjustment of life years. *Social Science and Medicine* 1992; 34(5): 559–569.

Nord E., Undrum Enge A., Gundersen V. QALYs: Is the value of treatment proportional to the size of the health gain? *Health Economics* 2010; 19: 596–607.

Oppe M., Devlin N.J., Hout van B., Krabbe P.F.M., De Charro F. A program of methodological research to arrive at the new international EQ-5D-5L valuation protocol. *Value in Health* 2014; 17(4): 445–453.

Patrick D.L., Bush J.W., Chen M.M. Methods for measuring levels of well-being for a health status index. *Health Services Research* 1973; 8(3): 228–245.

Porter M.E. What is value in health care? *The New England Journal of Medicine* 2010; 363(26): 2477–2481.

Rasch, G. An item analysis which takes individual differences into account. *British Journal of Mathematical and Statistical Psychology* 1966; 19: 49–57.

Rasch G. *Probabilistic Models for Some Intelligence and Attainment Tests*. Expanded edition with foreword and afterword by B.D. Wright. Chicago: University of Chicago Press, 1980.

Ratcliffe J., Brazier J., Tsuchiya A. et al. Using DCE and ranking data to estimate cardinal values for health states for deriving a preference-based single index from the sexual quality of life questionnaire. *Health Economics* 2009; 18: 1261–1276.

Revicki D.A., Cella D.F. Health status assessment for the twenty-first century: Item response theory, item banking and computer adaptive testing. *Quality of Life Research* 1997; 6: 595–600.

Richardson J. Cost utility analysis: What should be measured? *Social Science and Medicine* 1994; 39: 7–21.

Rose J.M., Bliemer M.C.J., Hensher D.A., Collins A.T. Designing efficient stated choice experiments in the presence of reference alternatives. *Transportation Research Part B: Methodological* 2008; 42(4): 395–406.

Salomon J.A. Reconsidering the use of rankings in the valuation of health states: A model for estimating cardinal values from ordinal data. *Population Health Metrics* 2003; 1: 12.

Salomon J.A., Vos T., Hogan D.R. et al. Common values in assessing health outcomes from disease and injury: Disability weights measurement study for the Global Burden of Disease Study 2010. *Lancet* 2012; 380: 2129–2143.

Stolk E., Oppe M., Scalone L., Krabbe P.F.M. Discrete choice modeling for the quantification of health states: The case of the EQ-5D. *Value in Health* 2010; 13: 1005–1013.

Strauss D. Some results on random utility models. *Journal of Mathematical Psychology* 1979; 20: 35–52.

Streiner D.L., Norman G.R. Item response theory. In: Norman, G. R. (Ed), *Health Measurement Scales: A Practical Guide to Their Development and Use* (4th edn.), pp. 77–95. New York: Oxford University Press, 2008.

Ten Klooster P.M., Taal E., Laar van de M.A. Rasch analysis of the Dutch Health Assessment Questionnaire disability index and the Health Assessment Questionnaire II in patients with rheumatoid arthritis. *Arthritis Rheumatology* 2008; 59(12): 1721–1728.

Thurstone L.L. A law of comparative judgments. *Psychological Review* 1927; 34: 273–286.

Torrance G.W. Social preferences for health states: An empirical evaluation of three measurement techniques. *Socio-Economic Planning Sciences* 1976; 10: 129–136.

WHO, Preamble to the Constitution of the World Health Organization as adopted by the International Health Conference, New York, 19–22 June 1946; signed on 22 July 1946 by the representatives of 61 States (Official Records of the World Health Organization, no. 2, p. 100) and entered into force on 7 April 1948, 1946.

chapter thirteen

Evolution of large-scale dimensional metrology from the viewpoint of scientific articles and patents

Fiorenzo Franceschini, Domenico Maisano, and Paola Pedone

Contents

13.1 Introduction

Over the last 20 years there has been a rapid growth of large-scale dimensional metrology (LSDM) instruments for measuring medium- to large-sized objects (i.e., *objects with linear dimensions ranging from tens to hundreds of meters* (Puttock, 1978)), where accuracy levels of a few tenths of a millimeter are generally tolerated (Franceschini et al., 2009, 2011). These systems usually support the assembly phase and/or dimensional compliance test on large volume objects, *in situ*. Table 13.1 sketches the history of the major LSDM instruments with a brief description of their characteristics. For more information on LSDM instruments and their technologies (e.g., technical features, usual applications, metrological performance, cost, etc.), we refer the reader to the extensive reviews (Estler et al., 2002; Peggs et al., 2009; Muelaner et al., 2010; Franceschini et al., 2011).

The aim of this chapter is a wide-range analysis of the existing scientific literature, from the dual perspective of (i) *scientific publications*, which reflect the interest of the (academic) scientific community and (ii) *patents*, which reflect the well-established (at industrial level) technologies or the emerging ones (Jun and Lee, 2012). The analysis tries to delineate the technological and scientific evolution of LSDM systems in the last 20–30

Table 13.1 List of some of the major LSDM instruments, with a brief description of their features

Instrument	Introduction time	Short description
Tapes/Sticks	Thousands of years ago	Reference artifacts used for measuring the length of an object directly.
Theodolite	≈1700	Consists of a movable telescope mounted within two perpendicular axes—a horizontal and a vertical one. When the telescope is pointed at a target point of interest, the angles with respect to each of these axes can be measured with great precision.
Contact-probe Coordinate Measuring Machine (CMM)	≈1960s	Used for measuring the geometrical characteristics of an object. It is composed of three moving axes, which are orthogonal to each other in a typical three-dimensional coordinate system. Measurements are defined by a mechanical contact-probe attached to the third moving axis.
Total station	Early 1970s	A sort of theodolite integrated with an Electronic Distance Meter (EDM) to read angles and distances from the instrument to a point of interest. System set-up is partially automated, and when measuring, there is no need for an assistant staff member as the operator may control the total station from the observed point.
3D scanners based on structured light, laser or photogrammetry	≈1990s	Instruments measure the three-dimensional shape of an object by processing images—taken by two or more cameras—of (i) encoded light patterns, (ii) laser beams, or (iii) targets projected on the object surface. Measurements are characterized by a high level of automation.
Optical-probe CMM	≈1990s	A classical CMM equipped with a contactless optical-probe, which allows the localization of a high density cloud of points on the object surface. Measurements are less accurate but considerably faster than those carried out with a contact-probe.
Laser tracker, tracer, and radar	≈1990s and early 2000s	Instruments allow the accurate localization of a target on the surface of the object of interest, by measuring mutual angles and distances. Distance measurements are generally based on the laser-interferometric and/or Absolute Distance Meter (ADM) technology.
Various distributed systems	≈2000s	Unlike the previous instruments, which can be classified as "single-station," these consist of a network of devices distributed around the measurement volume. The spatial coordinates of targets, in contact with the object of interest, are determined by two localization techniques: (i) multilateration (based on distances between targets and network devices) and (ii) triangulation (based on angles between targets and network devices). Distributed systems may rely on different technologies (e.g., ultrasonic, photogrammetric, optical, etc.).

Note: Instruments are sorted chronologically according to the approximate time of their introduction.

years by comparing these two types of outcomes and highlighting similarities and—more interestingly—differences and counter-tendencies. We emphasize that the academic literature—which generally consists of journal articles, conference proceedings, and book chapters and monographs—and the patent literature are not necessarily dependent on each other, as also evidenced by the relatively low incidence of patent citations in scientific publications and vice versa (Glänzel and Meyer, 2003; Seeber, 2007). This is the key reason for which combining these two analysis perspectives may lead to interesting results.

In the literature, there are many cases in which scientific publications and patents are investigated in conjunction; for instance, the contributions by Van Looy et al. (2006), Czarnitzki et al. (2007), Breschi and Catalini (2010), Franceschini et al. (2012), and many others. However, this type of analysis is a novelty in the LSDM field.

Our study focuses on the scientific publications and patents in the LSDM field and the citations that they obtained, from other publications and patents, respectively. While the citations that a scientific publication obtains from other publications depict its impact/diffusion within scientific community (De Bellis, 2009), the citations that a patent obtains from other patents can be indicative of its technological importance or even (potential) market value and profitability (Cheng et al., 2010). Furthermore, this study makes it possible to identify the LSDM articles and patents of greater impact, the major scientific journals and conference proceedings, and the major patent assignees.

The rest of the chapter is structured in three sections. Section 13.2 focuses on data collection and data cleaning aimed at selecting a portfolio of documents as comprehensive as possible and of the state of the art on LSDM. Section 13.3 briefly describes the methodology of analysis of the selected portfolio and presents the analysis results in detail. Analysis can be split into three parts: (i) evaluation of the dominant technologies, (ii) study of the evolution of LSDM systems in the last 20–30 years from the double perspective of scientific publications and patents, and (iii) identification of the main scientific journals, conference proceedings, and patent assignees in this field. The final section summarizes the results of the study highlighting the original contributions, implications, and limitations. An extended version of this chapter was published in the *International Journal of Advanced Manufacturing Technology* (Franceschini and Maisano, 2014).

13.2   Data collection

The objective of data collection is to identify a comprehensive set of scientific articles and patents from LSDM literature. Articles were collected through the Scopus database (Scopus-Elsevier, 2013). We chose this database for two reasons: (i) in the field of Engineering Science, Scopus' coverage is superior to that of Web of Science (Bar-Ilan, 2010) and (ii) Scopus is much more accurate than Google Scholar database (Jacsó, 2008). A limitation of Scopus is that it does not index books or book chapters but only articles from leading journals and conference proceedings, which, however, contain the vast majority of the relevant LSDM publications. Patents were collected using the Questel-Orbit database, which integrates patent statistics from more than 95 national and international patent authorities (Questel-Orbit, 2013), for example, EPO (European Patent Office), USPTO (United States Patent and Trademark Office), WIPO (World Intellectual Property Organization), JPO (Japan Patent Office), CPO (China Patent & Trademark Office), etc. For both articles and patents, databases were queried on March 8, 2013 with the following string: «("large scale*metrology" OR "large scale*measur*" OR "large volume*metrology" OR "large volume*measur*") OR {("large scale" OR "large volume") AND ["laser tracker" OR "photogrammetr*" OR

"CMM" OR "coordinate measur*" OR "structured light scan*"]}», searching into title, abstract or keywords of each document. Please note that "*" is a wildcard character, while "AND" and "OR" are basic Boolean operators. The search string was deliberately general, so as to reduce so-called "false negatives"—that is, documents that deal with LSDM but are ignored by databases because they do not meet the search query. The price to pay to reduce "false negatives" is identifying a large number of "false positives"—that is, documents that do not concern LSDM but meet the search query (Yu et al., 2004). To reduce "false positives," the documents returned by databases were cleaned manually; after examining title, abstract, and contents, we excluded those documents not concerning LSDM and the uncertain ones—for example, some unclear articles/patents in Russian or Chinese language only.

Regarding articles, Scopus returned 555 articles, which were reduced to 180 after data cleaning. Regarding patents, Questel-Orbit returned 334 patents, which were reduced to 53 after data cleaning. For duplicated records, that is, patents filed with multiple authorities, we considered the ones with the oldest "priority date."

One way to further reduce "false negatives" could be to expand the portfolio of cleaned documents by including their "neighbours," that is, those documents that cite or are cited by them, and then cleaning these additional documents manually. Of course, this operation would be significantly time-consuming.

During the manual cleaning activity, documents were also classified based on two criteria: (i) *dominant technology* and (ii) *typology of output*. The dominant-technology classification is illustrated in Table 13.2. While not claiming to be exhaustive, it includes the main technologies of the vast majority of LSDM systems in agreement with the literature review by Peggs et al. (2009).

Documents concerning minor technologies (e.g., moiré patterns or capacitive probes) or independent of a specific technology (e.g., documents about general measurement procedures or literature reviews) fall into the class *h-N/A*.

Table 13.3 illustrates the typology-of-output classification. Likewise the classification in Table 13.2, this other classification adapts to both articles and patents. It is reasonable to expect that most of the patents will fall into the first three classes (*A-Description of the measuring system*, *B-New hardware component(s)*, and *C-New application/working procedure*), while scientific articles will be distributed more uniformly among the seven classes.

13.3   Data analysis

13.3.1   Overall data analysis

Analyzing the two portfolios of articles and patents, it is possible to identify the most popular technologies both from the point of view of the total number of documents (P) and that of the total number of citations obtained (C). P is a proxy for the amount of research in the scientific literature while C reflects its impact on the scientific community or industry. For each article, we considered the citations obtained from other articles while for each patent, those obtained from other patents up to the moment of data collection (March 8, 2013). In the following subsections, articles and patents will be analyzed separately.

13.3.1.1   Dominant-technology classification

From Chart 1 in Figure 13.1, we note that articles are quite evenly distributed among the dominant technology classes. For example, the number of articles is not very different for classes *a-Laser-interferometry/ADM*, *b-Photogrammetry*, and *e-Theodolite/RLAT*.

Table 13.2 Dominant-technology classes for classifying documents (articles and patents)

Dominant technology		Description
a	*Laser-interferometry/ ADM*	Documents about measuring systems based on the laser-interferometric and/or Absolute Distance Meter (ADM) technology, or similar ones
b	*Photogrammetry*	Documents about measuring systems based on photogrammetric technology, in which one or more targets are localized using at least two cameras
c	*Ultrasound*	Documents about measuring systems in which one or more targets are localized by using the distances estimated by the time-of-flight of ultrasonic signals from/to three or more "landmarks," with known spatial position and orientation
d	*CMM*	Documents about Coordinate Measuring Machines (CMMs) for LSDM applications (e.g., gantry or horizontal arms CMMs)
e	*Theodolite/R-LAT*	Documents about systems based on classic theodolites or rotary laser automatic theodolites (R-LATs), such as the Nikon iGPS
f	*Structured light scanning*	Documents about measuring systems based on the projection of structured light (e.g., encoded patterns) on the surface of the object to be measured and analysis of the images taken by two or more cameras
g	*Hybrid*	Documents about measuring systems equipped with components derived from multiple technologies implementing multi-sensor data fusion techniques
h	*N/A*	Not applicable: documents concerning minor technologies or independent of a specific technology (e.g., documents about generic measurement procedures or literature reviews)

Regarding citations (see Chart 2), over half of the total citations are in the class *h-N/A*, being mainly captured by literature reviews, which are independent of a specific technology and tend to get more citations than standard research articles. Excluding the class *h-N/A*, citations are divided into the remaining technology classes in a fairly consistent way with articles.

Chart 3 illustrates the number of patents in the dominant technology classes and shows a clear predominance of the class *a-Laser-interferometry/ADM*. Analyzing the patents in this class, we noticed that most of them concern laser-trackers. Inventions in the class *b-Photogrammetry* are relatively more cited than the others (see Chart 4). This is symptomatic of the great industrial interest for this technology, even in contexts outside of LSDM (e.g., video game and home-entertainment). The class *h-N/A* includes many citations most of which (i.e., 113) come from a single outstandingly cited patent which is independent of a specific technology.

13.3.1.2 *Typology-of-output classification*

As regards articles, they seem quite evenly distributed among classes (see Chart 1 in Figure 13.2) with the predominance of documents concerning classes *C-New application/ working procedure* and *F-Performance analysis*. There is a relatively low fraction of articles in the classes *A-Description of the measuring system* and *B-New hardware component(s)*, which concerns the construction of new hardware equipment. Not surprisingly, nearly half of the total citations fall in the class *G-Literature review* (see Chart 2). Excluding this class, the distribution of citations among the remaining classes is quite in line with that of the articles (see Chart 1).

Table 13.3 Typology-of-output classes for classifying documents (articles and patents)

Typology of output		Description
A	Description of the measuring system	Detailed description of a new LSDM system, its technical features, functionality, measurement procedure, and (dis)advantages with respect to other systems.
B	New hardware component(s)	Development and characterization of new hardware components, which replace or complement those of an existing measuring system, improving its functionality and performance.
C	New application/ working procedure	Description of novel measurement procedures or applications, aimed at expanding and improving the functionality and/or performance of an existing measuring system. These procedures/applications generally require additional external hardware equipment.
D	Development of system set-up/calibration	Illustration of a new procedure/algorithm aimed at enhancing the system set-up/calibration stage of an existing measuring system.
E	Optimization of measurement operations	Improvement of the efficiency/effectiveness of measurement operations and data management. This optimization is typically software-based and does not imply any change in the measuring system's hardware or the introduction of external hardware.
F	Performance analysis	Analysis of the performance of an existent measuring system (e.g., evaluation of metrological characteristics such as repeatability, accuracy, measurement uncertainty, sampling rate, etc.) based on empirical data or simulations. It may include performance verification according to standards and/or comparison with other measuring systems.
G	Literature review	Lliterature review of the LSDM measuring systems or those based on a specific technology.

Regarding patents, the first three classes (*A-Description of the measuring system*, *B-New hardware component(s)*, and *C-New application/working procedure*) predominate, both from the viewpoint of *P* and *C* (see Charts 3 and 4). The typical industrial sectors in which LSDM systems are used can be identified by examining in more detail the documents in the class C-New application/working procedure. Almost all the patents examined concern the assembly phase of aircrafts. As for articles, nearly half of the applications concern the aircraft industry; other active sectors are those related to the assembly phase of ship hulls and railway equipment, measurement and control of telescope components, part alignment, and measurement in specific manufacturing processes (e.g., forming, welding).

13.3.1.3 "Technology–typology" maps

The maps in Figure 13.3 aggregate the results of the dominant-technology and typology-of-output classifications. Consistent with the results presented in Sections 13.3.1.1 and 13.3.1.2, it can be noticed that—as regards the patents—the densest part of the maps is the upper left-hand side whereas maps related to articles look more uniform.

13.3.2 Temporal evolution

13.3.2.1 Evolution according to the dominant-technology classification

The four diagrams in Figure 13.4 depict the temporal collocation of documents and their citations, according to the dominant-technology classification. The reference year for

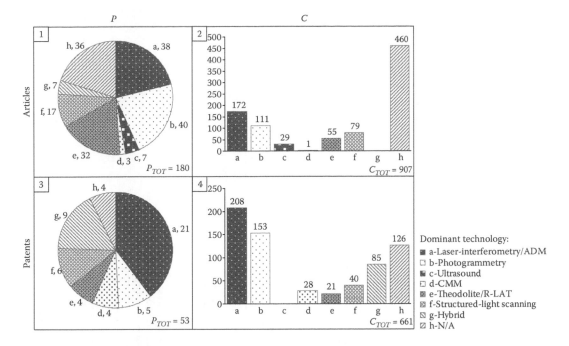

Figure 13.1 Charts concerning the distributions of the number of documents (*P*) and that of the corresponding citations (*C*) on the basis of the dominant-technology classification. Charts are constructed both for articles (Charts 1 and 2) and patents (Charts 3 and 4). For each chart, the number of documents/citations in each class is reported.

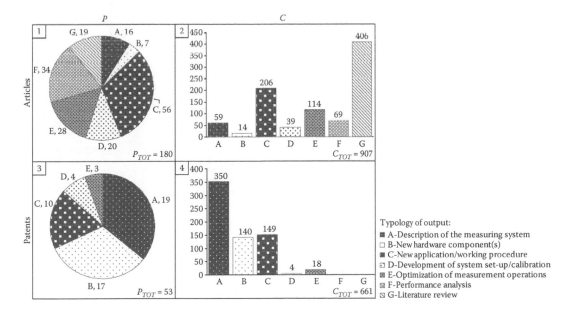

Figure 13.2 Charts concerning the distributions of the number of documents (*P*) and that of the corresponding citations (*C*) on the basis of the typology-of-output classification. Charts are constructed both for articles (Charts 1 and 2) and patents (Charts 3 and 4). For each chart, the number of documents/citations in each class is reported.

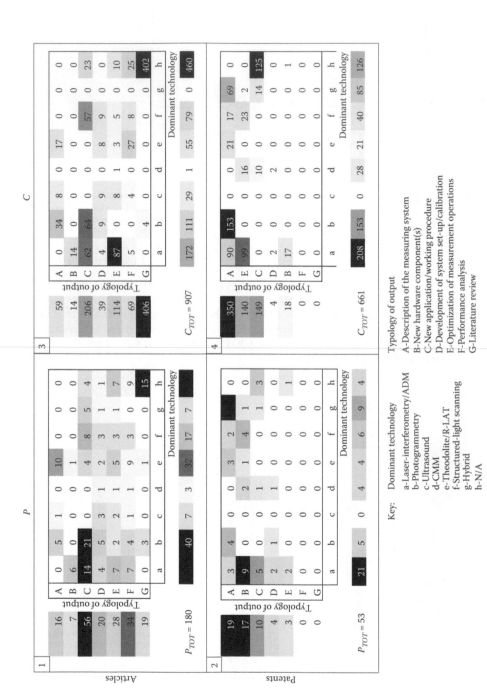

Figure 13.3 Maps concerning the partition of documents (P) and citations obtained (C) on the basis of dominant-technology (horizontal axis) and typology-of-output (vertical axis) classes—both for articles (maps (1) and (3)) and patents (maps (2) and (4)). To ease readability of each map, the gray level of cells is proportional to their numeric value, which represents the number of documents/citations.

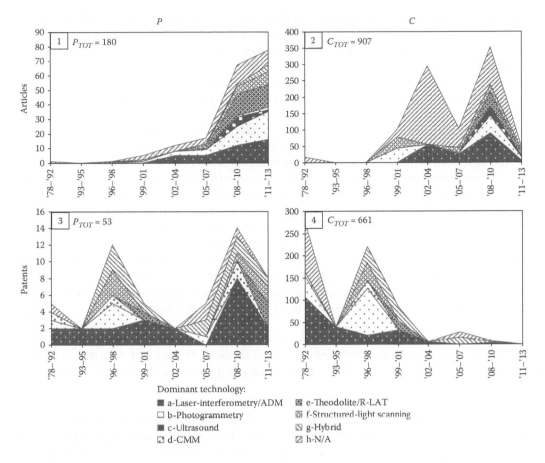

Figure 13.4 Temporal distribution of the number of documents (*P*) and the corresponding citations (*C*) on the basis of the dominant-technology classification. Charts are constructed both for articles (Charts 1 and 2) and patents (Charts 3 and 4).

patents is the oldest priority date, which is generally close to the date of the invention (Hinze and Schmoch, 2004). Time classes include three years; the only exception is represented by the broader first class (i.e., from 1978 to 1992) because of the relatively low number of documents issued in this period.

Chart 1, relating to the number of articles, shows that there is almost no article before 1996, apart from one published in 1978. Then, there is a gradual rise until 2008; this year signals a sort of explosion in terms of productivity. Consistently, with Figure 13.1(1), it can be noticed that the article distribution over the dominant-technology classes is relatively even. Of course, the 2008 explosion in the number of articles is partly affected by the general growth of the entire scientific literature in recent years (Larsen and von Ins, 2010). Nevertheless, so robust is this growth that it seems fairly symptomatic of a real increase of the scientific community's interest towards LSDM research.

Chart 2 in Figure 13.4 shows the number of citations obtained up to the moment of data collection (*C*), by articles issued in different years. We note that the pattern reflects that of the curves in Chart 1, except for two features:

- The huge number of citations received by some articles in the class *h-N/A* and issued in 2002–2004 (see the first local peak in Chart 2). Curiously, most of the citations (i.e., 229) come from three highly cited literature reviews, which are independent of a specific technology.
- The fact that relatively recent articles (issued in 2011–2013) obtained a few citations, due to the still not complete maturation period of the citation impact (Rogers, 2013).

Charts 3 and 4 are similar to the previous ones, but refer to patents. Chart 3 shows that a significant amount of patents were issued in the last 15 years of the twentieth century, and they are mainly related to the class *a-Laser-interferometry/ADM*. In the first 4–5 years of the twenty-first century there is a slight decrease in patent production followed by a powerful growth. This sort of "rebound" effect seems to be in synergy with the impressive growth of scientific articles in the last decade. Also, the numerous patents issued at the end of the twentieth century and the subsequent market entry of new patented technologies may have somehow stimulated the recent growth of scientific articles.

Chart 4, relating to the citations obtained by patents, is quite consistent with Chart 3, just before 2005. Instead, the recent growth phase (from 2005 onwards) is almost absent. This behavior is not surprising; regarding patents, the maturation time of citation impact is generally much longer than for scientific articles (Guan and Gao, 2009).

13.3.2.2   *Evolution according to the typology-of-output classification*

The four diagrams in Figure 13.5 are similar to those in Figure 13.4, but they are about the typology-of-output classification. Consistently with the results presented in Section 13.3.1.2, we note that

- Articles are quite uniformly distributed among the classes, over the years.
- Patents are almost exclusively related to the first three classes (*A-Description of the measuring system, B-New hardware component(s),* and *C-New application/working procedure*). And, in the last 10 years there has been a growth of patents in class *B* and *C*, to the detriment of those in class *A*.

A possible interpretation is that innovative LSDM systems were developed about 15–20 years ago, but due to the engineering-design-process time (Hara et al., 2007), they began to enter the market and started to be studied at academic level in the last 10–12 years only. In this recent period, there has been an explosion of new scientific articles and patents aimed at improving and refining these systems.

13.3.2.3   *Trying to predict the near future of LSDM research*

After studying the past evolution of the LSDM research, a question may arise: How will it evolve? Answering this is not easy because the documents we examined inevitably reflect the past. This applies to articles and especially patents since their pending-period (i.e., time lapse between the filing date and issue date) is very rarely shorter than three years. Also, extrapolating future trends can be hasty because LSDM is a relatively young and unpredictable discipline. Nevertheless, we will try to outline possible future scenarios, taking into account the results of this study, but also the current level of diffusion of LSDM equipment within industrial environments.

The analysis of the documents showed that the major technological innovations stem from a first "wave" of patents, issued in the last decade of the twentieth century. Around ten years later, many of these patented technologies were turned into real measuring

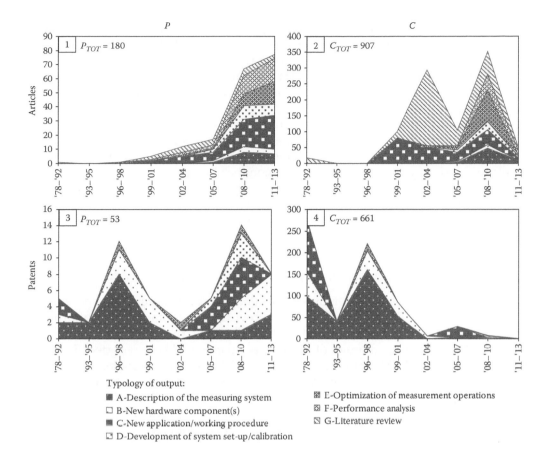

Figure 13.5 Temporal distribution of the number of documents (P) and the corresponding citations (C) on the basis of the typology-of-output classification. Charts are constructed both for articles (Charts 1 and 2) and patents (Charts 3 and 4).

systems, which began to be used in industrial contexts and analyzed in scientific articles. This confirms it is unlikely that disruptive innovations are introduced to the outside world in the form of scientific articles. Given the obvious strategic and commercial implications, it is customary to protect the most promising innovations with patents or even to keep them secret until they are ready to be transformed into well-established products (Mohan-Ram, 2001).

The second "wave" of patents—that is, those deposited roughly in the last 10–12 years and the simultaneous explosion of scientific articles—suggests that the current trend is to gradually improve and refine the available technologies. More precisely, some of the possible research directions are

1. Development of new hardware components, which enhances the functionality of existing measuring systems (e.g., innovative photogrammetric cameras, 6-DOF probes).
2. Optimization of measurement procedures such as enhancing and speeding up the set up/calibration stage of existing measuring systems.
3. Development of new standards aimed at assessing the functionality and performance of existing measuring systems and allowing their verification and comparison.

Nowadays, standards on LSDM equipment are very few; for instance, the relatively recent ASME B89.4.19 for laser trackers (ASME B89.4.19-2006, 2006), the German guideline VDI/VDE 2634 for photogrammetry (VDI/VDE 2634-1, 2002), and the relatively well-established ISO 10360 series standards for CMMs (ISO 10360-2, 2001). For most of the other systems standards are absent (e.g., there still is no standard for the emerging R-LAT systems).

4. Development of measurement techniques based on the combined use of LSDM systems of different nature. This requirement originates from a practical reason: In industrial contexts, complex measurements often require the combined use of multiple measuring systems. As a result, common (hardware and software) platforms that enable sharing and integration of information are essential. In the recent scientific literature, there are several researches aimed at achieving them (e.g., development of systems based on hybrid technologies); however, our impression is that this is just the beginning.

13.3.3   Study of the major scientific journals and patent assignees

Table 13.4 contains the list of the main sources (journals and conference proceedings) of the articles examined. For each source, the abbreviation, the total number of articles on LSDM (P), the total number of citations they have obtained up to the moment of data collection (C), and the average number of citations per paper (CPP) are reported. The map in Figure 13.6a displays the bibliometric positioning of the sources according to their P and C values. Of course, the most influential sources in this field are those that tend to be positioned in the upper right corner. The dominant source is Src1 (*CIRP Annals—Manufacturing Technology*), which contains articles presented in a traditional world-class conference in the field of manufacturing (CIRP, 2013). The vast number of citations is likely because seven

Table 13.4 List of the sources (journals and conference proceedings) that contain at least four LSDM articles

Abbr.	Source title	P	C	CPP
Src1	*CIRP Annals—Manufacturing Technology*	14	377	26.9
Src2	*Proceedings of the Institution of Mechanical Engineers, Part B: Journal of Engineering Manufacture*	5	56	11.2
Src3	*International Journal of Computer Integrated Manufacturing*	4	35	8.8
Src4	*Optics and Lasers in Engineering*	4	29	7.3
Src5	*Guangxue Xuebao/Acta Optica Sinica*	10	71	7.1
Src6	*Guangdianzi Jiguang/Journal of Optoelectronics Laser*	6	19	3.2
Src7	*Precision Engineering*	7	22	3.1
Src8	*Measurement Science and Technology*	6	14	2.3
Src9	*International Journal of Advanced Manufacturing Technology*	8	10	1.3
Src10	*Hongwai yu Jiguang Gongcheng/Infrared and Laser Engineering*	5	6	1.2
Src11	*Guangxue Jingmi Gongcheng/Optics and Precision Engineering*	4	2	0.5
Src12	*Proceedings of SPIE—The International Society for Optical Engineering*	15	5	0.3
Src13	*Applied Mechanics and Materials*	4	0	0.0

Note: For each source, its abbreviation ("Abbr."), total number of documents (P), total number of citations accumulated up to the moment of data collection (C), and average citations per document (CPP) are reported. Sources are sorted according to their CPP values.

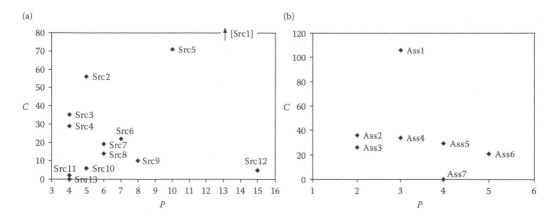

Figure 13.6 Scatter maps representing the bibliometric positioning of the sources/assignees reported in Tables 13.4 and 13.5 according to their *P* and *C* values. The corresponding *CPP* values are reported in brackets. Abbreviations of sources and assignees are shown in Tables 13.4 and 13.5, respectively. Given the high number of citations (377), Src1 falls above the quadrant of the map (a). (a) Scatter map concerning publication sources. (b) Scatter map concerning patent assignees.

of the 14 articles published by this source are literature reviews with a relatively large diffusion. Src12 is the source with the largest number of articles (i.e., 15), but their citation impact is dramatically low (just five total citations).

The same analysis can be extended to patents, that is, just replacing article sources with patent assignees. Table 13.5 shows the list of the assignees of at least two patents. For each assignee, the total number of patents (*P*), the total number of citations they have obtained (*C*), and the average number of citations per patent (*CPP*) are also reported. The map in Figure 13.6b displays the bibliometric positioning of the patent assignees, according to their *P* and *C* values. Curiously, the most cited patents are those relating to video game and home-entertainment applications based on photogrammetry or structured light scanning. We also notice that the list includes important companies in the field of metrology (e.g., Faro and Nikon) and aerospace (Boeing). Finally, it can be seen that some important universities (e.g., MIT and Zhejang University) are assignees of a portion of the patents.

Table 13.5 List of the assignees of at least two LSDM patents

Abbr.	Assignee	*P*	*C*	*CPP*
Ass1	Motion Games	3	106	35.3
Ass2	MIT	2	36	18.0
Ass3	Lawrence Livermore National Security	2	26	13.0
Ass4	Faro	3	34	11.3
Ass5	Boeing	4	29	7.3
Ass6	Nikon	5	21	4.2
Ass7	Zhejiang University	4	0	0.0

Note: For each assignee, its abbreviation ("Abbr."), total number of documents (*P*), total number of citations accumulated up to the moment of data collection (*C*), and average citations per document (*CPP*) are reported. Assignees are sorted according to their *CPP* values.

13.4   Conclusions

The recent evolution of LSDM systems is evidenced by the gradual growth of scientific articles and patents, particularly in the last 10–15 years. Our study isolated a platform of documents that could be of interest to the reader who approaches LSDM. The combined analysis of scientific articles and patents showed kind of dualism, summarized in the following points:

- As expected, almost all patents deal with new measurement systems, hardware components, and working procedures. As for the articles—apart from literature reviews—the main contributions concern working procedures, performance analysis, development of the system set-up/calibration stage, and optimization of measurement operations, for existing measuring systems.
- Regarding the temporal collocation, a substantial portion of patents was issued in the last fifteen years of the twentieth century. On the other hand, most of the articles were published after 2004. It appears clear that articles followed the evolution of patents, probably after the new (patented) technologies became available in the market.
- Referring to the dominant-technology classification, it is observed that scientific articles are distributed quite uniformly among the major available technologies. On the contrary, patents are polarized primarily on the laser-interferometry technology, followed by other more affordable ones but with lower accuracy, that is, photogrammetry and structured-light scanning. It may be true that the future evolution of these emerging technologies will make them more competitive. Another impression we experienced is that academic research is probably more focused on developing affordable measurement systems, while the most expensive and accurate ones are mainly designed and refined by companies who have the resources for developing, engineering, and marketing them.

The main limitation of this work is the data collection procedure, which, although being meticulous and transparent, is not free from the risk of omitting a portion of LSDM documents (i.e., relevant documents that are not indexed by the databases in use or do not meet the search query). This risk is even larger for patents since their language is sometimes captious and can complicate the searching by keywords (Foglia, 2007). Future research could be aimed at extending the portfolio of documents potentially related to LSDM by analyzing those that cite or are cited by other ones.

References

ASME B89.4.19-2006. 2006. Performance evaluation of laser-based spherical coordinate measurement systems, http://www.asme.org (March 8, 2013).

Bar-Ilan, J. 2010. Citations to the introduction to informetrics indexed by WOS, Scopus and Google Scholar. *Scientometrics*, 82(3): 495–506.

Breschi, S., Catalini, C. 2010. Tracing the links between science and technology: An exploratory analysis of scientists' and inventors' networks. *Research Policy*, 39(1): 14–26.

Cheng, Y.H., Kuan, F.Y., Chuang, S.C. 2010. Profitability decided by patent quality? An empirical study of the U.S. semiconductor industry. *Scientometrics*, 82(1): 175–183.

CIRP 2013. Organization website: http://www.cirp.net (March 8, 2013).

Czarnitzki, D., Glänzel, W., Hussinger, K. 2007. Patent and publication activities of German professors: An empirical assessment of their co-activity. *Research Evaluation*, 16(4): 311–319.

De Bellis, N. 2009. *Bibliometrics and Citation Analysis: From the Science Citation Index to Cybermetrics*. Scarecrow Press, Lanham, MD.

Estler, W.T., Edmundson, K.L., Peggs, G.N., Parker, D.H. 2002. Large-scale metrology—An update. *CIRP Annals—Manufacturing Technology*, 51(2): 587–609.

Foglia, P. 2007. Patentability search strategies and the reformed IPC: A patent office perspective. *World Patent Information*, 29(1): 33–53.

Franceschini, F., Galetto, M., Maisano, D., Mastrogiacomo, L. 2009. On-line diagnostics in the mobile spatial coordinate measuring system (MScMS). *Precision Engineering*, 33(4): 408–417.

Franceschini, F., Maisano, D., Turina, E. 2012. European research in the field of production technology and manufacturing systems: An exploratory analysis through publications and patents. *International Journal of Advanced Manufacturing Technology*, 62(1–4): 329–350.

Franceschini, F., Galetto, M., Maisano, D., Mastrogiacomo, L., Pralio, B. 2011. *Distributed Large-Scale Dimensional Metrology*. Springer, London.

Franceschini, F., Maisano, D. 2014. The evolution of large-scale dimensional metrology from the perspective of scientific articles and patents. *International Journal of Advanced Manufacturing Technology*, 70(5–8): 887–909.

Glänzel, W., Meyer, M. 2003. Patents cited in the scientific literature: An exploratory study of reverse citation relations. *Scientometrics*, 58(2): 415–428.

Guan, J.C., Gao, X. 2009. Exploring the h-index at patent level. *Journal of the American Society for Information Science and Technology*, 60(1): 35–40.

Hara, T., Arai, T., Shimomura, Y., Sakao, T. 2007. Service/product engineering: A new discipline for value production, in *19th International Conference on Production Research (ICPR-19)*, July 29 to August 2, 2007, Valparaiso, Chile.

Hinze, S., Schmoch, U. 2004. Opening the black box, in: Moed, H.F., Glänzel, W., Schmoch U. (eds.), *Handbook of Quantitative Science and Technology Studies*, 215–236. Kluwer Academic Publishers, Dordrecht.

ISO 10360-2 2001. Geometrical product specifications (GPS). *Acceptance and Reverification Tests for Coordinate. Measuring Machines (CMM), Part 2: CMMs Used for Measuring Size*. International Organization for Standardization, Geneva.

Jacsó, P. 2008. Testing the calculation of a realistic h-index in Google Scholar, Scopus, and Web of Science for F. W. Lancaster. *Library Trends*, 56(4): 784–815.

Jun, S., Lee, J. 2012. Emerging technology forecasting using new patent information analysis. *International Journal of Software Engineering and its Applications*, 6(3): 107–116.

Larsen, P.O., von Ins, M. 2010. The rate of growth in scientific publication and the decline in coverage provided by Science Citation Index. *Scientometrics*, 84(3): 575–603.

Mohan-Ram, V. 2001. Patent first, publish later: How not to ruin your chances of winning a patent. *Science Career*, October 26, 2001, http://sciencecareers.sciencemag.org (March 8, 2013).

Muelaner, J.E., Cai, B., Maropoulos, P.G. 2010. Large-volume metrology instrument selection and measurability analysis. *Proceedings of the Institution of Mechanical Engineers, Part B: Journal of Engineering Manufacture*, 224(6): 853–868.

Peggs, G.N., Maropoulos, P.G., Hughes, E.B., Forbes, A.B., Robson, S., Ziebart, M., Muralikrishnan, B. 2009. Recent developments in large-scale dimensional metrology. *Proceedings of the Institution of Mechanical Engineers, Part B: Journal of Engineering Manufacture*, 223(6): 571–595.

Puttock, M.J. 1978. Large-scale metrology. *Ann. CIRP*, 21(1): 351–356.

Questel-Orbit 2013. www.questel.com/prodsandservices/orbit.htm (March 8, 2013).

Rogers, E. 2013. *Diffusion of Innovations*. 5th edn., Free Press, New York.

Scopus-Elsevier 2013. www.info.scopus.com (March 8, 2013).

Seeber, F. 2007. Patent searches as a complement to literature searches in the life sciences–a "how-to" tutorial. *Nature Protocols*, 2(10): 2418–2428.

Van Looy, B., Callaert, J., Debackere, K. 2006. Publication and patent behavior of academic researchers: Conflicting, reinforcing or merely co-existing? *Research Policy*, 35(4): 596–608.

VDI/VDE 2634–1 2002. *Optical 3-D Measuring Systems and Imaging Systems with Point-by-Point Probing*. Beuth Verlag, Berlin.

Yu, J., Chong, Z., Lu, H., Zhou, A. 2004. False positive or false negative: Mining frequent itemsets from high speed transactional data streams. *Proceedings of the Thirtieth International Conference on Very Large Data Bases (VLDB '04)*, 30: 204–215.

chapter fourteen

Gaussian-smoothed Wigner function and its application to precision analysis

Hai-Woong Lee

Contents

14.1   Introduction

The phase-space formulation of quantum mechanics, which originates from the classic work of Wigner (1932), has enjoyed wide popularity in all areas of physics (Lee 1995, Schleich 2001). As there is no unique way of assigning a quantum-mechanical operator to a given classical function of conjugate variables, there can exist many different quantum phase-space distribution functions, the best known of which are the Wigner function (Wigner 1932), Husimi function (Husimi 1940), and P and Q functions (Glauber 1963, Sudarshan 1963). All these distribution functions are equivalent to one another, in the sense that any of them can be used to evaluate the expectation value of any arbitrary operator. Only the rule of ordering of noncommuting operators is different for a different function.

In this work, we study a class of distribution functions that results from a Gaussian convolution of the Wigner function in phase space. This class of functions, which we call the Gaussian-smoothed Wigner function, has physical significance in that it represents a probability distribution resulting from simultaneous measurements of position and momentum, where the widths in position and momentum space, respectively, of the smoothing Gaussian function can be identified as the measurement uncertainties of the devices used to measure the position and momentum. We find the rule of ordering for the Gaussian-smoothed Wigner function, which allows an accurate determination of expectation values of phase-space operators. Based on this, we show that the Gaussian-smoothed Wigner function can be used to determine accurately the expectation values from measurement data obtained even with realistic, nonideal detectors. This finding has immediate application in quantum optics to the problem of detecting the state of a radiation field, because there exists the method of eight-port homodyne detection, which performs simultaneous measurements of two quadrature amplitudes of the radiation field.

14.2   Gaussian-smoothed Wigner function

The starting point of our study is the equation

$$H(q,p) = \frac{1}{\pi\hbar} \int dq' \int dp' e^{-(q'-q)^2/2\sigma_q^2} e^{-(p'-p)^2/2(\hbar/2\sigma_q)^2} W(q',p') \tag{14.1}$$

which defines the Husimi function H in terms of a Gaussian convolution of the Wigner function W in phase space (Lee 1994, 1995). Note that the width σ_q of the Gaussian function in q space and the width $\sigma_p = \hbar/2\sigma_q$ in p space satisfy the Heisenberg minimum uncertainty relation

$$\sigma_q \sigma_p = \frac{\hbar}{2} \tag{14.2}$$

It is straightforward to generalize Equation 14.1 and define the Gaussian-smoothed Wigner function G as

$$G(q,p) = \frac{1}{2\pi\sigma_q\sigma_p} \int dp' \int dp' e^{-(q'-q)^2/2\sigma_q^2} e^{-(p'-p)^2/2\sigma_p^2} W(q',p') \tag{14.3}$$

The Husimi function is everywhere real and nonnegative (Cartwright 1976) and is entitled to probability interpretation. It has indeed been shown that the Husimi function represents a proper probability distribution associated with ideal simultaneous measurements of position and momentum (Arthurs and Kelly 1965, Braunstein et al. 1991, Stenholm 1992), where the ideal measurement refers to a measurement performed with a perfect measurement device satisfying the Heisenberg minimum uncertainty relation. We note that the widths σ_q and σ_p of the smoothing Gaussian function in Equation 14.1 are identified with the measurement uncertainties in q and p, respectively. In other words, the act of simultaneous measurement is modeled by phase-space Gaussian smoothing, with the widths of the smoothing Gaussian function identified as measurement uncertainties (Wodkiewicz 1984). The physical significance of the function G defined by Equation 14.3 should now be clear. It represents a probability distribution resulting from simultaneous measurements of position and momentum, where the measurements are performed with a device characterized by measurement uncertainties σ_q and σ_p. One may consider a large number of identically prepared systems on each of which a simultaneous measurement of position and momentum is performed. Each time, the measurement is performed with an identical measurement device of measurement uncertainties σ_q and σ_p. The probability distribution in q and p resulting from such measurements is the function G of Equation 14.3, where the widths σ_q and σ_p of the smoothing Gaussian function are given by the measurement uncertainties in q and p, respectively, of the measurement device used. According to the Heisenberg uncertainty principle, the measurement uncertainties must satisfy

$$\sigma_q \sigma_p \geq \frac{\hbar}{2} \tag{14.4}$$

The function G, with $\sigma_q\sigma_p \geq \hbar/2$, is thus physically measurable through simultaneous measurements of position and momentum.

14.3 Rule of ordering

It should be emphasized that, at least in principle, the function G is as good a quantum phase-space distribution function as the Husimi function or the Wigner function. The expectation value of any arbitrary operator can be calculated using the function G as well as the Husimi function or the Wigner function. Only the rule of ordering of noncommuting operators is different. In order to find the rule of ordering associated with the function G, we begin with the equation (Lee 1995)

$$\text{Tr}\{\hat{\rho}e^{i\xi\hat{q}+i\eta\hat{p}}f(\xi,\eta)\} = \int dp \int dp\, e^{i\xi q+i\eta p}G(q,p) \tag{14.5}$$

It can be shown that the function $f(\xi,\eta)$ that determines the rule of ordering for the function G is given by

$$f(\xi,\eta) = e^{-\sigma_q^2\xi^2/2-\sigma_p^2\eta^2/2} \tag{14.6}$$

At this point, we find it convenient to introduce two parameters κ and s defined as

$$\kappa = \frac{\sigma_p}{\sigma_q} \tag{14.7}$$

and

$$s = -\frac{\sigma_q\sigma_p}{\hbar/2} - \frac{\kappa\sigma_q^2}{\hbar/2} \tag{14.8}$$

The parameter κ has a dimension of $m\omega$ [mass/time]. The parameter s is real and negative, and its absolute value measures the product of the widths σ_q and σ_p associated with the function G being considered with respect to that of the minimum uncertainty Gaussian wave packet. Once σ_q and σ_p are given, κ and s are determined, and vice versa. Equation 14.6 can be rewritten, in terms of κ and s, as

$$f(\xi,\eta) = e^{(s\hbar\xi^2/4\kappa)+(s\hbar\kappa\eta^2/4)} \tag{14.9}$$

We further introduce dimensionless parameters v and β and an operator $\hat{b}$ as

$$v = i\xi\sqrt{\frac{\hbar}{2\kappa}} - \eta\sqrt{\frac{\hbar\kappa}{2}} \tag{14.10}$$

$$\beta = \sqrt{\frac{\kappa}{2\hbar}}q + \frac{i}{\sqrt{2\hbar\kappa}}p \tag{14.11}$$

$$\hat{b} = \sqrt{\frac{\kappa}{2\hbar}}\hat{q} + \frac{i}{\sqrt{2\hbar\kappa}}\hat{p} \tag{14.12}$$

Equation 14.5 with $f(\xi, \eta)$ given by Equation 14.6 can then be rewritten as

$$\mathrm{Tr}\{\hat{\rho}e^{v\hat{b}^\dagger - v^*\hat{b}}e^{s(|v|^2/2)}\} = \mathrm{Tr}\{\hat{\rho}e^{-v^*\hat{b}}e^{v\hat{b}^\dagger}e^{(s+1)(|v|^2/2)}\} = \int d^2\beta e^{v\beta^* - v^*\beta}G(\beta,\beta^*) \qquad (14.13)$$

The rule of ordering for the function G can now be determined from Equation 14.13 using the same method that Cahill and Glauber (1969) adopted for their s-parameterized distribution function. The final result is

$$\{\hat{b}^{\dagger n}\hat{b}^m\} = \sum_{k=0}^{(n,m)} k!\binom{n}{k}\binom{m}{k}\left(-\frac{s}{2}-\frac{1}{2}\right)^k \hat{b}^{(m-k)}\hat{b}^{\dagger(n-k)} \qquad (14.14)$$

where $\{\hat{b}^{\dagger n}\hat{b}^m\}$ represents the rule of ordering for the function G, the symbol (n, m) denotes the smaller of the two integers n and m, and $\binom{n}{k}$ is a binomial coefficient. Equation 14.14 yields, for example, $\{\hat{b}^\dagger\} = \hat{b}^\dagger$, $\{\hat{b}\} = \hat{b}$, $\{\hat{b}^\dagger\hat{b}\} = \hat{b}\hat{b}^\dagger - (1/2)(s+1)$, $\{\hat{b}^\dagger\hat{b}^2\} = \hat{b}^2\hat{b}^\dagger - (s+1)\hat{b}$, $\{\hat{b}^{\dagger 2}\hat{b}\} = \hat{b}\hat{b}^{\dagger 2} - (s+1)\hat{b}^\dagger$, and $\{\hat{b}^{\dagger 2}\hat{b}^2\} = \hat{b}^2\hat{b}^{\dagger 2} - 2(s+1)\hat{b}\hat{b}^\dagger + (1/2)(s+1)^2$.

As an illustration, let us find the expectation value of $\hat{q}\hat{p}^2$ with the function $G(q, p)$. We first need to use Equations 14.7 and 14.11 and perform change of variables from (q, p) to (β, β^*) to obtain $G(\beta, \beta^*)$. Expressing $\hat{q}\hat{p}^2$ in terms of $\{\hat{b}^{\dagger m}\hat{b}^n\}$, we obtain

$$\hat{q}\hat{p}^2 = -\sqrt{\frac{\hbar^3\kappa}{8}}[\{\hat{b}^3\} + \{\hat{b}^{\dagger 3}\} - \{\hat{b}^\dagger\hat{b}^2\} - \{\hat{b}^{\dagger 2}\hat{b}\} - (s+2)\{\hat{b}\} - (s-2)\{\hat{b}^\dagger\}] \qquad (14.15)$$

Equation 14.15 leads immediately to

$$\hat{q}\hat{p}^2 = -\sqrt{\frac{\hbar^3\kappa}{8}}\int d^2\beta G(\beta,\beta^*)[\beta^3 + \beta^{*3} - \beta^2\beta^* - \beta\beta^{*2} - (s+2)\beta - (s-2)\beta^*] \qquad (14.16)$$

14.4 Application to precision analysis

14.4.1 General consideration

We now discuss a possible application of the function G to precision analysis. When performing measurements, one faces a realistic problem of having to deal with imperfect detectors of efficiencies lower than 1. For precision analysis, one needs to find a reliable way of correcting unavoidable errors arising from imperfect measurements. An attractive feature of the function G introduced here is that it provides a definite recipe, in the form of the rule of ordering, that allows one to obtain precise quantitative information about the system being considered from measurement data collected by imperfect detectors. Let us suppose that simultaneous measurements of two conjugate variables are performed with realistic detectors of efficiencies less than 1, from which the joint count probability distribution is obtained. One can identify exactly the function G representing this distribution, as long as the efficiencies of the detectors are known. Since the rule of ordering of non-commuting operators for the function G is exactly known, the expectation value of any operator can be evaluated accurately using the function G. Hence, at least in principle, a high degree of accuracy comparable to that with near-perfect detectors is within reach, even if measurements are performed with imperfect detectors. We illustrate this below by considering the eight-port homodyne detection experiment (Noh et al. 1991, Schleich 2001) that performs simultaneous measurements of two quadrature amplitudes of a radiation field. The issue of obtaining precise information from measurements performed with realistic detectors in the eight-port homodyne detection experiment was investigated by Paris (1996a,b) for the case when the measurement uncertainties are described by the s-parameterized function of Cahill and Glauber (1969). Here, we wish to provide a general theory applicable to arbitrary values of measurement uncertainties.

14.4.2   Eight-port homodyne detection experiment

When applied to a radiation field, which mathematically is equivalent to a harmonic oscillator of mass $m = 1$, Equation 14.3 translates into

$$G(\alpha,\alpha^{*}) = \frac{1}{2\pi\sigma_1\sigma_2}\int d^2\alpha' e^{-(\alpha_1' - \alpha_1)^2/2\sigma_1^2} e^{-(\alpha_2' - \alpha_2)^2/2\sigma_2^2} W(\alpha',\alpha^{*\prime}) \tag{14.17}$$

where α_1 and α_2, real and imaginary parts of α, respectively, refer to two quadrature amplitudes of the radiation field. When $\sigma_1\sigma_2 = 1/4$, the function G becomes the Husimi function. In particular, when $\sigma_1 = \sigma_2 = 1/2$, the function G is reduced to the Q function. Furthermore, when $\sigma_1 = \sigma_2 = \sigma$, the function G becomes the s-parameterized function of Cahill and Glauber (1969) with $s = -4\sigma^2$.

Simultaneous measurements of two quadrature amplitudes of a radiation field, contingent upon the Heisenberg uncertainty principle $\sigma_1\sigma_2 \geq 1/4$, can be performed using the eight-port homodyne detection scheme proposed earlier (Noh et al. 1991). It has been shown (Freyberger et al. 1993, Leonhardt 1993) that the joint count probability of two quadrature amplitudes is given, in the limit of a strong local oscillator, by the Q function of the signal field provided that the detectors used for the measurements are perfect. In reality, however, detectors have nonunit efficiency $\eta < 1$. Assuming that the detectors have the identical nonunit efficiency η, Leonhardt and Paul (1993) have shown that the eight-port homodyne scheme measures the s-parameterized function of Cahill and Glauber (1969) where $s = -(2 - \eta)/\eta$. This result can be generalized in a straightforward way to the case where the detectors used to measure different quadrature amplitudes have different efficiencies η_1 and η_2. (We assume, however, that two detectors used to measure the same quadrature amplitude have the same efficiency.) In this case, the eight-port homodyne scheme measures the function G with σ_1 and σ_2 given by $4\sigma_1^2 = -s_1 = (2 - \eta_1)/\eta_1$ and $4\sigma_2^2 = -s_2 = (2 - \eta_2)/\eta_2$, respectively. For this case, the parameters κ and s are given by (with mass $m = 1$)

$$\kappa = \omega\frac{\sigma_2}{\sigma_1} = \omega\sqrt{\frac{(2 - \eta_2)\eta_1}{(2 - \eta_1)\eta_2}} \tag{14.18}$$

$$s = -4\sigma_1\sigma_2 = -\sqrt{\frac{(2 - \eta_1)(2 - \eta_2)}{\eta_1\eta_2}} \tag{14.19}$$

where ω is the angular frequency of the field. The corresponding rule of ordering is given by Equation 14.14, where the operator $\hat{b}$, the annihilation operator of a "squeezed" photon, is defined in Equation 14.12 with κ given by Equation 14.18.

14.4.3 Examples

Let us suppose that we wish to find the expectation value of the photon number, $\hat{a}^\dagger \hat{a}$, in the signal field. For this purpose, we first need to express $\hat{a}^\dagger \hat{a}$ in terms of $\{\hat{b}^{\dagger n}\hat{b}^m\}$ of Equation 14.14. A straightforward algebra yields

$$\hat{a}^\dagger \hat{a} = -(\{\hat{b}^{\dagger 2}\} + \{\hat{b}^2\})\frac{\sinh 2r}{2} + \{\hat{b}^\dagger \hat{b}\}\cosh 2r + \frac{s}{2}\cosh 2r - \frac{1}{2} \tag{14.20}$$

where the "squeeze" parameter r is defined as

$$e^r = \sqrt{\frac{\kappa}{\omega}} \tag{14.21}$$

We thus have

$$\hat{a}^\dagger \hat{a} = \int d^2\beta G(\beta,\beta^*)\left[-(\beta^{*2} + \beta^2)\frac{\sinh 2r}{2} + |\beta|^2 \cosh 2r\right] + \frac{s}{2}\cosh 2r - \frac{1}{2} \tag{14.22}$$

Here, the parameter β is related to the quadrature amplitude α by

$$\beta = \alpha \cosh r + \alpha^* \sinh r \tag{14.23}$$

The joint count probability of two quadrature amplitudes measured with (imperfect) detectors in the eight-port homodyne scheme leads directly to the identification of the function $G(\alpha, \alpha^*)$. One can then obtain $G(\beta, \beta^*)$ through change of variables from (α, α^*) to (β, β^*). The expectation value $\hat{a}^\dagger \hat{a}$ can then be calculated using Equation 14.22.

We emphasize that Equation 14.22 is valid for any arbitrary values η_1 and η_2 of the efficiencies of the detectors used. Hence, an accurate determination of $\hat{a}^\dagger \hat{a}$ can be achieved even from measurement data performed with imperfect detectors. One only needs to determine the function $G(\alpha, \alpha^*)$ accurately from the joint count probability obtained with (imperfect) measurements. The function to be determined here is the function G associated with the very detectors used, not the function $G(=Q)$, which would be obtained with the ideal detectors. To elaborate further on this point, let us consider the simple case when the detectors have the same efficiency $\eta_1 = \eta_2 \equiv \eta$. In this case, we have $\kappa = \omega$, $\beta = \alpha$, and $r = 0$, and Equation 14.22 is simplified to

$$\hat{a}^\dagger \hat{a} = \int d^2\alpha G(\alpha,\alpha^*)|\alpha|^2 + \frac{1}{2}(s - 1) \tag{14.24}$$

If the measurements were performed with the perfect detectors, the joint count probability would yield the Q function (Freyberger et al. 1993, Leonhardt and Paul 1993), and the expectation value $\hat{a}^\dagger \hat{a}$ would be calculated by

$$\hat{a}^{\dagger}\hat{a} = \hat{a}\hat{a}^{\dagger} - 1 = \int d^2\alpha Q(\alpha,\alpha^*)\,|\alpha|^2 - 1 \qquad (14.25)$$

The difference between Equations 14.24 and 14.25, namely, $(s+1)/2$, represents the "correction" factor that needs to be added to compensate for the use of imperfect detectors. This presents no problem because the correction factor can be determined exactly (see Equation 14.19) once the efficiency η is known. One can thus say that the expectation value $\hat{a}^{\dagger}\hat{a}$ can be determined from the eight-port homodyne detection experiment to a near-perfect degree of accuracy, regardless of the efficiencies of the detectors used.

Difficulty arises, however, when one wants to evaluate the expectation values $\hat{a}^{\dagger n}\hat{a}^m$ for large integers n and m. In general, the determination of $\hat{a}^{\dagger n}\hat{a}^m$ requires an accurate evaluation of the integrals $\int d^2\beta G(\beta,\beta^*)\beta^{*n}\beta^m$, $\int d^2\beta G(\beta,\beta^*)\beta^{*n-1}\beta^{m-1}$, etc. When n and/or m are large, the value of these integrals may vary widely with respect to small changes in the function G, and thus it is important to determine accurately the function G from the experiment. When low-efficiency detectors are used, the function G resulting from the joint count probability is a strongly smoothed function and thus exhibits a relatively flat distribution. In such a case, an accurate determination of the function G requires an accurate determination of a large number of significant figures of its values, which puts a heavy (and perhaps impossible if n and m are quite large and if the detector efficiencies deviate significantly from unity) burden on the experiment. Another difficulty arises from the fact that an accurate evaluation of $\hat{a}^{\dagger n}\hat{a}^m$ requires an accurate evaluation of lower-order expectation values, and thus a small error in $\hat{a}^{\dagger}\hat{a}$ and other low-order expectation values are magnified in the evaluation of high-order expectation values $\hat{a}^{\dagger n}\hat{a}^m$. One thus concludes that the greater the integers n and m are, the closer to unity the efficiencies η_1 and η_2 are required to be for an accurate evaluation of $\hat{a}^{\dagger n}\hat{a}^m$; that is, the requirement on detector efficiencies gets increasingly severe for larger integer values of n and m. In other words, the lower the efficiencies of the detectors used are, the more strongly limited the number of expectation values $\hat{a}^{\dagger n}\hat{a}^m$ that can be determined reliably. With imperfect detectors, it is practically impossible to accurately evaluate the expectation values $\hat{a}^{\dagger n}\hat{a}^m$ for all integers n and m. Hence, an accurate state reconstruction, for example, which requires $\hat{a}^{\dagger n}\hat{a}^m$ for all integers n and m, may be difficult, unless the signal state is given by a finite superposition of number states and thus only a finite number of $\hat{a}^{\dagger n}\hat{a}^m$ are different from zero (Paris 1996a,b). Nevertheless, if one is primarily interested in the expectation values $\hat{a}^{\dagger n}\hat{a}^m$ for small integers n and m, then the phase-space approach with the function G may provide a way to determine them with a high degree of accuracy, even if one is equipped only with moderately imperfect detectors.

14.5 Discussion and summary

The question arises how low the efficiencies can be if one can still hope to get reliable quantitative information about the system being measured. As a rough estimate, it may be expected that the function G with $\sigma_1^2 \approx \sigma_2^2 \approx 1/2$ is as good a phase-space distribution function as the Q function, in the same sense that the Q function with $\sigma_1^2 \approx \sigma_2^2 \approx 1/4$ is as good a function as the Wigner function. Taking $\sigma_1^2 \approx \sigma_2^2 \approx 1/2$, one obtains $\eta_1 = \eta_2 \approx 0.67$, the value well within reach of the present technology (Eisamann et al. 2011). If the detectors used in the eight-port homodyne detection experiment have efficiencies higher than ~0.67, then the function G constructed from the experiment may be expected to provide reasonably accurate information about the signal field.

The difficulty mentioned in the Section 14.4.3 associated with a strongly smoothed function G derived from low-efficiency detectors has its mathematical root in the fact that the Gaussian convolution operation of Equation 14.3 (or Equation 14.17) is, as has already been noted (Agarwal and Wolf 1970), the two-dimensional Weierstrass transform, which is an invertible point-to-point integral transform. As such, there is, in principle, no information loss when the convolution operation is performed, even if fine structures are inevitably smoothed. This is consistent with the fact that, regardless of the strength of smoothing, a definite rule of ordering exists in the form of Equation 14.14, which enables, in principle, an accurate evaluation of the expectation values $\hat{a}^{\dagger n}\hat{a}^{m}$ for all integers n and m. There, however, exists practical difficulty with the inverse Weierstrass transform, because a small error is magnified exponentially in the inverse transform. Hence, the requirement on the accuracy of the function G is increasingly severe, as the strength of smoothing is increased. Despite this practical difficulty, it is still encouraging that the expectation values $\hat{a}^{\dagger n}\hat{a}^{m}$ for small integers can be accurately evaluated, even if one has only imperfect detectors and is therefore provided with a strongly smoothed function G.

In this work, we have shown that the Gaussian-smoothed Wigner function G can be used in precision analysis to determine accurately the expectation values of phase-space operators from experimental data obtained with inefficient detectors. Another possible application of the function G is to the problem of experimental test of quantum nonlocality. It has already been shown (Banaszek and Wodkiewicz 1999, Lee et al. 2009, Lee et al. 2010) that Bell-type inequalities can be constructed with the s-parameterized function of Cahill and Glauber, as well as with the Wigner function and Q function, and can be used to experimentally test quantum nonlocality and witness entanglement. As the function G defined in Equation 14.17 is a generalized version of the s-parameterized function, in the sense that the former reduces to the latter when $\sigma_1 = \sigma_2$, it is straightforward to construct generalized Bell inequalities based on the function G, which will provide a variety of ways of demonstrating fundamental quantum properties.

In summary, we have found the rule of ordering of conjugate variables for the Gaussian-smoothed Wigner function G, which allows an accurate evaluation of the expectation values $\hat{q}^{n}\hat{p}^{m}$ (or $\hat{a}^{\dagger n}\hat{a}^{m}$). On the basis of the fact that the function G represents the joint count probability in simultaneous measurements of two conjugate variables q and p (or α_1 and α_2), we have shown that the data obtained from simultaneous measurements performed with realistic, nonideal detectors can be analyzed in such a way that a fairly accurate evaluation of the expectation values $\hat{q}^{n}\hat{p}^{m}$ (or $\hat{a}^{\dagger n}\hat{a}^{m}$) for low integers n and m can be achieved.

Acknowledgment

The author thanks Dr. Heeseung Zoe for extensive technical help. This work originally appeared in *Optics Communications*, 337:62–65, 2015 and has been reproduced with the permission from Elsevier.

References

Agarwal, G.S., Wolf, E., Calculus for functions of noncommuting operators and general phase-space method in quantum mechanics. I. Mapping theorems and ordering of functions of noncommuting operators. *Phys. Rev. D* 2:2161–2186, 1970.

Arthurs, E., Kelly, Jr. J.L., On the simultaneous measurement of a pair of conjugate observables. *Bell Syst. Tech. J.* 44:725–729, 1965.

Banaszek, K., Wodkiewicz, K., Testing quantum nonlocality in phase space. *Phys. Rev. Lett.* 82:2009–2013, 1999.

Braunstein, S.L., Caves, C.M., Milburn, G.J., Interpretation for a positive *P* representation. *Phys. Rev. A* 43:1153–1159, 1991.

Cahill, K.E., Glauber, R.J., Ordered expansions in Boson amplitude operators. *Phys. Rev.* 177(5):1857–1881, 1969.

Cartwright, N.D., A non-negative Wigner-type distribution. *Physica A* 83(1):210–212, 1976.

Eisamann, M.D., Fan, J., Migdall, A., Polyakov, S.V., Single-photon sources and detectors. *Rev. Sci. Instrum.* 82:071101, 2011.

Freyberger, M., Vogel, K., Schleich, W.P., From photon counts to quantum phase. *Phys. Lett. A* 176(1–2):41–46, 1993.

Glauber, R.J., Coherent and incoherent states of the radiation field. *Phys. Rev.* 131:2766–2788, 1963.

Husimi, K., Some formal properties of the density matrix. *Proc. Phys. Math. Soc. Japan* 22:264–314, 1940.

Lee, H.W., Generalized antinormally ordered quantum phase-space distribution functions. *Phys. Rev. A* 50(3):2746–2749, 1994.

Lee, H.W., Theory and application of the quantum phase-space distribution functions. *Phys. Rep.* 259(3):147–211, 1995.

Lee, S.W., Jeong, H., Jaksch, D., Testing quantum nonlocality by generalized quasiprobability functions. *Phys. Rev. A* 80:0220104, 2009.

Lee, S.W., Jeong, H., Jaksch, D., Witnessing entanglement in phase space using inefficient detectors. *Phys. Rev. A* 81:021302, 2010.

Leonhardt, U., Paul, H., Phase measurement and Q function. *Phys. Rev. A* 47:R2460, 1993.

Leonhardt, U., Paul, H., Realistic optical homodyne measurements and quasiprobability distributions. *Phys. Rev. A* 48:4598, 1993.

Noh, J.W., Fougeres, A., Mandel, L., Measurement of the quantum phase by photon counting. *Phys. Rev. Lett.* 67(11):1426–1429, 1991.

Paris, M.G.A., Quantum state measurement by realistic heterodyne detection. *Phys. Rev. A* 53(4):2658–2663, 1996a.

Paris, M.G.A., On density matrix reconstruction from measured distributions. *Opt. Commun.* 124(3–4):277–282, 1996b.

Schleich, W.P., *Quantum Optics in Phase Space*, Wiley-VCH, Berlin, 2001.

Sudarshan, E.C.G., Equivalence of semiclassical and quantum mechanical descriptions of statistical light beams. *Phys. Rev. Lett.* 10(7):277–279, 1963.

Stenholm, S., Simultaneous measurement of conjugate variables. *Ann. Phys.* 218(2):233–254, 1992.

Wigner, E., On the quantum correction for thermodynamic equilibrium. *Phys. Rev.* 40:749–759, 1932.

Wodkiewicz, K., Operational approach to phase-space measurements in quantum mechanics. *Phys. Rev. Lett.* 52(13):1064–1067, 1984.

chapter fifteen

Measurement issues in performance-based logistics

Kenneth Doerr, Donald R. Eaton, and Ira A. Lewis

Contents

15.1 Introduction

Performance-Based Service Acquisition is a Department of Defense (DoD) acquisition reform that has had noted success in reducing cost and streamlining the management of noncore government service capabilities (Office of the Deputy Under Secretary of Defense—Defense Acquisition Reform [OSD-DAR], 2000). The guiding principle in Performance-Based Service Acquisition is that when an outside vendor exists who can perform a service more effectively than a government user could organically, that user should specify measurable outcomes to a service vendor and allow the vendor to best determine the appropriate processes (how) of delivering the service. In adopting this reform, DoD has been influenced by the perceived success of outsourcing in the private sector. Firms have witnessed gains from devolving noncore activities to suppliers, while achieving high levels of transparency so that visibility of inventory and information is maintained throughout the supply chain (Spekman and Davis, 2004).

Performance-Based Logistics (PBL) is an extension of Performance-Based Service Acquisition aimed at the logistic services for major weapon systems. The Quadrennial Defense Review mandated DoD to implement PBL in order to "compress the supply chain and improve readiness for major weapons systems and commodities" (Office of the Secretary of Defense, 2001: 56). PBL is intended to reduce life-cycle cost, increase readiness, improve reliability, and reduce the logistical footprint of weapon systems. A number of case studies of successful PBL initiatives are available (e.g., Candreva et al., 2001).

This chapter takes for granted the ubiquitous nature of PBL initiatives, and takes as its starting point the question of how best to measure the degree of their success. In support of our prescriptions for measurement, we will draw not only on successful best practice but also on the underlying logic and justification of outsourcing as laid out in the economics and management literature. While PBL prescriptions from OSD are always careful to explain that a PBL initiative may result in the selection of an "organic contractor" (i.e., another DoD command), actual instances of "organic contractors" are fairly rare, and in any event, some of the same measurement issues arise regardless of the blend of private sector and organic resources.

While measuring the performance of ongoing PBL initiatives is our starting point, we also intend this chapter to inform valuation questions. From the initial question of whether to bring forward a weapon system or a major component of a weapon system as a candidate for PBL to later design questions of "what form" of PBL is best applied to that candidate, measurement issues are endemic. After all, the logistic services to be outsourced will be priced contractually, and for some services there is no clear market to determine that price. When discussing CONUS transportation, prices are perhaps not difficult to determine by reference to a market. However, when discussing something like intermediate-level maintenance of a deployed weapon system on which the DoD has a monopsony and the number of qualified bidders is quite limited (and may indeed be only one or two), the market paradigm clearly breaks down and is perhaps best understood in the context of game theory (Shubik and Levitan, 1980). Unlike the simple solutions of monopolistic games, however, the monopsonistic game of buying weapon systems logistics is hampered by the difficulty of measuring the value of the services to be obtained.

In discussing whether a case could be made for the privatization of a particular governmental service, Bendick (1984) said it was important to compare private to "nonmarket" (i.e., organic) alternatives and that the private sector should only be employed if it could reasonably be expected to be more efficient. He listed four aspects of market efficiency [that] are important to examine.

- In producing the services ... do the private sector's production processes and input costs allow it to generate output at a lower total cost than could the public sector?
- Are the administrative costs incurred by government to mobilize and control the private sector less than the cost savings from more efficient production?
- Is the supply side of the market sufficiently responsive that private firms enter markets rapidly and smoothly?
- Are purchasers sufficiently rational and careful, and the quality of the service sufficiently definable and measurable, that effective, informed consumer sovereignty can be exercised? (Bendick, 1984, pp. 153–154)

Each of these considerations is potentially problematic when examining PBL initiatives. When considering the first of the above factors, the existence of PBL contracts in which the private sector vendor has hired back organic resources as subcontractors to do the manual labor puts in question exactly what services are being outsourced—logistics or management? When considering the third of his factors, the consolidation of the defense industry and the decline of the number of independent companies that might act as potential bidders raise concern. However, this chapter will primarily concern itself with the second and especially the fourth of these factors. We will discuss how an excess of measurement can make administration of comprehensive PBL contracts more costly, while the difficulty with defining and measuring some logistic services make consumer sovereignty difficult to establish.

The rest of the chapter will be organized as follows. First, we will lay out a structural framework upon which measurement issues will be developed. Upon that framework, we will then develop questions about how measurement informs which sorts of candidates are best suited for PBL. Finally, we will discuss how measurement issues should be considered when deciding on the form of PBL to be adopted for a particular candidate and the management of ongoing PBL contracts. We are not attempting to clearly delineate between good and bad measures, or good or bad candidates for PBL. Rather, we are attempting to surface imbedded measurement-related issues that may make the difference between a problematic implementation and an easy one. Thus, this chapter is not intended as a guidebook for implementation, but rather as a guidebook for further investigation.

15.2   A hierarchical bridge framework of measurement for PBL

When describing logistics service acquisition for a weapon system as an economic game, it is important at the outset to note the dissimilarity between the two players. The vendor has a clear objective of maximizing the wealth of its owners and a clear profit incentive (again, we assume throughout the chapter that we are dealing with a private sector vendor). The objective of the user acquiring the service is not so easy to state and far more difficult to assess. Ambiguities in goals and a lack of linkage between services acquisition goals and strategic objectives are intrinsic aspects of the services acquisition process (Ausink et al., 2002; Camm et al., 2004). Maximizing national security would be one way to state the objective, and the incentive (at least at the organizational level) might be understood in the same terms—to gain more security for the nation. At the outset then, the game has a measurement and a translation problem—measuring the services in terms of their contribution to the objectives and incentives of the DoD and translating that measure into the dollar measurement used by the private sector.

Of course, it might be claimed that business does not really have such clear objectives and incentives either. There is venerable literature pointing out that the maximization of shareholder wealth should not be (and is not in practice) the sole aim of a public corporation. Stakeholder analysis has its roots in this observation (Donaldson and Preston, 1995). But even stakeholder analysis (in narrow form at least) does not deny the centrality of profit as a corporate incentive, rather the discussion centers on rights of resource holders and equitable distribution of profits.

The management fashion of balanced scorecards has demonstrated the willingness of corporate executives to look beyond profit in analyzing performance (Kaplan and Norton, 1992). But it would be a mistake to take the current proliferation of balanced scorecards as evidence that corporations suffer under the same sorts of fundamental measurement problems with their objectives and incentives as the DoD. The balanced scorecard is clearly meant to be a *diagnostic tool* to inform management decisions beyond retrospective financial figures about the long-term viability of the firm (i.e., it is meant in part to help predict and control *future* financial performance). Kaplan and Norton (1992) discuss the shortcomings of financial performance measures in terms of their ability to guide (1) the innovation necessary to obtain future profitability, (2) the diagnosis of internal process problems that limit current and future profitability, and (3) the relationship with the customer necessary to sustain future profitability. Their main criticisms of current financial measures (which are a part of the balanced scorecard) are that they are historical and external to operations. They tell a firm how well it has performed, not why, or what to do next to maintain or improve future performance.

But measurement-related differences between the DoD and the corporate world exist not only in the incentives and objectives of each, but also in the process capabilities that are important in developing logistics tactics to meet those objectives. In reviewing essential dimensions to be considered in logistic performance analysis in the commercial sector, Mentzer and Konrad (1991) developed a matrix in which five core logistics functions (transportation, warehousing, inventory, order processing, and administration) could be measured along six dimensions (cost, labor, facilities, equipment, time, and energy). Distinguish between those dimensions and the four "overarching goals of PBL … to compress the supply chain, eliminate non-value added steps, reduce Total Ownership Cost and improve readiness for weapons systems …" (Department of Defense—Defense Contract Management Agency [DoD-DCMA], undated) to which one should add "increased reliability and reduced logistics footprint" (Office of the Secretary of Defense [OSD], 2003). Aside from cost, these sets of six factors seem to have little in common. But all of the commercial sector factors can be translated into cost and can be understood as the essential dimensions that must be managed efficiently and effectively in order to facilitate logistics support of the firm's profitability objective. The DoD factors, on the other hand, do not all translate so readily into cost and fall into three categories of dimensions that logistics improves warfighting capability: improved readiness (facilitated both directly by a focus on readiness and indirectly by a focus on reliability), increased agility (reducing logistical footprint, eliminating non-value-added steps, supply chain compression, and improved reliability), and reducing cost (by freeing capital for other warfighting priorities).

This is a significant difference in how logistics is viewed. The concept of readiness shows up as "equipment" to commercial firms, who view the maintenance and functioning (and depreciation) of their operating capital primarily as a financial question—when will it become so expensive to maintain that I will have to replace it? Since DoD weapon systems are often quite old, very expensive, and difficult to recapitalize (lacking a depreciation mechanism, recapitalization is often driven by technological obsolescence or budget constraints), readiness is a much more central issue. Improvements in readiness, of course, improve warfighting capability, but marginal improvements are quite difficult to value in dollar terms. The idea of "agility" is increasingly important to commercial firms, but agility in a commercial operation means, for example, the flexibility to quickly change production volumes or quickly change production technology. It shows up in the list above as "time," which is also translatable to dollars. DoD operations on the other hand are mobile, and mobility directly impacts their effectiveness. Agility is not a newly discovered competitive dimension—it has always been an operational necessity. Once again, however, the operational effectiveness derived from a marginal improvement in logistics agility is very difficult to translate into dollars.

These differences in organizational objectives and the consequent logistics objectives further devolve into differences in process measurement. Caplice and Sheffi (1994), in a classification and review of corporate logistic process metrics, developed three categories: utilization, productivity, and effectiveness (see Table 15.1). Utilization measures simply address the question of how much of a resource is used, compared to what has been made available. While these sorts of measures may be useful in assessing the efficiency of a narrow segment of a process (e.g., space utilization may be useful in assessing the efficiency of a facilities layout manager), they have virtually no contribution to the understanding of logistics contribution to organizational objectives, primarily because they do not measure outputs at all.

It might be claimed that they measure waste, but even this is not true—all they measure is activity, not whether that activity is directed toward some valued outcome. What Caplice and Sheffi (1994) have called effectiveness measures, on the other hand, "beg the question" in an essential way—those measures are only as good as the norms one establishes for outputs. They may be useful for historical comparison of a single process, but their value in comparing across processes or in guiding resource allocation decisions is quite limited.

Table 15.1 Corporate logistics metrics

Dimension	Form of metric	Logistics examples
Utilization	Actual input/ input norm	Labor hour used/labor hours budgeted Area of warehouse occupied/total area Hours machine used/machine capacity
Productivity	Actual output/ actual input	Ton-miles delivered/costs incurred Orders processed/hours of labor Pallets unloaded/hour of dock time
Effectiveness	Actual output/ output norm	Items filled/items requested Shipments on time/shipments sent Transactions w/o error/total transactions

Source: Adapted from Caplice, C. and Sheffi, Y. 1994. *The International Journal of Logistics Management,* 5(2), 11–28.

Productivity measures, on the other hand, incorporate both outputs and inputs. For the corporation, assessing the contribution of an activity to its objectives is a matter of relating those inputs-outputs to profits. While of course this is not necessarily easy (e.g., single factor productivity measures do not capture a comprehensive cost picture), at least the examples given by Caplice and Sheffi (1994) can be measured or translated to dollars (e.g., dollars paid for orders processed, or shipments made), and this is broadly true of metrics proposed in other reviews of corporate logistic performance measurement systems as well (e.g., Mentzer and Konrad, 1991; Chow et al., 1994; Lambert and Burduroglu, 2000), with the important exception of customer satisfaction metrics.

The importance of the "customer view" has already been mentioned in relation to balanced scorecards and is often mentioned by authors on logistic performance measurement (e.g., Mentzer et al., 1999). However, it is worth noting that Lambert and Burduroglu (2000) list "reliance on management outside of logistics to identify the impact [of customer satisfaction] on revenues, which typically does not happen" as a primary disadvantage of customer satisfaction measurement. Hence, beyond simple utilization measures, corporate logistic performance measures can, or are desired to be, understood in terms of their impact on profitability.

Compare those corporate logistic measures to what might be proposed as a productivity ratio for weapons systems logistics:

$$A_o = \frac{A_{mch}}{T_{dh}}$$

where

A_o = Operational availability
A_{mch} = Fully mission-capable hours available
T_{dh} = Total deployed hours

For example, if 10 aircrafts are deployed in a squadron in a given month, and nine of them are fully mission-capable for the whole month, while the 10th is down for maintenance the whole month, that squadron would report an A_o of 90%. At first glance, this looks like a utilization measure, not a productivity measure, but A_o is often used as a surrogate for readiness, which is typically given as a primary *outcome* objective of military logistics (it would be a utilization measure if, e.g., flight hours were in the numerator). The denominator translates to dollars in a budget (whether or not they could be translated to an actual cost is another issue), but the numerator is not and should not be translatable to dollars because profit is not the objective. While measurable, it is difficult to value in terms of the dollars that might be spent to increase it or relinquished in order to pursue other priorities.

Another problem is that A_o is only a surrogate for readiness because it is a "single factor" measure. It is also not fine grained enough for many resource allocation decisions we wish to make (hence the distinction between "mission-capable" and "fully mission-capable" systems). And finally, readiness itself, after all, is only a surrogate for the organizational objectives of the DoD (i.e., ready for what?). Note that if A_o were really the objective, it could be maximized by parking equipment. Hence, logistic performance of weapon systems is more difficult to measure than commercial logistics (at least in terms of productivity) and is perhaps more attenuated from DoD objectives than are commercial logistic measurement systems (Figure 15.1).

How do these measurement issues inform the decision to bring forward a weapon system or component as a PBL candidate? First, again considering only outsourced PBL solutions, we must consider the economic logic behind outsourcing. One basic economic justification of outsourcing is the trade-off of economies of scale with a reduction in transaction costs. If the outsourced service can be performed by an organization that offers similar services to a number of other customers, that organization gains economies of scale and should be able to offer the service more cheaply than if it were done by the outsourcing organization in house. The price that is usually paid for such outsourced services is usually in terms of increased transaction costs to negotiate price and services, and monitor performance (Gustafson et al., 1996).

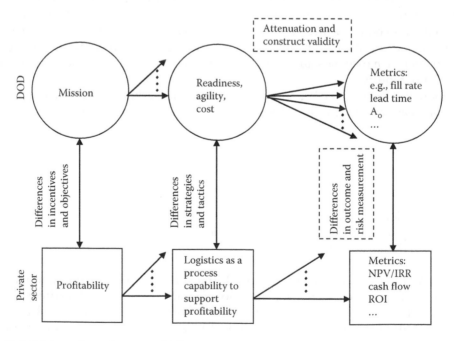

Figure 15.1 A hierarchical framework of measurement issues.

When economies of scale are difficult to obtain, as with a unique weapon system requirement, some of the underlying justification for outsourcing disappear. On the other hand, high internal transaction costs, due to, for example, high reporting requirements, or inefficient internal controls make outsourcing relatively more attractive. If high internal transaction costs are part of the justification for outsourcing a PBL contract, then it is important that the system or component being outsourced avoids some of those transaction costs. When the measurement of logistic outcomes (readiness, agility, and cost) is more difficult, it will mean higher transaction costs, because performance monitoring systems will have to be more elaborate, and fair prices will be more difficult to determine and negotiate.

One way to make pricing and performance monitoring easier is by reference to a market for similar services. Hence, in prescribing a methodology for the analysis of performance-based contracts for contract managers, market research is indicated as a required step (OSD-DAR, 2000). For comprehensive weapon system logistical support or for weapon system-unique components, there will likely be no ready market for maintenance or many other logistical support functions. In those cases, the implementation of an outsourced PBL solution will require more cost and effort to develop appropriate metrics and negotiate appropriate prices.

In summary, measurement issues are endemic to the relationship between commercial sector vendors and the DoD. From the point of view of measurement, the best PBL candidates are those with external markets for services, and clear outcomes that are easy to relate to mission objectives. This is not to place a definite boundary on the systems where PBL ought to be applied, but only to point out that measurement issues may make some PBL implementations far more difficult and expensive, and may affect the form of the PBL solution.

15.3 Measurement, the PBL support spectrum, and the management of ongoing contracts

One of the characteristics of PBL is that general characterizations are hard to make. The top-level guidance for the initiative always has caveats such as

> There is no one-size-fits-all approach to PBL. Several programs have started the move to PBL under initiatives designed to meet the programs' specific requirements. Each program has tailored the PBL application to its unique circumstances taking into account cost, schedule, or product integrity to meet warfighter capability. (DoD, 2001, p. 2)

In reviewing implementations, a wide variety of approaches can be found, in terms of measurement and incentives and in terms of the level at which the PBL contract is written—from a complete weapon system as with the DDG-51 (class of guided missile destroyers) to component-level stock support as with the Aegis (an integrated naval weapons system). The spectrum of choices is usually described in terms of the degree of commercial support involved, and a frequently encountered graphic (which we have been unable to track to its original source) is shown in Figure 15.2.

While examples of systems are often given in association with this chart, and definitions of the various anchor points (e.g., mini-stock point) are offered, very little guidance

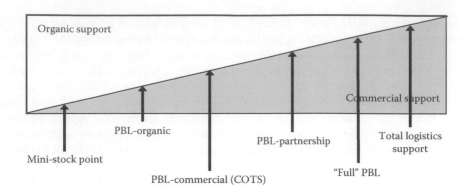

Figure 15.2 The PBL support spectrum.

can be found for the program manager as to what characteristics of a weapon system should inform the choice of the degree of commercial involvement, and whether the contract should be offered at the system or component level. It is our position that measurement issues should inform this choice.

A primary aspect of measurement informing the choice of commercial involvement, which we have not discussed yet, is risk. As should be clear by reference to Figure 15.1, vendors will be primarily interested in reductions of financial risk, while the DoD is entirely concerned with operational risk. The trade-off of these two kinds of risk is central to the logic of PBL outsourcing. Contracts are almost always offered across multiple years (lowering financial risk for the vendor), with the expectation that the vendor will assume some degree of operational risk. Figure 15.3 shows the expected trade-off of operational risk by the vendor.

Although risk is clearly indicated as a factor to consider when developing a PBL strategy (ASN-RDA, 2003; Office of the Assistant Deputy Under Secretary of Defense—Logistics

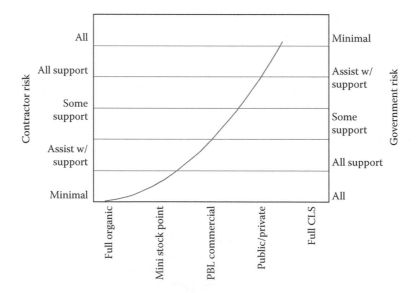

Figure 15.3 Intended risk transfer under PBL. (From DoD-Defense Contract Management Agency, 2002. *Performance Based Logistics Support Guidebook.* Unpublished manuscript dated October 2002, downloaded June 10, 2015 from https://acc.dau.mil/adl/en-US/54825/file/18745/PBL-GUIDE.doc.)

and Material Readiness [OSD-LMR], undated), this factor is rarely mentioned as a candidate for measures of ongoing performance in PBL contracts. Indeed, it has been said that

> minimal contract management involvement is anticipated as long as the contractor meets contractually specified performance metrics. However our involvement may increase if the contractor systems and processes are not functioning correctly and end users are not appropriately supported. (DoD-DCMA, undated, pp. 28–29)

This is a curious form of risk transfer. Operational risk, as we understand it, involves the variance associated with outcomes. The assumption of operational risk by a contractor then would entail accountability for control of that variance and assurance (with appropriate remedy in the case of failure) of the mitigation of its impact. Here, rather, it seems that vendors are being asked to assume some aspects of risk (not clearly defined), but that the DoD will "increase involvement if the contractor" fails to meet requirements. We do not claim that this is risk transfer in name only, but that the form of risk being assumed by vendors is not what is commonly understood as operational risk, and the degree of risk they are assuming is apparently quite limited.

It is our view that the degree of operational risk a contractor can assume is limited in many cases by the nature of military operations. It is unrealistic, for example, to assume that contractors will be able to perform operational-level maintenance on a ground combat weapon system; difficult issues relating to the physical risk, insurance, and liability of non-DoD personnel in or near combat need to be addressed. These sorts of operational risks are difficult to measure and even more difficult to value. We think it likely therefore that commercial sector vendors will be reluctant to undertake these risks. Depot-level maintenance and operation of CONUS inventory control points involve less operational risk, risk in a form that is easier to measure and less costly for a vendor to assume. Hence, we think it likely that the more operational risk involved in the logistical support of a particular system, the more organic resources will need to be involved.

Proposition 15.1

When operational risk is high or difficult to measure, PBL strategies should seek less commercial sector involvement.

Within the context of a price negotiation, it is also key to understand the benefit the DoD provides by eliminating financial risk, as this is part of what we are paying to potential vendors. Especially if interest rates and rise and the difference between the cost of capital and risk-free rates increase, what the DoD offers in terms of financial risk mitigation is highly valuable. This valuable benefit is not free for the government to offer, and should be incorporated into pricing and contract negotiations. If less operational risk is assumed by the vendor (or if that risk is difficult to assess), less financial risk should be mitigated—meaning contract terms should be reduced.

Proposition 15.2

When commercial sector vendors assume less (measurable) operational risk under a PBL contract, the term of that contract should be less.

On the other hand, the outcomes of PBL strategies involving only certain components, or only depot-level support, are more difficult to tie to weapon system

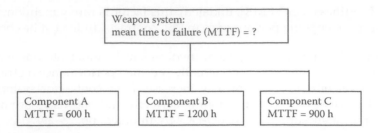

Figure 15.4 Need for integrated system model to judge impact of component outcomes.

outcomes. Consider Figure 15.4, which shows a highly stylized and simplified version of a weapon system and its major components, along with the failure rates (mean time to failure) of each of the components. Assuming failure of any of the components causes the weapon system to become non-mission-capable, the failure rate of the overall weapon system is then an order statistic, formed of the *distributions* of the time to failure of *all* of the components. Now consider the problem faced by a program manager who has decided that his best PBL strategy involves outsourcing only component A (the one with the highest failure rate). To properly value the impact of, for example, a proposed incentive to improve the reliability of component A by 10%, the program manager would need not only distributional information about the time to fail of all the other components, but also a working model, which imbeds that system in mission requirements. After all, the final value of an improvement in reliability of a component (to readiness—of course there are other benefits in terms of reduced life cycle cost of spares, and improved agility through reduced footprint) rests in the increased likelihood of mission success in the deployed weapon system. The sort of integrated simulation model needed to properly assess the impact of improved component reliability would be expensive, and more importantly, time consuming to build. Given the time pressures put upon Program Managers, it is easy to see that the situation is problematic.

Proposition 15.3

PBL strategies involving less than comprehensive logistical support of a weapon system (e.g., for a component) should nonetheless have integrated weapons system models in support of their business case analysis.

In summary, measurement issues exist across the PBL spectrum, but present different sorts of challenges at either end. Ultimately, there are at least two core measurement issues that should be referred to when deciding on an appropriate level of support within the PBL spectrum. The first is the valuation of outcome-related performance, and the second is valuation of operational and financial risk. While outcomes are easier to measure at the right end of the spectrum, one is less likely to find a relevant market to support price and value decisions. On the left hand of the spectrum, markets may well exist that essentially duplicate, for example, the services of a mini-stock point. However, the valuation of those isolated services in terms of weapon systems performance is even more difficult.

Aside from risk, another distinction important to develop when discussing management of ongoing PBL contracts is the difference between process and outcome measures. It is our position that, while PBL is clearly intended to buy outcomes and

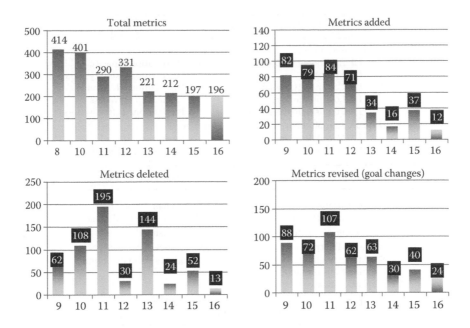

Figure 15.5 A measure of PBL measures used at one DoD command.

relieve management of the necessity of monitoring the details of "how" performance is obtained, a great deal of effort is still being devoted to process measurement. It has recently been said that "too many metrics" is a major problem with PBL implementation (Office of the Secretary of Defense, Undersecretary for Acquisition, Technology and Logistics [OSD-ATL], 2004). If the DoD is buying outcomes, not processes, then it may be that much of the process measurement is unnecessary.

An example of "too many metrics" is shown in Figure 15.5. This is a slide shown in a brief to a base commander to provide an overview of the PBL contracts at his base. This is a small base, with only a handful of PBL contracts. Clearly, the commander understands that there are too many metrics, and is tracking them quarterly in order to push for their reduction. Here, the number of metrics itself has become a metric, with visibility to the top operating officer at a command.

Exactly how the superabundance of metrics arises is an open question. PBL is a process meant to streamline management concern with the details of a logistical process. In part, it may arise from a broader context within DoD, of understanding the systemic relationships of which logistics is only a part. Under various titles, including Integrated Logistics Support, the last several years has seen an increasing awareness of the embedded nature of logistical support, and the interrelationships involved between, for example, manpower, maintenance, and supply. Figures 15.6 and 15.7 were taken from a presentation to a PBL "tiger team" concerned with establishing metrics. Figure 15.6 lists the "balanced scorecard" of top-level factors for weapon system support.

Figure 15.7 translates the scorecard into logistics measures by mapping between the high-level scorecard factors, and the primary factors (process elements) of another management fashion called integrated logistical support (ILS). The details of ILS process elements are not germane here, only to note that they are indeed *detailed* and *process* oriented. If one starts with a multidimensional balanced scorecard, and works

Warfighter

L1: Force readiness
    L2: Equipment readiness
    L2: MC rates
    L2: Commodity availability
    L2: Logistics units readiness
    L2: Prepositioned readiness
L1: Force sustainment
    L2: Materiel support
    L2: Services support

L1: Force closure
    L2: Distribution capacity
    L2: Distribution visibility and control
    L2: Distribution effectiveness
    L2: Combat effectiveness
    L2: Prepositioned effectiveness
L1: Reset
    L2: Reset
    L2: Operational drawdown

Innovation and learning

L1: Innovation realization
    L2: Rate of transportation
    L2: Rate of improvement
L1: Workforce capability
    L2: Workforce shaping
    L2: Learning transformation
    L2: Employee climate

Logistics process

L1: Logistics chain reliability
    L2: Perfect order fulfillment
L1: Logistics chain effectiveness
    L2: Logistics chain agility
    L2: Logistics chain cycle time

Resource planning

L1: Logistics chain predictability
    L2: Planned cost to actual cost
    L2: Demand forecast accuracy

L1: Logistics chain cost-effectiveness
    L2: Total logistics chain cost percentage
    L2: Total logistics chain cost
    L2: Value added productivity

Figure 15.6 DoD balanced scorecard for weapon system support. (From Doerr, K., Eaton, D.R. and Lewis, I.A. 2004. In *Proceedings of 2nd Annual Acquisition Research Symposium of the Naval Postgraduate School*, Monterey, CA, May 13, 2004. Approved for public release, distribution unlimited. With permission.)

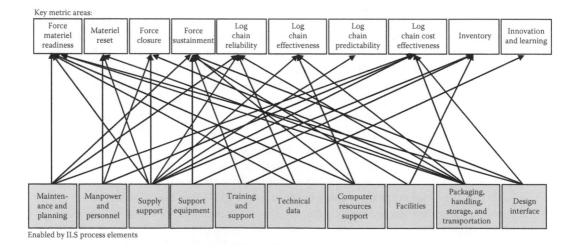

Figure 15.7 Metric areas informed by balanced scorecard and ILS process elements. (From Doerr, K., Eaton, D.R. and Lewis, I.A. 2004. In *Proceedings of 2nd Annual Acquisition Research Symposium of the Naval Postgraduate School*, Monterey, CA, May 13, 2004. Approved for public release, distribution unlimited. With permission.)

through these process elements, it is easy to see how the number of metrics that must be tracked could be numbered in the hundreds—in fact, it would be hard to avoid. One must ask why, however, if we are engaged in an initiative to buy performance, we are starting with a detailed map of the internals of the process. Would it not make more sense to measure only key outcomes, and measure them well?

When we first presented this idea at a conference, we were met with the objection that an abundance of measures do not necessarily distract a decision maker from key tasks. The analogy was drawn to a pilot in a jet, where the cockpit has a superabundance of meters and instruments, almost all of which can be ignored, except in the case of an emergency. The analogy is a telling one, in that most of the people making decisions about metrics for PBL have themselves been pilots, or ship captains, or in charge of some complex process in the past. However, PBL is not supposed to present the DoD with a complex process to manage—it is supposed to take one off the hands of the DoD. We are not supposed to be flying the plane—we are supposed to be passengers. When you are paying someone else to get you to your destination, you care about the price of the ticket and arriving on time.

Of course this is a simplification. When discussing a PBL contract with one deployed squadron, we were met with a complaint about the difficulty of obtaining requisition status for parts that were backordered. The maintenance officer for that squadron was not impressed that the average lead time to get the parts had been reduced, because his primary decision when confronted with a backorder was whether or not to devote the manpower to strip a spare part from another downed aircraft. It was a decision that was difficult to make, without knowing how long it would be before the spare part arrived. Clearly, in this case, some visibility into the process was needed, *but only because a key operational decision rested on the measurement of that process.* Given that at least part of the justification for PBL is meant to be a reduction in transaction costs, we think this should probably be a general rule.

Proposition 15.4

In the management of ongoing PBL contracts, metrics should primarily concern themselves with valued outcomes, and should be related to weapon system cost, readiness, and agility. Process measures should only be applied when key operating decisions depend on the status of the process itself.

Finally, we turn to the measurement of risk in ongoing PBL contracts. Operational risk is always difficult to assess. In the context of support for a weapon system, it can be understood as variance in the logistics-related readiness of that system. A common measure of readiness is A_o. To see how variability, or risk, affects A_o, consider Table 15.2. The table shows the availability of two squadrons of 10 aircrafts over a 20-day period. Over that period, both squadrons would report operational availability of 95%. But consider that a mission requiring 10 aircrafts would be degraded (more likely to fail), only 5% of the time in the first case, but 50% of the time in the second case. The difference is the variance in A_o. To our knowledge, in spite of the exhortations of the centrality of risk assumption and readiness to PBL, there are no programs currently tracking this measure.

In a very thoughtful document, the office of the Assistant Secretary of the Navy for Research, Development and Acquisition (ASN-RDA, 2003) listed factors to consider when deciding whether to use a PBL strategy: life cycle stage, acquisition program strategy (including programmatic risks), organic impact (e.g., maintaining

Table 15.2 Distributional differences in a 95% A_o

Day	FMC aircraft in squadron A (10 aircrafts total)	FMC aircraft in squadron B (10 aircrafts total)
1	10	10
2	10	09
3	10	10
4	10	09
5	10	10
6	10	09
7	10	10
8	10	09
9	10	10
10	10	09
11	10	10
12	10	09
13	10	10
14	10	09
15	10	10
16	10	09
17	10	10
18	10	09
19	10	10
20	0	09

engineering expertise), commercial base (including additional risk required of industry partners), design considerations (including risk associated with incentives and performance thresholds), and technology considerations (including supportability risks). Although risk is mentioned in four of the six factors, there is no mention of the sort of operational risk discussed above and shown in Figure 15.3. Indeed, there is little mention of risk for ongoing contracts in this, or any other guidance documents for PBL. We find this curious. If part of what we are buying is operational risk mitigation (in key performance dimensions), it seems to us that it ought to be measured.

15.4   Conclusion

Organic acquisition and sustaining logistics in the DoD of today is seriously challenged in its ability to provide significantly higher levels of operational availability and effectiveness commensurate with the complexity and high costs of weapon systems. In addition, we exacerbate the pressure on logistics support as we prosecute our national security strategies with ships, airplanes, and equipment that are aging when a sufficient replenishment rate to counter that aging is unaffordable. Any change in the Defense Department Logistics discipline that can produce greater utility at lower costs should be the focus of an energetic pursuit. In order to bring about the necessary improvements, we must mitigate cultural impediments and concurrently align measurement and rewards with good logistics performance in the acquisition process and operation of weapons systems in the field. It is also important to fully fund logistics requirements, safeguard the funding, and engage in long-term contracts.

The DoD faces major challenges in establishing a coherent logistics support strategy for its weapon systems. While PBL has been decreed as a preferred implementation strategy, real questions remain unanswered about objectives and measurability. The disaggregation of logistics support and the emphasis on an increased role for the private sector in logistics also raise major issues related to the assumption of risk. In making sourcing decisions, DoD must act as a "knowledgeable client" and avoid simplistic decisions that may lead to critical systems being left unsupported.

In summary, this chapter has presented a framework and propositions about the impact of measurement on PBL. None of the propositions have the status of fully supported hypotheses or fully developed theorems. All need further investigation. Some of the propositions are empirical and need to be investigated in the field. Others are prescriptive and need to be supported by modeling and analysis. Our hope is that we have furthered the discussion of metrics for PBL and added to the momentum for improved implementations of PBL.

Acknowledgments

The authors would like to acknowledge the support of RADM Jim Greene whose broad approach to innovating DoD acquisition enabled this effort. Dr. Keith Snider has also been instrumental in organizing this research effort and bringing it to a timely completion. We also wish to thank Ms. Karey Shaffer, whose patience and labor over our drafts were invaluable. Finally, we want to thank Tom Najarian of Lockheed Martin, Lawrence Fitzpatrick of NAVSEA, and Jeffery Heron of NAVAIR, whose insight and background provided the foundation for our effort.

This work has been published in the *Journal of Public Procurement*, 5(2):164–196, 2005. It has been reproduced here with permission from the *Journal of Public Procurement*.

References

Assistant Secretary of the Navy—Research, Development and Acquisition [ASN-RDA]. 2003. *Performance Based Logistics Guidance Document*. Unpublished manuscript.

Ausink, J., Baldwin, L.H., Hunter S. and Shirley, C. 2002. *Implementing Performance-Based Services Acquisition (PBSA): Perspectives from an Air Logistics Center and a Product Center*. Santa Monica, CA: RAND Corporation (Report No. DB-388).

Bendick, M. 1984. Privatization of public services: Recent experience. In: *Public–Private Partnership*, Brooks, H., Liebman L. and Schelling, C. (eds), Cambridge, MA: Published for the American Academy of Arts & Sciences by Ballinger Publishing Company.

Camm, F.A., Blickstein, I. and Venzor, J. 2004. *Recent Large Service Acquisitions in the Department of Defense: Lessons for the Office of the Secretary of Defense*. Santa Monica , CA: RAND Corporation (Report No. MG-107).

Candreva, P., Hill, B., Marcinek, B., Sturken, B. and Vince, B. 2001. *Auxillary Power Units Total Logistics Support Case Study*. Philadelphia, PA: Working Paper, Tench Francis School of Business.

Caplice, C. and Sheffi, Y. 1994. A review and evaluation of logistics metrics. *The International Journal of Logistics Management*, 5(2), 11–28.

Chow, G., Heaver, T.D. and Henriksson, L.E. 1994. Logistics performance: Definition and measurement. *International Journal of Physical Distribution and Logistics Management*, 24(1), 17–28.

Department of Defense [DoD]. 2001. *Product Support: A Program Managers Guide to Buying Performance*. November 2001. Unpublished manuscript.

Department of Defense—Defense Contract Management Agency [DoD-DCMA]. 2002. *Performance Based Logistics Support Guidebook.* Unpublished manuscript dated October 2002, downloaded June 10, 2015 from https://acc.dau.mil/adl/en-US/54825/file/18745/PBL-GUIDE.doc

Doerr, K., Eaton, D.R. and Lewis, I.A. 2004. In *Proceedings of 2nd Annual Acquisition Research Symposium of the Naval Postgraduate School,* Monterey, CA, May 13, 2004.

Donaldson, T. and Preston, L.E. 1995. The stakeholder theory of the corporation: Concepts, evidence, and implications. *Academy of Management Review,* 20(1), 65–91.

Gustafson, K., Aubert, B.A., Rivard, S. and Patry, M. 1996. A transaction cost approach to outsourcing behavior: Some empirical evidence. *Information and Management,* 30, 51–90.

Kaplan, R.S. and Norton, D.P. 1992. The balanced scorecard: Measures that drive performance. *Harvard Business Review,* 70(1), 71–79.

Lambert, D.M. and Burduroglu, R. 2000. Measuring and selling the value of logistics. *The International Journal of Logistics Management,* 11(1), 1–17.

Mentzer, J.T., Flint, D.J. and Kent, J.L. 1999. Developing a logistics service quality scale. *Journal of Business Logistics,* 20(1), 9–32.

Mentzer, J.T. and Konrad, B.P. 1991. An efficiency/effectiveness approach to logistics performance analysis. *Journal of Business Logistics,* 12(1), 33–62.

Office of the Assistant Deputy Under Secretary of Defense—Logistics & Material Readiness [OSD-LMR]. (undated). *DoD Template for Application of TLCSM and PBL in the Weapon System Life Cycle.* Unpublished manuscript.

Office of the Deputy Under Secretary of Defense—Defense Acquisition Reform [OSD-DAR]. 2000. *The Guidebook for Performance-Based Services Acquisition (PBSA) in the Department of Defense.* Unpublished manuscript.

Office of the Secretary of Defense [OSD]. 2001. Quadrennial Defense Review Report, September 30. Retrieved July 6, 2004 from www.defenselink.mil/pubs/qdr2001.pdf.

Office of the Secretary of Defense [OSD]. 2003. *Designing and Assessing Supportability in DoD Weapon Systems.* Unpublished manuscript, dated May 29, 2003.

Office of the Secretary of Defense, Undersecretary for Acquisition, Technology and Logistics [OSD-ATL]. 2004. PBL Purchasing Using Performance Based Criteria. *Proceedings of the 2004 DoD Procurement Conference,* Orlando, FL.

Shubik, M. and Levitan, R. 1980. *Market Structure and Behavior.* Cambridge, MA: Harvard University Press.

Spekman, R.E. and Davis, D.E. 2004. Risky business: Expanding the discussion on risk and the extended enterprise. *International Journal of Physical Distribution and Logistics Management,* 34(5), 414–433.

chapter sixteen

Data processing and acquisition systems

Livio Conti, Vittorio Sgrigna, David Zilpimiani, and Dario Assante

Contents

16.1 Introduction

In many applications, the amplitude and frequency content of analog signals to be detected is *a priori* unknown and can lie in a very large range of values. This is a particularly challenging problem to be managed, especially when the data acquisition and signal analysis must be carried out in extreme environments (such as in space exploration) where large data-processing resources are requested together with a low-power consumption, radiation-tolerant electronics, and high efficiency of the device. The method described in this chapter (Conti et al. 2014) is specifically useful when the power spectrum of measured signals exhibits a very large variability in different frequency bands (typically associated with several coexisting different parameters or fields), thus making the runtime multichannel signal analysis of the phenomenon under study particularly difficult (see, for instance, Proakis 1995). Examples can be found in the laboratory, medical applications (neurology, cardiology, etc.), space science, aviation, telecommunications, etc. (Baciarello et al. 1987, Biagi et al. 1990, Parrot 2002, Sgrigna et al. 1993, 2005, 2007, 2012). Nowadays, there are many methods and devices available for spectral analysis such as *general-purpose* laboratory instruments (conceived to be used in the widest range of signal analysis) (Agilent 2013, Rohde and Schwarz 2013) or *processing devices* made for specific needs (Kudeki and Munson 2008, Rybin 2012, and references therein). These two classes of instruments have obvious drawbacks with respect to a device based on our method. Thus, for example, laboratory spectrum analyzers, while allowing very sophisticated frequency analyses (but redundant with respect to specific uses for which our method is proposed), are oversized and generally not suitable for many *in situ* measurements and application areas, being bulky complex instruments of considerable overall dimensions and power consumption. In this chapter, Section 16.2 describes the proposed method to sample an analog signal and to reconstruct its spectrogram by using an original three-stage scheme:

conditioning module (Section 16.2.1), acquisition module (Section 16.2.2), and processing module (Section 16.2.3). Section 16.3 introduces the electronic board of the signal-conditioning module that has been constructed and tested for the case of eight-frequency bands. Finally, the discussion and conclusions are reported in Section 16.4.

16.2   Description of the method

The method is based on three (conditioning, acquisition, and processing) interconnected modules performing the following three main functions of data-analysis procedures as illustrated in Figure 16.1:

1. To perform signal conditioning. The analog input signal is preamplified and subdivided into a fixed number of frequency channels, at each one of which a different tunable amplification is applied, defined as a function of weights estimated by a calibration procedure.
2. To sum up the conditioned channel signals. The resulting signal is digitized by a single analog-to-digital converter (ADC) to perform an appropriate spectrum analysis.
3. To execute the calibration procedure mentioned in item 1. The calibration is accomplished on the basis of the spectrum analysis of the ADC output signal (with the aim at separately optimizing the amplification of each frequency channel) after which the spectrum of the initial input signal is reconstructed.

The method may be applied to any analog signal $S(t)$ to reconstruct its spectrogram (i.e., the spectrum $S(\omega)$ as a function of time) through the continuous optimization of the differential amplification of each individual signal frequency band, so that all the spectral

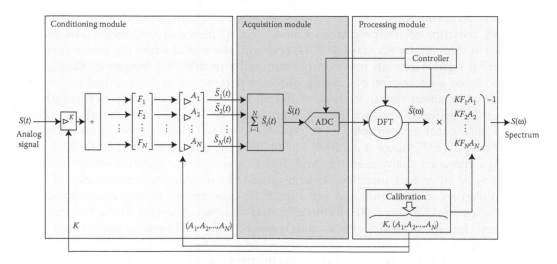

Figure 16.1 Scheme of the signal conditioning, data acquisition, and data-processing modules. (Reprinted from *Nucl. Instrum. Meth. Phys. Res. A*, 756, Conti, L. et al., Method for signal conditioning and data acquisition system, based on variable amplification and feedback technique, 23–29, Copyright 2014, with permission from Elsevier.)

amplitudes of the various frequency bands are made comparable between them. Once the amplification gains (that in general vary over time) are known, the true $S(\omega)$ spectrum can be reconstructed. The possibility to separately vary the gain of each frequency channel allows exploiting the resolution of a single ADC. The change in amplification, through a calibration procedure, optimizes the dynamic resolution of the ADC sampling in all frequency bands without "wasting bits" (Hadji-Abdolhamid and Johns 2003, Haykin 1994, Johns and Martin 1997). This flexibility of the method is particularly useful for signals in which the spectral content varies considerably from one frequency band to another. It follows that the actual dynamic resolution of the digitization procedure, achieved by the proposed method, is higher than that of the ADC static procedure.

Concerning the configuration, it must be underlined that the subdivision into modules, schematically shown in Figure 16.1, is to be considered only as an example, since the functional units can be placed and grouped in different configurations depending on the application or technological needs. The functional scheme described for each module may be constructed through both physical devices (hardware) and processing and control algorithms (software). All modules can be installed on a single electronic board or on separated interconnected units. In the basic scheme (described below), the method is applied to an analog input signal to produce its spectrogram. However, it is possible to exploit the modularity of the method to achieve different architectures. In particular, in case one wants to simultaneously analyze more analog input signals, it is possible to build an architecture consisting of a single processing module that simultaneously handles many subsystems. Each subsystem will consist of a conditioning module and an acquisition module and will be able to separately acquire several incoming analog signals. The single processing module will analyze all the various signals simultaneously and return all the spectra of the input signals in parallel outgoing.

16.2.1   The conditioning module

The conditioning module (Figure 16.1) consists of an amplifier K (including one or more stages with programmable gain), a divider (with N outputs), N filters with transfer functions $(F_1, ..., F_N)$, and N blocks of amplification $(A_1, ..., A_N)$. The conditioning module is connected to the acquisition module and the processing module. The output signal of K is divided by the divider block into N separate signals $S_i(t)$ $(i = 1, ..., N)$ of equal amplitude and identical frequency content to the input signal: $S_1(t) = S_2(t) = \cdots = S_N(t) = S(t)$. The N signals are sent in parallel and separately to the N filters, each one of which is chosen so as to select a particular frequency range (for its input signal), which is relevant for the purpose of the signal analysis and the specific application of the device. Also, the number N of filters depends on the number of frequency bands needed to analyze the $S(t)$ signal. The limit of $N = 1$ is possible and corresponds to an implementation of the method with only one frequency range analyzed. The N filters may be passive or active elements (with different stages depending on the requested filtering quality factor) and with characteristic parameters (cutoff frequencies, attenuation factor, bandwidth, etc.) fixed or variable depending on the desired configuration. A possible choice for the filters is constituted by N band-pass filters centered on adjacent and consecutive bands so as to cover the whole frequency range of the input signal $S(t)$. Another possibility is to choose filters that suppress specific intervals of unwanted frequencies in the input signal. Each of the N channels is sent separately to one of the N blocks of amplification (with variable gains $A_1, ..., A_N$) that can be constituted by one or more stages.

The output signal $\tilde{S}(t)$ from the *i*th block of amplification is given by $\tilde{S}_i(t) = S(t) \times K \times F_i \times A_i$. The gains $(A_1, ..., A_N)$ are set and varied (as described below) during the calibration phase by the processing module, which is activated at the beginning of the acquisition and can be triggered cyclically during the acquisition, depending on the temporal evolution of the $S(t)$ input signal or on a defined schedule. Once the calibration of the amplification is performed, the blocks maintain their settings until the next calibration cycle and the N-conditioned signals $\tilde{S}(t)$ $(i = 1, ..., N)$ are sent from the conditioning module to the acquisition module.

16.2.2   The acquisition module

The acquisition module includes an adder and an ADC. The N-conditioned analog signals $\tilde{S}(t)$ are summed up by the adder into a single analog signal given by $\tilde{S}(t) = \sum_1^N \tilde{S}(t)$. By setting the gains of all but one channel to zero, it is possible to select a single *r*th channel through the calibration procedure among the N-added signals: $\tilde{S}(t) = \tilde{S}_r(t)$. This mode allows to analyze the original signal $S(t)$ in only one of the N frequency ranges selected by filters. Similarly, by setting groups of gain amplifiers to zero, it is possible to select groups of channels. The output signal $\tilde{S}(t)$ from the adder is sampled by a single ADC. The characteristics of the ADC, such as sampling frequency, resolution, external trigger, serial or parallel architecture, etc., are chosen according to the input signal characteristics and the needs of analysis required by the application. In particular, the ADC sampling frequency should be large enough to acquire the highest frequency channel of the conditioning module. From the acquisition module, the digitized signal is sent to the processing module, which manages and sets all the acquisition parameters.

16.2.3   The processing module

The processing module includes a programmable and reprogrammable controller unit, a block that calculates $\tilde{S}(\omega)$ that is the discrete Fourier transform (DFT) of the conditioned signal $\tilde{S}(t)$, a calibration unit, and a unit to reconstruct the spectrum $S(\omega)$ of the input $S(t)$ signal under study. The processing module can be based on a single data processor or on separate devices, such as central processing unit (CPU), digital signal processing (DSP), field-programmable gate array (FPGA), etc. The controller unit manages the other processing blocks (DFT, calibration, and spectrum reconstruction), synchronizes the ADC with any external signal trigger, oversees cycles of start/stops of calibration and acquisition, and interfaces the entire three modules of the data-acquisition system with the external environment. The processing module computes the DFT and produces the spectrum $\tilde{S}(\omega)$ of the conditioned and digitized signal $\tilde{S}(t)$. The calculation of the spectrum $\tilde{S}(\omega)$ is repeated continuously for data packets of a suitable time duration and according to the need for the signal analysis. The sequence of spectra $\tilde{S}(\omega)$ as a function of time produces the spectrogram. The number M of the DFT lines determines the resolution of the spectrum $\tilde{S}(\omega)$ of the conditioned signal (as well as that of the spectrum $S(\omega)$ of the input $S(t)$ signal) and is chosen as a function of the number N and bandwidths of the channels in the conditioning module. Generally, the greater the number N of channels, the greater will be the resolution necessary to know the spectral content in each channel and therefore, the size M of the DFT should be greater than N. By selecting the number N of channels, the bandwidth (and the spectral content) can be reduced for each channel so as to optimize the gain to be applied to each channel. The size M of the DFT, the time between a DFT and the subsequent one (dead time), as well as the frequency with which the spectrum $\tilde{S}(\omega)$ is evaluated are fixed or tuned via the controller.

The calibration unit is devoted to calculate the overall gain K, applied to the entire input signal $S(t)$, and the gains $(A_1,..., A_N)$, specific of each of the N blocks of amplification, to optimize the effective dynamic resolution of the acquisition method. The calibration is performed on the basis of the $\tilde{S}(\omega)$ spectrum by setting first of all the amplification K and then by varying each of the amplifications $(A_1,..., A_N)$ individually as a function of the $S(t)$ spectral content. For each of the N channels, the relative *spectral* power is calculated by integrating the DFT rows within the corresponding interval of frequencies. Generally, depending on the frequency content of the signal $S(t)$, the spectrum amplitudes $\tilde{S}_1(\omega),...,\tilde{S}_N(\omega)$ of the N channels can be very different from each other, even for several orders of magnitude. As known, sampling a signal with strong amplitude variations in the various frequency ranges would mean that part of the static ADC resolution is "wasted" to take account of the range of variation of the amplitudes in the different bands. In the proposed method, the calibration block calculates $\tilde{S}_1(\omega),...,\tilde{S}_N(\omega)$ and individually varies gains $(A_1,..., A_N)$ in such a way that all the N spectral amplitudes are comparable. Overall, gain K is varied as a scaling factor. This allows to maximally optimize the dynamic range of the ADC, significantly increasing the effective resolution of the digitization of the method.

The value chosen for N, the distribution of the frequencies of the filters, and the value M of the DFT size determine the resolution at which the processing module can set the amplifier gains $(A_1,..., A_N)$. A possible configuration is to vary the filters $(F_1,..., F_N)$, to select the channel bandwidths, and to choose the size $M \geq N$ to know at least one spectral line for each of the N frequency channels so as to set each of the gains $(A_1,..., A_N)$ separately. If $M < N$, there would not be at least one spectral line in each range of filter frequencies $(F_1,..., F_N)$; nevertheless, it is still possible to set the gains $(A_1,..., A_N)$ for groups of channels. The possibility to independently vary (or in groups) the gain of the amplification blocks allows for completely exploiting the resolution of the ADC. The entire ADC static resolution is dynamically optimized to sample the range of variability of the signal in each frequency band without "wasting bits." This extreme flexibility of the method is particularly useful for signals in which the spectral content changes considerably from one frequency band to another.

The calibration can be performed one time at the beginning of acquisition or repeated cyclically, according to the $S(t)$ time variability, the environmental acquisition conditions, and depending on the specific device applications. The time duration of the calibration phase depends on the size M of the DFT as well as on the processing module performances and can be set or repeated cyclically to continuously optimize the amplification of each frequency channel. At the end of each calibration cycle, the gains are updated and the acquisition continues with the new setting. Once the spectrum $\tilde{S}(\omega)$ of the conditioned signal, the gain K, the filters transfer functions $(F_1,..., F_N)$ (each of them is a function of the frequency in the hth interval for $h = 1,..., N$), and the gains $(A_1,..., A_N)$ are known, it is possible to reconstruct the original spectrum $S(\omega)$ of the input signal $S(t)$. This reconstruction is made by multiplying the spectral lines $\tilde{S}(\omega)$ (for harmonics ω in the frequency range $\Delta\omega_h$ of the hth filter ($\omega \in \Delta\omega_h$) by the factor $(KF_h(\omega)A_h)^{-1}$ that is $S(\omega) = \tilde{S}(\omega) \times (KF_h(\omega)A_h)^{-1}\delta_{\omega h}$, with $\delta_{\omega h} = 1$ if $\omega \in \Delta\omega_h$, or 0 if $\omega \notin \Delta\omega_h$ for $h = 1,..., N$. It is noted that in general, the gains K and $(A_1,..., A_N)$ will change time by time at each calibration cycle, and therefore, the coefficients needed to reconstruct the $S(\omega)$ spectrum will also change dynamically over time. The spectrum $S(\omega)$ of the signal thus reconstructed constitutes the output of the entire processing method.

16.3   Construction and testing of an electronic board of the conditioning module

An electronic board of the signal-conditioning module has been constructed according to the diagram of Figure 16.1 for the case of $N = 8$ frequency bands. Following the general scheme of the proposed conditioning module, the board has been specifically designed for measuring the relative variation (instead of the absolute value) of each channel spectral content. In particular, the developed board is useful in the analysis of the input analog signal having a "smooth" spectral content and not monochromatic harmonics or thin-bandwidth signals, such as the natural electromagnetic signals detected by satellites in the near-Earth space for remote sensing and in general for investigation of natural (not artificial) electromagnetic phenomena. The analysis of these natural signals does not require thin-bandwidth and highly selective filters in the conditioning module. By using filters of low order, we expect an overlap between the spectral content of the eight-frequency channels, but, for the specific application purposes we planned, even a first-order filtering is enough to follow the time evolution and the spectrogram of the input signal to be analyzed.

Figure 16.2 shows the electric scheme of the ith filtering and amplification block of the built board. All eight filters are passive first order, with each of them constituted by the sequence of a resistance–capacitance (RC) low-pass filter and a capacitance–resistance (CR) high-pass filter. The values of resistances and capacitors, shown in Figure 16.2, are the following: $R_1^{(i)} = R_2^{(i)} = 1.62 \times 10^{(3+int((i-1)/2))}\,\Omega$, $R_3^{(i)} = R_5^{(i)} = 475\,\Omega$, $R_4^{(i)} = R_6^{(i)} = R_7^{(i)} = 4.3\,\Omega$, and $C_1^{(i)} = C_2^{(i)} = 10^{(-9+int((i-2)/2))}\,F$ with i = 1,...,8 (e.g., for $i = 1$ and 2, the values are $R_1^{(1)} = R_2^{(1)} = 1.62\,\Omega$, $R_1^{(2)} = R_2^{(2)} = 16.2\,\Omega$, $C_1^1 = C_2^1 = 100\,pF$, and $C_1^2 = C_2^2 = 1nF$). Consequently, the eight-frequency bandwidths are centered at values ranging from 0.1 Hz up to 1 MHz with a step of one decade. The choice of the RC–CR combination is just an example of easy-to-implement band-pass filters that are well controllable (with respect to the central frequency and attenuation factor), stable (with respect to the temperature, aging, failures induced by radiation, fabrication tolerance, etc.), and less noisy than active filters (based on the operational components). As mentioned in Section 16.2.1, the scheme of the conditioning board can also be implemented by adopting filters of any (higher) order and distance between adjacent frequency bands as well as of a different type and performances. A photo of the electronic board is reported in Figure 16.3.

The board accepts the input supply voltages, an analog signal S(t) to be analyzed, and a set of digital control signals (in general variable over time) that serve to fix the gains of the amplifiers. The electrical circuit of the device was built on a four-layer PC104 form-factor PCB (3.550" × 3.775") that allows easy installation in a variety of electronic architectures. The board is constituted by an amplifier K (LT1028 by Linear Technology) and eight blocks of filtering and amplification, each of them consisting of a passive band-pass RC–CR filter, two amplifiers ($A_I^{(i)}$ and $A_{II}^{(i)}$), and a selector CD4053B (by Texas Instruments). The input signal S(t) is preamplified (K); the resulting signal $\tilde{V}_{in}$ is divided into blocks of filtering and amplification on eight channels. The specific amplifiers installed on board are $A_I^{(i)} = A_{II}^{(i)} = OPA627$ (by Texas Instruments) for $i = 1, 2$; $A_I^{(i)} = A_{II}^{(i)} = OP97$ (by Analog Devices) for $i = 3, 4, 5$; and $A_I^{(i)} = A_{II}^{(i)} = OP1177$ (by Analog Devices) for $i = 6, 7, 8$. Each block separately amplifies the corresponding frequency channel with a variable gain set by the external calibration procedure. By varying the digital signals that control the selector CD4053B, the overall gain of the two amplifiers of each block can be set to the values 1 or 10 or 100. The eight conditioned channels $\tilde{V}_{out}^i$ can be sent out separately or added

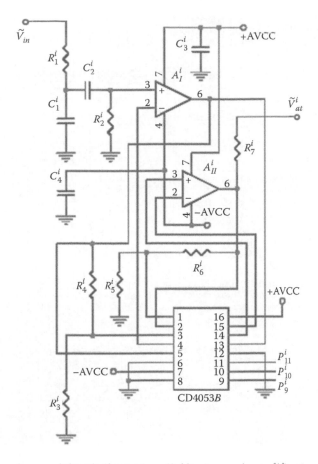

Figure 16.2 Electric scheme of the *i*th (for $i = 1,..., 8$) filtering and amplification block of the signal-conditioning board. The values of the power bypass capacitors are $C_3^i - C_4^i = 0.1\,\mu F$. See the text for details. (Reprinted from *Nucl. Instrum. Meth. Phys. Res. A*, 756, Conti, L. et al., Method for signal conditioning and data acquisition system, based on variable amplification and feedback technique, 23–29, Copyright 2014, with permission from Elsevier.)

together in various groups and combinations thanks to several jumpers. In particular, it is also possible to "turn off" one or more channels completely excluding them from the output. The signals of the output channels can be acquired (individually, in groups, or all together) for subsequent digitization, calculation of spectrograms, or to reconstruct the waveform.

To verify the frequency response of the board and its capability to selectively amplify specific spectral components (then avoiding significant contamination from neighboring frequency windows), we have performed several simulations of the circuit by the LTspice IV software (LTspice 2015). Figure 16.4 (panel (a)) shows the frequency response of each separated block of the board, with gain amplification set to ×10 for all the blocks. Owing to the frequency response of the OP627 amplifier at high frequencies (shown in panel (b) of Figure 16.4), the block with the filter frequency centered at 1 MHz shows an output signal that is lower than those of the other blocks. The frequency response of the board as a whole is reported in the bottom panels of Figure 16.4 (i.e., when all the output signals are added), for an input signal of flat spectral content, in two different configurations of

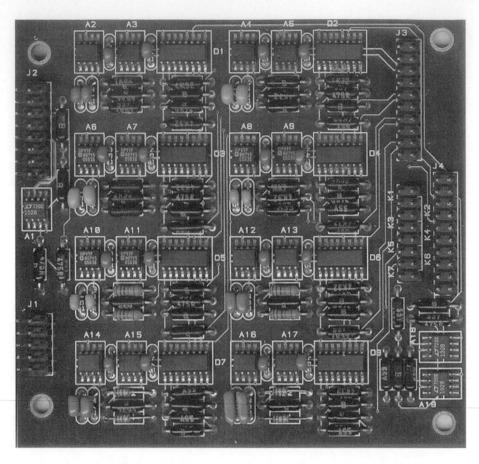

Figure 16.3 Photo of the signal-conditioning electronic board. The four-layer PCB has dimensions of 3.550″ × 3.775″ as the PC104 form factor. (Reprinted from *Nucl. Instrum. Meth. Phys. Res. A*, 756, Conti, L. et al., Method for signal conditioning and data acquisition system, based on variable amplification and feedback technique, 23–29, Copyright 2014, with permission from Elsevier.)

the blocks amplification gains: (×10, ×100, ×1, ×10, ×10, ×10, ×100, ×10) (panel (c)) and (×100, ×10, ×1, ×100, ×1, ×100, ×10, ×10) (panel (d)). Finally, an example of the performances of the board obtained in the case of a specific input signal is shown in Figure 16.5. The spectral content of the input signal (top panel) is constituted by a broad low-frequency band and three sharp and very-low-amplitude harmonic signals of medium and high frequencies (i.e., 1.6, 160, and 500 kHz). The middle panel of Figure 16.5 shows the board frequency response, whereas the output signal is plotted in the bottom panel. The example reported in Figure 16.5 shows that the board allows to amplify some specific spectral components of the input signal separately that might be of interest to the required application and which could have too low amplitude in the original input signal. At the same time, the frequency response of the board points out its capability to exclude (also by using first-order band-pass filters) the low-frequency component and, in general, to significantly reduce the contamination from frequency bands contiguous to those of interest.

In the real working configuration, the signal-conditioning board must be connected to both an acquisition module and a processing module. The gains of the conditioning board may be fixed one-off at the beginning of the acquisition process or varied cyclically with

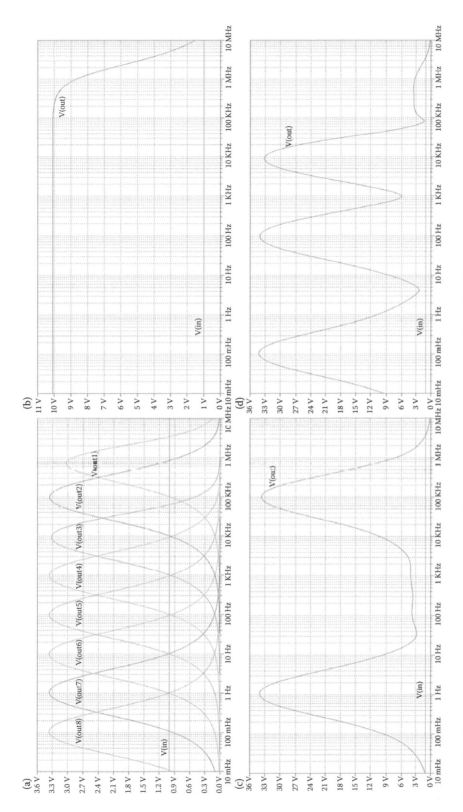

Figure 16.4 Panel (a) shows the frequency response V(out_i) of each single block ($i = 1,..., 8$) of the board. Owing to the transfer function of the OP627 amplifier in the high-frequency range (about 1 MHz) shown in panel (b), the frequency response of the first block is lower than those of the other blocks. In panels (c) and (d), the frequency response (V(out)) of the board as a whole (by adding all the blocks output signals), for two different configurations of the amplification gains, is shown when an input signal (V(in)) with flat spectral content is applied. (Reprinted from *Nucl. Instrum. Meth. Phys. Res. A*, 756, Conti, L. et al., Method for signal conditioning and data acquisition system, based on variable amplification and feedback technique, 23–29, Copyright 2014, with permission from Elsevier.)

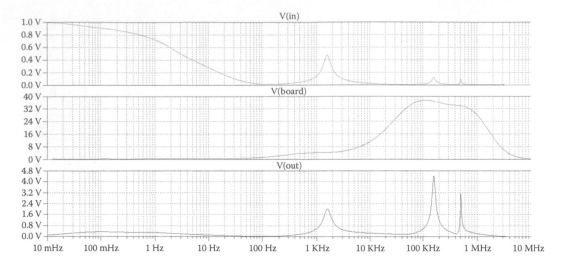

Figure 16.5 Example of the board output obtained by applying the input signal when the board frequency response is that shown in the middle panel. (Reprinted from *Nucl. Instrum. Meth. Phys. Res. A*, 756, Conti, L. et al., Method for signal conditioning and data acquisition system, based on variable amplification and feedback technique, 23–29, Copyright 2014, with permission from Elsevier.)

a feedback process during acquisition as described above. To control the amplifications of the board, the processing unit will determine the optimum value of the amplifications ("calibration") and set the parameters of the selector CD4053B. Once the calibration phase is complete, the blocks of amplification of the board retain their settings until the next calibration cycle. The flexibility of the board allows that these control functions can be performed by devices such as CPU, DSP, FPGA, or on a suitable dedicated processor that manages and synchronizes the calibration and acquisition. The simplicity of the design of the board, in particular due to the use of stable and robust passive filters, also makes the determination of the transfer function of the entire board easier. Once the waveform of the ith conditioned channel, the transfer functions of the filters, and the gains of the amplifiers are known, it is possible to easily reconstruct the waveform or the spectrum of the original signal (without conditioning). It does not appear that such a board is currently in the market.

16.4   Discussion and conclusions

In this chapter, we have proposed an original signal-processing method based on variable amplification and feedback technique with the aim of optimizing the spectrum resolution in analog acquisition systems. The method can be included among the dynamic multiband compression procedures. The outline design consists of three interconnected modules for the conditioning, digitizing, and processing of an input analog signal (generally *a priori* unknown and that can be measured in a large variety of different phenomena on the ground, in space, or in the laboratory) designed to determine the spectrum of the signal with great accuracy. An example of the electronic board of the conditioning module for a specific configuration and application has been constructed and tested as described in the chapter. The method consists of dividing the investigated analog input signal into different frequency bands and making each of the different channel amplitudes comparable with one another, through an appropriate calibration process (by assigning a suitable weight to each channel), and digitizing all the channels with a single analog–digital converter. Such weights are determined on the basis of the spectral contents of the input signal. This approach optimizes the performances of the ADC by maximizing its dynamic range. Therefore, the method exhibits all its versatility and adaptivity when applied to a signal with a largely variable spectral content in different frequency bands. The output of the processing module produces the optimized spectrograms of the analog input signal, for example, the spectra as a function of time. On the other hand, when compared to the category of existing processing devices, developed for specific needs (Kudeki and Munson 2008, Rybin 2012 and references therein), the method proposed here is clearly superior, in particular for a better adaptive feedback amplification, the use of a single ADC to scan the multichannel signal, the optimization of the ADC dynamic range, and the high device performance with reduced weights and costs. Moreover, the method is particularly suitable when it is necessary to carry out a multichannel analysis of an analog signal, that is, when the input signal is the result of a superimposition of many components of different frequency bands, characterized by a significant difference in amplitudes, and that flow simultaneously in the acquisition system. It should also be considered that the proposed approach is particularly useful for all the satellite applications and data analyses in space, which typically ask for extremely reduced power consumption and electronic components resistant to radiations so to survive at the severe conditions occurring in space, such as in the solar wind, Earth magnetosphere, ionosphere, etc. For data-acquisition systems designed to operate in not-so-demanding environments or conceived just for laboratory applications, the up-to-date technologies allow oversampling, digital high-order filtering, and fast data processing on a reprogrammable architecture, all via highly integrated ICs (such as FPGAs or application-specific integrated circuits [ASICs] with millions of gates). On the contrary, in space conditions, this "extreme" and fully programmable approach must be carefully revised. In fact, the ADCs with high-frequency sampling and high resolution, as well as the most-performing digital processing chipsets, are also generally very power consuming and extremely sensitive to radiation damages (such as single-event upset [SEU], single-event latchup [SEL], single-event transient [SET], etc.) induced by ionizing radiations and energetic particles. In space, a more sophisticated but more expensive and fragile data acquisition based on radiation-tolerant and very power-consuming electronic components is less reliable than the simpler but robust and cheaper strategy of our method (based on analog/passive filtering, optimized sampling, low-power, and less radiation-sensitive devices).

Acknowledgments

This chapter has been reprinted from Conti, L., Sgrigna, V., Zilpimiani, D., Dario, A., 2014, Method for signal conditioning and data acquisition system, based on variable amplification and feedback technique, *Nucl. Instrum. Meth. Phys. Res. A*, 756, 23–29, doi:10.1016/j.nima.2014.04.009, Copyright (2014), with permission from Elsevier.

References

Agilent, 2013, Spectrum analyzer and signal analyzer, http://www.agilent.com/find/sa, accessed July 17, 2015.

Baciarello, G., Bella, R., Di Maio, F., Sgrigna, V., Villani, M., Villatico Campbell, S., Sciacca, A., 1987, Analysis of high-resolution atrial activation: Report on 403 cases, *Am. Heart J.*, 113, 307–315.

Biagi, P.F., Manjgaladze, P., Sgrigna, V., Zilpimiani, D., 1990, Preliminary laboratory observations of non-linear electromechanic effects, *Nuovo Cimento*, 13C, 631–638.

Conti, L., Sgrigna, V., Zilpimiani, D., Dario, A., 2014, Method for signal conditioning and data acquisition system, based on variable amplification and feedback technique, *Nucl. Instrum. Meth. Phys. Res. A*, 756, 23–29, doi:10.1016/j.nima.2014.04.009

Hadji-Abdolhamid, A., Johns, A.D., 2003, A 400-MHz 6-bit ADC with a partial analog equalizer for coaxial cable channels, *Proceedings of the 29th European Solid-State Circuits Conference, ESSCIRC '03*, September 16–18 2003, Estoril, Portugal, 0-7803-7995-0, pp. 237–240, doi: 10.1109/ ESSCIRC.2003.1257116.

Haykin, S., 1994, *Communication Systems*, 3rd ed. New York: Wiley.

Johns, D.A, Martin, K.W., 1997, *Analog Integrated Circuit Design*, John Wiley & Sons, Inc., ISBN 0471144487.

Kudeki, E., Munson, D.C. Jr. 2008, *Analog Signals and Systems*, Illinois Ece Series, Urbana-Champaign, ISBN 0131435063.

LTspice IV, 2015, High performance SPICE simulator, http://www.linear.com/designtools/ software/, accessed February 18, 2015.

Parrot, M., 2002, The micro-satellite DEMETER, *J. Geodyn.*, 33, 535–541.

Proakis, J., 1995, *Digital Communications*, 3rd ed. Singapore: McGraw-Hill.

Rohde & Schwarz GmbH & Co. KG, 2013, Signal and spectrum analyzers, New York, NY: Rohde & Schwarz GmbH & Co. KG, www.rohde-schwarz.com

Rybin, Y.K., 2012, *Electronic Devices for Analog Signal Processing, Springer Series in Advanced Microelectronics*, Springer, Dordrecht, ISBN 9400722040.

Sgrigna, V., Della Monica, G., Villani, M., Iannucci, G., Alessandri, N., Bella, R., Gallo, R., Massa, E., Baciarello, G., 1993, Automatic analysis of high-resolution atrial activation in patients with isolated mitral valve stenosis, *Int. J. Cardiol.*, 42, 63–70.

Sgrigna, V., Carota, L., Conti, L., Corsi, M., Galper, A., Koldashov, S.V., Murashov, A.M., Picozza, P., Scrimaglio, R., Stagni, L., 2005, Correlations between earthquakes and anomalous particle bursts from SAMPEX/PET satellite observations, *J. Atm. Sol.-Terr. Phys.*, 67, 1448–1462.

Sgrigna, V., Altamura, F., Ascani, S., Battiston, R., Bencardino, R., Blasko, S., Buzzi, A. et al., 2007, First data from the EGLE space experiment onboard the ISS, *Microgr. Sci. Technol.*, 19(2), 45–49.

Sgrigna, V., Conti, L., 2012, A deterministic approach to earthquake prediction, *Int. J. Geophys.*, 20, doi:10.1155/2012/406278.

chapter seventeen

Part A: Visualization of big data
Current trends

Isaac J. Donaldson, Sandra C. Hom, Thomas Housel,
Johnathan Mun, and Trent Silkey

Contents

17.1   Introduction

The world is exploding in digital data. IDC Corporation predicts that from 2005 to 2020, the digital universe will grow by a factor of 300, from 130 to 40,000 exabytes or 40 zettabytes. Moreover, the digital universe will about double every 2 years from now to 2020, a 50-fold growth in 10 years (Gantz and Reinsel, 2012).

More than five billion people are calling, texting, tweeting, and browsing on mobile phones worldwide, and 350 million tweets are sent per day (Gantz and Reinsel, 2012). Companies around the world are capturing trillions of bytes of information on customers, suppliers, and operations. The McKinsey Global Institute (MGI; Manyika et al., 2011) estimates that global enterprises stored more than 7 exabytes of new data on disk drives in 2010, while consumers stored more than 6 exabytes of new data on devices such as PCs and notebooks. The U.S. government produced 848 petabytes of data in 2009. Data collected by the U.S. Library of Congress as of April 2011 totals 235 terabytes.

For the purposes of our research, we will use MGI's definition of big data as "datasets whose size is beyond the ability of typical database software tools to capture, store, manage, and analyze" (Manyika et al., 2011) . There are many challenges with big data including the ability to capture, store, curate, search, transfer, share, analyze, and visualize the data. This section focuses on the big data eco-structure. It begins with a discussion of the market size, then discusses some of the tools and technologies used in big data analysis, and looks at federal government initiatives involving big data.

The total big data market reached $11.59 billion in 2012 and was estimated to grow at an annual growth rate of 61% to $18.1 billion in 2013, according to Wikibon (Kelly et al., 2013). Big data consists of software, hardware, services, and storage. In addition, Wikibon

(Kelly et al., 2013) predicted the big data market to exceed $47 billion by 2017, growing at a 31% compound annual growth rate over the 5-year period from 2012 to 2017.

17.2   The big data ecosystem

Fueling the growth in big data sales are several factors as follows:

- Increased awareness of the benefits of big data as applied to industries beyond the web, most notably financial services, pharmaceuticals, and retail.
- Maturation of big data software such as Hadoop, NoSQL (not only structured query language), data stores, in-memory analytic engines, and massively parallel processing analytic databases.
- Increasingly sophisticated professional services practices that assist enterprises in practically applying big data hardware and software to business use cases.
- Increased investment in big data infrastructure by massive web properties—most notable Google, Facebook, and Amazon—and government agencies for intelligence and counter-terrorism purposes (Kelly et al., 2013, Growth Drivers and Adoption Barriers, para. 3).

Big data is generated by a variety of sources. Big data originates from sources including industry specific transactions, machine/sensor indications, web applications, and text (Ferguson, 2013). Industry specific transactions can include call records and geographic location data. Machines generate extremely large volumes of information every day and can range in complexity from simple temperature readings to the performance parameters of a gas-turbine engine. Big data on the web also ranges in format from machine language to customer comments on social networks and is also produced in considerably sizeable portions. Text sources can include archived documents, external reports, or customer account information (Ferguson, 2013).

Because big data comes from a variety of sources, it also possesses characteristics that distinguish it from data in the traditional context. Common terms used to define the qualities of big data include volume, variety, velocity, and value (Dijcks, 2013). From this listing of sources, one can understand that the *volume* of data generated on a daily basis is enormous. For example, Dijcks (2013) stated that just a single jet engine produces 10 terabytes of data in 30 min. Extrapolate that example to include all the aircraft currently airborne, and then include all the factory infrastructure around the globe collecting data on production, service life, and maintenance requirements, and the enormity of big data volumes begins to emerge. Another characteristic of big data, *variety*, can be directly translated from the various sources into the variety of data formats. Various data formats require additional consideration to ensure the ability of all systems to share data. *Velocity*, which is related to volume, is the frequency with which big data is created. To illustrate velocity, consider the relative size of a single twitter feed (140 characters) to the large number of feeds generated in a given period (Dijcks, 2013). Finally, *value* is the feature of big data that is important to any enterprise.

17.3   Big data technologies and tools

Many techniques can be used to analyze datasets. These techniques, which often draw upon statistics and computer science, can be applied to big data to generate insights into large and diverse datasets, as well as smaller, diverse datasets. Table 17.1 summarizes some techniques (Manyika et al., 2011).

Table 17.1 Big data analysis techniques

Technique name	Technique description
A/B testing	• The technique in which a control group is compared with a variety of test groups in order to determine what treatments (i.e., changes) will improve a given objective variable. • Big data enables huge numbers of tests to be executed and analyzed, ensuring that groups are of sufficient size to detect meaningful (i.e., statistically significant) differences between the control and treatment groups.
Association rule learning	• Set of techniques for discovering interesting relationships (i.e., "association rules") among variables in large databases. • These techniques consist of a variety of algorithms to generate and test possible rules. • An application is market basket analysis, in which a retailer can determine which products are frequently bought together and use this information for marketing (a commonly cited example is the discovery that many supermarket shoppers who buy diapers also tend to buy beer). • Used for data mining.
Classification	• Set of techniques to identify the categories in which new data points belong, based on a training set containing data points that have already been categorized. • One application is the prediction of segment-specific customer behavior (e.g., buying decisions, churn rate, and consumption rate) where there is a clear hypothesis or objective outcome. • These techniques are often described as supervised learning because of the existence of a training set; they stand in contrast to cluster analysis, a type of unsupervised learning. • Used for data mining.
Cluster analysis	• A statistical method for classifying objects that splits a diverse group into smaller groups of similar objects, whose characteristics of similarity are not known in advance. • An example of cluster analysis is segmenting consumers into self-similar groups for targeted marketing. • This is a type of unsupervised learning because training data are not used. • Used for data mining.
Crowdsourcing	• The technique for collecting data submitted by a large group of people or community (i.e., the "crowd") through an open call, usually through networked media such as the web. • This is a type of mass collaboration and an instance of using Web 2.0.
Data fusion and data integration	• Set of techniques that integrate and analyze data from multiple sources in order to develop insights in ways that are more efficient and potentially more accurate than if they were developed by analyzing a single source of data. • Signal processing techniques can be used to implement some types of data fusion. • One example of an application is sensor data from the *Internet of Things* being combined to develop an integrated perspective on the performance of a complex distributed system such as an oil refinery. • Data from social media, analyzed by natural language processing, can be combined with real-time sales data, in order to determine what effect a marketing campaign is having on customer sentiment and purchasing behavior.

(Continued)

Table 17.1 (Continued) Big data analysis techniques

Technique name	Technique description
Data mining	• Set of techniques to extract patterns from large datasets by combining methods from statistics and machine learning with database management. • These techniques include association rule learning, cluster analysis, classification, and regression. • Applications include mining customer data to determine segments most likely to respond to an offer, mining human resources data to identify characteristics of most successful employees, or market basket analysis to model the purchase behavior of customers.
Ensemble learning	• Using multiple predictive models (each developed using statistics and/or machine learning) to obtain better predictive performance than could be obtained from any of the constituent models. • This is a type of supervised learning.
Genetic algorithms	• Technique used for optimization that is inspired by the process of natural evolution or "survival of the fittest." • Potential solutions are encoded as "chromosomes" that can combine and mutate. • These individual chromosomes are selected for survival within a modeled "environment" that determines the fitness or performance of each individual in the population. • Often described as a type of "evolutionary algorithm," these algorithms are well-suited for solving nonlinear problems. • Examples of applications include improving job scheduling in manufacturing and optimizing the performance of an investment portfolio.
Machine learning	• Subspecialty of computer science (within a field historically called "artificial intelligence") concerned with the design and development of algorithms that allow computers to evolve behaviors based on empirical data. • A major focus of machine learning research is to learn automatically to recognize complex patterns and make intelligent decisions based on data. Natural language processing is an example of machine learning.
Natural language processing (NLP)	• Set of techniques from a subspecialty of computer science (within a field historically called "artificial intelligence") and linguistics that uses computer algorithms to analyze human (natural) language. • Many NLP techniques are types of machine learning. • One application of NLP is using sentiment analysis on social media to determine how prospective customers are reacting to a branding campaign.
Neural networks	• Computational models, inspired by the structure and workings of biological neural networks (i.e., the cells and connections within a brain), that find patterns in data. • Neural networks are well-suited for finding nonlinear patterns. • Can be used for pattern recognition and optimization. Some neural network applications involve supervised learning, and others involve unsupervised learning. • Examples of applications include identifying high-value customers that are at risk of leaving a particular company and identifying fraudulent insurance claims.

(Continued)

Table 17.1 (Continued) Big data analysis techniques

Technique name	Technique description
Network analysis	• Set of techniques used to characterize relationships among discrete nodes in a graph or a network. • In social network analysis, connections between individuals in a community or organization are analyzed (e.g., how information travels) or who has the most influence over whom. • Examples of applications include identifying key opinion leaders to target for marketing and identifying bottlenecks in enterprise information flows.
Optimization	• Portfolio of numerical techniques used to redesign complex systems and processes to improve their performance according to one or more objective measures (e.g., cost, speed, or reliability). • Examples of applications include improving operational processes such as scheduling, routing, and floor layout, and making strategic decisions such as product range strategy, linked investment analysis, and R&D portfolio strategy. • Genetic algorithms are an example of an optimization technique.
Pattern recognition	• Set of machine learning techniques that assign some kind of output value (or label) to a given input value (or instance) according to a specific algorithm. • Classification techniques are an example.
Predictive modeling	• A set of techniques in which a mathematical model is created or chosen to predict best the probability of an outcome. • Example of an application in customer relationship management is the use of predictive models to estimate the likelihood that a customer will "churn" (i.e., change providers) or the likelihood that a customer can be cross-sold another product. • Regression is one example of the many predictive modeling techniques.
Regression	• Set of statistical techniques to determine how the value of the dependent variable changes when one or more independent variables are modified. • Often used for forecasting or prediction. • Examples of applications include forecasting sales volumes based on various market and economic variables or determining what measurable manufacturing parameters most influence customer satisfaction. • Used for data mining.
Sentiment analysis	• Application of natural language processing and other analytic techniques to identify and extract subjective information from source text material. • Key aspects of these analyses include identifying the feature, aspect, or product about which a sentiment is being expressed, and determining the type, "polarity" (i.e., positive, negative, or neutral), and the degree and strength of the sentiment. • Examples of applications include companies applying sentiment analysis to analyze social media (e.g., blogs, microblogs, and social networks) to determine how different customer segments and stakeholders are reacting to their products and actions.
Signal processing	• Set of techniques from electrical engineering and applied mathematics originally developed to analyze discrete and continuous signals (i.e., representations of analog physical quantities [even if represented digitally] such as radio signals, sounds, and images).

(Continued)

Table 17.1 (Continued) Big data analysis techniques

Technique name	Technique description
	• This category includes techniques from signal detection theory, which quantifies the ability to discern between signal and noise. • Sample applications include modeling for time series analysis or implementing data fusion to determine a more precise reading by combining data from a set of less precise data sources (i.e., extracting the signal from the noise).
Spatial analysis	• Set of techniques, some applied from statistics, which analyze the topological, geometric, or geographic properties encoded in a dataset. • Often the data for the spatial analysis come from geographic information systems (GIS) that capture data including location information (e.g., addresses or latitude/longitude coordinates). • Examples of applications include the incorporation of spatial data into spatial regressions (e.g., how is consumer willingness to purchase a product correlated with location?) or simulations (e.g., how would a manufacturing supply chain network perform with sites in different locations?).
Statistics	• The science of the collection, organization, and interpretation of data, including the design of surveys and experiments. • Statistical techniques are often used to make judgments about what relationships between variables could have occurred by chance (the "null hypothesis") and what relationships between variables likely result from some kind of underlying causal relationship (i.e., that are "statistically significant"). • Statistical techniques are also used to reduce the likelihood of Type I errors ("false positives") and Type II errors ("false negatives"). • An example of an application is A/B testing to determine what types of marketing material will most increase revenue.
Supervised learning	• Set of machine learning techniques that infer a function or relationship from a set of training data. • Examples include classification and support vector machines.
Simulation	• Modeling the behavior of complex systems, often used for forecasting, predicting, and scenario planning. Monte Carlo simulations, for example, are a class of algorithms that rely on repeated random sampling (i.e., running thousands of simulations, each based on different assumptions). • The result is a histogram that gives a probability distribution of outcomes. • One application is assessing the likelihood of meeting financial targets given uncertainties about the success of various initiatives.
Time series analysis	• Set of techniques from both statistics and signal processing for analyzing sequences of data points, representing values at successive times, to extract meaningful characteristics from the data. • Examples of time series analysis include the hourly value of a stock market index or the number of patients diagnosed with a given condition every day. • Time series forecasting is the use of a model to predict future values of a time series based on known past values of the same or another series. • Some of these techniques (e.g., structural modeling) decompose a series into trend, seasonal, and residual components, which can be useful for identifying cyclical patterns in the data. • Examples of applications include forecasting sales figures or predicting the number of people who will be diagnosed with an infectious disease.

(Continued)

Table 17.1 (Continued) Big data analysis techniques

Technique name	Technique description
Unsupervised learning	• Set of machine learning techniques that find hidden structure in unlabeled data. • Cluster analysis is an example of unsupervised learning (in contrast to supervised learning).
Visualization	• Techniques used for creating images, diagrams, or animations to communicate, understand, and improve the results of big data analyses.

There are a growing number of technologies used to aggregate, manipulate, manage, and analyze big data. Some of the more widely used technologies, used to aggregate, manage, and analyze big data are found in Table 17.2 (Manyika et al., 2011).

In working with massive amounts of data, displaying summary data and using visualization is critical to finding connections and relevance among millions of parameters and variables to convey linkages, hypotheses, metrics, and project future outcomes. Taken one level further, interactive visualization moves visualization from static spreadsheets and graphics to images capable of drilling down for more details, and immediately changing how data are presented and processed.

Examples of visualization methods include the following

- Bar charts are commonly used for comparing the quantities of different categories or groups.
- Box plots represent a distribution of data values. They display five statistics of minimum, lower quartile, median, upper quartile, and the maximum values that summarize the distribution of a set of data. Extreme values are represented by whiskers extending from the edges of the box.
- Bubble plots are variations of a scatter plot in which the data markers are replaced with bubbles, with each bubble representing an observation (or group of observations). They are useful for datasets with many values or when values differ by orders of magnitude.
- Correlation matrices combine big data with fast response times to identify quickly which variables among millions/billions are related. They also show the relationship strength between variables.
- Cross-tabulation charts show frequency distributions or other aggregate statistics for the intersections of two or more category data items. Crosstabs enable examination of data for intersections of hierarchy nodes or category values.
- Clustergrams display how individual members of a dataset are assigned to clusters as the number of members increases.
- Geo maps display data as a bubble plot overlaid on a geographic map. Either each bubble is located at the center of a geographic region or at location coordinates.
- Heat maps display a distribution of values for two data items using a table with colored cells. Colors are used to communicate relationships between data values.
- Histograms are variations of bar charts using rectangles to show the frequency of data items in successive numerical intervals of equal size. They are often used to quickly show the distribution of values in large datasets.
- History flow charts show the evolution of a document edited by multiple contributing authors. Time appears on the horizontal axis, while contributions to the text are

Table 17.2 Big data analysis technologies

Technique name	Technique description
Big table	• Proprietary distributed database system built on the Google File System. • Inspiration for HBase.
Business intelligence	• A type of application software designed to report, analyze, and present data. • Often used to read data previously stored in a data warehouse or data mart. • Also used to create standard reports that are generated on a periodic basis, or to display information on real-time management dashboards (i.e., integrated displays of metrics that measure the performance of a system).
Cassandra	• An open source (free) database management system designed to handle huge amounts of data on a distributed system. • The system was originally developed at Facebook and is now managed as a project of the Apache Software Foundation.
Cloud computing	• A computing paradigm in which highly scalable computing resources, often configured as a distributed system, are provided as a service through a network.
Data mart	• A subset of a data warehouse used to provide data to users usually through business intelligence tools.
Data warehouse	• Specialized database optimized for reporting, often used for storing large amounts of structured data. • Data uploaded using ETL (extract, transform, and load) tools from operational data stores, and reports are often generated using business intelligence tools.
Distributed System	• Multiple computers, communicating through a network, used to solve a common computational problem. • Problem is divided into multiple tasks, each of which is solved by one or more computers working in parallel. • Benefits of distributed systems include higher performance at a lower cost (i.e., because a cluster of lower end computers can be less expensive than a single higher-end computer), higher reliability (i.e., because of a lack of a single point of failure), and more scalability (i.e., because increasing the power of a distributed system can be accomplished by simply adding more nodes rather than completely replacing a central computer).
Dynamo	• Proprietary distributed data storage system developed by Amazon.
Extract, transform, and load (ETL)	• Software tools used to extract data from outside sources, transform them to fit operational needs and load them into a database or data warehouse.
Google File System	• Proprietary distributed file system developed by Google; part of the inspiration for Hadoop.31
Hadoop	• Open source software framework for processing huge datasets on certain kinds of problems on a distributed system. Its development was inspired by Google's MapReduce and Google File System. It was originally developed at Yahoo! and is now managed as a project of the Apache Software Foundation.
HBase	• Open source, distributed, non-relational database modeled on Google's Big Table. • Originally developed by Powerset and is now managed as a project of the Apache Software Foundation as part of the Hadoop.
MapReduce	• Software framework introduced by Google for processing huge datasets on certain kinds of problems on a distributed system. • Also implemented in Hadoop.

(*Continued*)

Table 17.2 (Continued) Big data analysis technologies

Technique name	Technique description
Mashup	• An application that uses and combines data presentation or functionality from two or more sources to create new services. • Applications are often made available on the Web and frequently use data accessed through open application programming interfaces or from open data sources.
Metadata	• Data that describes the content and context of data files (e.g., means of creation, purpose, time and date of creation, and author).
Non-relational database	• A database that does not store data in tables (rows and columns).
R	• Open source (free) programming language and software environment for statistical computing and graphics. • R language has become a de facto standard among statisticians for developing statistical software and is widely used for statistical software development and data analysis.
Relational database	• Database made up of a collection of tables (relations; i.e., data are stored in rows and columns). • Relational database management systems (RDBMS) store a type of structured data. • SQL is the most widely used language for managing relational databases.
Semi-structured data	• Data that do not conform to fixed fields but contain tags and other markers to separate data elements. • Examples include XML- or HTML-tagged text.
SQL	• Originally, an acronym for a structured query language, SQL is a computer language designed for managing data in relational databases. • The technique includes the ability to insert, query, update, and delete data, as well as manage data schema (database structures) and control access to data in the database.
Stream processing	• Technologies designed to process large real-time streams of event data. • Enables applications such as algorithmic trading in financial services, RFID event processing applications, fraud detection, process monitoring, and location-based services in telecommunications.
Structured data	• Data that reside in fixed fields. • Examples include relational databases or data in spreadsheets.
Unstructured data	• Data that do not reside in fixed fields. • Examples include free-form text (e.g., books, articles, body of e-mail messages) and untagged audio, image, and video data.
Visualization	• Technologies for creating images, diagrams, or animations to communicate a message often used to synthesize the results of big data analyses.

on the vertical axis; each author has a different color code and the vertical length of a bar indicates the amount of text written by each author.

- Line charts show the relationship of one variable to another by using a line that connects the data values. They are most often used to track changes or trends over time.
- Pareto charts are a specialized type of vertical bar chart where values of the dependent variables are plotted in decreasing the order of frequency from left to right. They are used to identify quickly when certain issues need attention.

- Scatter plots are two-dimensional plots showing the joint variation of two (or three) variables from a group of table rows. They are useful for examining the relationships, or correlations, between numeric data items.
- Tag clouds are weighted visual lists in which words appearing most frequently are larger and words appearing less frequently are smaller.
- Tree maps are a variation of heat maps using rectangles (tiles) to represent data components. The largest rectangle represents the dominant division of the data, and smaller rectangles represent subdivisions.

17.4   Government spending on big data

The U.S. federal government is fueling the growth of big data spending on national security and military applications. According to the Biometrics Research Group (King, 2013), federal agencies spent approximately US$5 billion on big data resources in fiscal year (FY) 2012, and they estimated annual spending would grow to US$6 billion in 2014. By 2017, that figure will reach US$8 billion, growing at a compound annual growth rate of 10%.

During the near to midterm, Biometrics Research Group (King, 2013) predicts that most of the spending will be on military applications of the U.S. government with federal agencies pursuing more than 150 big data projects (grants, procurements, or related activities). The agency leading big data research is the U.S. DoD (Department of Defense) with more than 30 projects and, in particular, the Defense Advanced Research Projects Agency (DARPA) with nine major projects (King, 2013).

In a recent study sponsored by EMC (King, 2013) that surveyed 150 U.S. government information technology (IT) executives, 70% respondents stated that big data will be critical to all government operations within 5 years. Big data, according to the survey, has the potential to save nearly $500 billion, or 14%, of agency budgets across the federal government by increasing efficiency, enabling smarter decisions, and deepening insight. However, only 31% of respondents said their agency has an adequate big data strategy (King, 2013).

Government agencies are seeking to make big data a greater part of their mission. The Department of Homeland Security (DHS) posted a solicitation July 24, 2013 (DHS, 2013) requesting additional information from industry in order to identify transformational opportunities to improve mission and operational efficiencies and lower costs through advanced analytic automation for the DHS and Homeland Security Enterprise (HSE). The request for information (RFI) read as follows:

> The purpose of this RFI is to ascertain available sources to provide widely used big data infrastructure, computing, storage, analytics, and visualization capabilities that are based on open source or commonly available commercial technologies and represent technology options of high value to the future of homeland security.

17.5   Big data projects in government

In 2012, the Obama administration announced the *Big Data Research and Development Initiative* to help solve challenges by improving the ability to extract knowledge and insights from large and complex collections of digital data (Office of Science and Technology Policy, 2012). The initiative's objective is to analyze big data and achieve advances in several sectors such as healthcare, security, the environment, education, and the sciences. Six federal departments and agencies launched the initiative with more than $200 million in commitments that promise to greatly improve the tools and techniques needed to access, organize, and glean discoveries from huge volumes of digital data.

The Big Data Research and Development Initiative was created to achieve the following

- Advance state-of-the-art core technologies needed to collect, store, preserve, manage, analyze, and share huge quantities of data.
- Harness these technologies to accelerate the pace of discovery in science and engineering, strengthen our national security, and transform teaching and learning.
- Expand the workforce needed to develop and use big data technologies (Office of Science and Technology Policy, 2012, p. 1).

The DoD announced plans to invest approximately $250 million annually across the military departments in a series of programs that will

- Harness and utilize massive data in new ways and bring together sensing, perception, and decision support to make truly autonomous systems that can maneuver and make decisions on their own.
- Improve situational awareness to help warfighters and analysts and provide increased support to operations. The department is seeking a 100-fold increase in the ability of analysts to extract information from texts in any language and a similar increase in the number of objects, activities, and events that an analyst can observe (Office of Science and Technology Policy, 2012, pp. 2–3).

According to King (2013), DoD big data programs include XDATA, Cyber-Insider Threat (CINDER), Anomaly Detection at Multiple Scales (ADAMS), Insight, Mind's Eye, Machine Reading, Mission-Oriented Resilient Clouds, Programming Computation on Encrypted Data (PROCEED), and Video and Image Retrieval and Analysis Tool (VIRAT).

- XDATA is a 4-year, $25 million-per-year program to develop computational techniques and software tools for analyzing large volumes of data, both semi-structured (e.g., tabular, relational, categorical, meta-data) and unstructured (e.g., text documents, message traffic). Some core challenges include developing scalable algorithms for processing imperfect data in distributed data stores and effective human–computer interaction tools that are rapidly customizable to facilitate visual reasoning for diverse missions. XDATA envisions open source software toolkits for flexible software development, enabling processing of large volumes of data for use in targeted defense applications (King, 2013, para. 13).
- The CINDER program seeks to develop innovative approaches to detect activities consistent with cyber espionage in military computer networks. CINDER will apply various models of adversary missions to *normal* activity on internal networks as a method to expose hidden operations. The program also intends to increase the accuracy, rate, and speed with which cyber threats are detected (King, 2013, para. 6).
- The ADAMS program addresses the issue of anomaly detection and characterization in massive datasets. Data anomalies are intended to cue the collection of additional, actionable information in a wide variety of real-world contexts. Initially, ADAMS will focus on insider threat detection, in which anomalous actions by an individual are detected against a background of routine network activity (King, 2013, para. 5).
- The Insight program addresses key shortfalls in current intelligence, surveillance, and reconnaissance systems. Automation and integrated human-machine reasoning enable operators to analyze greater numbers of potential threats ahead of time-sensitive

situations. This program seeks to develop a resource management system that automatically identifies threat networks and irregular warfare operations by the analysis of information from imaging and nonimaging sensors and other sources (King, 2013, para. 7).

- The Mind's Eye program seeks to develop a capability for *visual intelligence* in machines. Unlike the traditional study of machine vision where progress has been made in recognizing a wide range of objects and their properties or the nouns in the description of a scene, Mind's Eye seeks to add the perceptual and cognitive underpinnings needed for recognizing and reasoning about the verbs in those scenes. Collectively, these technologies could enable a more complete visual narrative (King, 2013, para. 9).
- The Machine Reading program seeks to realize artificial intelligence applications by developing learning systems that process natural text and insert the resulting semantic representation into a knowledge base rather than relying on expensive and time-consuming current processes for knowledge representation that require expert and associated knowledge engineers to handcraft information (King, 2013, para. 8).
- The Mission-Oriented Resilient Clouds program aims to address security challenges inherent in cloud computing by developing technologies to detect, diagnose, and respond to attacks (King, 2013, para. 10).
- The PROCEED research effort targets a major challenge for information security in the cloud computing environments by developing practical methods and associated modern programming languages for computation on data that remains encrypted the entire time it is in use. An interception by an adversary would be more difficult if users had the ability to manipulate encrypted data without first decrypting (King, 2013, para. 11).
- The VIRAT program aims to develop a system to provide military imagery analysts with the capability to exploit the vast amount of overhead video content being collected. If it is successful, VIRAT will enable analysts to establish alerts for activities and events of interest as they occur. Tools will also be developed to enable analysts to retrieve rapidly, with high precision and recall, video content from extremely large video libraries (King, 2013, para. 12).

17.6   *Government big data case studies*

Government agencies have implemented big data projects to transform agencies' processes and procedures. The U.S. Army, for example, is already leveraging big data technologies in conjunction with cloud computing (Conway, 2012). Started in April 2009, the U.S. Army's Big Data Cloud program extends to forward operating bases, which can double as local nodes that collect data from various sources. The private cloud, which went live in March 2011, conveys the latest intelligence information to U.S. troops in Afghanistan in real or near-real time (Conway, 2012).

The National Archive and Records Administration (NARA) challenge is to digitize a huge volume of unstructured data to provide quick access while maintaining the data in both classified and unclassified environments (TechAmerica Foundation, 2012). NARA is charged with providing the Electronic Records Archive (ERA) and online, public access systems for U.S. records and documentary heritage. In January 2012, NARA managed approximately 142 terabytes of information, consisting of more than seven billion objects and incorporating records from across the federal agencies, Congress, and several presidential libraries. There are more than 350 million annual hits on its website. In addition to managing the ERA, NARA must digitize more than four million cubic feet of traditional archival holdings, including about 400 million pages of classified information scheduled for declassification, pending review with the intelligence community (TechAmerica Foundation, 2012).

NARA used big data tools to address those challenges. In conjunction with traditional data capture, digitizing, and storage capabilities, advanced big data capabilities were used for search, retrieval, and presentation, all while supporting strict security guidelines. Faster result ingestion and categorization of documents, improved end user experience, and dramatically reduced storage costs were the results (TechAmerica Foundation, 2012). Other big data cases involving government agencies are summarized in Table 17.3 (adapted from TechAmerica Foundation, 2012).

17.7   Lessons learned

To better understand big data, it is useful to consider this somewhat ambiguous concept by taking advantage of lessons learned by other organizations dealing with similar problems. The TechAmerica Foundation Big Data Commission released a study in October 2012 on how big data can move beyond the tidal wave of data and transform government. The Commission's mandate was to demystify the term *big data* by defining its characteristics, describing the key business outcomes it will serve, and providing a framework for policy discussion. Its goal was to provide guidance to federal government's senior policy and decision makers.

The Commission identified a number of lessons learned from early government big data initiatives (TechAmerica Foundation, 2012) as follows:

- The path toward becoming big data "capable" will be iterative and cyclical.
- Successful big data initiatives seem to begin with a burning business or mission requirement that government leaders are unable to address with traditional approaches.
- Successful big data initiatives commonly start with a specific and narrowly defined business or mission requirement, and not a plan to deploy a new and universal technical platform to support perceived future requirements.
- Successful initiatives seek to address the initial set of use cases by augmenting current IT investments, but do so with an eye to leveraging these investments for inevitable expansion to support far wider use cases in subsequent phases of deployment.
- Once an initial set of business requirements has been identified and defined, the leaders of successful initiatives then assess the technical requirements, identify gaps in their current capabilities, and plan the investments to close those gaps.
- Successful initiatives tend to follow three patterns of deployment underpinned by the selection of one big data "entry point" that corresponds to one of the key characteristics of big data—volume, variety, and velocity.
- After completing their initial deployments, government leaders typically expand to adjacent use cases, building out a more robust and unified set of core technical capabilities. These capabilities include the ability to analyze streaming data in real time; the use of Hadoop or Hadoop-like technologies to tap huge, distributed data sources; and the adoption of advanced data warehousing and data mining software (TechAmerica Foundation, 2012, p. 7).

The Commission made the following recommendations for government agency leaders to adopt when implementing big data solutions:

- Understand the "Art of the Possible" by reviewing case studies of prior implementations to understand practical examples.

Table 17.3 High-level summary of case studies

Agency/org./co. big data project name	Underpinning technologies	Big data metrics	Initial big data entry point	Public/user benefits
NARA ERA	Metadata, submission, access, repository, search, and taxonomy applications for storage and archival systems	Petabytes, terabytes/s, semi-structured	Warehouse optimization, distributed info management	Provides ERA and online public Access systems for U.S. records and documentary heritage
National Aeronautics and Space Administration (NASA) Human Space Flight Imagery	Metadata, archival, search, and taxonomy applications for tape library systems, government off-the-shelf (GOTS)	Petabytes, terabytes/s, semi-structured	Warehouse optimization	Provide industry and the public with iconic and historic human spaceflight imagery for scientific discovery, education, and entertainment
National Oceanic and Atmospheric Administration (NOAA) National Weather Service	HPC modeling; data from satellites, ships, aircraft, and deployed sensors	Petabytes, terabytes/s, semi-structured, ExaFLOPS, PetaFLOPS	Streaming data and analytics, warehouse optimization, distributed info management	Provide weather, water, and climate data, and forecasts and warnings for the protection of life and property and enhancement of the national economy.
Internal Revenue Service (IRS) Compliance Data Warehouse	Columnar database architecture; multiple analytics applications; descriptive, exploratory, and predictive analysis	Petabytes	Streaming data and analytics, warehouse optimization, distributed info management	Provide taxpayers top quality service by helping them understand and meet their tax responsibilities and enforce the law with integrity and fairness.
Centers for Medicare & Medicaid Services (CMS) Medical Records Analytics	Columnar and NoSQL databases, Hadoop being looked at, EHR on the front end, with legacy structured database systems (including DB2 and COBOL)	Petabytes, terabytes/day	Streaming data and analytics, warehouse optimization, distributed info management	Protect the health of all Americans and ensure compliant processing of insurance claims

- Identify 2–4 key business or mission requirements that big data can address for the government agency, and define and develop underpinning use cases that would create value for both the agency and the public.
- Take inventory of the "data assets." Explore the data available both within the agency enterprise and across the government ecosystem within the context of the business requirements and the use cases.
- Assess current capabilities and architecture against what is required to support goals, and select the deployment entry point that best fits your big data challenge, whether it is volume, variety, or velocity.
- Explore which data assets can be made open and available to the public to help spur innovation outside the agency (TechAmerica Foundation, 2012, p. 8).

17.8 Big data in the U.S. Navy

The U.S. Naval Air Systems Command (NAVAIR) has optimized its resources with big data. NAVAIR implemented the Decision Knowledge Programming for Logistics Analysis and Technical Evaluation (DECKPLATE) system to centralize and streamline management of aircraft fleet and aircraft carriers deployed around the world (Sverdlik, 2012). DECKPLATE is used to manage fleet resources during both military and humanitarian missions. When the Fukushima Daiichi nuclear power plant was leaking radiation, DECKPLATE was used to determine the readiness of the fleet operating in the area. It also provided real-time data on the danger of radiation exposure to the Navy's assets during this time (Sverdlik, 2012).

DECKPLATE provides the following

- Enterprise-wide visibility: DECKPLATE uses about 23 years of trend analysis of aircraft readiness, checking data in areas such as aircraft maintenance, flight usage and inventory, configuration baseline management, engine total asset visibility, technical directives, and supply cost.
- Daily reporting: Daily readiness reporting is provided with messages going out every day from an aircraft carrier deployed at sea concerning aircraft status. In 2004, these reports would be correlated on a monthly basis, put on a DVD, and sent to commanders with the readiness status.
- Constant process optimization: DECKPLATE provides ongoing improvements of its processes. It can provide data to address problems proactively before they occur, which the traditional reporting process did not allow for.
- Changing logistics philosophy: Historically, the military wanted 100% of its assets up 100% of the time, and that required expenditures to fix things that were not really necessary. With DECKPLATE, an initiative was created to optimize the logistics process to have the right assets with the right configuration in the right place at the right time (Sverdlik, 2012, Enterprise-Wide Visibility, para. 6).

The next phase for DECKPLATE is *binning* in which data would be evaluated on a more granular level (Sverdlik, 2012). In the binning project, a history of some 200 million maintenance actions would be broken down into the individual maintenance actions required. The historical maintenance actions would then be further broken down into every 15 min. This process will answer the question, "Was the aircraft awaiting maintenance during that time or was it awaiting supply?" The final objective of identifying exactly how and where time was spent on the aircraft during the maintenance period requires a massive amount of data to be collected and analyzed over a 5-year period on approximately 5000 aircraft (Sverdlik, 2012).

References

Conway, S. 2012. Big data cloud delivers military intelligence to U.S. Army in Afghanistan (Press release). Retrieved from http://www.datanami.com/datanami/2012-02-06/big_data_cloud_delivers_military_intelligence_to_u.s._army_in_ afghanistan.html. Accessed July 17, 2015.

Department of Homeland Security (DHS). 2013. Big data storage and analytics technology (Solicitation No. HSHQDC-13-BIG_DATA_RFI). Retrieved from https://www.fbo.gov/index ?s=opportunity&mode=form&id=ca5d230b7454 615562a026466264acc2&tab=core&_cview=1. Accessed July 17, 2015.

Dijcks, J. P. 2013. *Oracle: Big Data for the Enterprise* (White Paper). Redwood Shores, CA: Oracle. Retrieved from http://www.oracle.com/us/products/database/big-data-for-enterprise-5191 35.pdf. Accessed July 17, 2015.

Donaldson, I. 2013. Big data implications for enterprise architecture. Unpublished manuscript, Department of Information Sciences, Naval Postgraduate School, Monterey, CA.

Ferguson, M. 2013. *Enterprise Information Protection: The Impact of Big Data* (White Paper). Cheshire, England: Intelligent Business Strategies. Retrieved from http://www-01.ibm.com/software/os/systemz/pdf/Info_Security_and_Big_Data_on_Z_White_Paper_Final.pdf. Accessed July 17, 2015.

Gantz, J. and Reinsel, D. 2012. *The Digital Universe in 2020: Big Data, Bigger Digital Shadows, and Biggest Growth in the Far East*. Framingham, MA: IDC Corporation.

Housel, T. J. and Bell, A. 2001. *Measuring and Managing Knowledge*. Boston, MA: McGraw-Hill.

Kelly, J., Floyer, D., Vellante, D. and Miniman, S. 2013. Big data vendor market revenue and forecast. Retrieved from http://wikibon.org/wiki/v/Big_Data_Vendor_Revenue_and_Market_Forecast_2012-2017. Accessed July 17, 2015.

Kenney, M. E. 2013. Cost reduction through the use of additive manufacturing (3D printing) and collaborative product life cycle management technologies to enhance the Navy's maintenance programs, Master's thesis, Naval Postgraduate School. Retrieved from http://acquisitionresearch.net. Accessed July 17, 2015.

King, R. 2013. U.S. government spending on big data to grow exponentially. Retrieved from http://www.biometricupdate.com/201308/u-s-government-spending-on-big-data-to-grow-exponentially. Accessed July 17, 2015.

Knowledge Value-Added (KVA) Methodology. (n.d.). International Engineering Consortium. Retrieved from http://cmapspublic3.ihmc.us/rid=1G9L62WTW-NJ5G5D-C84/KVAmethodology.pdf. Accessed July 17, 2015.

Komoroski, C. L. 2005. Reducing Cycle Time and Increasing Value through the Application of Knowledge Value Added Methodology to the U.S. Navy Shipyard Planning Process, Master's thesis, Naval Postgraduate School. Retrieved from http://acquisitionresearch.net/. Accessed July 17, 2015.

Manyika, J., Chui, M., Brown, B., Bughin, J., Dobbs, R., Roxburgh, C. and Byers, A. H. 2011. *Big Data: The Next Frontier for Innovation, Competition, and Productivity*. McKinsey Global Institute. Retrieved from http://www.mckinsey.com/insights/business_technology/big_data_the_next_frontier_for_innovation. Accessed June 28, 2015.

Office of Science and Technology Policy. 2012. Obama administration unveils big data initiative: Announces $200 million in new R&D investments (Press release). Retrieved from http://www.whitehouse.gov/sites/default/files/microsites/ostp/big_data_press_release_final_2.pdf. Accessed July 17, 2015.

Sverdlik, Y. 2012. *Big Data Gives Real Time Logistics to U.S. Navy: A Data Analytics Platform Changes the Way Navy Thinks About Logistics* (Press release). Retrieved from http://www.datacenterdynamics.com/focus/archive/2012/02/big-data-gives-real-time-logistics-us-navy. Accessed July 17, 2015.

TechAmerica Foundation. 2012. *Demystifying Big Data: A Practical Guide to Transforming the Business of Government*. Washington, DC. Retrieved from http://www.techamerica.org/Docs/fileManager.cfm?f=techamerica-bigdatareport-final.pdf. Accessed July 17, 2015.

chapter eighteen

Part B: Visualization of big data
Ship maintenance metrics analysis

**Isaac J. Donaldson, Sandra C. Hom, Thomas Housel,
Johnathan Mun, and Trent Silkey**

Contents

18.1 Introduction

There are between 150 and 200 parameters for measuring the performance of ship maintenance processes in the U.S. Navy. Despite this level of detail, budgets and timelines for performing maintenance on the Navy's fleet appear to be problematic. Making sense of what these parameters mean in terms of the overall performance of ship maintenance processes is clearly a "big data" problem.

A team from the Naval Postgraduate School (NPS) was requested by Program Executive Office (PEO) Ships to work with naval ship maintenance metrics groups to provide additional options regarding how large datasets could be optimized. The current process for presenting data on more than 150 parameters measuring ship performance maintenance costs and processes, containing billions of data points, is still done by static, cumbersome spreadsheets. The central goal of this project was to provide a means to aggregate voluminous maintenance data in such a way that ship maintenance leadership can better understand the causal factors contributing to cost and schedule overruns. By providing this kind of information in an intuitively visual form, leadership could be assisted in budget and scheduling decision making.

18.2 Ship maintenance vignettes

Maintenance is crucial to the Navy's fleet readiness and ensures that the fleet reaches its expected service life. This section uses vignettes to show three aspects of ship maintenance that provide a framework for understanding these types of activities within the Navy. It begins with a general discussion on maintenance and modernization budgets and then provides specific ship case examples.

18.2.1 Maintenance and modernization spending

Maintenance and modernization are essential to derive full benefits of DoD assets and, more importantly, they enable the United States to respond quickly to security challenges and offer humanitarian assistance around the world. In FY2010, the DoD spent approximately $83.7 billion to maintain strategic material readiness for 13,900 aircraft, 800 strategic missiles, 350,000 ground combat and tactical vehicles, 283 ships, and myriad other DoD weapon systems (Office of the Assistant Secretary of Defense for Logistics and Material

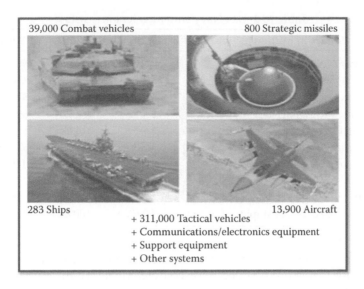

Figure 18.1 Systems supported by DoD maintenance.

Readiness [OASD (L&MR)], 2011). Figure 18.1 shows the systems supported by the DoD (OASD [L&MR], 2011). Maintenance was provided through the efforts of approximately 657,000 military and civilian maintainers and thousands of commercial firms.

Performed at several levels, DoD material maintenance ranges in complexity from daily system inspections to rapid removal and replacement of components, to complete overhauls or rebuilds of a weapon system. The three levels of maintenance are as follows: depot-level maintenance for the most complex and extensive work; intermediate-level maintenance for less complex maintenance activities performed by operating unit back-shops, base-wide activities, or consolidated regional facilities; and field-level maintenance, a combination of organizational depot and intermediate levels (OASD [L&MR], 2011).

In early 2011, the DoD operated 17 major depot activities and expended more than 98 million direct labor hours (DLHs) annually (Avdellas et al., 2011). DoD depots' property, plants, and equipment were valued at more than $48 billion with an infrastructure consisting of more than 5600 buildings and structures (Avdellas et al., 2011).

To maintain readiness and ensure that the fleet reaches its expected service life, the Navy spent $8.5 billion on ship maintenance in FY2011. Table 18.1 shows the Navy's maintenance budget in recent years (adapted from Department of the Navy [DoN], 2012a).

18.2.2 Maintenance vignettes

Each of the three vignettes describes an aspect of ship maintenance work: new work (NW), deferred maintenance (DM), and modernizations. Although there is another category, *original work* (OW), maintenance for which planning has been completed and is included in the maintenance package before the evolution (an *availability*, or *avail*), this section focuses on NW, DM, and modernizations.

NW is maintenance added to a specific ship's availability after planning has been completed (i.e., not part of the original maintenance package). NW can result from discrepancies that have not yet been discovered or from work that was not added to the availability work package until after, planning was complete. DM refers to the status of maintenance

Table 18.1 U.S. Navy ship maintenance costs

(U.S. dollars in millions)	FY2011	FY2012	FY2013
Active forces			
Ship maintenance	$4726	$4533	$5090
Depot operations support	$1326	$1296	$1315
Baseline ship maintenance (O&M, N)	$6052	$5829	$6405
Overseas contingency operations	$2484	$1493	$1310
Total ship maintenance (O&M, N)	$8536	$7322	$7715
Percentage of projection funded	100%	97%	100%
Annual DM	$0	$217	$0
CVN refueling overhauls (SCN)	1664	530	1683
Percentage of SCN estimates funded	100%	100%	100%

rather than the time of its inclusion in the maintenance package and may be either OW or NW. DM is work that is rescheduled to be completed later in the current availability or as part of a future maintenance period. Modernizations (or *mods*) are system upgrades. A modernization can range in scope from a short-term software upgrade to a long-term ship infrastructure remodeling. Generally, the planning for all the modernization work is completed before the availability begins, and is, therefore, classified as OW. However, in the modernization vignette in this section, two cases demonstrate that situations can arise that require modernization work to become NW. Figure 18.2 shows the relationships among the different categories.

The three ships used in the vignettes under the cognizance of Norfolk Ship Support Activity (NSSA) are, the United States Ship (USS) *Wasp* (LHD-1), the USS *Bataan* (LHD-5), and the USS *Iwo Jima* (LHD-7). First, LHD-7 is a case study to describe NW. Second, to depict DM, both LHD-5 and LHD-7 are examples. Finally, LHD-1 and LHD-7 are used to illustrate modernizations, as shown in Figure 18.3.

The three vignettes that follow were derived from two phone conversations with David J. Furey, a civilian employee of the NSSA, on September 9 and 11, 2013.

18.2.2.1 New work vignette: USS Iwo Jima

The USS *Iwo Jima* is an example requiring NW, DM, and modernization. In addition, this case examines NW and how complications from NW can impact schedules. In this vignette, the focus is on the rudder and the bilge. The rudder, a critical portion of the

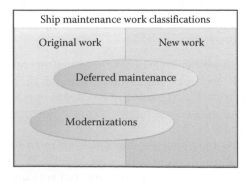

Figure 18.2 Ship maintenance work classifications.

	Vignette		
	New work	Deferred maintenance	Modernizations
USS *Wasp* (LHD-1)			X
USS *Bataam* (LHD-5)		X	
USS *Iwo Jima* (LHD-7)	X	X	X

Figure 18.3 Vignette overview.

ship's steering system, caused a schedule extension due to degradation that was not readily apparent. All appropriate assessments, checks, and leakage tests were conducted by maintenance technicians, and the results indicated the rudder was in good condition. All the tests associated with the rudder were within specified parameters, and the rudder passed the preliminary inspection. Unfortunately, *bearing clearance testing*, tests that analyze rudder performance over the entire range of operation (full left to full right), exposed inconsistencies prior testing did not reveal. Results from the test were irregular and upon examination of the rudder bearings, metal debris and rust were discovered. The NSSA ultimately made the decision to remove and replace the rudder, which resulted in the availability schedule being extended by 14 days.

NW was also required on the bilge of the USS *Iwo Jima*. As part of the entire availability, high pressure washing was required in the bilge. While performing this evolution, fuel piping was damaged, and a leak developed. *Ship's force*, a term that describes the active duty sailors onboard the ship, repaired the damage by using a *soft patch*. A soft patch is a temporary repair method for low pressure piping. However, the NSSA was constrained by more restrictive requirements and was required to replace the faulty piping. To determine the extent of the damage, *ultrasonic testing* (UT) was used, which uses sound wave properties to determine the remaining amount of pipe wall thickness. If less than 50% of the pipe wall remains, the NSSA is required to replace the pipe. UT was performed and revealed 40 ft of fuel piping and an additional 20 ft of oily waste piping required replacement. The availability schedule was extended by 40 days to replace the identified piping.

18.2.2.2   *Deferred maintenance vignette: USS* Bataan *and USS* Iwo Jima

In these vignettes, the USS *Bataan* example relates to cost cutting while the USS *Iwo Jima* example relates to prioritization. The overall magnitude of work to be accomplished during the USS *Bataan* availability made it a target of cost cutting during shrinking fiscal budgets in 2012. A common item to be deferred is paintwork, and the USS *Bataan* was not an exception. Much of the tank paintwork was deferred from the 2012 availability to the 2015 availability as a result of fiscal cutbacks.

The USS *Iwo Jima* also experienced DM, but the maintenance was deferred because higher priorities required the ship to be waterborne. Specifically, the 7-K-O-W tank, the forward feed tank for the ship's ballast system, was due for preservation and required the ship to remain in drydock. The tank had not been opened since commissioning as this was the ship's first drydock availability. Inspection revealed the tank to be in Tank Condition 4, which means that a profound failure had been discovered. UT showed that no more than 17% surface wastage had occurred, and, therefore, the tank had become a candidate for deferral. Higher priority maintenance necessitated that the ship be waterborne, so the drydock was flooded and the 7-K-O-W tank preservation was deferred.

While the effect on a ship's availability schedule of the addition of NW can be directly measured, the consequences of deferring maintenance are a matter of risk. The USS *Iwo Jima* added NW to its availability and incurred schedule delays, or *lost operating days* (LODs); 14 days were attributed to work on the rudder and 40 days to the replacement of the pipe. In both cases, the impact can be easily measured.

As for DM, the impact can range from minimal to substantial. For instance, the tank paintwork for the USS *Bataan* was deferred until the next planned availability in 2015. The paintwork would have cost a certain dollar amount in 2012 and would have provided the tank a level of preservation protection. In 2015, the paintwork will cost more not only because of inflation and the degradation of the paint associated with time, but also because corrosion will have developed at a higher rate than it would have with a fresh application of paint. The difference between the costs of paintwork in 2015 versus in 2012 (including corrosion correction) is the impact of this DM example and would be comparatively minimal. However, the possibility of a larger effect exists. Perhaps the development and growth of corrosion on the 7-K-O-W tank is underestimated. If the corrosion progresses significantly faster, then the likelihood of structural failure increases. Should the structural failure occur outside the maintenance environment of the shipyard, and then the impact would be far greater, and the costs associated with unscheduled maintenance much higher. The decision to defer the preservation of the tank must consider both the likelihood and severity of all the possible outcomes. In other words, the decision maker must consider all the associated risks before deferring maintenance.

18.2.2.3 Modernizations vignettes: USS Iwo Jima and USS Wasp

Modernizations have the most potential to impact the schedule of the three classifications of shipyard maintenance examined in this section. In the cases of the USS *Iwo Jima* and USS *Wasp*, modernizations may affect the timetable because not all the required drawings had been completed prior to the start of work. For the USS *Iwo Jima*, a single modernization is presented, whereas the USS *Wasp* serves as a more general example. However, a brief overview of the shipyard planning evolution is presented first to explain the importance of timely drawings.

Before the shipyard period starts, the plan for a scheduled availability must be completed. To complete the plan for availability, NAVSEA (Naval Sea Systems Command) must approve the contractor-provided estimate (Department of the Navy [DoN], 2012b). To generate the estimate, however, the contractor must review all the drawings (first-tier and second-tier) associated with the work to be performed (D. Furey, personal communication, September 9 and 11, 2013). First-tier drawings are the main focus of the modernization, whereas second-tier drawings involve infrastructure and subsystems related to the work. For a particular modernization, if all the drawings are not completed, the contractor cannot create the estimate and an approved plan will not exist. In addition, availabilities must sometimes commence on a partial solution; otherwise, all work would be completed late. In the situation without an approved plan, the project completion date (PCD) has a larger margin of error, and schedule changes are more likely to occur.

This was the case with the CANES (Consolidated Afloat Networks and Enterprise Services) installation in the USS *Iwo Jima* availability. CANES, or, as its name implies, is a program created to consolidate many networks and services aboard ships into a single information technology system. Although not all the drawings were received, the maintenance period started anyway. There was another work to perform; CANES was not the only reason for the USS *Iwo Jima* to visit the shipyard. As drawings for CANES were completed, they were then provided to the contractor. However, the plan for CANES could not be approved until all the drawings were received, the contractor generated the estimate, and NAVSEA accepted the plan.

In the case of the USS *Wasp*, the estimated modernization cost was extremely high at $250 million to $300 million. The high cost was partially due to modernizations needed to accommodate the F-35 Joint Strike Fighter (JSF) because the USS *Wasp* was to be the first ship to test the JSF, and part of the flight deck had to be strengthened. Not only was the structural reinforcement of the after flight deck a large package, but the ship was also undergoing many other modernizations. Unfortunately, the USS *Wasp* also started its availability without a complete plan. Twenty modernization packages were not included in the plan, including the structural reinforcement of the after flight deck because the drawings had not yet been delivered.

In addition, the NSSA erroneously included one large work item in the plan for which second-tier drawings had not yet been received. The contractor brought the discrepancy to the NSSA's attention explaining that they, the contractors, would not be able to complete an estimate before the plan was completed (also known as 100% *lock*). The NSSA had two options: either extend the lock or pull the work item out and add it back in later as NW. They chose the latter.

In both these vignettes, modernizations had significant potential to affect severely the scheduled PCD because the drawings were not completed. Two questions arise that are associated with the implications of missing PCD on ship maintenance costs.

- Is there a cost premium to NW? In other words, do costs increase because a modernization was added after 100% lock?
- Are LODs caused by planning or scope? In other words, is it the planning process or the scope of work that is to blame for missing PCD?

18.3   Ship maintenance simulations

A team from the NPS was tasked by PEO Ships to work with naval ship maintenance metrics groups to provide additional options regarding how large datasets could be optimized. In particular, presentation methods were requested succinctly showing a ship's maintenance status including all operational costs and schedule deviations from planned maintenance. Project sponsors also sought suggestions for improving how key information could be summarized and ultimately used in making critical maintenance allocation decisions. The current process for presenting data on more than 150 parameters measuring ship performance maintenance costs and processes, containing billions of data points, is still done with static, cumbersome spreadsheets.

The project was conducted in three distinct phases, as seen in Figure 18.4. First, data were collected on 19 U.S. Navy guided missile destroyers (DDG) with maintenance periods spanning a few years, 2010 to mid-2013. Data were collected on 21 maintenance availabilities for those DDGs and included definitized estimates prepared by SMEs in the planning process along with the actual cost and availability data on three maintenance categories. In Phase 2, a hypothesis was tested, and two simulations were run using the knowledge value added (KVA) methodology. In Simulation 1, we tested the potential impact of incorporating three-dimensional printing (3DP) on ship maintenance programs while in Simulation 2 we evaluated the combination of 3DP plus two more technologies (3D laser scanning technology [3D LST] and collaborative product lifecycle management [CPLM]). In Phase 3, a visualization tool offered by an independent software vendor was selected to show how large volumes of data could be shown in a succinct manner.

> **Phase 1: Data collection**
> - Definitized estimates for 19 guided missile destroyers (DDG)
> - Twenty-one maintenance availabilities from 2010 to mid-2013
> - Actual costs from surface team one metrics system (ST1MS)
> - Cost categories of growth, new growth, new work, original

> **Phase 2: Simulations**
> - Simulation 1 Three-dimensional printing technology (3DP)
> - Simulation 2 Three-dimensional printing technology (3DP)
>   - Three-dimensional laser scanning technology (3D LST)
>   - Collaborative product lifecycle management (CPLM)

> **Phase 3: Analysis and results**
> - Definitized cost estimates for maintenance work ($313.7 million)
> - Actual costs for maintenance work ($435.5 million)
> - Cost estimates after simulations incorporating technologies ($271.1 million)
> - Potential cost savings of 37.7% ($164.4 million)

Figure 18.4 Project phases.

The visualization software provides a higher level of visual clarity, enabling faster and more intuitive interpretation of ship maintenance data by presenting the data relationships in diagrams, graphs, and charts. Relationships among variables are more readily discoverable and, more importantly, those relationships can be used in forecasting to develop more accurate maintenance data, estimates that are based on historical data. Decision makers can see analytical results quickly with visualization software, which allows them to find relevance among millions of variables, communicate concepts and hypotheses to others, and even forecast possible scenarios.

This section of the report is divided into several topics. First, maintenance categories and the data collection process are reviewed. Final simulation results are highlighted to provide a framework for understanding the power of visualization software, followed by a general discussion on the original definitized cost estimate. Actual costs are then compared with the definitized cost estimates and discrepancies between the two are discussed. An analysis of the potential effect on ship maintenance costs by incorporating specific technologies in Simulations 1 and Simulation 2 is discussed in greater detail. Alternative presentation methods, which drill down into specific detail, are then explored. Next, a description and analysis of a common ship maintenance metric, LODs, are given along with a recommendation of a more useful metric, availability density. This section concludes with further examples of visualization tools' abilities to drill down into specific details.

18.3.1   Maintenance cost categories

There are several cost categories for ship maintenance: OW, growth (G), NW, and new growth (NG). OW is the estimated ship maintenance cost (shipyard or contractor, labor, and material costs) at the completion of planning and is also known as the definitized cost estimate. The definitized cost estimate is a figure provided by an SME in the planning process.

G is an expansion of OW and can result from many factors including undiscovered discrepancies or an increase in scope. For example, the OW plan for a hypothetical ship called for preservation work on the ship's hull. While conducting the preservation work, the maintenance technician discovered hull damage that required minor repair. The minor repair work would be classified as G.

NW is maintenance that is added to a ship's availability after planning has been completed (i.e., not part of the OW maintenance package). NW can result from discrepancies that have not yet been discovered and are unrelated to previously planned maintenance or from work that was not added to the availability work package until after planning was complete. For example, while conducting preservation work on the hypothetical ship, the maintenance technician discovered damage to a communication antenna. The resulting repair work would be classified as NW.

NG is the growth resulting from an expansion in NW, similar to the relationship between G and OW. For example, the antenna maintenance technician conducting antenna repair work discovered that the antenna was beyond repair and needed to be replaced. Replacement of the antenna would be considered NG.

18.3.2   Data collection

Data for this analysis were derived from the ST1MS website (https://mfom-shipmain.nmci.navy.mil). In particular, ship availabilities were selected for examination based on several factors designed to establish a proof of concept for the use of big data to shape executive-level decisions. The availabilities were restricted to only U.S. Navy DDGs whose maintenance period started by 2010 and whose final reports were closed and completed by the time this study began in 2013. Ships whose close-out reports were incomplete or missing data were not included in the analysis.

The figures in this *section* are screenshots of solar graph results that were captured while using the visualization software program to process the ship maintenance data obtained from the ST1MS website. The data consist of 21 maintenance availabilities for the DDGs.

18.3.3   Final simulation results incorporating different combinations of technologies into U.S. Navy ship maintenance programs

Two simulations were run to show the potential cost savings of incorporating specific technologies. In Simulation 1, only 3DP technology was evaluated while, in Simulation 2, three combined technologies were evaluated. Tables 18.2 and 18.3 (both based on J. Kornitsky, personal communication, November 2013) reflect the differences between definitized costs, actual costs, and projected costs for Simulations 1 and 2. The definitized cost estimate was $313.7 million, compared to the actual cost of $435.5 million. If 3DP, 3D LST, and CPLM technologies combined were incorporated into the ship maintenance processes, the costs would have been reduced to an estimated $271.1 million.

Table 18.2 Cost comparison by ship (U.S. dollars represented in millions)

Ship	Definitized cost estimates	Actual costs	% versus definitized	Simulation 1 (3DP)	% versus definitized	% versus actual	Simulation 2 radical (3DP, 3D, LST, CPLM)	% versus definitized	% versus actual	% versus 3DP
Barry	$48.0	$70.1	46.0	$65.8	37.1	-6.1	$43.9	-8.5	-37.4	-33.3
Arleigh Burke	$46.9	$58.0	23.7	$56.4	20.3	-2.8	$35.7	-23.9	-38.4	-36.7
Ramage	$46.3	$57.2	23.5	$55.9	20.7	-2.3	$35.7	-22.9	-37.6	-36.1
Donald Cook	$21.4	$36.3	69.6	$36.2	69.2	-0.3	$22.9	7.0	-36.9	-36.7
Stout	$45.4	$63.2	39.2	$64.1	41.2	1.4	$38.6	-15.0	-38.9	-39.8
All other	$105.6	$150.5	42.5	$147.5	39.7	-2.0	$94.1	-10.9	-37.5	-36.2
Total	$313.6	$435.3	38.8	$425.9	35.8	-2.2	$270.9	-13.6	-37.8	-36.4

Table 18.3 Cost comparison by work (U.S. dollars represented in millions)

Work	Definitized cost estimates	Actual costs	% versus definitized	Simulation 1 (3DP)	% versus actual	Simulation 2 radical (3DP, 3D, LST, CPLM)	% versus actual
Original	$313.7	$313.7	0.0	$307.3	−2.0	$195.4	−37.7
Growth	$0.0	$47.1	100.0	$45.7	−3.0	$28.1	−40.2
New work	$0.0	$66.8	100.0	$65.5	−1.9	$43.0	−35.6
New growth	$0.0	$7.7	100.0	$7.4	−3.9	$4.5	−41.6
Total	$313.7	$435.5	38.7	$426.2	−2.1	$271.1	−37.7

18.3.4   Visualization software analysis of U.S. navy ship maintenance

18.3.4.1   Visualization model

The visualization model (Figure 18.5, J. Kornitsky, personal communication, November 2013) is an overview of how the DDG spreadsheet data was mapped into the software. It shows four cost categories on top, all 19 ships by name in the middle, and their combined availabilities at the bottom. The lines between the boxes depict connection relationships.

The 24 boxes referred to in the model have a number above each that represents the aggregate cost. For example, the box on the middle left side of Figure 18.5, labeled *Stout*, indicated $28.1 million of aggregate cost attributed to the availability. In addition, the horizontal bar between the cost number and the box represented the relative portion of cost attributed to that availability when compared with all availabilities. The box in the top left corner, labeled Growth, indicated a relative cost that resulted in the length of the bar shown.

At the bottom of each box, the number of connections to all other variables was depicted in two ways. The number displayed on the bottom right of each box and the number of ovals displayed on the bottom left of each box. At the box at the bottom of Figure 18.5, labeled avail indicated 21 connections to all other variables with both the numeral, "21," and the number of ovals displayed, 21.

At the top of Figure 18.5, four boxes are depicted and represent one category of cost, type of work. The labels on each box indicate a particular type of work, G, NG, NW, and OW. Each particular type of work accounted for cost indicated.

In the middle of Figure 18.5, the 19 boxes labeled with ship names indicate the maintenance cost each ship incurred. For 17 of the ships, the ship maintenance cost was attributed to a single availability. For the *Arleigh Burke* and the *Donald Cook*, the ship maintenance cost was attributed to two availabilities. For example, the box labeled *Arleigh Burke* in Figure 18.5 indicated $35.7 million in ship maintenance cost, but for two unique availability periods. This can be verified by referencing the number in the lower-right portion of the ship name boxes. For most of the ships, this number was 4, and the number of ovals was four. This represented the number of connections to the kinds of cost. In any single availability, there were four types of work (cost) identified (OW, G, NW, and NG). In the cases of the *Arleigh Burke* and *Donald Cook* ships, there were two availabilities recorded, and, therefore, eight connections to the four types of work (cost) as was indicated by the number, 8, and the eight ovals indicated in either box in Figure 18.5.

The single box depicted at the bottom of Figure 18.5 represented the aggregate forecasted cost of all availabilities, $271.1 million.

Figure 18.5 Visualization model.

18.3.4.2   Definitized estimate, all ships

Definitized estimates are the total projected costs of an availability upon completion of the planning phase of ship maintenance, provided by SMEs in the planning process. According to the *Joint Fleet Maintenance Manual* (DoN, 2012b), the planning phase for an availability for a DDG begins 720 days before the first day of maintenance (A-720). By this day, A-720, an availability must be added to the U.S. Navy surface ship availability schedule. The next milestone, a letter of authorization, occurs on or before A-360 and obligates the stakeholders to the specific cost of prorate schedules. Through the next three milestones, A-240 (50%), A-120 (80%), and A-75 (100%), progressively, more of the budgeted funds must be allocated, or locked, to specific work items. By A-60, the overall plan for maintenance must be finalized to allow the detailed work schedule to be formulated, and cost estimates completed. The final cost estimate, or definitized work package must be finished by A-35 and represents all costs attributed to OW. After definitization, all additional work items are considered to be G, NW, or NG (DoN, 2012b).

Figure 18.6 (J. Kornitsky, personal communication, November 2013) shows how each ship contributed to the total expected cost of all the availabilities analyzed. The total of $313.7 million is greater than the total presented in the previous image, $271.1 million. As explained earlier, this is because the first screenshot shows the total costs after the combined incorporation of three different technologies into the ship maintenance process.

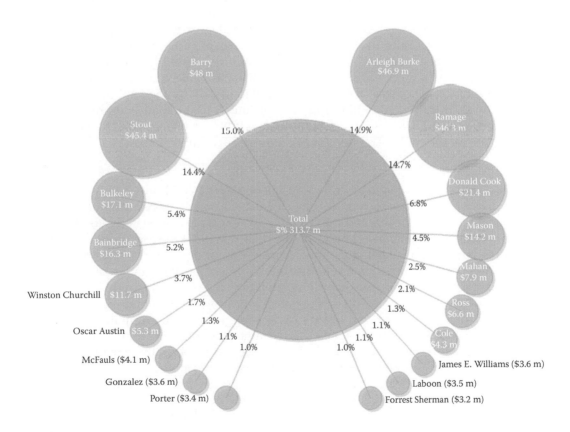

Figure 18.6 Definitized estimate, all ships screenshot.

All the figures shown in this section present a parent–child type of relationship hierarchy, similar to object-oriented programming. In Figure 18.6, there exists only a simple relationship with each instance having assumed a single role. The total definitized estimate of $313.7 million in the center is the parent while all the ships and their total maintenance costs, are the children. Multi-role instances, where the single solar graph screenshots can be both parent and child, are presented in later figures, beginning with Figure 18.8.

Each ship contributed to the total definitized estimate of $313.7 million. The amount each contributed is presented in three different ways. First, the size of each bubble signifies its cost relative to the total cost bubble in the middle of the screenshot. The larger the relative cost of the ship identified the larger the bubble. Second, the relative impact of each ship on cost is also identified by a percentage written on the line connecting each ship with the total. Finally, the actual dollar amount of each ship's impact upon the definitized estimate is shown either inside the instance for larger contributors or near the instance for smaller ones.

The *Winston Churchill*, for example, which is located at the eight o'clock position on Figure 18.6, was not the largest contributor to the total definitized estimate. However, a brief visual analysis of the entire figure shows it was not the least significant either because many of the ship solar graph screenshots are smaller. The relative sizes and organization of all the instances enable an intuitive understanding to be quickly developed. The *Winston Churchill* screenshot is larger than the four instances directly below it, but it is also smaller than the four instances directly above it. The relative location of the *Winston Churchill* instance enables a decision maker to quickly identify that the ship's relative contribution to the overall definitized estimate lies somewhere in the middle of the pack.

If further understanding of the relative contribution is needed, the decision maker would then refer to the percentage indicated along the line connecting the *Winston Churchill* to the total estimate. The *Winston Churchill* accounted for 3.7% of the total definitized estimate. However, if the actual dollar amount contributed to the total is desired, then the decision maker could refer to the number located within the instance. In the case considered, the *Winston Churchill* accounted for $11.7 million in absolute terms.

18.3.4.3 *Definitized estimates of the top five ships*

Figure 18.7 (J. Kornitsky, personal communication, November 2013) is nearly identical to Figure 18.6 except that it has been modified to identify the largest cost contributors. The five largest contributors are shown, and the remaining 14 ships are aggregated into "all other."

Consider the decision maker analyzing the presentation. If the executive is only interested in the largest cost contributors, then the addition of the other 14 ships only makes interpretation of the information more difficult. However, the aggregation of the remaining ships into a single instance also provides another view of the data. In this example, the total definitized estimate of the other ships is $105.6 million and represents 33.6% of the entire sum. This view may be significant to a decision maker who originally thought that the largest cost contributors represented a much larger portion of the total. In this figure, a decision maker would easily be able to determine that the impact of the remaining 14 ships is much greater than the impact of any single large cost contributor.

Alternatively, if the decision maker were more interested in determining the sources of the expenses, then an additional level of detail would be necessary. Figure 18.7 provided cost information, and the costs were aggregated at the ship level. An executive interested in determining the primary drivers of cost would need more detailed information that can be found in Figure 18.8.

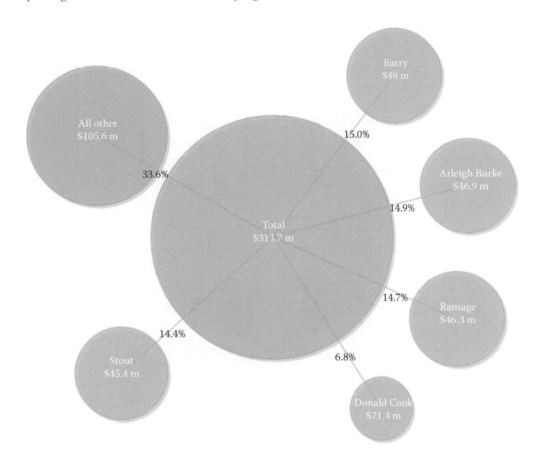

Figure 18.7 Definitized estimate, top five ships solar graph.

18.3.4.4   Definitized estimates of top five ships by expense details

Figure 18.8 (J. Kornitsky, personal communication, November 2013) adds one level of detail to the figure previously discussed. These added dimensions are two cost categories of labor and material, which can be seen radiating further from the graph's center and labeled with the availability's identification number from which it originated. These additional details to the definitized estimate of the top five ships increased the complexity of the parent–child hierarchy and produced different numbers of children among the ship level instances. For the executive using this solar graph to make important ship maintenance decisions, it is important to understand the changes.

First, the parent–child relationship hierarchy has increased in complexity. With the addition of another level of detail, or another layer of children, the ship name solar screenshots have become both parent and child. The ship names are still children to the parent, total definitized estimate but are now also parents to the expense details. For example, located at the one o'clock position in Figure 18.8, the *Barry* solar graph instance has spawned two children, Labor, and Material. The *Barry*, originally only a child to the total definitized estimate, is now also a parent to its two children. However, this concept has produced ship name parents with varying numbers of children, and their causes may not be initially intuitive.

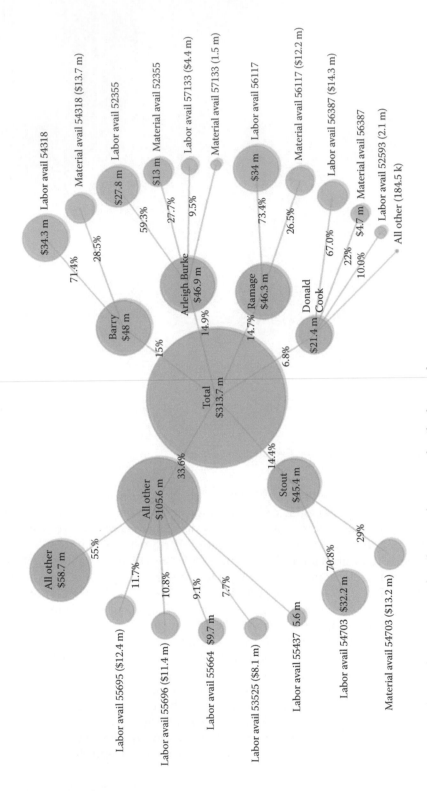

Figure 18.8 Definitized estimate, top five ships, and expense detail solar graph.

Earlier, both the *Arleigh Burke* and the *Donald Cook* were identified as being irregular because they represented multiple availabilities. The addition of expense detail has further demonstrated the presence of two separate maintenance periods within each. Just above the three o'clock position in the solar graph, the *Arleigh Burke* shows four children. Two are labeled as Labor, and two are labeled as Material. However, each one labeled Labor is identified by a unique availability identification number, and each one labeled Material has the same unique numbers. The *Arleigh Burke* and *Donald Cook* multiple availability instances produced four children as opposed to the two children generated by the single availability instances of the *Barry, Ramage,* and *Stout* ships.

To an executive, the additional level of detail in the solar graph begins to remove ambiguity and provide clear relationships among the sources of cost. However, if the manner and method in which the detail is presented is confusing, then the additional information will only further confound the decision maker. Understanding why ships produced varying numbers of Labor and Material children is important for the executive to make appropriate decisions regarding ship maintenance based on the solar graph. However, the six children subordinate to the "all other" instance at the 10-o'clock position in Figure 18.8 also requires explanation.

The reason the "all other" instance produced six children is two-fold. First, the "all other" instance includes 14 ships and, therefore, 14 availabilities (since the *Arleigh Burke* and *Donald Cook* have already been accounted for). The definitized cost estimate for each availability has been categorized into Labor and Material. Therefore, 14 availabilities should have generated 28 expense detail children. There are more than just two or four children available to display. This leads to the second part of the two-fold explanation. In Figure 18.8, the number of children to be displayed was arbitrarily chosen. The top five largest contributors retained their individual solar screenshots, and the remaining were aggregated into the "all other instance." The choice to display the top five ships in the screenshot with less detail has also affected this graph. The biggest five individual contributors, all which happen to be Labor instances, are displayed while the remaining are aggregated into the "all other" instance. Again, the implication for the executive using this solar graph to form ship maintenance policy decisions is that if the manner and method of solar graph creation are not known, then the insight derived from the graph will be erroneous. For example, if decision makers assumed that the "all other" category displayed all its children, then they would misunderstand the graph and believe that only labor costs were incurred for those 14 ships.

From Figure 18.7, previously seen, the decision maker was interested in finding more about the cost sources. Now in Figure 18.8, with an added level of detail, the decision maker could make more observations and gain a deeper understanding of cost drivers. For instance, the top five ships all demonstrated that for a given availability, labor impacted cost more than material. Specifically, consider the *Barry, Ramage,* and *Stout.* The labor costs accounted for percentages ranging from 70.8% to 73.4%. In this small sample, the decision maker could develop cost baselines indicating that for a given availability, labor accounted for about 70% of the cost and material accounted for about 30%. Given that the small sample size is an accurate screenshot of DDG ship maintenance, then the definitized cost estimates of future availabilities could be compared to the baseline and predictions generated about how the cost profile might change before work is completed.

18.3.5 *Actual costs of the top five ships by type expense*

While the definitized cost estimate solar graphs do produce valuable information, they represent only well-educated guesses of the actual cost. The next screenshot (Figure 18.9,

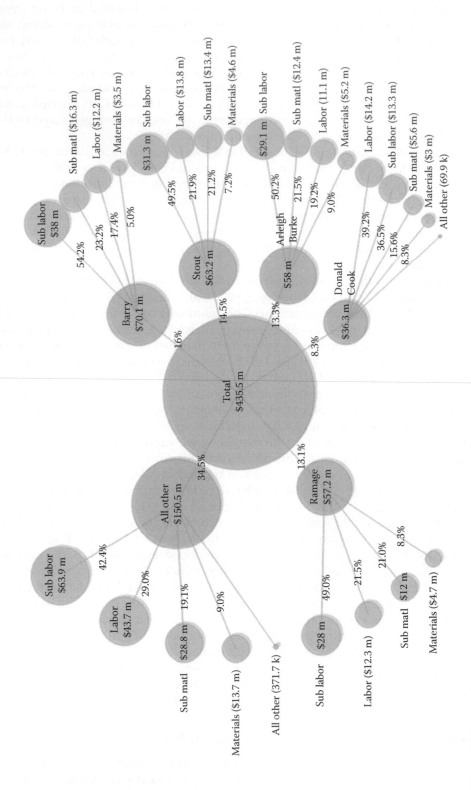

Figure 18.9 Actual cost, top five ships, type expense solar graph.

J. Kornitsky, personal communication, November 2013) provides actual cost and is organized by the top five ships with an additional level of detail. Figure 18.9 also provides the
additional information on "type of expense." comment

Most noticeably, the total actual cost, represented by the largest solar instance in the
center of the graph, has increased to $435.5 million. However, referring to the previous fig-
ure (Figure 18.8), definitized cost was estimated to be $313.7 million, so the costs actually
increased by 38.8%. A visualization tool enables the decision maker to drill down further
to identify the largest cost drivers.

The types of expenses figure provides the ability to drill down further into the cost
sources. Whereas expense detail was broken down into only labor and material categories,
type expense splits those into (shipyard) labor, sub (contractor) labor, (shipyard) material,
and sub (contractor) material. From here forward, the additional description in parenthe-
ses will be excluded, but the terms retain their definitions. Labor and material, in the
context of type expense, refer to the labor and material costs associated with the shipyard
hosting the availability. Sub labor and sub material refer to the same costs, but those associated with the expense incurred by subcontractors.

In the *Arleigh Burke*, at the four o'clock position on the figure, the definitized estimate
for this ship was $46.9 million, and the actual cost was $58 million. That represents an
increase of 23.7%. However, a decision maker, knowing that labor is a larger contributor to
cost than material, would want to know what type of labor expense is more responsible,
the shipyard or the subcontractors. In the case of the *Arleigh Burke*, sub labor accounted
for 50.2%, whereas labor represented only 19.2% of total availability cost. Representing a
majority of the cost for the *Arleigh Burke*, perhaps sub labor should be examined for cost-
reduction opportunities.

The bubble charts of either definitized estimates or actual costs provide decision makers with valuable insight. However, the size difference between estimates and actual costs
would provide an understanding of the sources of cost growth. For instance, an executive
is interested in determining the primary driver of increased costs. While the previous
solar graphs possess the necessary information, further calculations are needed to figure
changes in cost. If the relative and actual changes in cost were displayed on the same
graph, then the decision maker would be able to identify easily the primary drivers of cost
growth and cost savings. The next four figures (Figures 18.10 through 18.13) demonstrate
the concept of representing both the definitized estimates and actual costs, simultaneously.

18.3.6   Definitized estimate versus actual of the top five ships by type expense

Figure 18.10 (J. Kornitsky, personal communication, November, 2013) displays a zoomed-in
look at the comparison format to provide an introduction to the new characteristics and
to review some old ones. Starting at the nine o'clock position on the bubble chart (Figure
18.10), the first characteristic examined is the shell. The shell thickness represents the difference in the amount of change and whether the change was cost growth or cost savings.

Proceeding clockwise, the terms are familiar, but their presentation is new. Definitized
cost estimate and actual cost refer to the estimated cost at the end of planning and the cost
incurred upon completion of the availability, respectively. In this figure, the definitized
estimate is represented by the inner layer of the shell and the actual cost by the outer
layer. For example, the largest bubble represents total cost. The inner layer shows how
large the instance would be if only the total definitized estimate, $313.7 million, were
displayed. The outer layer shows how large the instance would be if only the total actual
cost, $435.5 million, were displayed. The difference between the layers, or the thickness

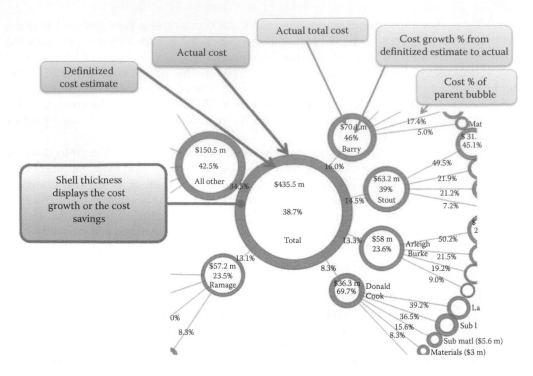

Figure 18.10 Definitized estimate versus actual, top five ships, type expense, solar graph. Close-up of center elements.

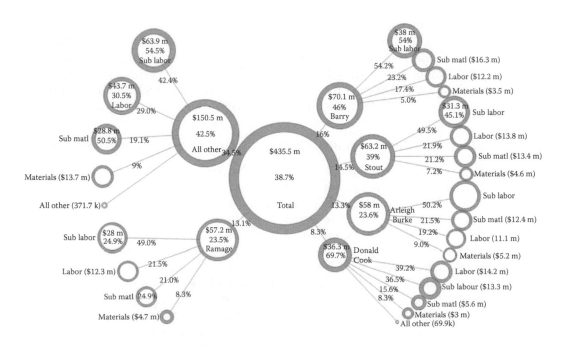

Figure 18.11 Definitized estimate versus actual, top five ships, type expense solar graph.

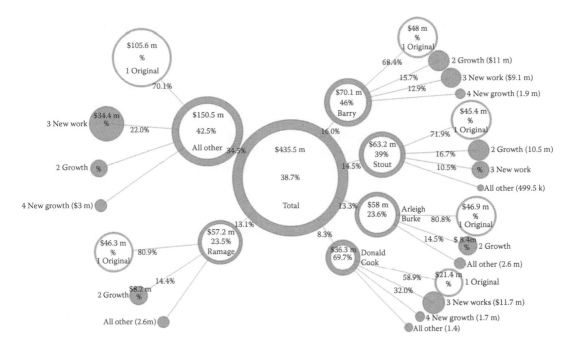

Figure 18.12 Definitized estimate versus actual, top five ships, work solar graph.

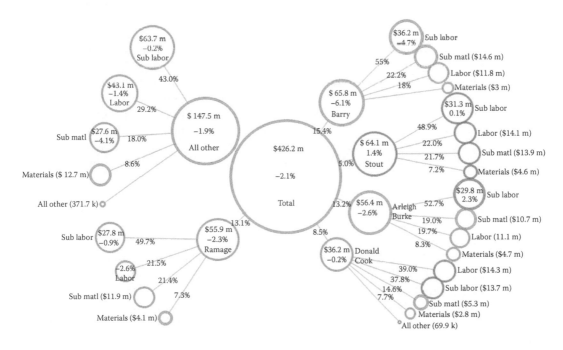

Figure 18.13 Actual versus 3DP, top five ships, type expense solar graph.

of the shell, represents the change in cost and is numerically indicated by the percentage shown, 38.7%. The definitized estimate was less than the actual cost, which means that there was cost growth.

Although the next two aspects of the bubble chart are familiar, they require further clarification. First, the number represented in millions of dollars is the final state of the instance. For this comparison between definitized estimate and the actual cost of the *Barry*, located at the one o'clock position in Figure 18.10, the final state is the actual cost, which was $70.1 million. Second, the percentage immediately below the actual cost value indicates the change from the initial state (definitized estimate) to the final state (actual cost). In the case of the *Barry*, the cost grew by 46% from the definitized estimate to the actual cost.

The final characteristic identified on the close-up is another percentage. Whereas the percentage within the instance represented cost growth, the percentage on the line between parent and child represented the proportion of the parent's cost that the child contributed. In the figure, the dialog box arrow points at the percentage the child instance accounted for with regard to its parent, the *Barry*, or 17.4% of the total actual cost incurred by the *Barry*.

18.3.7 *Definitized estimate versus actual of the top five ships by type expense*

The most distinguishing feature of Figure 18.11 (J. Kornitsky, personal communication, November 2013), which has been organized to display the top five ships to show an additional level of detail according to type expense, is the cost growth. The two instances of cost savings, both titled "all other" and located at about the nine and five o'clock positions on the outermost ring, are relatively insignificant, representing only 0.1% of the total cost, $435.5 million. Examples of cost growth are abundant, but an examination of the largest contributor to total cost may produce valuable insight for the executive-level decision maker.

The "all other" instance at the 10 o'clock position represents 14 ships. Those 14 ships accounted for $150.5 million, or 34.5%, of the total actual cost. The shell and the percentage inside the "all other" instance together indicate 42.5% aggregate cost growth for the 14 ships. These numbers reveal that the "all other" category would be an area for a decision maker to examine more closely in an attempt to identify the drivers of cost growth. A cursory glance at the children of the "all other" instance shows that subcontractors both sub labor and sub material, experienced more than 50% cost growth. Therefore, subcontractors are a primary driver of cost growth for at least the 14 ships represented by the "all other" instance.

The visualization software provides the ability to delve into the data to discover more detail. For example, if personnel are preparing a presentation based on definitized estimate versus actual type expense data for top 5 ships (Figure 18.11, and the decision maker asks the question, "What was the definitized estimate for the Barry?" then the answer can be found readily. Rather than regressing to previous screenshots, the presenter can simply select the *Barry* instance and pull up a bar chart that, among other information, displays the definitized estimate. Perhaps the decision maker may request even finer details. The software possesses the ability to drill down five levels of detail and can reproduce the data located on the original spreadsheet. Therefore, more detail is available than just what is displayed on the static screenshots presented. Refer to Figures 18.23 and 18.24 near the end of this section for examples.

18.3.8 Definitized estimate versus actual of the top five ships by work

Figure 18.12 (J. Kornitsky, personal communication, November, 2013), is the last of the three figures showing simultaneous display of both definitized estimates and actual costs and provides the additional detail of work instead of type of expense. As previously discussed, maintenance work is broken into four types: OW, G, NW, and NG. Changing the detail to allocate cost by work creates a couple of peculiarities, both related to the definitions of the work, and important for the executive-level decision maker to understand.

There are two anomalies when the data are changed to show work details. The first peculiarity is that there are now a significant number of instances that possibly indicate cost savings. Unfortunately, all the percentages within the cost savings shelled instances are left blank revealing that no change (0%) has taken place. That is because the instances are representing OW, which does not change after the completion of planning, making the percentage within the instance irrelevant. For example, refer to the *Arleigh Burke* solar graph instance at the four o'clock position in Figure 18.12. The cost savings shelled child instance attached to the *Arleigh Burke* is labeled "Original" for OW. The percentage displayed is blank that indicates 0% change in cost has occurred because any change in cost is recorded by the other categories of work. The percentage that is important for the decision maker to acknowledge, though, is indicated along the line connecting the child to parent. The 80.8% for OW indicated what portion of the total actual cost, for the *Arleigh Burke* that OW accounted for.

The second peculiarity, also a result of definitions, is that the instances for the other three categories of work the baseline, or definitized estimate in this case, was $0 and the actual cost is all cost growth. That is because the other three categories of work (G, NW, and NG) all result from work needed in addition to the OW, and are, therefore, cost growth by definition. Continuing with the examination of the *Arleigh Burke*, its larger child is labeled Growth. The percentage within the instance is blank, but again, it is less important. The significant values important to the decision maker are the actual cost of G, $8.4 million, and the proportion of the *Arleigh Burke*'s total actual cost that G work accounted for, 14.5%. With the two peculiarities defined and understood, reconsider the previous figure to identify a cost driver.

The decision maker examined the largest "all other instance" more closely and determined that subcontractors were a primary driver of cost growth. The decision maker might then ask to see the additional detail organized by work to expand further his or her understanding. Again looking at the "all other" instance located at the 10 o'clock position in Figure 18.12, the largest driver of cost growth is NW, which accounted for $34.4 million, or 22%, of the actual costs for "all other" 14 ships. Combine the knowledge derived from examining both graphs (Figures 18.11 and 18.12, respectively) and the keen decision maker might direct staff personnel to investigate NW performed by subcontractors for cost-savings opportunities.

Figure 18.12 demonstrates how costs aggregate from the bottom up. Costs are created at the operational level and occur in different forms. Here, the forms are categorized according to the classification of work that created the cost. As the costs move from the outer rings of the solar graph, they are aggregated into ship instances that provide less cost detail but are still useful as another way of looking at cost. Finally, all the ships' actual costs are aggregated into the center solar instance "Total." The visualization software offers the opportunity to view the cost data at many levels of detail, each of which delivers valuable information for decision makers.

18.3.9   Simulations 1 and 2: Introduction of 3DP and additive manufacturing radical

Visualization tools provide decision makers with insights into historical data, and more importantly, offer forecasting capabilities. Before implementing process changes, which involve risk and uncertainty, an executive could use bubble charts to forecast the effects of such changes. Consider the following example.

The executive-level decision maker has analyzed the figures previously presented and has concluded that changes to the ship maintenance process are necessary to control cost growth. Three technologies have been identified to reduce costs: 3DP, 3D LST, and CPLM. To test this hypothesis, two simulations were conducted with differing implementation strategies. In Simulation 1, 3DP technology only was applied while in Simulation 2, all three technologies (3DP, 3D LST, and CPLM combined) were applied to the ship maintenance process. Simulation results, which could identify potential cost savings, are discussed further in this section.

To quantify the potential benefits of those technologies, the KVA methodology was applied. KVA assigns a value to the knowledge assets of an organization (Housel and Bell, 2001) and was used to forecast the effect that 3DP, 3D LST, and CPLM technologies would have on U.S. Navy ship maintenance programs. In one prior study, the researchers found that 3DP and CPLM could result in cost savings of as much as 81% (Kenney, 2013). Another study determined that cost savings of as much as 84% could result from the use of 3D LST and CPLM in U.S. Navy ship maintenance programs (Komoroski, 2005). The potential impact of these three technologies has been determined to be substantial. Therefore, they were used to demonstrate the ability of the software program to create intuitive screenshots of the cost savings generated by their implementation.

In the previous set of comparison figures, the definitized estimate was the baseline, and the actual cost was the value compared. In the next set of four comparison figures (Figures 18.13 through 18.16), the baseline and the value compared are changed to examine the effect of three different technologies on ship maintenance actual cost. In the first two figures to follow (Figures 18.13 and 18.14), the actual cost is the baseline, and the forecasted effect of 3DP only is the compared value. The next two figures (Figures 18.15 and 18.16) visualize the effect that the combination of 3DP, 3D LST, and CPLM, labeled as additive manufacturing (AM) radical, has on ship maintenance costs.

18.3.9.1   Actual versus 3DP for the top five ships by type expense

Figure 18.13 (J. Kornitsky, personal communication, November, 2013), visualizes the effect on the actual cost of implementing 3DP into the ship maintenance process. The familiar top five ship format is maintained, and the additional level of detail is organized by type expense. The baseline is the actual cost incurred, and the compared value is the backcasted effect that 3DP would have had on actual cost.

To the executive-level decision maker analyzing the effect of 3DP on U.S. Navy ship maintenance, this figure provides two important pieces of information. The first is that, overall the actual cost of ship maintenance can be reduced with the implementation of 3DP technology. The center instance in Figure 18.13 shows that the effect of 3DP on the ship maintenance process could have reduced the total cost by 2.1%, as is indicated by the percentage and the cost savings shell. The cost of ship maintenance with the incorporation of 3DP is now $426.2 million versus the original $435.5 million for savings of $9.3 million. The *Barry*, again located at the one o'clock position, is the ship that demonstrates the largest percentage cost savings at 6.1% and reduced costs across all types of expense.

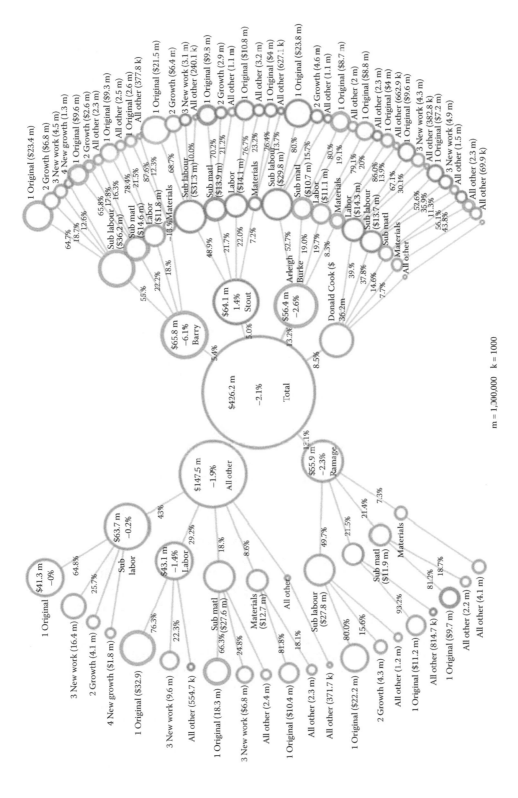

Figure 18.14 Actual versus 3DP, top five ships, type expense, work solar graph.

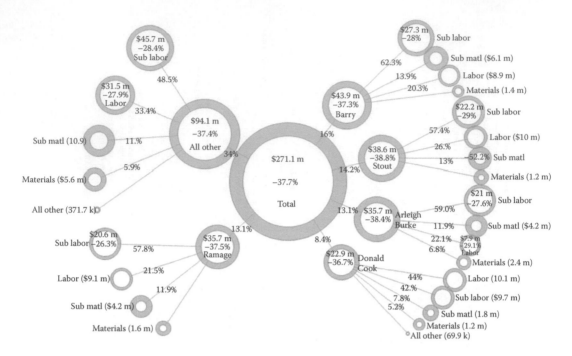

Figure 18.15 Actual versus AM radical, top five ships, type expense solar graph.

Second, not every ship may benefit from the use of 3DP technology. Just above the three o'clock position on Figure 18.13, the *Stout* indicates 1.4% cost growth for a backcasted total cost of $64.1 million or $0.9 million greater than the original cost. Drilling down one level of detail into expense type, the decision maker can easily determine that every category of expense contributed to the cost growth for the *Stout*. However, additional levels of detail are available, and the Executive may request that more information be displayed to help identify the primary drivers of cost growth for the *Stout* and/or the leading sources of cost savings for the *Barry*. Therefore, Figure 18.14, organized by work, adds another layer of detail.

18.3.9.2   Actual versus 3DP of the top 5 ships by type expense, work

Figure 18.14 (J. Kornitsky, personal communication, November 2013) is the second in the series of comparison figures. It allows the decision maker to visually drill down into the data even further. In Figure 18.13, the *Barry* displayed cost savings across all types of expense, and the *Stout* indicated cost growth with the implementation of 3DP into the ship maintenance process. The addition of the classification of work detail, however, indicated where each ship derived its savings or growth with 3DP.

The executive drilling down into the 3DP backcasted cost data for the *Barry* can quickly identify one classification of work, in one type of expense, which produced cost growth. The only shelled solar graph instance subordinate to the *Barry* in Figure 18.14 is the NG instance, subordinate to sub labor, which has been backcasted to account for $1.3 million dollars of sub labor cost. However, the percentage growth is not displayed because the software limits the presence of information to reduce clutter and increase clarity. However, the executive requiring more information can simply select the shelled NG instance, and more information is available immediately including the percentage of cost

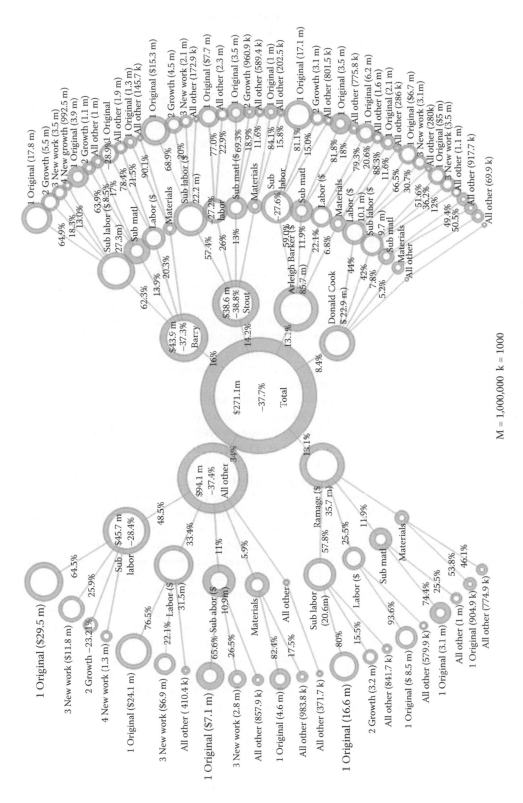

Figure 18.16 Actual versus AM radical, top five ships, type expense, work solar graph.

growth. If the decision maker decided to implement the 3DP-only strategy, then the NG work attributed to sub labor could be an aspect that should be looked at for improvement.

The executive examining the *Stout*, at the two o'clock position in Figure 18.14 more closely, can quickly see that even though, the aggregate change in cost is cost growth, there are indications of possible cost savings. Immediately subordinate to the *Stout*, sub labor is backcasted to account for $31.3 million. Again, the percentage increase in cost is not displayed but is available by selecting the solar graph instance. Even though the sub labor instance indicates cost growth, there are children subordinate to sub labor that signify cost savings. For example, the Growth instance is cost savings shelled and is backcasted to account for $6.4 million. To the decision maker, this figure is forecasting the possible effect of implementing 3DP into ship maintenance using historical data, and it provides the ability to examine the effect a particular technology might have on cost without the risk and uncertainty involved with actual implementation.

18.3.9.3   Actual versus AM radical of the top 5 ships by type expense

Figures 18.15 and 18.16 compare the baseline, actual cost, to the backcasted effect that the implementation of all three combined technologies might have had on cost. The structure of the graph remains familiar, but the increase in cost savings demonstrates the ability of the software to produce intuitive screenshots that easily communicate the differences in effect on cost. Figure 18.15 (J. Kornitsky, personal communication, November 2013) reverts back to the format of Figure 18.13 with less detail, but easily demonstrates the difference in cost savings. The substantial increase in cost savings is communicated by, most intuitively, the thickness of the solar graph instance shells but is also indicated by the absolute and relative values displayed in or near the instance.

To the executive-level decision maker concerned with cost, the most evident display of cost savings is the center instance. The total backcasted cost of ship maintenance for all 19 ships, had 3DP, 3D LST, and CPLM technologies been implemented, and was $271.1 million or 37.7% cost savings under actual cost. The difference, $164.4 million, could have been used to finance other needs such as system upgrades, structural improvements, or reduction of the number of maintenance jobs deferred until the next availability due to shrinking fiscal budgets. The decision maker analyzing the change in cost might also be interested in understanding the difference in cost savings of individual ships.

In contrast to the 3DP-only implementation strategy, which slightly increased cost for one of the top five ships, AM Radical decreased costs for all top five ships. There appear to be substantial cost savings in the "all other" solar graph instance located at the 10-o'clock position in Figure 18.15 as well, but current settings prevent concluding that all 19 ships incurred cost savings. In the case of the *Barry*, cost savings is significantly increased. With the implementation of 3DP only, the backcasted cost was $65.8 million, or 6.1% cost savings. With the use of all three technologies or AM Radical implementation, the backcasted cost for the *Barry* is $43.9 million, a cost savings of 37.3% when compared with the actual cost. Drilling down one layer of detail, two of the type expense children subordinate to the *Barry* have thicker cost savings shells than the others, an intuitive indication of substantial cost savings. In fact, sub labor and sub material account for almost 80% of the increase in the cost savings of AM Radical over the 3DP-only implementation strategy for the *Barry*.

Earlier, in the description of Figure 18.11 solar graph, the executive-level decision maker identified subcontractor labor and material as primary drivers of cost growth. The keen decision maker might begin to formulate that a possible solution to subcontractor labor and material cost growth is the implementation of all three technologies. However, the addition of another layer of detail is available, and it could provide either supporting or contradictory evidence.

18.3.9.4 Actual versus AM radical of the top 5 ships by type expense, work

Figure 18.16 (J. Kornitsky, personal communication, November 2013), the fourth and final screenshot of this comparison series, allows the executive-level decision maker to drill down visually into the data even further. The additional layer of detail is organized by work and provides more information about the sources of cost savings.

An executive analyzing this figure could notice the most obvious aspect first, the fact that AM Radical implementation creates cost savings throughout the entire dataset. Whereas 3DP-only implementation indicated cost growth in one ship, various type expenses and classifications of work, the backcasted effect AM Radical implementation could have produced cost savings in every instance. For example, with 3DP-only implementation, Figure 18.14 identified cost growth in one classification of work, NG, which accounted for $1.3 million of sub labor. However, with AM Radical implementation, the NG instance subordinate to the *Barry* on this solar graph, Figure 18.16 indicates cost savings and now accounts for $0.99 million. As stated before, the percentage change is not displayed to reduce clutter; however, it is available by simply selecting the instance. Possibly more interesting to the executive-level decision maker is the case of the *Stout*, which changed from a source of cost growth to a significant driver of cost savings.

In the previous solar graph, Figure 18.14 showing the backcasted effect of 3DP, the *Stout* displayed an absolute cost of $64.1 million and cost growth of 1.4%. The classification of work that contributed most to the cost of the *Stout* was OW, a child of sub labor and indicated an absolute cost of $21.5 million. But, with AM Radical implementation, this solar graph (Figure 18.16) backcasted the cost to $15.3 million for the OW associated with sub labor, a cost savings of $6.2 million when compared to 3DP-only implementation.

The *Stout*, as well as the other top five ships, could have produced significant cost savings had the AM Radical approach been implemented. However, the actual costs have already been incurred. The significance of this series of figures is that a decision maker can visualize the effect the technology implementation strategies might have had on historical data and then make predictions about the effect on future costs. The decision maker, armed with the predictions derived from the solar graphs, weighs additional executive-level organizational considerations, can make better cost-control choices for the future of U.S. Navy ship maintenance.

18.3.10 Alternative figures

The final series of solar graph figures (Figures 18.17 through 18.19) demonstrate the flexibility of visualization tools, enabling drilling down into specific details. All but the first of the figures described thus far have used the top five ships structuring concept for the first level of detail. While the use of the single method of organizing the first layer of detail made comprehension of the graphs easier, it limited the appearance of the flexibility of the third-party software. Therefore, other methods of organizing and presenting the data are explored in the following three comparison figures.

18.3.10.1 Definitized estimate versus actual of the type expense by work

Figure 18.17 (J. Kornitsky, personal communication, November 2013) is useful for the decision maker interested in analyzing cost growth without discriminating by ship. This figure reverts back to using the definitized estimate as the baseline and the actual cost

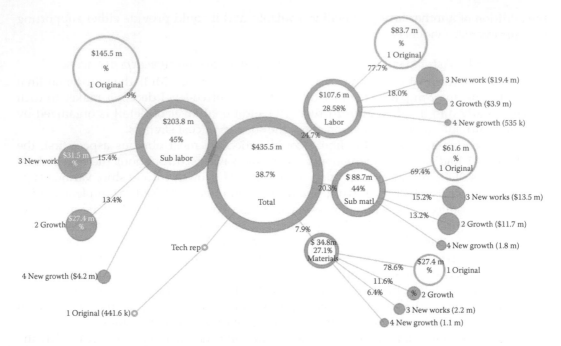

Figure 18.17 Definitized estimate versus actual, type expense, work solar graph.

as the comparison, as in Figures 18.11 and 18.12, respectively. However, the top five ships are not used as an organizing concept. Instead, the first layer of detail is grouped by type expense, and the additional layer is organized by work.

Consider the theory arrived at by the executive during the analysis of Figure 18.11. The decision maker noted that subcontractor labor and material appeared to be primary drivers of cost growth. In this screenshot, Figure 18.17, the sub labor instance appears at 10 o'clock and the sub material instance at three-o'clock. The indicated percentages of cost growth are 45% and 44%, respectively. Compared to the cost growth of labor and materials associated with the shipyard, 28.5% and 27.1%, respectively, subcontractors also appear here to be primary drivers of cost growth. The decision maker is interested in understanding the causes of subcontractor cost growth at a deeper level of detail. Therefore, the executive might analyze the graph further and discover that NW is the largest absolute contributor to both sub labor, at $31.5 million, and sub material, at $13.5 million.

The same information was derived from the analysis of two sequential solar graphs described earlier, Figures 18.11 and 18.12. The same understanding was derived from two unique presentations, one with two graphs and the other with this one graph. Arriving at the same conclusion from different presentations of the data builds confidence in the decision maker that the data are accurate and the visualization methods are valid.

18.3.10.2   *Definitized estimate versus actual of the work by ship*

The remaining two alternative figures are complementary. Figure 18.18 (J. Kornitsky, personal communication, November 2013) is a figure that a decision maker could use to identify problem areas of cost growth based on classification of work. Figure 18.19 (J. Kornitsky, personal communication, November, 2013) keeps the same organization format but enables the decision maker to analyze how the implementation of the three technologies could have created cost savings.

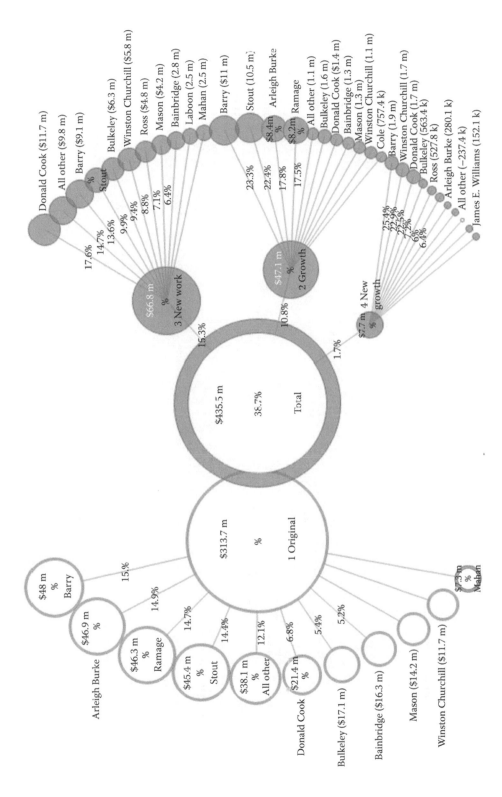

Figure 18.18 Definitized estimate versus actual, work, ship solar graph.

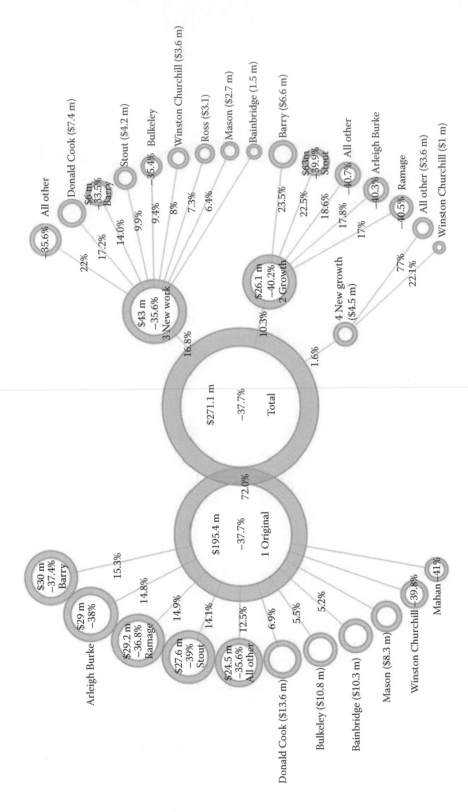

Figure 18.19 Actual versus AM radical, work, ship solar graph.

In Figure 18.18, which demonstrates cost growth, there is only one peculiarity that has already been explained. All the thin, cost savings shelled instances on the left side of the graph represent 0% growth because of the definition of OW, which cannot grow in expense. In addition, the solid, shelled instances on the right side represent only cost growth that occurred and are classified as NW, G or NG because of their definitions.

Figure 18.18 is important to the executive-level decision maker because it exhibits data already presented in another format, Figure 18.12, which organized the first level of detail by ship and the second level by work. In this graph, the organizing concepts have been reversed. If the same deduction can be derived from this screenshot, then the decision makers' confidence in their ability to make accurate and valid choices for the future of U.S. Navy ship maintenance processes, increases.

The deduction already made by the decision maker was that NW, over the other classifications of work, accounted for the largest portion of cost growth. Refer to Figure 18.18, a quick visual scan over the classification of work instances creates an intuitive understanding. The NW instance is the largest indicator of cost growth. Further examination by the decision maker provides the dollar values that support the intuitive perception. The NG instance, located at the five o'clock position, accounted for $7.7 million. The G instance, located just below the three o'clock position, represented $47.1 million. Finally, the NW instance, located at the two o'clock position, produced $66.8 million in cost growth. Even though the data were organized and presented differently, the same deduction was reached: NW was the primary driver of cost growth.

If the executive were interested in determining the ships that produced the largest cost growth, then simply referring to the additional level of detail would provide the answer. For example, since NW was the primary driver of cost growth, identification of the largest contributing ship may provide a specific case for further analysis of cost growth. Referring to the NW instance, located at the two o'clock position in Figure 18.18, the child ship that represents the largest portion of cost growth is the *Donald Cook*. The decision maker, remembering that the *Donald Cook* represents two availabilities, would drill down into the next level of detail by selecting the *Donald Cook*. Then, the determination would be made whether either one of the *Donald Cook* availabilities or the next largest individual ship (the *Barry*) was the ship representing the most cost growth for NW. Once the ship was identified, the executive could direct further study into the causes of cost growth.

18.3.10.3   *Actual versus AM radical of the work by ship*
Figure 18.19 shows the backcasted effect that the implementation of all three technologies combined might have had on U.S. Navy ship maintenance costs. Figure 18.19 maintains the organizing structure of the immediately previous solar graph to provide an easy comparison for the executive-level decision maker.

For example, the decision maker is interested in figuring out the overall effect that AM Radical implementation has compared to definitized cost. The center instance in Figure 18.19, total, indicates the bottom-line cost savings that may have occurred had the AM Radical implementation strategy been employed. At $271.1 million, AM Radical implementation might have resulted in 37.7% cost savings, but that is compared to actual cost. Referring to the OW instance located at the nine o'clock position on the previous solar graph (Figure 18.4), the value is $313.7 million. Because of the definition of OW and the position of the OW instance at the first level of detail, it also represents the total definitized estimate. Simple math shows that AM Radical implementation might have caused the ships analyzed to come under budget by $42.6 million or 13.6%. Cost growth could have been turned into cost savings through the backcasted effect that AM

Radical implementation might have had on the ships studied. To the executive-level decision maker, this is important because if the three technologies (3DP, 3D LST, and CPLM combined) were selected for implementation, then future U.S. Navy ship maintenance budgets might be reduced and result in reallocation of funding to higher priority projects.

18.3.11 LOD and availability density bubble charts

LODs have long been considered by the U.S. Navy ship maintenance metrics groups to be a valuable indication of the performance of the ship maintenance process. The LOD metric is often included in reports made by regional maintenance centers (RMCs) to NAVSEA as an indication of the effect on ship's schedule caused by delays (M. Leftwich, personal communication, September 4, 2013). However, the LOD metric has been linked to the quality of the definitized estimate, and that quality has been determined to be random (T. Laverghetta, personal communication, November 26, 2013). To the executive-level decision maker, the important aspect of ship maintenance is cost. Availability density is considered a better metric for predicting cost and was provided to this study for further analysis (P. Pascanik, personal communication, November 21, 2013). Of the following three figures (Figures 18.20 through 18.22, all from J. Kornitsky, personal communication, November 2013), the first two highlight the lack of correlation between the LOD metric and actual cost. The third demonstrates the validity of using the availability density metric to indicate actual maintenance cost.

18.3.11.1 LOD versus expense (actual cost)

Figure 18.20 is presented first to introduce the structure of the chart. Figure 18.21 is then to be considered as the chart important to the U.S. Navy ship maintenance executive-level decision maker interested in controlling costs.

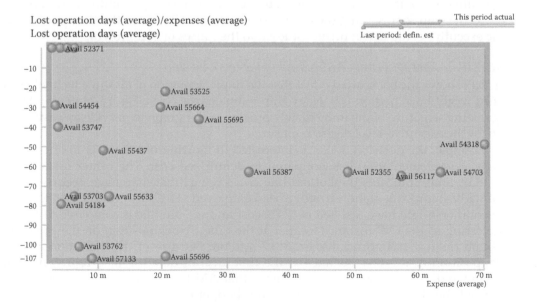

Figure 18.20 LOD versus expense (actual cost) bubble chart.

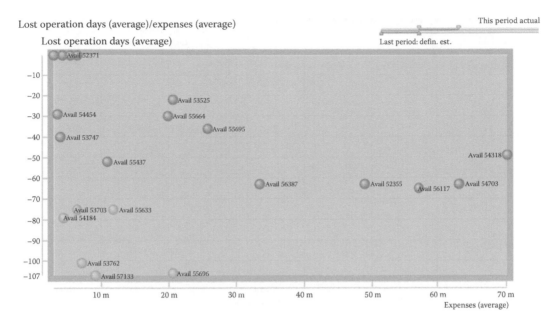

Figure 18.21 LOD versus expense (actual cost)—highlighted bubble chart.

Figure 18.20 is structured as an XY scatter plot. The X-axis represents expense or actual cost of ship availability and ranges from $0 to $70 million. The Y-axis represents the total LODs incurred during availability and ranges from 0 to (–107); negative numbers represent operating days *lost*. The data points scattered throughout the chart represent the LOD and expense values for the individual availabilities and are labeled with their unique availability identification numbers. For example, the data point labeled "avail 56387" near the center of the bubble chart represents one of the availabilities for the *Donald Cook*. The LODs incurred during that availability totaled 63, and the total expense was $33.4 million.

18.3.11.2   *LOD versus expense (actual cost): highlighted*

Figure 18.21 is important to the executive-level decision maker because it demonstrates that the LOD metric is not useful for forecasting the actual cost of availability. This is shown by both a visual analysis of the chart and by a mathematical calculation of the correlation factor.

Visually, the data points show that smaller availabilities, under $30 million, can result in either the highest number or the lowest number of LODs. For example, the data point labeled "avail 52371" in Figure 18.21 is for the *James E. Williams* and indicates an actual cost of $4.2 million with a total of zero LODs. Meanwhile, the data point labeled "avail 57133" is for one of the *Arleigh Burke* availabilities and indicates an actual cost of $9.1 million with a total of 107 LODs. In fact, the six data points highlighted in the lower left corner of the bubble chart all represent availabilities of relatively small cost that incurred relatively high numbers of LODs, which prevent the appearance of a linear relationship. Therefore, the LOD metric is not a good indicator of availability cost.

The mathematical calculation also demonstrates the lack of connection between the LOD metric and expense. The expense, or actual cost, of each availability was totaled, to

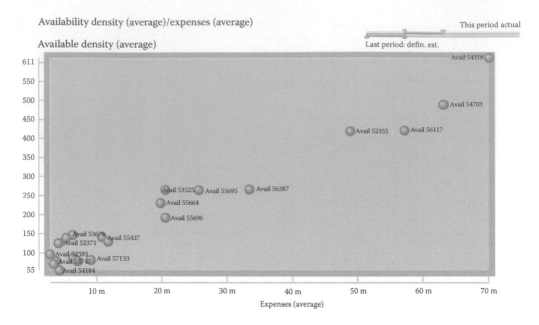

Figure 18.22 Availability density versus expense (actual cost) bubble chart.

include OW, G, NW, and NG. Then, a correlation factor was calculated between the cost of each availability and the number of LODs incurred during each availability. The correlation factor is (−0.14). This number shows that, mathematically, the LOD metric is not a good indicator of cost. For the executive-level decision maker, LODs have been visually and mathematically shown not to correlate well with cost. However, the metric that correlates well with cost is the metric provided to this study for further analysis.

18.3.11.3   *Availability density versus expense (actual cost)*

The metric of availability density was provided to this study and is defined as the total actual man days divided by the total availability duration days. In other words, the availability density number represents the average number of man-days performed each calendar day of the availability. For example, in Figure 18.22, the *Stout* (Avail 54703) used 134,254 man-days to complete its availability, which lasted 275 calendar days. The availability density for the *Stout* is 488. For each calendar day of the *Stout*'s availability, an average of 488 man-days were performed.

Figure 18.22 changes one axis to represent the new metric. The X-axis remains as an expense, but the Y-axis now represents availability density and ranges from 55 to 611. For example, the *Stout* (Avail 54703) data point near the top-right corner of the chart indicates an average of 488 man-days per availability calendar day and an actual cost of $63.2 million.

Availability density is a better indicator of cost (Figure 18.22), and proves that visually and mathematically. Visually, availability density correlates with the expense. For example, the data point labeled Avail 57133 near the bottom-left portion of the chart represents one of the *Arleigh Burke* availabilities and indicates an availability density of 85 and expense of $9.1 million. In the diagonally opposite corner, avail 54318 represents the *Barry* and indicates an availability density of 611 and expense of $70.1 million. Visually, availability density provides a good indication of availability expense through its linear response.

Mathematically, availability density and cost correlate very well. The expense of each availability was again totaled. Then, a correlation factor was calculated between the cost and availability density of each availability. The correlation factor is 0.98. This number shows that, mathematically, the availability density metric is a strong indicator of cost. Availability density is visually and mathematically an accurate indicator of cost.

For the executive-level decision maker, predicting the actual cost of events in progress is extremely valuable. The metric availability density shows such a strong correlation to cost that it may be able to predict whether a particular current availability is expected to meet or exceed the definitized estimate. The ability to predict the ending actual cost of a ship maintenance evolution in progress would enable decision makers to avert large cost growth by implementing changes earlier in the U.S. Navy ship maintenance process.

18.3.12 Drill-down spreadsheets

Figures 18.23 and 18.24 (both from J. Kornitsky, personal communication, November 2013) are examples of drill-down spreadsheets that can be produced from any solar graph instance. The ship selected as the target for drill down was the *Barry*, and time analysis spreadsheet was produced at two different levels of detail. The time analysis covers the actual cost, 3DP-only implementation backcasted cost, and AM Radical implementation backcasted cost. The titles of each spreadsheet indicate the levels of detail shown; Figure 18.23 shows three levels of detail and Figure 18.24 shows four.

These spreadsheets would be valuable to the executive-level decision maker who wanted to see the numbers that the visualization software translates into intuitive solar graphs. For example, consider the executive analyzing solar graph, Figure 18.15. The *Barry* instance, located at the one o'clock position on that graph, indicates a backcasted absolute cost of $43.9 million if all three technologies had been implemented into the ship maintenance process. However, the decision maker wants to see the absolute values for the actual cost, the 3DP-only backcasted cost, and the AM Radical backcasted cost together for comparison. Then, simple selection of the *Barry* instance would produce the option to generate detailed spreadsheets at varying levels of detail. If the decision maker only wanted to see a little additional detail, then Figure 18.23 option might be selected. If the decision maker really wanted to drill down into the data, then Figure 18.24 spreadsheet could be generated. Either way, these spreadsheets provide the executive-level decision maker with drill-down capability sufficient to meet the needs of the most detail-oriented executive.

Barry/time analysis

Account Code	Name	Level	Totals Actual Am.		% Revenue	% Cost	% Growth		Actual Amount	3DP Amount	Am radical Amount
TEx	Expenses	1	43,941,901			100.00%	−37.35%		70,145,759	65,829,780	43,941,901
LBR	Labor	2	36,325,647			82.66%	−27.79%		50,310,106	48,128,487	36,325,647
LTG	Labor total growth (G + NW + NG)	3	11,522,533			26.22%	−27.95%		15,994,526	15,336,894	11,522,533
TOL	Total original labor	3	24,803,113			56.44%	−27.72%		34,315,580	32,791,593	24,803,113
MTR	Materials	2	7,616,254			17.33%	−61.60%		19,835,653	17,701,293	7,616,254
MTG	Material total growth (G + NW + NG)	3	2,357,236			5.36%	−61.52%		6,126,742	5,381,784	2,357,236
TOM	Total original material	3	5,259,017			11.96%	−61.63%		13,708,911	12,319,508	5,259,017
TG	Total increase (G + NW + NG)	1	13,879,769			31.58%	−37.25%		22,121,268	20,718,678	13,879,769
To	Total original	1	30,062,131			68.41%	−37.40%		48,024,491	45,111,101	30,062,131

*** % Growth is AM radical vs. actual

Figure 18.23 Barry drill down, 3 levels of detail.

Barry/time analysis

Account Code	Name	Level	Totals Actual Am.		% Revenue	% Cost	% Growth		Actual Growth Amount	3DP Amount	Am radical Amount
TEx	Expenses	1	43,941,901		100.00%	−37.35%		70,145,759	65,829,780	43,941,901
LBR	Labor	2	36,325,647		82.66%	−27.79%		50,310,106	48,128,487	36,325,647
LTG	Labor total growth (G+NW+NG)	3	11,522,533		26.22%	−27.95%		15,994,526	15,336,894	11,522,533
TLG	Total growth labor (original)	4	5,389,107		12.26%	−29.80%		7,677,244	7,298,394	5,389,107
TLNG	Total labor new growth	4	1,026,852		2.33%	−26.03%		1,388,275	1,414,158	1,026,852
TLNW	Total labor new work	4	5,106,572		11.62%	−26.30%		6,929,007	6,624,341	5,106,572
TOL	Total original labor	3	24,803,113		56.44%	−27.72%		34,315,580	32,791,593	24,803,113
OL	Original labor	4	7,000,423		15.93%	−27.00%		9,589,621	9,301,932	7,000,423
OSL	Original sub labor	4	17,802,690		40.51%	−28.00%		24,725,959	23,489,661	17,802,690
MTR	Materials	2	7,616,254		17.33%	−61.60%		19,835,653	17,701,293	7,616,254
MTG	Material total growth (G+NW+NG)	3	2,357,236		5.36%	−61.52%		6,126,742	5,381,784	2,357,236
TMG	Total growth material (original)	4	1,240,520		2.82%	−63.08%		3,360,462	2,882,044	1,240,520
TMNG	Total material new growth	4	164,808		0.37%	−71.73%		583,025	531,539	164,808
TMNW	Total material new work	4	951,907		2.16%	−56.39%		2,183,255	1,968,201	951,907
TOM	Total original material	3	5,259,017		11.96%	−61.63%		13,708,911	12,319,508	5,259,017
OM	Original material	4	1,338,166		3.04%	−57.00%		3,112,015	2,676,332	1,338,166
OSM	Original submaterial	4	3,920,851		8.92%	−63.00%		10,596,896	9,643,175	3,920,851
TG	Total increase (G+NW+NG)	1	13,879,769		31.58%	−37.25%		22,121,268	20,718,678	13,879,769
To	Total original	1	30,062,131		68.41%	−37.40%		48,024,491	45,111,101	30,062,131

*** % Growth is AM radical vs. actual

Figure 18.24 Barry drill down, 4 levels of detail.

18.4 Conclusions and recommendations

18.4.1 Conclusions

PEO Ships asked the team from NPS to work with U.S. Navy ship maintenance metrics groups to provide additional options regarding the optimization of large datasets. Static, cumbersome spreadsheets are no longer suitable for executive-level decision makers to make strategic choices regarding ship maintenance budgeting and scheduling. The visualization software used to present ship maintenance big data provides a means to aggregate voluminous data in visually intuitive ways to understand better cost drivers and factors that lead to schedule over-runs. Big data visualization allows decision makers to identify trends quickly, develop a better understanding of the problem space, establish defensible baselines for monitoring activities, perform forecasting, and determine the usefulness of metrics.

Visualization software provides decision makers with a tool that makes quick identification of trends possible. Referring to Figures 18.11 and 18.12 or Figure 18.17 in the example scenarios presented, an executive-level decision maker was interested in identifying the largest cost contributor. Visual analysis of the figures led the decision maker to quickly identify that subcontractor labor resulting from NW caused a trend of higher costs.

Better understanding of the problem space is also provided by the visualization of big data. Before decision makers can make choices about the future of U.S. Navy ship maintenance, they must be able to understand the characteristics of the problem as a whole. Charts, diagrams, and solar graphs enable executives to visualize how all the data points relate to each other, to define which categories of data are of particular interest, and to forecast the impact of policy changes. Through the manipulation of big data visualization tools, decision makers can develop a better understanding of their specific problem space.

Continued collection of ship maintenance, big data would provide for the creation of defensible cost and schedule performance baselines. The sample data analyzed in this project represented a limited number of availabilities and is, therefore, limited in its ability to represent U.S. Navy ship maintenance as an industry. However, expanded and continued collection of ship maintenance big data would provide data that more accurately reflect the industry. If the collection of data were expanded to include all types of ships and continued to provide for the analysis of many years of data, then the visualization software could be used to create defensible cost and schedule baselines.

Executive-level decision makers are often concerned with the future impact of their current policy change choices. Historically, executives relied upon the advice of experts and instincts developed over several years of personal experience to select which policy changes would create the effects desired. Through big data visualization software, manipulation of the data is possible to allow for forecasting. In the simulations, which examined the implementation of either 3DP technology only or the combination of multiple technologies (3DP, 3D LST, and CPLM), cost-savings trends, derived from previous research on those technologies, were applied to historical ship maintenance data. The results were then presented as screenshots from visualization software in a manner that allowed a decision maker to understand intuitively the forecasted effect without the need for expensive test cases or extensive research by experts.

Metrics provide an indication of performance as long as they represent a causal relationship. LODs have long been a metric used to indicate ship maintenance performance, but their validity was questioned. Availability density was offered as an alternative metric, but proof of its validity was necessary before being considered as a real substitute for the LOD metric. Using bubble charts, the visualization software created a visually intuitive display that demonstrated the correlation to the expense of each of the metrics. The LOD metric was shown to be a poor indicator of cost, and the availability density metric was shown to be a good indicator of cost.

Through the use of big data visualization tools, executive-level decision makers can identify trends quickly, develop a better understanding of the problem space, establish defensible baselines for monitoring activities, perform forecasting, and determine the usefulness of metrics.

18.4.2   Recommendations

Big data visualization tools are beneficial to executive-level decision makers responsible for implementing policy throughout their enterprise. For U.S. Navy ship maintenance decision makers considering the use of visualization software in their industry, the following recommendations are made:

- Continue collection of data: Data that reflect ship maintenance over time will provide greater value and more defensible baselines.
- Expand collection of data: Data that reflect all types of ships in the U.S. Navy would better reflect the industry and better characterize the problem space.
- Identify performance accounting software for tracking: Software packages are available that would provide for a systematic, common, and seamless method for collecting, storing, and analyzing performance data.
- Begin forecasting once performance baselines are established: Forecasting the effects of policy decisions is only as accurate, and, therefore, valuable, as the baselines used to derive the forecast. Continued and expanded collection of data in a common software package over a period must be accomplished before value can be obtained through forecasting.
- Develop a meaningful numerator for evaluating ship maintenance performance: Return on investment (ROI) is calculated by dividing the output by the input. U.S.

Navy ship maintenance collects troves of data on the input, the denominator, in the form of dollars of cost. However, there is no output, or benefit, derived from ship maintenance, which is collected as a metric and represented in generic units of output. Without a numerator, the ROI of U.S. Navy ship maintenance cannot be determined.

Through the implementation of these recommendations, U.S. Navy ship maintenance executive-level decision makers would be well on their way to deriving the benefits of big data visualization. Those benefits include the ability to identify trends quickly, develop a better understanding of the problem space, establish defensible baselines for monitoring activities, perform forecasting, and determine the usefulness of metrics.

References

Avdellas, N., Berry, J., Disano, M., Oaks, D. and Wingrove, W. 2011. *Future capability of DOD main-tenance depots,* Report LG901M2. Retrieved from http://armedservices.house.gov/index.cfm/files/serve?File_id=5ea27615-e4fe-41ad-8884-d205988dd772.

Department of the Navy (DoN). 2012a. *Highlights of the DON FY2013 budget.* Retrieved from http://www.finance.hq.navy.mil/fmb/13pres/OMN_Vol1_book.pdf.

Department of the Navy (DoN). 2012b. *Joint Fleet Maintenance Manual.* Retrieved from http://www.submepp.navy.mil/.

Office of the Assistant Secretary of Defense for Logistics and Material Readiness (OASD [L&MR]). 2011. *DOD Maintenance Fact Book.* Washington, DC: Author. Retrieved from http://www.acq.osd.mil/log/mpp/factbooks/2011_Fact_Book_ final.pdf.

chapter nineteen

Defining and measuring the success of services contracts

Patrick Hagan, Joseph Spede, and Trisha Sutton

Contents

19.1 Introduction

Department of Defense (DoD) spending on services has been trending upwards for over a decade and as of 2011 accounted for 56% of total contract spending compared to 48% in 2000 (Berteau et al., 2012). The Center for Strategic and International Studies modified the standard DoD portfolio grouping by adjusting services codes to align more appropriately with their respective categories (Berteau et al., 2012). In fiscal year (FY) 2011, more than half of the DoD's $375 billion in contract obligations was for services (Government Accountability Office [GAO], 2012a). The DoD has steadily increased its reliance on services contractors to augment its critical capability shortfalls across the six services categories (GAO, 2011b).

As DoD spending on services has increased over the last 10 years, numerous published Government Accountability Office (GAO) and Department of Defense Inspector General (DoDIG) reports have cited deficiencies in multiple areas of the contracting process. Between 2001 and 2009, the GAO issued 16 reports citing deficiencies in the acquisition of services. Between FY2003 and FY2008, the DoDIG issued 142 reports citing weaknesses in the acquisition and contract administration process (DoDIG, 2009). Contract management has been on the GAO high-risk list since 1992, revealing the difficulties in meeting services procurement, cost, schedule, and performance objectives (Apte et al., 2010). Deficient process areas found by both the GAO and DoDIG included market research, contract type, requirements management, project management, contractor oversight, and personnel training (Apte et al., 2010). These cited problems were compounded by an acquisition workforce that has remained the same size since 2001 while spending services have doubled over the same period (GAO, 2009b). The DoD's contract management process capabilities have also been found to be lacking in all phases of the contracting process. The contract administration and the contract closeout phases, specifically, have even lower process capability than the other phases.

The DoD needs to focus on improving services contract management by first identifying how disparate stakeholders define and measure the success of services acquisitions (Gansler, 2011). Research is necessary to gain an increased understanding of differing stakeholders' goals and objectives, which could be used to develop a standardized definition of services acquisition success and to align stakeholders toward a common goal. In order to maximize the use of scarce acquisition resources and improve outcomes, research is necessary to identify and provide recommendations on the factors of successful services contracts.

The primary purpose of this study was to conduct further analysis of services contracting management practices within the U.S. Navy. The objective of this research was to build upon the understanding developed in prior research projects to generate metrics for defining and measuring successful services contracts. These developed metrics will later help identify factors that influence successful services contract outcomes. We designed a survey based on the exploratory findings of Miller et al. (2012) and distributed it to the primary stakeholders for services acquisitions to determine how different stakeholders define and measure success.

19.2   Deficiencies in services contracts

DoD services contract obligations rose from $92 billion in 2001 to over $200 billion in 2008 (GAO, 2009a). In FY2006, more funds were obligated for services contracts than for supplies and equipment combined (GAO, 2007a). This massive spending growth in services is not attributable solely to the sizable logistical support efforts required in Iraq and Afghanistan. In 2008, these operations accounted for $25 billion of the $200 billion spent on services, but the remaining growth is due to other factors (GAO, 2009a). These factors include contractors filling roles previously held by government employees through outsourcing and the DoD using services contracts for historically non-services acquisitions (GAO, 2007a). It is important to note that during the recent years of defense spending drawdown (2008–2011), spending on services has decreased at a lower percentage rate than spending on products (Berteau et al., 2012).

The significant growth in services spending discussed previously is accompanied by a contracting workforce that has remained relatively the same size since 2001 (GAO, 2009b). Prior to 2001, the acquisition workforce was slashed by nearly 50% from 1989 to 1999 (GAO, 2012c). A sufficiently trained and competent acquisition workforce is necessary for effective contract management and is needed to achieve successful defense acquisition outcomes (Apte et al., 2010). The GAO (2009b) reports that acquisition workforce capability shortfalls make it difficult to ensure that value is achieved and expose the DoD to unnecessary risk.

Government reports indicate insufficient or undocumented market research during the acquisition planning process for services (DoDIG, 2009; GAO, 2012b). Market research is required by the Federal Acquisition Regulation (FAR, 2012), and its purpose is to reveal the market's capability of meeting the government's specific acquisition requirement and to assist in determining the appropriate acquisition strategy to fulfill it (FAR, 2012). Market research is also necessary to enhance competition and to ensure government socioeconomic objectives are achieved. A primary government socioeconomic objective is to promote opportunities for small business concerns to compete for government acquisitions. The government uses small business set-asides to award certain acquisitions exclusively to small business concerns and more specific categories such as services-disabled veteran-owned small businesses (FAR, 2012). Market research determines if an acquisition is suitable for set-aside by determining if responsible small business concerns can satisfy the government's requirement.

The GAO and DoDIG have found inappropriate contract types used on services contracts, which lead the government to shoulder increased cost risk (DoDIG, 2009; GAO, 2009a). Specifically, the GAO found the overuse of high-risk time-and-materials contracts due to their ease of use, speed, and flexibility (GAO, 2009a). The use of time-and-materials contracts should be limited because they provide no incentive to the contractor for cost control and efficiency (GAO, 2009a). There is not a blanket good or bad type of contract, only the inappropriate use of a specific contract type in a procurement. The appropriate contract type is necessary to properly distribute contract risk between the government and contractor and to provide the contractor with the most incentive for efficient performance.

The use of a proven program management approach is considered the best practice for managing services contracts (Apte et al., 2010). A program management approach utilizes methods such as formal project managers, project teams, integrated processes, and a project life cycle to manage contracts. Despite these proven best practices, government reports have shown that the DoD does not have an adequate management structure to oversee services acquisitions at both the strategic and transactional levels (DoDIG, 2009; GAO, 2009a).

Government reports have repeatedly identified contractor oversight and contract administration as serious problems in services contract management (DoDIG, 2009; GAO, 2007b,c, 2009a). Proper oversight and administration ensure that contractors are providing timely and quality services in accordance with the terms of the contract and mitigate contractor performance risk throughout the period of performance.

Both the GAO and DoDIG have identified poorly defined services requirements as a deficient area of services contracts (DoDIG, 2009; GAO, 2009a). A clearly defined description of what the contractor is required to provide the government is necessary to hold effectively contractors accountable, meet customer needs, and ensure that the best value is achieved.

The noted deficiencies discussed in this section provide insight into why contract management has been on the GAO high-risk list for 20 years. Improvements in multiple contract management areas are required to effectively reduce the government's exposure to the risk of overpaying for services.

With DoD financial resources declining since 2009, the DoD must achieve optimal value for defense acquisitions. In his 2010 *Better Buying Power Guidance Memorandum* (Office of the Under Secretary of Defense for Acquisition, Technology, and Logistics OUSD[AT&L]), the Under Secretary of Defense (Acquisition, Technology, and Logistics; USD[AT&L]) acknowledged that DoD practices for services procurement are far less mature than for weapons systems, and he provided guidance to improve services acquisition efficiency. The USD(AT&L) directed each branch to appoint a flag level senior manager for services, adopt a standard taxonomy for types of services, and address the root causes of poor tradecraft in services (OUSD[AT&L], 2010). The component senior manager for services is to be responsible for governing the planning, execution, strategic sourcing, and management of services contracts (OUSD[AT&L], 2010). This USD (AT&L) initiative takes steps to address weaknesses cited by the GAO in strategic management of services acquisitions. A standard taxonomy of reporting categories will ensure consistency across and within the departments and improve visibility and the ability to measure productivity and success across the DoD. The USD (AT&L) cited the following examples of poor tradecraft that must be improved: mission creep, one-bid competitive procurements, misuse of time-and-materials and award fee contracts, and the need to incentivize productivity for large services contracts (OUSD[AT&L], 2010). The Secretary's strategic guidance is an important step in addressing the problems found by the GAO and DoDIG, and it provides a path for delivering better value to the taxpayers.

The deficiencies described previously are significant and by no means comprehensive. There are numerous other examples of insufficient oversight, lack of knowledge by acquisition personnel, and improper contract administration. Some of the deficiencies listed could be corrected if a program management approach were widely implemented by the DoD into the services contracting process. In Section 19.3, we give a brief overview of program management concepts and how they could be applied to services acquisitions.

19.3 Program management approach to services acquisitions

The lack of a mature program management infrastructure and a life-cycle approach to services acquisition project management exposes the DoD to the risk of not meeting cost, schedule, and performance objectives (Apte and Rendon, 2007). The DoDIG and GAO identified critical deficiencies when examining the DoD's existing management structure for acquiring services. In this section, we review basic program management concepts and the implications of research conducted by Apte and Rendon (2007), which examined the application of a program management approach and project management concepts to services acquisition.

19.3.1 Overview of the program management approach and project management concepts

Apte and Rendon (2007) use the term *program management* to describe the approach and methodology needed for the management of complicated projects. A program management approach includes the foundation that enables the attainment of cost, schedule, and performance objectives, and represents the coordinated centralized management of multiple projects to achieve the program's strategic objectives and benefits (Apte and Rendon, 2007; Project Management Institute [PMI], 2008). A systematic program management approach includes the following basic project management concepts: project life cycle, integrated processes, project team structure, a project manager, and a suitable organizational structure (Apte and Rendon, 2007).

The first project management concept we examine is the project life cycle. The Project Management Institute's (PMI) *Project Management Body of Knowledge* (PMBOK; 2008) defines *project life cycle* as a collection of generally sequential phases whose name and number are determined by the control needs of the organization(s) involved in the project. By dividing the project into phases, management is able to control activities more effectively within each phase and the overall progress of the project (Apte and Rendon, 2007). The *Defense Acquisition Guidebook* (DAU, 2012) breaks the project life cycle into the following phases: material solution analysis, technology development, engineering and manufacturing development, production and deployment, and operations and support.

Integrated processes are an essential element of a systematic program management approach and are vital to project success. The PMI *PMBOK* (2008) identifies five project management process groups required for any project. These groups are initiating, planning, executing, monitoring and controlling, and closing processes (PMI, 2008). Each process group includes functional phases, such as cost and schedule management, that are part of a respective process group (PMI, 2008). A structured program management approach integrates these processes to ensure coordination and unity of the total program effort.

The project team structure, with cross-functional expertise across various disciplines, is necessary to integrate effectively project management efforts to achieve the project's objective. In collaboration with the project manager, these integrated subject-matter experts must coordinate and determine which, and to what degree, respective integrated processes are appropriate for the effort (PMI, 2008).

For an effective program management approach, a project manager must be designated by the organization to synchronize the project activities of the various functional team members toward the overall project objectives. Complex projects require the project manager to oversee activities and determine applicable resource applications to these activities (PMI, 2008).

The appropriate organizational structure is a vital element of a disciplined program management approach. Organizational structures influence how projects are conducted and range from functional to project specific, with various matrix types in between (PMI, 2008). Selection of a suitable organizational structure that supports the integrated processes, project teams, and project manager will contribute to project success substantially (Apte and Rendon, 2007). In Section 19.3.2, we look at how program management concepts are currently applied to weapons systems acquisitions.

19.3.2 Application of program management concepts to weapons systems acquisitions

Program management concepts are well established for defense weapons systems acquisitions and are essential practices for complex high-technology weapons systems projects (Apte and Rendon, 2007). The defense acquisition life cycle is a disciplined management process that takes acquisition programs through a series of phases, milestones, decision points, and reports. Control gates assist with keeping projects within the three major constraints of cost, schedule, and performance (Rendon and Snider, 2008). DoD Instruction 5000.2 (OUSD[AT&L], 2008) establishes the Defense Acquisition Management System as the project life cycle for Major Defense Acquisition Programs and was updated in 2008 to include services, specifically. (See Figure 19.1 [DAU, 2012] for an illustration of the milestones and phases of the DoD Acquisition Management Framework.) The program manager (PM) is the designated individual responsible for program objectives; is accountable for cost, schedule, and performance; and reports to the Milestone Decision Authority (MDA; OUSD[AT&L], 2003). Integrated processes and integrated product teams (IPTs), which are established by DoD 5000 regulations, enable PMs to maintain continual and effective communication throughout project execution (Apte and Rendon, 2007). The DoD relies substantially on tailored organizational structures to enhance the integration of project processes and project teams for weapons systems acquisition programs (Apte and Rendon, 2007). The DoD typically uses matrix and projectized (project specific) organizational structures for weapons systems acquisitions. The type of tailored organizational structure used for a specific project depends on factors such as the number of functional areas involved in the project, the degree of integration required within the functional areas and between the organization and

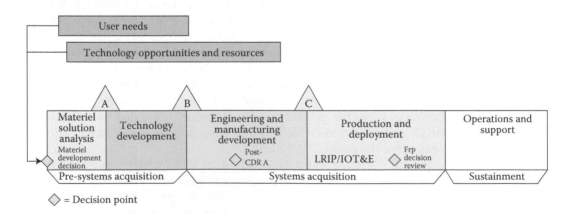

Figure 19.1 The defense acquisition management framework.

customer, and the organization's experience level for the work that the project requires (Apte and Rendon, 2007). Integrated project teams are cross-functional teams with subject-matter experts from multiple functional areas including engineering, contracting, financial management, logistics, and legal (Apte and Rendon, 2007). These teams are led and managed by the designated project manager who ensures coordination and integration to accomplish the project's objective (Apte and Rendon, 2007). In the next section, we examine how these proven program management concepts for weapons systems can be applied to services acquisitions.

19.4 Application of program management concepts to services acquisitions

Apte and Rendon (2007) assessed current services acquisition practices at various activity levels of the Air Force and examined how a program management approach and concepts can be applied to manage services acquisition programs successfully. The findings showed that at the installation level, the acquisition of services was managed ad hoc as opposed to using a program management approach (Apte and Rendon, 2007). The Air Force used some program management concepts at the installation level, but they were not institutionalized throughout the organization and were inconsistent in application (Apte and Rendon, 2007). Apte and Rendon (2007) found that the traditional method used for the acquisition of services does not utilize a program management approach. The traditional method lacked the disciplined use of project life cycles, integrated processes, designated PMs, integrated cross-functional teams, and an appropriate organizational structure.

However, Apte and Rendon (2007) observed two organizations at the major-command level that are applying innovative program management approaches in terms of organizational structure, project life cycle, integrated processes, and project teams for the successful acquisition of services. Air Education and Training Command (AETC) in the U.S. Air Force integrates critical processes by utilizing a disciplined and structured project life cycle for services acquisitions (Apte and Rendon, 2007). Figure 19.2 depicts the various phases within the AETC services project life cycle (Apte and Rendon, 2007). In addition, the AETC utilizes formal project teams for pre- and post-award activities and an integrated and matrixed organizational structure for the acquisition of support services (Apte and Rendon, 2007). Apte and Rendon (2007) found the U.S. Air Force's Air Combat Command (ACC) Acquisition Management and Integration Center (AMIC) to be a fully integrated organization that employs the critical program management concepts discussed previously and includes all essential elements of an acquisition program office. The AMIC provides integrated cradle-to-grave services acquisition support, which enables a resource-efficient process-oriented approach for acquisitions (Apte and Rendon, 2007). Both AETC and the AMIC successfully demonstrated a disciplined program management

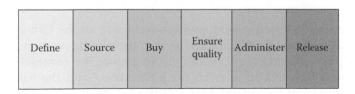

Figure 19.2 Services project life cycle.

approach for the acquisition of services (Apte and Rendon, 2007). The research clearly demonstrated the viability and applicability of a program management approach to services acquisitions. This program management approach provides a process-oriented framework for improved services acquisition outcomes.

19.5   Services contracting process

Contract management is an integral part of the defense acquisition management system and is defined as "the art and science of managing a contractual agreement throughout the contracting process" (Rendon and Snider, 2008). It is the means through which progress is made in the different phases of the acquisition life cycle. The acquisition of all systems, goods, and services is done through contracting and is guided by the FAR. The contracting process involves six primary phases conducted by both a buying organization (government) and a selling organization (contractor), each with a unique contract management perspective. The buyer's perspective involves procurement planning, solicitation planning, solicitation, source selection, contract administration, and contract closeout/termination (Rendon and Snider, 2008). The seller's perspective involves similar phases that align with the buyer's phases. They are presales activities, bid/no-bid decision making, bid/proposal preparation, contract negotiation and formation, contract administration, and contract closeout/termination (Rendon and Snider, 2008). Since this research is about the problems of services contracts within the U.S. government, we discuss only the buying (government) side of the contracting process.

Phase 1 for the buyer is the procurement planning process and involves identifying which business needs can be met by reaching outside the organization for products and services (Rendon and Snider, 2008). The government decides if procurement is necessary and, if so, what, how much, and when to procure (Rendon and Snider, 2008). Part of the planning process includes determining and defining the requirement and conducting market research (Rendon and Snider, 2008). Budgets and cost estimates are developed by both the buyer and seller, and preliminary work documents that delineate what services will be performed, are written (Rendon and Snider, 2008).

Phase 2 is solicitation planning. The process involves determining which procurement method the government will use to procure the goods or services such as through negotiated proposals or sealed bids. The type of contract (cost versus fixed price) and the contract/award strategy (lowest price versus price trade-off) are determined, criteria are developed for evaluating proposals, and the solicitation documentation is drawn up with contract terms and conditions (Rendon and Snider, 2008).

Phase 3 consists of the solicitation process. Solicitation involves receiving bids and proposals from potential sellers (contractors). In order to receive bids and increase competition, the government must advertize that there is a procurement opportunity (Rendon and Snider, 2008). To do this, the contracting officer (CO) transmits a notice though the Government Point of Entry (Rendon and Snider, 2008). After the government issues the solicitation, contractors prepare offers and attempt to persuade the government to accept their bid (Rendon and Snider, 2008).

Phase 4 is the source selection and involves receiving all the bids or proposals from the sellers. The proposals are then evaluated against the criteria previously set in the solicitation planning phase (Rendon and Snider, 2008). Before the contract can be awarded, contract negotiations must occur between the government and the contractor to ensure that there is full agreement on cost, schedule, terms and conditions, and contractor performance (Rendon and Snider, 2008).

Phase 5, contract administration, is the most critical phase in the contracting process. Both the government and the contractor conduct contract administration, each ensuring that performance from both sides meets the contractual obligations. Both parties attend a pre-performance conference to ensure that everyone understands the contract requirements and to discuss protocols for performance management, communication, and contract change management (Garrett and Rendon, 2005). Any changes or modifications to the contract are executed, by authorized individuals only, through a formal process (Rendon and Snider, 2008). The government closely monitors the contractor for quality, cost control, and performance. Best practices in contract administration include, but are not limited to, establishing a system to verify contractual conformance, assigning responsibility to check actual performance against requirements, ensuring that someone takes appropriate corrective action to significant variances, and maintaining all contract documentation (Garrett, 2011).

Phase 6 is contract closeout or termination. A government contract can end when the contractor successfully completes the job, when the government deems that it is convenient to terminate the contract, or when the contractor is in default (Rendon and Snider, 2008). Regardless of why the contract is closed, all final administrative and legal matters must be completed (e.g., price adjustments and final payments made, claims settled, final products or services accepted, and past performance reports documented; Rendon and Snider, 2008).

Recent studies on the contract management process in the DoD have indicated that, on the average, the process capability for the procurement planning, solicitation planning, solicitation, and source selection phases are at the Structured level (Level 3), meaning that these processes are not fully integrated throughout the other functional areas of the agency, even though they are institutionalized. They are also not measured or continuously improved. Additionally, on the average, the process maturity level for the contract administration and contract closeout phases are at the Basic level (Level 2), meaning that these processes are not sufficiently institutionalized within the contracting department, even though they are somewhat established with some documentation, nor are these processes fully integrated throughout the other functional areas of the agency (Rendon, 2009, 2010, 2011).

Each of the phases in the contracting process has activities performed by different individuals, each of whom has a stake in making sure that a project/program is performed in accordance with his/her organization's regulations. These *stakeholders* can affect or be affected by the accomplishment of their organization's objectives (Freeman, 1984). Stakeholder theory, next section, describes and discusses this concept.

19.6   Stakeholder theory

In the private sector, stakeholder theory is described as a corporation that identifies the people who have a stake or interest in that corporation and then acts accordingly to further the interests of those stakeholders (Cleland, 1986). The main assumption of this theory is that the purpose of the corporation is to generate and dispense some form of wealth to various stakeholders, and that in order to achieve that purpose, all of the stakeholders cooperate (Freeman, 1984). The word *stakeholder* originated in the 1960s at the Stanford Research Institute (SRI) with respect to the view that corporate management should only be concerned with the corporation's stakeholders (Parmar et al., 2010). Stakeholder theory, applied to government contracting, includes all stakeholders concerned with a certain project. The project is the corporation, and the contractor and the various government roles on the acquisition team are the stakeholders. That is a very broad view because both the contractor and the government have numerous individual stakeholders. On the government side, there are the CO, the PM, the contracting officer representative (COR), the financial managers, and the customers who use the final product. On the contractor side, there are the contracts manager, the contract specialist, the PM, all subcontractors conducting business with that company, all suppliers who provide goods or services to that company, and various executives who have a vested interest in the success of the company. This is a serious challenge for most managers because the stakeholder management approach views success as assessing the impact of every decision made by all stakeholders through the services acquisition process (Cleland, 1986).

As stated earlier, the CO, PM, COR, financial manager, and customer are all government stakeholders on various projects that fall under their sphere of influence. Because of their differing roles and responsibilities, their definitions of success cover a wide range and are often in conflict with each other in terms of the importance of the factors measured. These stakeholders are described in detail in Section 19.8.

19.7 Stakeholders in DoD services contracts

19.7.1 Contracting officer

The government obtains defense systems, equipment, and services from private-sector sources. Since a contract is the primary means of acquisition, contract management is a set of important skills and knowledge that is of great value in defense acquisition projects (Rendon and Snider, 2008). The FAR (2012) defines the CO as an agent of the government with the authority to enter into, administer, and/or terminate contracts, and make related determinations and findings. Not only is the CO responsible for performing all contract functions, but the CO must also be able to aid in requirements development (DoD, 2011). The CO is the principal business advisor to the acquisition team, and the role's specific responsibilities include forming a business strategy, participating in source selection, and possibly administering the resultant contract (DoD, 2011). He or she is the individual responsible for making sure that contracts are planned, executed, and closed out in accordance with agency regulations and statutory requirements (Rendon and Snider, 2008). COs must provide support to help achieve the contract objectives of cost, schedule, and performance. COs are the only agents with the ability to bind the government, and any individual who attempts to do so without delegated authority will make an unauthorized commitment (Rendon and Snider, 2008). A detailed listing of CO responsibilities is presented in the FAR (2012).

19.7.2 Program/project manager

The PM, also called a program director or a project manager, is the designated person responsible for accomplishing program objectives and ensuring that the desired results are achieved (Brown, 2010). The PM has a very involved and important role in acquisition and procurement because he or she is held accountable for overall cost, schedule, and performance. The PM identifies, plans, and controls various aspects of the project/program. These areas include, but are not limited to, delivery requirements, scheduling, conducting market research, and, normally, participating in source selection (DoD, 2011). The PM should be the one most familiar with the program requirements (DoD, 2011).

The PM must also have an aptitude in contracting, financial management, and cost estimating. In addition, more important, his or her management and leadership skills come into play when dealing with management challenges in day-to-day program execution (Wood, 2010). The PM's role is not an easy one, and the GAO has historically identified his or her lack of program management expertise as the primary cause of cost and schedule overruns in major acquisitions (GAO, 2005). Major weapons systems typically establish a program office; however, this is not always the case with services acquisitions (GAO, 2011a). When there is no PM designated in a service contract, the contracting teams work directly with the requiring organization (GAO, 2011a).

19.7.3 CO representative

Complex contracts for services require the addition of a technical expert, one who is intimately involved in the contract and has expert technical knowledge of the system or service being procured—this person is the COR. The COR is the onsite technical specialist who assesses performance against standards and then records and reports this information to the CO (GAO, 2011a). Only a U.S. government employee (civilian or military) or a North Atlantic Treaty Organization (NATO) partner can fill the role of a COR; the CO should never fill this role (FAR, 2012). The COR is formally appointed in writing by the CO and must have specific qualifications and experience appropriate for the responsibility delegated to him or her (FAR, 2012). The COR should be deeply involved in the entire acquisition and procurement process and should assist the CO in developing the quality assurance plan, the technical requirements in the contract, and other pre-award activities. The COR does not have the same authority as the CO, so he or she cannot make any commitments or changes that affect the terms of the contract (FAR, 2012). The COR assists the CO and, along with the PM, becomes the focal point of the contract by monitoring all of its day-to-day aspects, as well as inspecting and accepting services (DAU, 2012). A comprehensive list of the specific responsibilities of a COR can be found in the FAR (2012).

19.7.4 Finance manager

The finance manager is another critical individual in contract management and is well versed in the financial management regulation (FMR). His or her role is to serve as the fiscal and budgetary advisor to the acquisition team (DoD, 2011). The government finance manager is responsible for ensuring compliance with the statutory requirements of fiscal law (e.g., that proper authorization is granted for expending funds [purpose], the contract obligations occur during the time limits prescribed by appropriation [time], and adequate funding is available [amount]; Rendon and Snider, 2008).

19.7.5 Customer or end-user

The customer's role in the acquisition process is to have a detailed knowledge of the requirement so that it can be clearly conveyed to contractors (DoD, 2011). The customer helps to determine whether trade-offs are available for a requirement and what these trade-offs are (DoD, 2011). The customer plays a vital role in the acquisition process because, ultimately, he or she is a member of the team that drives how well the requirements document reflects their needs.

The roles and responsibilities of each of the stakeholders differ, and, as a result, their definitions of what constitutes a successful contract cover a wide range. Often these definitions are in conflict with each other in terms of the importance of the factors measured. Identifying how each of the stakeholders determines success is vital to contract administration. In Section 19.8, we discuss a study that identified seven factors that contribute to contract management success.

19.8   Success factors

As described earlier in this chapter, there are six phases in the contract management process. Of these six phases, only one is dedicated to contract administration. GAO audits have consistently stated that contract administration needs to be improved. Successful contract administration is dependent on effective contract management. The question then becomes, what determines the success of a contract? Rendon (2012) conducted a study asking this question specifically of DoD contracting. In the study, Rendon surveyed eight defense agencies and two defense contractors over the course of 4 years. Since the survey was designed for defense procurement, it was administered only to warranted COs and individuals fully qualified in government contracting. The results of the survey identified seven critical success factors (CSF) in defense procurement: Workforce, Relationships, Processes, Resources, Leadership, Policy, and Requirements (Rendon, 2012).

Rendon analyzed and summarized over 2000 responses to the survey into the seven CSF categories. The Workforce factor in the study is related to using proper staffing, hiring, and recruitment processes; having the right number of personnel; and having experienced, trained, and competent people. Relationship responses involved communication, cooperation, and coordination at all levels within and between agencies. The Processes category involved having a consistent, efficient, standardized, enforced, streamlined, and documented contracting process. Resource responses included the need for contract tracking tools, automated contract writing systems, technical support, and adequate travel funds. Leadership responses related to the need for clear lines of authority, strong management support, and an empowered leadership. The Policies category included the need for clear and concise guidance and regulations. The final CSF, Requirements, related to timely procurement request packages, well-written statements of work, proper technical reviews, and adequate procurement funding (Rendon, 2012).

A comparison of the DoD and industry/contractor responses shows some interesting differences. The DoD considers workforce-related elements to be the most important success factor and requirements-related elements to be the least important. Industry responses showed that processes were the most important factor, and policies were the least important. The overall results of the study provide some thought-provoking insights into the differences between organizations and the disconnection that can occur when measuring and defining the success of a contract. Agency Theory, described in Section 19.9, explains how and why this disconnection can occur.

19.9   Agency theory

Agency theory is aimed at the relationship that arises when one party (the principal) engages another party (the agent) to perform a specific effort focusing on a certain outcome (Eisenhardt, 1989). The theory mainly discusses these relationships and describes how the principal and the agent are engaged in cooperative behavior but have differing goals and attitudes (Eisenhardt, 1989). Agency Theory states that the principal can limit this difference by creating suitable incentives for the agent or creating sufficient means to monitor the agent (Hill and Jones, 1992). When applying this theory to services contracts, the government is the principal, and the contractor is the agent. The government employs the contractor to perform specific tasks that are defined in great detail within the contract.

As described in Section 19.8, the government and contractor place differing importance on each of the success factors—and this is one of the cornerstone assumptions of Agency Theory (Hill and Jones, 1992). The government and the contractor clearly have different goals in mind when executing a service contract. The government's goal is to have the services performed to a certain standard in the most effective way for a fair and reasonable price (FAR, 2012). The contractor wishes to stay in business, so its ultimate goal is to generate profit. The government incentivizes the contractor by providing a fee—and in some cases, a price premium—in order to facilitate desired behaviors. Agency Theory describes the complex government–contractor relationship and how a principal and an agent can be involved in a cooperative effort but have differing goals and attitudes (Eisenhardt, 1989).

19.10 Survey development

Our research methodology included the development of a survey instrument to collect empirical data for answering our research questions. Following verification of validity and cohesiveness, the survey was deployed to the various stakeholders (PMs, COs, CORs, contractors, and end users) at the participating commands (Fleet Logistics Centers [FLC] Philadelphia, FLC Jacksonville, FLC Norfolk, FLC Puget Sound, FLC San Diego, Naval Sea Systems Command [NAVSEA], Military Sealift Command [MSC], and Space and Naval Warfare Systems Command [SPAWAR]). We then analyzed the data using descriptive statistics to provide recommendations and conclusions.

The focus of the survey was on answering the following core research questions:

- How do different stakeholders define successful services contracts within the Navy?
- How do different stakeholders measure services contracts within the Navy?

We took the qualitative results from previous research conducted by Miller et al. (2012) and then identified four metrics—process, cost, schedule, and performance—about which we could ask further detailed questions.

The survey questions that addressed our two core research questions consisted of seven demographic questions and 12 research questions such as

1. What is your branch of service or service affiliation?
2. What organization are you affiliated with?
3. What is your current, primary functional role?
4. What is your DAWIA (Defense Acquisition Workforce Improvement Act) level certification ?
5. How many years of acquisition experience do you have?
6. What type of services do you predominantly procure?
7. What broad category do the majority of your contracts fall into?
8. How do you define a successful service contract?
9. How do you measure the success of a service contract?

The purpose of the demographic questions was so that we could differentiate our results, and compare, and contrast to determine trends across different areas such as functional role, DAWIA level, and type of service provided.

The purpose of the core research questions was to establish the importance of different factors when defining and measuring the success of services contracts. We asked several questions related to the contracting process, as well as questions concerning different outcomes such as cost, schedule, and performance. Our process questions involved but were not limited to, the level of administrative load, the occurrence of protests, and levels of communication between all stakeholders. The survey questions associated with cost dealt with over-runs, fair and reasonable pricing and profit, as well as cost control. The schedule questions were related to meeting major milestones and a timely completion of the contract. Performance questions were connected to customer satisfaction, adherence to the statement of work (SOW), and reliance on COR reports. We also used these factors to differentiate responses in order to determine trends.

The survey provided specific questions related to how commands define the success of a service contract. The first two questions ask participants to rank various definitions relating to the four metrics in order of most important (1) to least important (5). The next three main questions ask participants to rate definition statements relating to process, schedule, cost, and performance. These questions use a Likert scale asking level of agreement, importance, and the amount of time devoted by the participants. The Likert scale had a range of (1)–(5), with (1) representing a negative response and (5) representing a positive response.

The survey also asked specific questions related to how commands measure the success of a service contract. The first two questions ask participants to rank various measurements relating to the four metrics in order of most important (1) to least important (5). The last question in the section asks participants to rate on a Likert scale how often the organization conducts certain actions that pertain to the measurement of success concerning process, schedule, cost, and performance.

The survey included a final question soliciting any general comments that the participants may wish to share regarding the topic of defining and measuring successful services contracts. Figure 19.3 contains a diagram of the survey questions.

One of the participating commands was NAVSEA whose mission is to engineer, build, buy, and maintain ships, submarines, and their combat systems that meet the U. S. Naval Fleet's current and future operational requirements. As the largest of the Navy's five system commands, the NAVSEA has a $30 billion FY budget, accounting for 25% of the Navy's entire budget. To accomplish their mission, the NAVSEA manages billions of dollars in

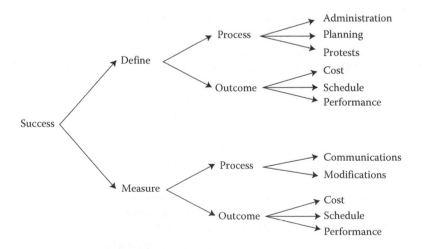

Figure 19.3 Diagram of survey questions.

annual foreign military sales and 150 acquisition programs. It plays a critical role in the Navy Enterprise and strives to be an efficient provider of defense resources for the United States (NAVSEA, n.d.).

SPAWAR is one of the three major Departments of Navy acquisition commands. Its mission is to design, develop, and deliver advanced communications and information dominance systems to the fleet. It supports the full life cycle of product and services delivery, which includes research, engineering, acquisition and deployment, and operations and logistics support services. The SPAWAR's products and services, "transform ships, aircraft and vehicles from individual platforms into integrated battle forces, enhancing information dominance and awareness among Navy, Marine, joint forces, federal agencies and international allies" (SPAWAR, n.d.). In order to accomplish their mission, SPAWAR partners and contracts with industry, including small businesses, to get the best value for information technology. In 2010, SPAWAR obligated $1.21 billion to small businesses and $4.84 billion to large businesses (Esaias, 2011).

The MSC is composed of over 100 noncombatant, civilian-crewed ships. Its mission is to support our nation by replenishing U.S. naval ships, moving military cargo and supplies, strategically prepositioning combat cargo at sea around the world, and conducting specialized missions. The MSC is organized around five mission areas: combat logistics force, special mission, propositioning, services support, and sealift. These worldwide operations are financed through the Navy Working Capital Fund and the Transportation Working Capital Fund. Its budget of about $3 billion is reimbursed by direct appropriations or by funds transfers by MSC customers (www.msc.navy.mil).

The FLCs Philadelphia, Jacksonville, Norfolk, Puget Sound, and San Diego are just some of the 12 total subordinate commands of the Naval Supply Systems Command (NAVSUP). NAVSUP as a whole provides 25 distinct products and services, ranging from supply chain management, warehousing, and foreign military sales, to postal services and quality of life programs (NAVSUP, n.d.). Each subordinate command is responsible for providing logistics, business and support services, and products to U.S. naval activities and other joint, civilian, and allied forces within their respective areas of responsibility. They "deliver combat capability through logistics by teaming with regional partners and customers to provide supply chain management, procurement, contracting and transportation services, technical and customer support, defense fuel products and worldwide movement of personal property" (NAVSUP, n.d.). They are given contracting authority by NAVSUP, contracting for over $4 billion annually in supplies, services, and equipment, and make about 120,000 individual purchases (GlobalSecurity.org, n.d.).

19.11   Survey instrument, results, and analysis

19.11.1   Defining the success of a service contract

One hundred and sixty-eight respondents were asked to rank different factors related to defining the success of a service contract. These questions also dealt with different aspects of processes and outcomes. Of the 172 respondents, 40% felt that process-related factors were the most important. Sixty percent felt that outcome-related factors were the most important. The distribution of highest ranked responses is displayed in Figure 19.4.

Breaking down the outcome-related factors further, 15% of respondents felt that cost-related factors were the most important, 19% felt that schedule-related factors were most important, and 26% felt that performance-related factors were most important.

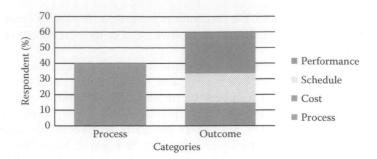

Figure 19.4 Aggregate stakeholder ranking of definitions of success.

19.11.2   Measuring the success of a service contract

Our survey also had participants rate on the Likert scale the various degrees of importance and the extent to which they agreed or disagreed; various factors related to how they measure the success of a service contract. Again, these factors related to either processes or outcomes. Outcomes were more important to our participants as a whole. Of the 168 respondents, 46% felt that process-related factors were the most important. Fifty-four percent felt that outcome-related factors were the most important. The distribution of highest ranked responses is displayed in Figure 19.5.

Breaking down the outcome-related factors further, 19% of respondents felt that cost-related factors were the most important, 12% felt that schedule-related factors were most important, and 23% felt that performance-related factors were most important.

19.11.3   Findings by stakeholder

We obtained the most responses from certain functional roles. PMs, CORs, and COs/contract specialists accounted for 87% of our total responses. In this section, we examine and compare how each of these functional areas defines success with regards to service contracts. COs and contract specialists were grouped under one category and from now on the term "CO" encompasses both demographics. We differentiated by functional role and made no other demographic distinctions. Our data include functional roles across all DAWIA levels, contract types, and organizations.

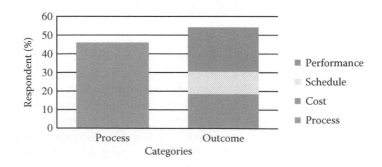

Figure 19.5 Aggregate stakeholder ranking of measurements of success.

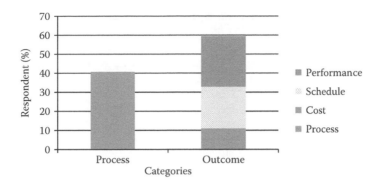

Figure 19.6 PM ranking of definitions of success.

19.11.3.1   PMs', CORs', and COs'/contract specialists' definitions of success

19.11.3.1.1   PMs' definition of success   With regard to the Likert scale ratings from PMs, we found, based on the 15 responses we received, that this group considers outcomes slightly more important than processes when defining success. Of the 15 respondents, 41% felt that process-related factors were the most important. Fifty-nine percent felt that outcome-related factors were the most important. The distribution of highest ranked responses is displayed in Figure 19.6.

Breaking down the outcome-related factors further, 11% of respondents felt that cost-related factors were the most important, 22% felt that schedule-related factors were most important, and 27% felt that performance-related factors were most important.

19.11.3.1.2   CORs' definition of success   We received 27 responses from CORs. The range of CORs definitions on the Likert scale is relatively low, and CORs also appear to favor outcomes over processes when defining the success of a service contract. Of the 27 respondents, 39% felt that process-related factors were the most important. 61% percent felt that outcome-related factors were the most important. The distribution of highest ranked responses is displayed in Figure 19.7.

Breaking down the outcome-related factors further, 18% of respondents felt that cost-related factors were the most important, 19% felt that schedule-related factors were most important, and 24% felt that performance-related factors were most important.

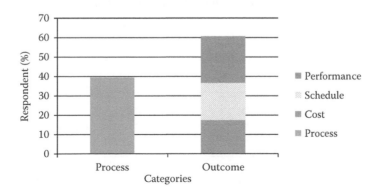

Figure 19.7 COR ranking of definitions of success.

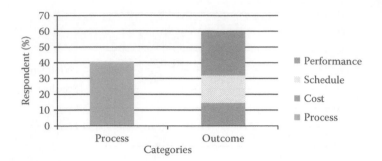

Figure 19.8 CO ranking of definitions of success.

19.11.3.1.3 COs' and contract specialists' definition of success We received responses from 126 self-identified COs/contract specialists who only slightly favored outcomes versus processes when defining the success of a service contract. Of the 126 respondents, 40% felt that process-related factors were the most important. Sixty percent felt that outcome-related factors were the most important. The distribution of highest ranked responses is displayed in Figure 19.8.

Breaking down the outcome-related factors further, 17% of respondents felt that cost-related factors were the most important, 15% felt that schedule-related factors were most important, and 28% felt that performance-related factors were most important.

19.11.3.2 PMs', CORs', and COs'/contract specialists' measurement of success

19.11.3.2.1 PMs' measurement of success PMs provided 15 responses that indicated they heavily rely on outcomes rather than processes to measure the success of a service contract. Of the 15 respondents, 43% felt that process-related factors were the most important. Fifty-seven percent felt that outcome-related factors were the most important. The distribution of highest ranked responses is displayed in Figure 19.9.

Breaking down the outcome-related factors further, 13% of respondents felt that cost-related factors were the most important, 14% felt that schedule-related factors were most important, and 30% felt that performance-related factors were most important.

19.11.3.2.2 CORs' measurement of success CORs provided 27 responses that suggested they also find outcomes significantly more important than processes when measuring the success of a service contract. Of the 27 respondents, 39% felt that process-related

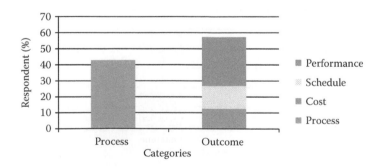

Figure 19.9 PM ranking of measurements of success.

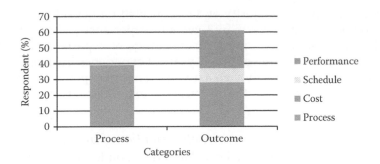

Figure 19.10 COR ranking of measurements of success.

factors were the most important. Approximately, 61% felt that outcome-related factors were the most important. The distribution of highest ranked responses is displayed in Figure 19.10.

Breaking down the outcome-related factors further, 28% of respondents felt that cost-related factors were the most important, 9% felt that schedule-related factors were most important, and 24% felt that performance-related factors were most important.

   19.11.3.2.3   COs' and contract specialists' measurement of success   COs and contract specialists provided 126 responses. COs and contract specialists were aligned with PMs and CORs when measuring the success of a service contract. All three functional roles agreed that outcome-related factors heavily outweighed process-related factors when measuring success. Of the 126 respondents, 49% felt that process-related factors were the most important. Fifty-one percent felt that outcome-related factors were the most important. The distribution of highest ranked responses is displayed in Figure 19.11.

Breaking down the outcome-related factors further, 17% felt that cost-related factors were the most important, 11% felt that schedule-related factors were most important, and 23% felt that performance-related factors were most important.

19.11.4   Findings on DAWIA level

This particular analysis is strictly divided among DAWIA levels and contains no other demographic differentiation. Of the 168 participants, 85% had a DAWIA certification level of I, II, or III. The education and experience requirements for certification rise in

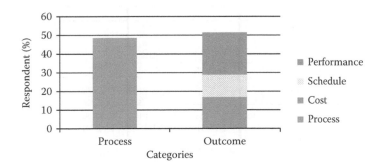

Figure 19.11 CO ranking of measurements of success.

conjunction with the level. A majority of data collected on Level I participants showed that these participants had fewer than 3 years of experience in an acquisition-related billet. Level II participants had a more diverse range of experience with the majority having more than 4 years, but fewer than 8 years of experience in an acquisition-related billet. Level III participants also showed a diverse range of experience, and, as expected, contained participants with more than 19 years of experience in an acquisition-related billet.

19.11.4.1 *DAWIA-certified participants' definition of success*

When differentiated by DAWIA level, Level I and II respondents showed a slightly higher rating than Level III participants. However, the responses concerning processes versus outcomes showed little difference within each level. While distinct levels may rate processes and outcomes differently, they felt that outcomes were slightly more important than processes. Yet, processes were still considered important when defining the success of service contracts.

One hundred forty-two respondents with a DAWIA certification provided responses when asked to rank definitions of success concerning a service contract. Each certification level provided similar answers within each of the categories of processes and outcomes. Of the Level I respondents, 42% felt that process-related factors were the most important while 58% felt that outcome-related factors were the most important. Of the Level II respondents, 39% felt that process-related factors were the most important while 61% felt that outcome-related factors were the most important. Of the Level III respondents, 40% felt that process-related factors were the most important while 60% felt that outcome-related factors were the most important. Again, this shows that these particular demographics tend to use outcomes when defining the success of a service contract. The distribution of the responses is shown in Figure 19.12.

19.11.4.2 *DAWIA-certified participants' measurement of success*

Our data suggest that when measuring the success of a service contract, the various DAWIA levels rely heavily on outcome-related factors rather than process-related factors. When asked to rank measurement factors of success concerning a service contract, there was a little more diversity among the respondents with various DAWIA levels, but all favored outcomes over processes. Forty-nine percent of Level I respondents felt that processes were most important, with 51% favoring outcomes. Forty-eight percent of Level II respondents felt that processes were most important, with 52% favoring outcomes. Forty-three percent

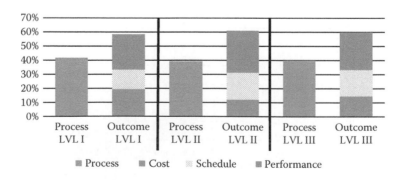

Figure 19.12 Ranking of definitions of success across DAWIA levels.

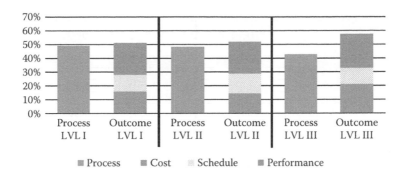

Figure 19.13 Ranking of measurements of success across DAWIA levels.

of Level III respondents felt that processes were most important, with 57% favoring outcomes. The distribution of the responses is shown in Figure 19.13.

19.11.5 Data analysis by type of service

We obtained the majority of responses from participants who procured certain types of services. Of the 168 total responses, the equipment-related service portfolio grouping accounted for 72 responses and the knowledge-based service portfolio groupings for 77 responses. Given the majority of responses from these two groups, we further analyzed them to identify trends across service types. The equipment-related service portfolio grouping included the procurement of maintenance, repair and overhaul, equipment modification, installation, and quality control services. The knowledge-based services portfolio grouping was composed of professional and administrative services, engineering management, program management, logistics management, and education and training. Further demographic breakdown of the two groups showed 40% of knowledge-based service participants were involved with cost reimbursement-type contracts compared to only 6% for equipment-related services. The higher percentage of cost reimbursement-type contracts for knowledge-based service participants was most likely due to the increased challenges and uncertainties in defining requirements associated with these types of services. Equipment-related services are generally more concrete in terms of requirements definitions and would be more suitable for a fixed-price contractual instrument.

19.11.5.1 Specific service type definition of success

In response to our questions asking participants to classify different factors related to defining the success of a service contract, we received 149 responses from participants who work in knowledge-based and equipment-related services. We received 72 responses from participants who worked on equipment-related services and 77 responses from those involved with knowledge-based services. When differentiating between types of service, we found that equipment-related service participants rated both processes and outcomes higher than knowledge-based service participants. Participants segregated by type of service may rate processes and outcomes slightly differently; however, they both indicated that outcomes are slightly more important than processes for defining success.

Seventy-two respondents involved with equipment-related services provided responses when asked to rank definitions of success concerning a service contract. The questions asked of participants dealt with different aspects that aligned with process- and

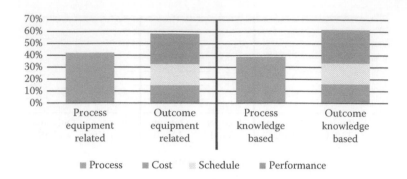

Figure 19.14 Ranking of definitions of success by service type.

outcome-related groupings. Of the 72 respondents, 42% believed that process factors were most important while 58% felt that outcomes more accurately defined the success of a service contract.

Breaking down the outcome-related factors further, 15% felt that cost-related factors were the most important, 18% felt that schedule-related factors were most important, and 25% felt that performance-related factors were most important.

Seventy-seven respondents involved with knowledge-based services provided responses when asked to rank definitions of success concerning a service contract. Of the 77 respondents, 39% believed that process factors are most important while 61% felt that outcomes more accurately defined the success of a service contract. The results show that equipment-related service participants rated processes higher and outcomes lower than knowledge-based participants, but both groups indicated that outcomes are more important than processes for defining success, based on their responses to ranking questions. The distribution of highest ranked responses is displayed in Figure 19.14.

Breaking down the outcome-related factors further, 16% of respondents felt that cost-related factors were the most important, 18% felt that schedule-related factors were most important, and 28% felt that performance-related factors were most important.

19.11.5.2 Specific service type measurements of success

In measuring the success for different types of services, the equipment-related service participants rated processes higher and outcomes lower than knowledge-based participants, but both groups indicated that outcomes are more important than processes for measuring success.

In response to our questions asking participants to classify different factors related to measuring the success of a service contract, we received 149 responses from participants who work in knowledge-based and equipment-related services. The questions dealt with process- and outcome-related groupings. Of the 72 equipment-related responses to our ranking questions, process-related factors were ranked most important 44% of the time while outcome-related factors were ranked as most important 56% of the time.

Breaking down the outcome-related factors further, 21% of respondents felt that cost-related factors were the most important, 12% felt that schedule-related factors were most important, and 21% felt that performance-related factors were most important.

Forty-eight percent of the 77 respondents associated with knowledge-based services felt that process-related factors were the most important when measuring success.

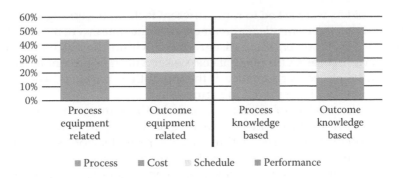

Figure 19.15 Ranking of measurement of success by service type.

Fifty-two percent of respondents felt that outcome-related factors were the most important. The results show that equipment-related service participants rated processes lower and outcomes higher than knowledge-based participants, but both groups indicated that outcomes are more important than processes for measuring success, based on their responses to our ranking questions. The distribution of highest ranked responses is displayed in Figure 19.15.

Breaking down the outcome-related factors further, 16% of respondents felt that cost-related factors were the most important, 11% felt that schedule-related factors were most important, and 25% felt that performance-related factors were most important.

19.11.6   Analysis

19.11.6.1   Analysis of aggregate survey responses

The research findings from our overall survey showed that different stakeholders find all aspects of processes and outcomes important when defining the success of a service contract. As a whole, our population favored neither process nor outcome when defining success. When asked if something such as cost overruns, major milestones, or a lack of protests is important, all stakeholders invariably said "yes." When respondents were forced to rank, the responses differed, and outcome-related responses received a high rank of one or two 60% of the time. This is because outcomes, such as keeping on schedule and adhering to a budget, are easy to define and understand. Process-related factors such as administration and communication are harder to quantify.

The findings also demonstrated that when measuring the success of a service contract, all stakeholders tend to focus on outcomes and do not take into consideration the processes. When ranking the different factors with respect to measuring success, the results were similar to defining success, with 56% of "most important" responses falling under the outcomes category.

19.11.6.2   Analysis across stakeholders

When examining how different stakeholders define the success of a service contract, we found that PMs, CORs, COs, and contract specialists all agreed that the outcome is slightly more important than process. Within outcome, performance-related factors received the highest average rating while schedule-related factors received the lowest average rating. All functional roles showed an upward trend from schedule, to cost, to performance.

When stakeholders were asked to rank different factors concerning their definition of success, we found that there was clear agreement that outcomes were more important than processes. There was, however, some disagreement within the outcome factors of cost, schedule, and performance. CORs felt that cost was the most important factor, while PMs, COs, and specialists placed performance at the top of their rankings. The distribution of highest ranked responses is displayed in Figure 19.16.

Stakeholders also tended to measure success in the same way. When asked to rate different factors related to stakeholders' measures of success, all respondents agreed that outcomes far outweighed processes. Within outcome-related factors, stakeholders showed an upward trend from performance to schedule, to cost. Our ranking data showed that, again, major stakeholders preferred outcome-related factors when measuring the success of service contracts. Of these responses, 43% of respondents felt process factors were most important while 57% favored factors related to outcomes. The distribution of highest ranked responses is displayed in Figure 19.17.

The data collected from the major stakeholders was similar to our cumulative findings. That is to be expected, considering they made up 87% of our total population. It is interesting that in both defining and measuring success, CORs ranked cost highest out of the three stakeholders. In previous research (Miller et al., 2012), CORs listed performance as more important than cost when defining and measuring success. This may be attributed to the open-ended nature of the questions asked in the previous research. Our survey may have brought to light issues or factors that CORs had never thought of before.

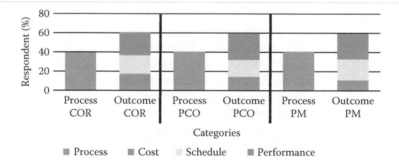

Figure 19.16 Major stakeholder ranking of definitions of success.

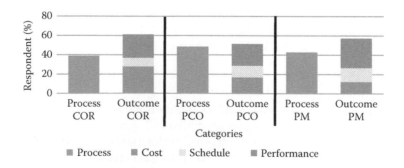

Figure 19.17 Major stakeholder ranking of measurements of success.

Another interesting result is that COs tended to place nearly equal importance on processes and outcomes when forced to rank factors concerning measuring success. This is probably due to the administrative nature of the COs' role. For example, their functional role has to deal with modifications, COR reports, and exercising options. PMs and CORs are not overly concerned with processes and are focused on the requirement and outcomes. The data reflect this fact.

It is of note that every demographic consistently rated processes significantly higher when defining success versus measuring success, possibly because stakeholders view measures as a tangible entity associated with post-award functions. Measures such as cost, schedule, and performance are straightforward in as much as either a goal is met or it is not. Processes such as communication flow and overall management are more obscure and subjective. The stakeholders rated processes higher for defining success because they are closely associated with mainly pre-award functions. Processes such as choosing the correct contract type and appropriate proposal evaluation are crucial for success. Because these are pre-award activities, it is easier to define success than to measure it.

We performed a statistical analysis across the major stakeholders to determine if there was a significant difference between the ratings on the Likert scale. We first performed an F-test two sample for variances to determine the appropriate t-test to perform. In all instances, we found an equal variance among stakeholders. The only significant difference we discovered was between the CORs and COs/specialists when measuring success. This could be because CORs viewed communication and other processes as a key factor when measuring the success of a service contract. CORs are also likely to view a protest as a serious issue when measuring success because it results in a delay of execution, and the CORs cannot perform their duties. There was no statistically significant difference between any other of the stakeholders on the Likert scale.

19.11.6.3   Analysis across DAWIA levels

We noted that a majority of the DAWIA level-certified personnel were mostly COs, so their results somewhat mirror that demographic. It is of note that Level III personnel tended to provide lower responses than Levels I and II when defining success. Level III personnel generally have more experience than Levels I, and II, and that may be the reason for this trend. Seventy-two percent of Level III-certified respondents reported having 10 or more years of experience, and this demographic had the most divergence from the CO role. The level of experience can explain the low Likert score for the process factors for participants with a Level III certification. Because those respondents have 10 or more years of experience, the processes have become routine and they view administration and communication as standard, whereas Level I and II respondents have relatively low experience and believe that processes are more important. Another reason that more experienced stakeholders view processes as less important is because, as a result of their experience, they have seen the outcomes of fully completed contracts whereas more junior personnel may have only dealt with award and administration. There is the possibility that they have never conducted close out on a contract to actually see the true outcome, and, as a result, Level I and II stakeholders believe that processes are more important than do Level III stakeholders.

19.11.6.4   Analysis by type of service

The findings based on the type of service showed no substantial deviation for defining the success of a service contract. Both equipment-related and knowledge-based groups rated outcomes slightly higher than processes. Ranking questions for definitions of success showed outcomes as more important than processes for both groups. The respondents indicated performance to be the most important component of outcomes. Ranking questions showed 25%–28% for performance and 18% for schedule across both groups. Perhaps ranking results, due to the limitation of choices, provided the most precise definitions of success. When participants were forced to rank, they emphasized the importance of outcomes (cost, schedule, and performance) over processes more definitively when defining the characteristics of successful service contracts.

The findings for measuring success also showed no substantial deviation based on the type of service. Both equipment-related and knowledge-based groups rated outcomes significantly higher than processes. Ranking questions for measuring success showed outcomes as more important than processes for both groups. Participants who worked on equipment-related services rated cost as more important, and performance as less important compared to knowledge-based participants. This finding seems to indicate that for knowledge-based services, stakeholders were more willing to trade cost for enhanced performance. This may be attributable to the highly specialized and technical nature of functions such as engineering management and program management, which the government lacks the crucial internal capacity to perform.

19.12   Conclusions and recommendations

19.12.1   Conclusions

Over the last 10 years, the DoD has steadily shown increases in the number of dollars spent on the acquisition of services, accounting for over half of total contract spending. This increased reliance on service contractors has prompted the GAO and DoDIG to increase surveillance and issue numerous reports citing deficiencies and weaknesses in the acquisition and contract management process. These flaws were noted in every aspect of the contracting process such as insufficient market research, choosing an inappropriate contract type, poor contractor oversight, and lack of properly trained government personnel.

Another major issue is that while these spending increases have occurred, the acquisition workforce has steadily decreased. The knowledge gap created by this decrease has contributed significantly to ineffective contract management over the last decade. Well-trained and capable personnel are essential for successful management of service contracts. It is imperative that the correct metrics are utilized to define and measure that success.

This research project was conducted to analyze further services contracting management practices within the Navy. The goal was to answer three research questions in order to strengthen the Navy's understanding of successful services acquisitions from disparate stakeholder perspectives.

- How do different stakeholders define successful services contracts within the Navy?
- How do different stakeholders measure services contracts within the Navy?
- How should Navy services contracts be defined and measured?

The objective of this research was to build upon the understanding developed in prior research projects to generate metrics for defining and measuring successful services contracts. These developed metrics could help identify factors that influence successful services contract outcomes.

We found that when defining success all stakeholders tend to utilize outcome-related factors over process-oriented factors. This is because outcomes tend to drive perceptions of success more than processes. Outcomes such as customer satisfaction or meeting major milestones are easily identified, whereas processes tend to be more ambiguous. A process, such as conducting market research in accordance with the FAR, does not necessarily define success. Based on our findings, processes are important factors in that they contribute to the overall success, but not necessarily the definition success. For example, public policy dictates that the government shall provide for full and open competition, award to small business when appropriate, and negotiate fair and reasonable prices. These are all statutory requirements. Based on our research findings, these processes do not necessarily define the success of a service contract; however, they do determine if a contract was awarded in accordance with federal law and public policy. Thus, it is conceivable that a contract may be defined as successful in terms of outcomes (cost, schedule, and performance) but, based on processes, may be in violation of public policy.

The findings in relation to measuring the success of a service contract were comparable to defining success. More participants favored outcomes, but this can also be explained by looking at the factors within the broad category. Cost and schedule are easy factors to track and the results can generally be quantified. If a schedule is not met, then the contract could be considered unsuccessful. However, when there is no appropriate contractor oversight of a process, success is not necessarily sacrificed.

Stakeholders were inclined to define and measure success in similar ways when asked about differing degrees of importance, agreement, and likely occurrence of events throughout the service contracting cycle. When asked to rate the factors on a Likert scale, all stakeholders agreed that process-related factors such as communication, administrative burden, and clear objectives are of high importance. They also agreed that outcomes relating specifically to cost, schedule, and performance are important when defining the success of a contract. However, when asked to rank these factors, stakeholders favor outcomes rather than processes when defining success.

We discovered that when measuring the success of a service contract all stakeholders felt that outcomes were better measures of success than processes. This was revealed because outcomes are easily quantifiable. It is easy to track costs, rate performance, and differentiate between a contract that is on schedule and one that is falling behind schedule. It is not as easy to determine an appropriate level of administrative work, a proper amount of communication among stakeholders, or the correct number of change orders or modifications that are necessary.

We also found that CORs viewed processes as more important than COs/specialists when measuring the success of a service contract. CORs, ideally, are in frequent communication with the COs/specialists to convey the progress and performance on a contract, so the CORs viewed processes as significantly more important. Another reason that CORs rated processes more favorably than *COs/specialists* is that protests completely stop work on a contract, and the CORs believed that a protest results in an unsuccessful contract.

Our research showed that the metrics used to measure success are typically related to cost, schedule, and performance. There is, however, no consensus among stakeholders as to what can be interpreted as the most important of these metrics. According to the data we collected, some demographics considered performance the most important measure, whereas others found the cost to be the most important measure of success.

19.12.2   Recommendations

Process-related factors need to be emphasized in stakeholders' definitions of success. Outcome-related factors of cost, schedule, and performance are dominant while processes are viewed with little importance when defining success. It is evident that this is an issue. Choosing the appropriate contract type, a step in the planning process, should be of great importance when defining the success of a service contract. Time-and-materials or cost-type contracts could result in overpayment, and the cost is considered an important factor when defining success. A well-written and clear SOW should also be highlighted when defining success. If the contractor or the acquisition team does not understand the requirement, modifications are necessary, and increased costs or a schedule slip could result. Process-related factors are inextricably tied to the definition of success of a service contract, and steps need to be taken so that stakeholders understand and act on their importance.

Internal control measures should be in place to ensure that proper processes are being followed and that all stakeholders place high importance on the value of these processes. If stakeholders are forced to take into consideration processes when measuring and defining the success of a service contract, some of the deficiencies in services procurement could be corrected. An example of an internal control measure is the Services Requirements Review Board, a program that NAVSEA implemented in 2011. Commonly referred to as "Services Court," the program is an annual review of the full range of NAVSEA service requirements to understand what services are required and ensure proper oversight is in place so that there is maximum value for dollars spent (NAVSEA, 2012). The board is chaired by the NAVSEA Commander and Executive Director in order to involve leadership at the top level. As a result of their findings, NAVSEA has modified their policies to ensure a clearer definition of service requirements, require more stringent contract file maintenance, and implement proper surveillance and proper performance reporting requirements (NAVSEA, 2012). All of these reforms are process related, and other commands could benefit by implementing similar internal control measures.

Another way to increase the emphasis placed on processes when defining and measuring success is to put in place an operational audit process. This tool can be used to determine the extent of use of process-related factors when defining and measuring success with relation to service contracts. If the correct processes are being followed in a proper way, then it is only natural that desired, or successful, outcomes will follow. A formal audit board should be staffed with their sole responsibility consisting of conducting audits and assessing the extent of proper use of processes in the service contracting process.

The DoD should implement the use of program management concepts to the services acquisition process. This would place a more rigid structure on how services are procured and the contracts subsequently administered. It would also ensure the involvement of PMs. Of the three major stakeholders, PMs made up the smallest portion or respondents. Previous research also notes the apparent lack of PMs involved in the process (Miller et al., 2012). The program management approach also forces the PM to be accountable to the MDA, who is briefed periodically throughout the life cycle. Program management concepts dictate the utilization of IPTs so that there is no lack of knowledge at any point in the acquisition process. The DoD would benefit from the application of program management concepts to service acquisition.

A final recommendation is that a standardized reporting process should be in place in order to track contractor performance related to both processes and outcomes. The COR should be intimately involved in this reporting process, and status needs to be regularly conveyed to the stakeholders. This report should include cost elements, as well as schedule and performance elements. However, it need not be limited to only those factors. If CORs are forced to report on adherence to a communication plan as well as customer satisfaction, the stakeholders could also track how effectively processes are being followed. Typically, customer satisfaction surveys deal with outcomes, but they could also refer to communications, planning, and administration. If a standardized reporting process were in place, it might be possible to capture metrics accurately in order to define and measure the success of a service contract.

References

Apte, A. and Rendon, R. 2007. *Managing the service supply chain in the Department of Defense: Implications for the program management infrastructure* (Technical Report, NPS-PM-07-126). Retrieved from Naval Postgraduate School, Acquisition Research Program website: http://www.acquisition-research.net.

Apte, U., Apte, A., and Rendon, R. 2010. *Services supply chain in the Department of Defense: A comparison and analysis of management practices in Army, Navy, and Air Force* (Technical Report, NPS-CM-10-161). Retrieved from Naval Postgraduate School, Acquisition Research Program website: http://www.acquisitionresearch.net.

Berteau, D., Ben-Ari, G., Sanders, G., Morrow, D., and Ellman, J. 2012. *U.S. Department of Defense Services Contract Spending and the Supporting Industrial Base, 2000–2011*. Washington, DC: Center for Strategic and International Studies.

Brown, B. 2010. *Introduction to Defense Acquisition Management (IDAM)*. Fort Belvoir, VA: Defense Acquisition University.

Cleland, D. I. 1986. Project stakeholder management. *Project Management Journal*, 17(4), 36–44.

Defense Acquisition University (DAU). 2012. *Defense Acquisition Guidebook (DAG)*. Retrieved from https://dag.dau.mil/.

Department of Defense (DoD). 2011. *Guidebook for the Acquisition of Services*. Retrieved from https://acc.dau.mil/adl/en-US/472568/file/60393/Services%20Acquisition%20Guidebook%20v5%207_20_2011.pdf.

Department of Defense Inspector General (DoDIG). 2009. *Summary of DoD Office of Inspector General Audits of Acquisition and Contract Administration* (DoDIG Report No. D-2009-071). Washington, DC: Author.

Eisenhardt, K. M. 1989. Agency theory: An assessment and review. *The Academy of Management Review*, 14(1), 57–74.

Esaias, F. (2011, January 25). *Contract opportunities at SPAWAR* [Presentation slides]. Retrieved from http://www.afcea.org/smallbusiness/files/West2011/SPAWAR_FayeEsaias_AFCEAWest2011.pdf.

Federal Acquisition Regulation (FAR). 48 C.F.R., ch. 1, 2012.

Freeman, R. E. 1984. *Strategic Management*. Boston, MA: Pitman.

Gansler, J. S. 2011. *Democracy's Arsenal: Creating A Twenty-First Century Defense Industry*. Cambridge, MA: MIT Press.

Garrett, G. A. 2011. *U.S. Government Services Contracting: Tools, Techniques, and Best Practices*. Riverwoods, IL: CCH.

Garrett, G. A. and Rendon, R. G. 2005. *Contract Management Organizational Assessment Tools*. McLean, VA: National Contract Management Association.

GlobalSecurity.org. (n.d.). Naval Supply Systems Command (NAVSUP). Retrieved from http://www.globalsecurity.org/military/agency/navy/navsup.htm.

Government Accountability Office (GAO). 2005, March. *Contract Management: Opportunities to Improve Surveillance on Department of Defense Services Contracts* (GAO-05-274). Washington, DC: Author.

Government Accountability Office (GAO). 2007a. *Defense Acquisitions: DoD Needs to Exert Management and Oversight to Better Control Acquisition of Services* (GAO-07-359T). Washington, DC: Author.

Government Accountability Office (GAO). 2007b. *Defense Acquisitions: Improved Management and Oversight Needed to Better Control DoD's Acquisition of Services* (GAO-07-832T). Washington, DC: Author.

Government Accountability Office (GAO). 2007c. *High Risk Series: An Update* (GAO-07-310). Washington, DC: Author.

Government Accountability Office (GAO). 2009a. *Defense Acquisitions: Actions Needed to Ensure Value for Services Contracts* (GAO-09-643T). Washington, DC: Author.

Government Accountability Office (GAO). 2009b. *Department of Defense: Additional Actions and Data Are Needed to Effectively Manage and Oversee DoD's Acquisition Workforce* (GAO-09-342). Washington, DC: Author.

Government Accountability Office (GAO). 2011a. *Better Identification, Development, and Oversight Needed for Personnel Involved in Acquiring Services* (GAO-11-892). Washington, DC: Author.

Government Accountability Office (GAO). 2011b. *High Risk Series: An Update* (GAO-11-34T). Washington, DC: Author.

Government Accountability Office (GAO). 2012a. *Competition for Services and Recent Initiatives to Increase Competitive Procurements* (GAO-12-384). Washington, DC: Author.

Government Accountability Office (GAO). 2012b. *Defense Acquisition Workforce: Improved Processes, Guidance, and Planning Needed to Enhance Use of Workforce Funds* (GAO-12-747R). Washington, DC: Author.

Government Accountability Office (GAO). 2012c. *Defense Contracting: Improved Policies and Tools Could Help Increase Competition on DoD's National Security Exception Procurements* (GAO-12-263). Washington, DC: Author.

Hill, C. W. and Jones, T. M. 1992. Stakeholder-agency theory. *Journal of Management Studies*, 29(2), 131–154.

Miller, F., Newton, J., and D'Amato, S. 2012. *Defining and Measuring Success of Service Contracts.* Master's thesis. Retrieved from Naval Postgraduate School, Acquisition Research Program website: http://www.acquisitionresearch.net.

Naval Sea Systems Command (NAVSEA). (n.d.). About NAVSEA. Retrieved from http://www.navsea.navy.mil/AboutNAVSEA.aspx.

Naval Sea Systems Command (NAVSEA). 2012. Challenges in contracting. Retrieved from http://www.navsea.navy.mil/OnWatch/aquisitioncontracts3.html.

Naval Supply Systems Command (NAVSUP). (n.d.). NAVSUP team. Retrieved from http://www.navsup.navy.mil/navsup/ourteam/navsup.

Office of the Under Secretary of Defense for Acquisition, Technology, and Logistics (OUSD[AT&L]). 2003. *The Defense Acquisition System* (DoD Directive 5000.1). Washington, DC: Author.

Office of the Under Secretary of Defense for Acquisition, Technology, and Logistics (OUSD[AT&L]). 2008. *Operation of the Defense Acquisition System* (DoD Instruction 5000.2). Washington, DC: Author.

Office of the Under Secretary of Defense for Acquisition, Technology, and Logistics (OUSD[AT&L]). 2010. *Better Buying Power: Guidance for Obtaining Greater Efficiency and Productivity in Defense Spending.* Washington, DC: Author.

Parmar, B. L., Freeman, R. E., Harrison, J. S., Wicks, A. C., Purnell, L., and De Colle, S. 2010. Stakeholder theory: The state of the art. *The Academy of Management Annals*, 4, 403–445.

Project Management Institute (PMI). 2008. *A Guide to the Project Management Body of Knowledge.* Newtown Square, PA: Author.

Rendon, R. G. 2009. *Contract Management Process Maturity: Empirical Analysis of Organizational Assessments* (Technical Report, NPS-CM-09-124). Retrieved from Naval Postgraduate School, Acquisition Research Program website: http://www.acquisitionresearch.net.

Rendon, R. G. 2010. *Assessment of Army Contracting Command's Contract Management Processes* (Technical Report, NPS-CM-10-154). Retrieved from Naval Postgraduate School, Acquisition Research Program website: http://www.acquisitionresearch.net.

Rendon, R. G. 2011. *Assessment of Army Contracting Command's Contract Management Processes* (Technical Report, NPS-CM-11-019). Retrieved from Naval Postgraduate School, Acquisition Research Program website: http://www.acquisitionresearch.net.

Rendon, R. G. 2012. Defense procurement: An empirical analysis of critical success factors. In G. L. Albano, K. F. Snider, and K. V. Thai (Eds.), *Charting a Course in Public Procurement Innovation and Knowledge Sharing*, pp. 174–208. Boca Raton, FL: PrAcademics Press.

Rendon, R. G. and Snider, K. F. 2008. *Management of Defense Acquisition Projects.* Reston, VA: American Institute of Aeronautics and Astronautics.

Space and Naval Warfare Systems Command (SPAWAR). (n.d.). About SPAWAR. Retrieved from http://www.public.navy.mil/spawar/Pages/default.aspx.

Wood, R. L. 2010. How well are PMs doing? Industry view of defense program manager counterparts. *Defense Acquisition Review Journal*, 206–218.

chapter twenty

Measurement of personnel productivity
During government furlough programs

Adedeji B. Badiru

Contents

20.1   Introduction

This chapter is a case study, based on Badiru (2014), of the adverse management approach used in the 2013 U.S. Federal government furlough program. No publication of research studies of the furlough program is currently available in the open literature. This chapter is intended to prompt potentially productive research investigations of the impact of personnel furloughs, particularly on defense acquisition programs. Pertinent analytical techniques are provided in this chapter to illustrate potential pathways for further research studies of furloughs. Defense acquisition programs are time-sensitive and systems-oriented. What appears to be a minor delay in one unit of an acquisition life cycle can lead to long-term encumbrances within the entire defense system resulting in enormous cost escalation. The analytical methodologies suggested in this chapter are useful for assessing how furloughs adversely impact organizational productivity. It is hoped that this chapter will spur research so that future furloughs can be better conceived, planned, executed, and managed, or avoided altogether. One takeaway from this chapter is that 100% of the work during a furlough cannot be done with fewer resources at the original work rate. In this chapter, we explore analytical methodologies for studying the impact of furloughs on employee work rate and organizational productivity.

High-dollar acquisition programs that suffer productivity impediments can lead to enormous cost escalations. A case example (Carey, 2012) is the 2012 revelation by the U.S. Government Accountability Office (GAO) that the U.S. Air Force would spend $9.7 billion over 20 years to upgrade the capabilities of F-22A Raptor as a result of the failure to anticipate the plane's long-term need for technology modernization. In high-cost and time-sensitive programs such as the F-22A Raptor, any additional slowdown and work rate decline in the acquisition process can result in adverse impacts on the overall readiness of the nation. Workforce work rate has a direct impact on overall organizational productivity. The very premise of the defense acquisition program is to ensure timely acquisition and deployment of critical technology to aid the warfighter. The purpose of this chapter is to provide thought-provoking research methodologies to analyze the management of furlough programs with respect to work productivity. Furlough-induced work slowdown in one segment of a defense organization can lead to overall work rate decline with a resulting decline in overall productivity and cost escalation. A furlough program takes both leadership and employees away from productive work because planning spans multiple weeks. Although the hypothesis of the chapter is anecdotal, it does present the basis for further empirical studies. It is intended to provoke more data-driven research on employee work rate analysis. Because 100% of the work cannot be done by fewer human resources working at the normal work rate during a furlough, a research study is needed to guide future decisions. The recommendations offered include the following:

- Ensuring that furloughs, if needed at all, be implemented in smaller manageable timelines. While changes are essential for organizational improvement, they should be implemented in incremental cost-cutting measures rather than one big furlough period.
- Focusing on gradual incremental fiscal adjustments rather than one big drastic implementation of budget cuts.
- Providing early and clear communication to clarify requirements and impacts in order to allay the fear of those affected and to minimize ambiguity.
- Ensuring the personal needs and welfare of employees are given priority in the execution of furlough programs.

20.2   Furlough programs impacts on acquisition systems

There are many possible sources of program delays including those caused by a lack of cohesive budget agreement and political discord, which result in the need for furloughs. Three leading sources of delays in acquisition programs are

- Technological limitations such as a sluggish maturation of new technology
- Externally-imposed limitations such as the prevailing global economic developments
- Self-induced procedural limitations such as political discord or procedural inefficiency

The ongoing federal budget sequestration is wreaking havoc on organizational productivity throughout the Department of Defense (DoD). A July 31, 2013, news headline (http://www.daytondailynews.com/news/news/new-af-center-to-lose-13m-work-hours-to-sequester/nY9Q5/) read, "New Air Force Center to lose 1.3 million hours to sequester." The news went on to affirm how the mid-year sequestration budget cuts adversely affected the Air Force Life Cycle Management Center (LCMC). A productivity loss of 1.3 million hours, depending on the base wage rate used, can translate to as much as $70 million. If we consider the 600,000-plus employees across DoD during a furlough program, it is obvious that 100% of the work cannot be fulfilled by the furlough-depleted workforce working at the original work rate. The economic impact of the reduction of work output is a good topic for future research. For a sequestration program that is supposed to be saving money, losing that much money is a move in the wrong direction. In addition to the serious financial impacts of furlough programs on family take-home disposable incomes, there are grave consequences on social well-being and community economic performance. Those personal impacts, coupled with organizational loss of productivity, make the net cost savings of furlough programs negligible.

20.3   Logistics and acquisition disaggregation

Stone (2013) emphasized how the civilian furlough period caused delays in moving and maintaining equipment at a time that the military cannot afford any operational disruption. The wartime drawdown is just one piece of the jigsaw complexity of military logistics and acquisitions. A poorly executed furlough program complicates an already complex undertaking. People and equipment have to be moved under a tight schedule with a shrinking base budget. The civilian workforce provides a key linkage between everything that has to be done. Reducing the availability of the workforce through a furlough program at a critical time impedes the overall goal of the DoD. Once again, to emphasize, 100% of the work cannot be done by a reduced workforce working at the original work rate.

20.4   Furloughs and loss of productivity

Employee furloughs, as a mechanism to save the Federal budget, do have deleterious effects on employee morale, functional coordination, and employee work rates. When morale is low, all other factors of productivity are adversely impacted. Thus, furloughs have several unintended consequences. In essence, employee furloughs do not offer much in the way of long-term benefits. Work backlogs that are caused by furloughs often take months to complete. To protect personnel-related data, hypothetical values are used in the computational examples. Organizations wishing to implement the computational methodologies presented in this chapter will use their own unit-specific data values. One anticipated benefit is that it will open up avenues for discussions and more rational decisions in advance for any future furlough programs. Hopefully, any future furlough programs can be better conceived, planned, executed, and managed, or avoided altogether. In the author's own furlough experience, the DoD furlough program in 2013 created protracted planning, execution ambiguity, disjointed implementations, uncertainty of expectations, inconsistent guidance, and disruption of workflow processes thereby causing adverse impacts on overall organizational productivity and impeding national defense preparedness.

For the specific case at Wright Patterson Air Force Base, the furlough period began in the week of July 8, 2013, for about 10,000 civilians. In the initial DoD implementation, it was expected that civilians who were affected by the furlough must endure a scheduled unpaid day off each week for a total of 11 furlough days Although this was later cut down to 6 furlough days, the productivity damage had already been done. Considering that the same amount of work must be accomplished, furloughed employees must prioritize tasks to determine what is done and what is compromised. In the absence of a standardized process, employees may inadvertently marginalize high-value tasks. Even flexibility for an employee to choose which day of the week to take a furlough has some unanticipated adverse impacts. In a normal workweek devoid of furlough or sequestration distractions,

Monday is typically the busiest (but not necessarily the most productive) day of the week. Tuesday is seen as the most productive day while Friday is the least busy day and, potentially, least productive. This phenomenon is a human cultural reaction to the progression of a work week that has been confirmed by several labor research studies (Dawkins and Tulsi, 1990; Pettengill, 1993, 2003; Weiss, 1996; Hill, 2000; Campolieti and Hyatt, 2006; Chandra, 2006; Taylor, 2006; Bryson and Forth, 2007a,b; Golden, 2011). One adverse impact of variable furlough days is the difficulty in synchronizing work across functional areas, which leads to overall diminished work output similar to what Figure 20.1 illustrates based on a study by Bryon and Forth (2007a). While the data in the study do not represent our DoD acquisition workforce of interest, the productivity ramp-up and ramp-down process is evident in every workforce, and the topic is fertile for future research.

Going by normal human nature in 80% of the population, according to Pareto distribution, some less-motivated workers, if given the option of picking a furlough day, will pick Monday. Monday, being the busiest day, is the day to opt out of work. The research literature has confirmed that Monday experiences the highest level of sick-day call-ins (Kronos, Inc., 2004). Friday, a normally slow day, is perceived as a day to come to work, knowing that typically not much work stress will occur on that day. These two bipolar behavioral observations will, thus, have greater adverse impact on overall productivity than what a normal furlough day might be expected to produce. The normally busy Monday suffers in two ways

- Reduced workforce due to furlough and
- Critical work pushed further down the week due to elective furlough day selection.

The situation can be compounded by some people taking Friday off one week, then taking Monday off the next week. Due to several subtle factors such as the above, getting

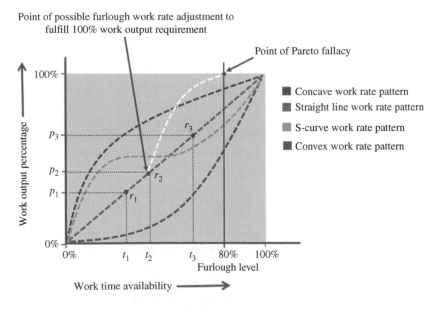

Figure 20.1 Workday-based ramp-up and ramp-down of worker productivity. (Adapted from Bryson, A. and Forth, J. 2007a. Are There Day-of-the-Week Productivity Effects? Manpower Human Resources Lab Report, Centre for Economic Performance, London School of Economics, July 2007.)

two full workday equivalents out of Monday and Friday proves fallacious in actual practice. The following paraphrased actual statement typifies the type of negative work impacts that the uncoordinated furlough program and sequestration caused. This statement is in response to a query following a critical task that went uncompleted and untracked for weeks.

> I apologize for the delay. While waiting for a response I put the request in a follow up folder; since I am part-time, and we have taken on the responsibilities of laid-off employees, not to mention the day of work we lose due to the furlough, it has taken me this long to get a moment to follow up on the task. Please know that I do not intend to make excuses but merely to explain the circumstances ...

Figure 20.2, based on a 2004 survey conducted by Harris Interactive for Kronos, Inc., illustrates that 61% of respondents report that "nothing gets done on their workload when they are absent from work." The population surveyed was a general office workforce. While this is not a DoD workforce, there are similarities in the office work environment of both populations.

Where human work is concerned, the psychology of work must be taken into account when deciding on new work practices either as a response to budgetary pressures or in pursuit of process improvement goals. The literature is replete with relevant research studies in this regard (Baltes et al., 1999; Hamermesh, 1999; Bailey and Kurland, 2002; Askenazy, 2004; Berg et al., 2004; Bertschek and Kaiser, 2004; Böheim and Taylor, 2004; Heisz and LaRochelle-Côté, 2006; Altman and Golden, 2007; Kelliher and Anderson, 2010). Unfortunately, technical workforce teams, such as those in defense acquisition programs, are rarely studied with respect to the best way to manage work schedules. Therein lies a flaw in the across-the-board implementation of the present furlough program.

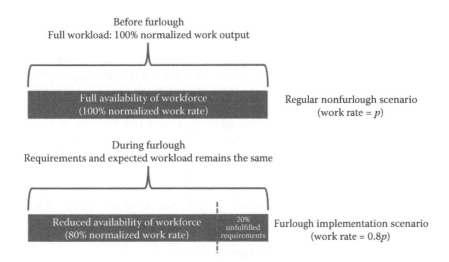

Figure 20.2 Adverse impacts of furloughs on productivity. (Adapted from Kronos, Inc. 2004. *Working in America: Absent Workforce Survey Highlights.* Kronos, Chelmsford, MA, http://www. workforceinstitute.org/wp-content/themes/revolution/docs/Working-in-Amer-Survey.pdf.)

20.5 Link between productivity and operational cost

The U. S. Government is using the SAVE (Securing Americans Value and Efficiency) program to solicit ideas from all federal employees to help identify areas where the nation can "cut wasteful spending." A review of the SAVE award website at www.whitehouse.gov shows that 89,000 ideas have been submitted over the past 4 years since the program started in 2009. It should be noted, however, that any cut of "wasteful spending" should be coupled with a mitigation of the subtle avenues of eroding productivity. Improving functional productivity can translate to lower operating cost (Blake, 2011) as stated below

> Industrial engineers make systems function better together with less waste, better quality, and fewer resources.

As it is with every organization, a major goal of the U.S. Air Force is to eliminate waste in consonance with the federal goal of cutting wasteful spending. In spite of its goal, some cost-cutting programs instead have the unintended consequence of reducing productivity, which increases operating cost. Consequently, the savings from cutting wasteful spending are nullified by the higher cost of lower productivity. An uncoordinated implementation of furlough programs is one glaring example of "robbing Peter to pay Paul." Efficiency, effectiveness, productivity, and cost reduction must be integrated analytically in order to get the desired composite organizational benefits. Organizational performance is defined in terms of several organization-specific metrics, which include efficiency, effectiveness, and productivity. The existing techniques for improving efficiency, effectiveness, and productivity (Badiru and Thomas, 2013, and all the references therein) are suitable for analyzing the impacts of furloughs. Efficiency refers to the extent to which a resource (time, money, effort, etc.) is properly utilized to achieve an expected outcome. The goal, thus, is to minimize resource expenditure, reduce waste, eliminate unnecessary effort, and maximize output. The ideal (i.e., the perfect case) is to have 100% efficiency. This is rarely possible in practice. Usually expressed as a percentage, *efficiency* (*e*) is computed as output over input.

$$e = \frac{Output}{Input} = \frac{Result}{Effort} \tag{20.1}$$

Equation 20.1 is also adapted for measuring productivity (Badiru and Thomas, 2013).

Effectiveness is primarily concerned with achieving the specific objectives, which constitute the broad goals of an organization. To model effectiveness quantitatively, we can consider the fact that an "objective" is essentially an "output" related to the numerator of the efficiency equation (20.1). Thus, we can assess the extent to which the various objectives of an organization are met with respect to the available resources. Although efficiency and effectiveness often go hand-in-hand, they are, indeed, different and distinct. For example, one can forego efficiency for the sake of getting a particular objective accomplished. Consider the statement, "If we can get it done, money is no object." The military, by virtue of being mission-driven, often operates this way. If, for instance, our goal is to go from point A to point B to hit a target and we do hit the target, no matter what it takes, then we are effective. We may not be efficient based on the amount of resources expended to hit the target. A cost-based measure of effectiveness is defined as

$$ef = \frac{S_0}{C_0}, \quad C_0 > 0 \tag{20.2}$$

where

    ef = measure of effectiveness on interval (0, 1)
    S_0 = level of satisfaction of the objective (rated on a scale of 0–1)
    C_0 = cost of achieving the objective (expressed in pertinent cost basis: money, time, measurable resource, etc.)

If an objective is fully achieved, its satisfaction rating will be 1. If not achieved at all, it will be zero. Thus, having the cost in the denominator gives a measure of achieving the objective per unit cost. If the effectiveness measures of achieving several objectives are to be compared, then the denominator (i.e., cost) will need to be normalized to a uniform scale. The overall system effectiveness can be computed as a summation as follows:

$$ef_c = \sum_{i=1}^{n} \frac{S_0}{C_0} \tag{20.3}$$

where

    ef_c = composite effectiveness measure
    n = number of objectives in the effectiveness window

Depending on the units used, the effectiveness measure may be very small with respect to the magnitude of the cost denominator. This may be handled by converting the measure to a scale of 0–100. Thus, the highest comparative effectiveness per unit cost will be 100 while the lowest will be 0. The above quantitative measure of effectiveness makes most sense when comparing alternatives for achieving a specific objective. If the effectiveness of achieving an objective is desired in non-comparative absolute terms, it would be necessary to determine the range of costs, minimum to maximum, applicable for achieving the objective. Then, we can assess how well we satisfy the objective with the expenditure of the maximum cost versus the expenditure of the minimum cost. By analogy, "killing two birds with one stone" is efficient. By comparison, the question of effectiveness is whether we kill a bird with one stone or kill the same bird with two stones, if the primary goal is to kill the bird nonetheless. In technical terms, systems that are designed with parallel redundancy can be effective, but not necessarily efficient. In such cases, the goal is to be effective (get the job done) rather than to be efficient. Productivity is a measure of throughput per unit time. Typical productivity formulas include the following:

$$P = \frac{Q}{q} \tag{20.4}$$

$$P = \frac{Q}{q}(u) \tag{20.5}$$

where

    P = productivity
    Q = output quantity
    q = input quantity
    u = utilization percentage

Notice that Q/q also represents efficiency (i.e., output/input) as defined earlier. Applying the utilization percentage to this ratio modifies the ratio to provide actual productivity yield. The acquisition workforce is composed primarily of knowledge workers whose productivity must be measured in alternate terms, perhaps through work rate analysis. Rifkin (2011) presents the following productivity equation suitable for implementation for the acquisition environment:

$$Product \text{ (i.e., output)} = Productivity \text{ (objects per person} - \text{time)}$$
$$\times Effort \text{ (person} - \text{time)} \tag{20.6}$$

where *Effort = Duration × Number of People*.

While changes are essential for organizational improvement, they should be implemented in smaller manageable chunks, possibly incrementally, with respect to cost-cutting measures rather than one big furlough period. Organizational focus should be on gradual incremental improvement rather than one-fell-swoop drastic implementation of budget cuts. These two points need to be addressed in further detail via further research studies that are based on live data collection and analysis. It is hoped that this chapter will spur research by pointing out some basic examples of productivity measurement through analytical computations.

20.6   Work rate computations

Work rate and work time availability are essential components of estimating the cost of specific tasks. Given a certain amount of work that must be done at a given work rate, the required time can be computed. Once the required time is known, the cost of the task can be computed on the basis of a specified cost per unit time. Work rate analysis is important for resource substitution decisions. The analysis can help identify where and when the same amount of work can be done with the same level of quality and within a reasonable time span by a less expensive resource. Potential future research may include learning curve analysis that may be used to predict the expected work rate. Although not generally applicable across the board for government work, learning curves are still useful for cases where work output accountability is tracked. The general relationship among work, work rate, and time is given by

$$Work \ done = (Work \ rate)(Time)$$
$$w = rt \tag{20.7}$$

where
   w = the amount of actual work done expressed in appropriate units. Example of work units are number of contract reviews completed, lines of computer code typed, gallons of oil spill cleaned, units of a product produced, and surface area painted
   r = the rate at which the work is accomplished (i.e., work accomplished per unit time)
   t = the total time required to perform the work excluding any embedded idle times

For simplification, work is defined as a physical measure of accomplishment with a uniform density. For example, cleaning 1 gallon of oil spill may be as desirable as cleaning any other gallon of oil spill within the same work environment. The production of one unit of a product is identical to the production of any other unit of the product. If uniform work density cannot be assumed for the particular work being analyzed, weighting factors must be applied to the elements contained in the relationship. Uniformity can be enhanced if the scope of the analysis is limited to discrete work elements of similar design. The larger the scope of the analysis, the more the variability from one work unit to another, and the less uniform the overall work measurement will be. For example, in a project involving the construction of 50 miles of surface road, the work analysis may be done in increments of 10 miles at a time rather than the total 50 miles. If the total amount of work to be analyzed is defined as one whole unit, then the following relationship can be developed for the case of a single resource performing the work, with the parameters as

Work rate: r
Time: t
Work done: 100% (1.0)

The work rate, r, is the amount of work accomplished per unit time. For a single resource to perform the whole unit (100%) of the work, we must have $rt = 1.0$.

For example, if an acquisition technician is to complete one work unit in 30 min, he or she must work at the rate of 1/30 of the work content per unit time. If the work rate is too low, then only a fraction of the required work will be performed. The information about the proportion of work completed may be useful for productivity measurement purposes. In the case of multiple technicians performing the work simultaneously, the work relationship is as presented in Table 20.1.

Even though the multiple technicians may work at different rates, the sum of the work they all performed must equal the required whole unit. In general, for multiple resources, we have the following relationship:

$$\sum_{i=1}^{n} r_i t_i = 1.0 \tag{20.8}$$

where
  n = number of different resource types
  r_i = work rate of resource type i
  t_i = work time of resource type i

For partial completion of work, the relationship is

$$\sum_{i=1}^{n} r_i t_i = p \tag{20.9}$$

where p is the proportion of the required work actually completed. In any furlough program, the expectation of 100% work completion does not match reality. Under a furlough program, only a fraction of the expected work will get done.

Table 20.1 Work rate tabulation for multiple technicians

Technician (i)	Work rate (r_i)	Time (t_i)	Work done (w)
1	r_1	t_1	$(r_1)(t_1)$
2	r_2	t_2	$(r_2)(t_2)$
...	...	...	...
n	r_n	t_n	$(r_n)(t_n)$
		Total	1.0

20.7   Employee work rate examples

Under a furlough program, there can be no expectation that 100% of the work can be accomplished with a 20% reduction of human resources operating at the pre-furlough work rate. Suppose technician A, working alone, can complete a task in 50 min. After he has been working on the task for 10 min, technician B is brought in to work with technician A to complete the job. Both technicians, working together as a team, finish the remaining work in 15 min. We are interested in finding the work rate for technician B, if the amount of work to be done is 1.0 whole unit (i.e., 100% of the job). The work rate of technician A is 1/50. The amount of work completed by technician A in the 10 min, working alone, is $(1/50)(10) = 1/5$ of the required total work. Therefore, the remaining amount of work to be done is 4/5 of the required total work. That is,

$$\frac{15}{50} + 15(r_2) = \frac{4}{5} \tag{20.10}$$

which yields $r_2 = 1/30$. Thus, the work rate for technician B is 1/30. That means technician B, working alone, can perform the same job in 30 min. A tabulated summary of this example is shown in Table 20.2.

In this example, it is assumed that both technicians produce an identical quality of work. If quality levels are not identical, we must consider the potentials for quality-time trade-offs in performing the required work. The relative costs of the different technician skills needed to perform the required work may be incorporated into the analysis as shown in Table 20.3.

By using the above relationship for work rate and cost, the work crew can be analyzed to determine the best strategy for accomplishing the required work, within the required time and within a specified budget, in a climate of a furlough program. For another simple example of possible acquisitions scenarios, consider a case where an acquisition IT technician can install new IT software at three work stations every 4 hours. At this rate, it is desired to compute how long it would take the technician to install the same software at five work stations. The proportion that "three stations" is to 4 hours is equivalent to the proportion that "five stations" is to x h, where x represents the number of hours the

Table 20.2 Work rate tabulation for technicians A and B

Technician (i)	Work rate (r_i)	Time (t_i)	Work done (w)
A	1/50	15	15/50
B	r_2	15	$15(r_2)$
		Total	4/5

Table 20.3 Incorporation of wage cost into work rate analysis

Technician (i)	Work rate (r_i)	Time (t_i)	Work done (w)	Pay rate (p_i)	Wage (P_i)
A	r_1	t_1	$(r_1)(t_1)$	p_1	P_1
B	r_2	t_2	$(r_2)(t_2)$	p_2	P_2
...	...	...	...	...	...
N	r_n	t_n	$(r_n)(t_n)$	p_n	P_n
		Total	1.0		Budget

technician would take to install software in the five stations. This gives the following work-and-time ratio relationship:

$$\frac{3\,\text{work stations}}{4\,\text{h}} = \frac{5\,\text{work stations}}{x\,\text{h}} \tag{20.11}$$

which yields $x = 6$ h, 40 min. Now consider a situation where the technician's competence with the software installation degrades over time for whatever reason, possibly due to furlough interruptions. We will see that the time requirements for the IT software installation will vary depending on the current competency level and the availability of the technician. Consider another example where an acquisition analyst can do contract checks at the rate of 120 contract line items per minute. A supervisor can inspect the checkmarks at the rate of three per second. How many supervisors are needed to keep up with 18 acquisition analysts? At the given work rate, one analyst can complete the task at the rate of two per second (i.e., 120 checkmarks every 60 s). Therefore, 18 analysts would complete 36 checkmarks per second. Now let x be the number of supervisors needed to keep up with the 18 analysts. Since one supervisor completes three inspections per second, x supervisors would inspect $3x$ checkmarks per second. That is, $3x = 36$, which yields $x = 12$ supervisors. Overall work slowdown will occur if, due to furloughs, the supervisors needed are not available to keep up with the workload. By the author's own estimation in his own furlough experience, as much as 25% of required work process checkmarks may be missed.

One more illustrative example is, suppose the work rate of team member 1 is such that she can perform a certain task in 30 days. It is desired to add team member 2 to the task so that the completion time of the task can be reduced. The work rate of team member 2 is such that she can perform the same task alone in 22 days. If team member 1 has already worked 12 days on the task before team member 2 comes in, we want to find the completion time of the task, if team member 1 starts the task at time 0. The amount of work to be done is 1.0 whole unit (i.e., the full task). The work rate of Team Member 1 is 1/30 of the task per unit time. The work rate of team member 2 is 1/22 of the task per unit time. The amount of work completed by team member 1 in the 12 days she worked alone is $(1/30)$ $(12) = 2/5$ (or 40%) of the required work. Therefore, the remaining work to be done is 3/5 (or 60%) of the full task. If we let T be the time for which both members work together, then we will have the following work-and-time equation:

$$\frac{T}{30} + \frac{T}{22} = \frac{3}{5} \tag{20.12}$$

which yields $T = 7.62$ days. Thus, the completion time of the task is $(12 + T) = 19.62$ days from time zero. It is assumed that both members produce identical quality of work and that the respective work rates remain consistent. The respective costs of the different resource types may be incorporated into the work rate analysis to determine where real cost savings can be achieved.

20.8 Furlough-induced work rate and productivity

The key benefit of doing an analytical work rate analysis is that the disconnection between employee work and the prevailing work load can be brought to the forefront. As a case example, the 2013 implementation of furlough at Wright Patterson Air Force Base required

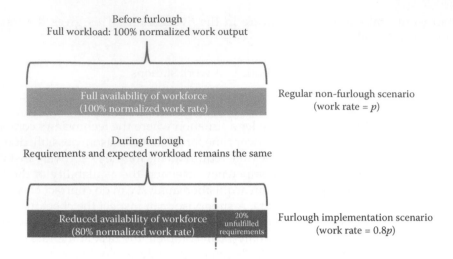

Figure 20.3 Pareto analysis of furlough work rate versus requirements workload.

each eligible employee to go on furlough 1 day each workweek for 11 weeks, which was later reduced to 6 weeks. For each week, this represented a 20% loss of availability to work. Meanwhile, the workload was not adjusted downward to account for the 20% loss of employee time availability. This resulted in an effort to do same (even more, in some cases) with less employee time. A simple Pareto plot of this work scenario quickly reveals a serious disconnect. To balance the equation, either the work rate of employees will have to increase or the expected work output (i.e., requirements) will need to be reduced. Figure 20.3 shows a pictorial representation of this disconnection. Figure 20.4 presents examples of furlough work rate adjustment curves. If employee time availability is cut to 80% (i.e., 1 workday

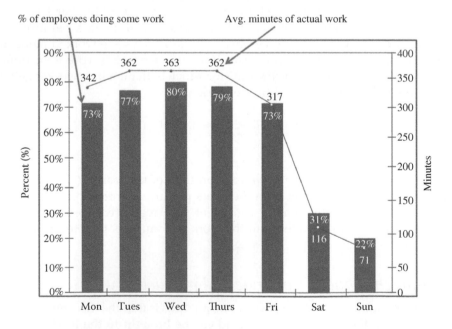

Figure 20.4 Furlough work rate adjustment curves.

furlough per workweek) as in the DoD furlough implementation, employee work rates must be adjusted upward if the expected workload is still to be accomplished. This is represented by the white curve starting at the point r_2 and ending at the intersection of 100% workload and 80% work time availability. The r_2 point was selected because it offers a mid-range point on the curve. In other words, we still accomplish most (if not all) of the required work using only 80% work time availability. However, this is at the expense of a higher mid-stream work rate of the employee. Something will have to be compromised if we expect the same work output from the same standard work rate under a reduction in work time availability. It should be pointed out, though, that the presumption of 100% completion of government work is not realistic. The concept may work analytically for countable units of production, but not directly for office type of work outputs. The concept, nonetheless, does provide guidance for rational thought about managing office work output under the condition of a furlough program.

We can now apply the above analysis to the previous example of teamwork rate analysis. If work rates remain the same, we must either reduce the work content or increase the duration (number of days) over which the task is accomplished. If the task duration is to be kept the same at $D = 19.62$ days, then work rates must be adjusted. Let us assume the following notations:

x_1 = normal work rate of Team Member 1
$x_{1(f)}$ = furlough work rate of Team Member 1
x_2 = normal work rate of Team Member 2
$x_{2(f)}$ = furlough work rate of Team Member 2
D = fixed expected task duration in days = 19.62 days
T = number of days remaining to due date = 19.62 − 12 days = 7.62 days.

Normally, team member 1 can complete the task in 30 days at a work rate of 1/30; so, $x_1 = 1/30$ and from the data given previously, $x_2 = 1/22$. Team member 1 works for 12 days before handing over to team member 2. Assuming that the work rate of team member 2 is the one to be adjusted while keeping $r_{1(normal)}$ constant, we see that team member 1 completes 1/30 work unit per day times 12 days, which yields 2/5 of the work content completed by team member 1 working alone. This leaves 3/5 of the work to be completed by member 2. This gives us the following relationship equation

$$T(x_1) + T(x_{2(f)}) = \text{Work content remaining to be done} \tag{20.13}$$

That is

$$7.62(1/30) + 7.62(x_{2(f)}) = 3/5 \tag{20.14}$$

However, member 1 hands over work to member 2 due to going on furlough. Therefore, the above equation reduces to member 2 working alone to complete the remaining 60% portion of the task in the 7.62 days before the due date. That is

$$7.62(x_{2(f)}) = 3/5 \tag{20.15}$$

which yields $x_{2(f)} = 0.0787$ work content per day. This is considerably higher than the normal work rate of 1/22 (i.e., 0.045) for member 2. In fact, it is 173% of the normal work rate for member 2, which is not practical to accomplish.

20.9   *Better management of furlough programs*

The DoD is made of teams of military personnel, government civilians, and contractors, who all are expected to work together seamlessly. Any furlough program that targets only one segment of the collaborative teams will create long-lasting disruptions that will nullify the intended benefits of defense teams. While one group is on furlough, the non-furlough groups cannot work at the best level of their potential. A prior analytical view of military–civilian work rate integration can help determine a better way to manage or avoid furloughs. Based on the analytical template presented above, it is recommended that future furlough programs, if there must be any, be managed with a consideration of the systems impact of employee absences. Systems engineering tools such as the V-model (DAU [Defense Acquisition University], 2013) and DEJI-model (Badiru, 2012) can be explored during the initial stages of furlough deliberations to determine how decision factors intermingled with respect to considerations for people, technology, and work process. Figure 20.5 illustrates some of the factors of consideration in applying the DEJI-model. Figure 20.6 presents a flowchart of performance sustainment, leading to possible performance optimization and resulting in performance enhancement. Once performance enhancement is achieved, it would be fed back as a sustainment goal for monitoring and coordinating functions. In such a flow process, the potential adverse impact of a furlough program can be identified earlier and in advance.

20.10   *Implementation strategy*

The simple process of communication, cooperation, and coordination can be used to get everyone on board for a furlough program. Projects are executed and accomplished through the collective efforts of *people*, *tools*, and *processes*. Communication is the glue that binds all these together. The author's own observations indicate that most project failures can be traced to poor communication at the beginning. Even in highly machined-controlled processes, the occasional human intervention can spell doom for a project if proper communication is not in effect. We often erroneously jump to the coordination

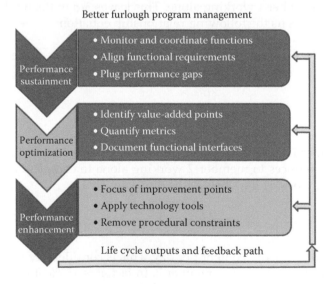

Figure 20.5 Furlough program design, evaluation, justification, and integration.

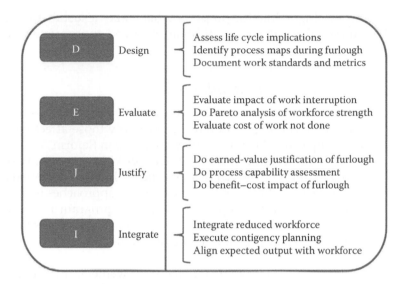

Figure 20.6 Life cycle feedback model for better furlough program management.

phase of a project believing that this is where project execution lies. But the fact is that a more fundamental foundation for project success lies well before the coordination phase. The author advocates building a structural project execution hierarchy, starting with communication, which facilitates cooperation, which paves the way for coordination, and which ends with the desired project success. That is, every project should build a project flow process as shown

Communication → Cooperation → Coordination → Program success

In the above process, investing in people communication is the easiest thing that an organization can do. Regardless of whatever technological tools, technical expertise, and enhanced processes are available in the project environment, basic human communication is required to get a project started right and moving forward efficiently and effectively. Communication highlights what must be done and when. It can also help to identify the resources (personnel, equipment, facilities, etc.) required for each effort. It points out important questions such as

- Does each project participant know what the objective is?
- Does each participant know his or her role in achieving the objective?
- What obstacles may prevent a participant from playing his or her role effectively?
- Does each person have "buy-in" into the project?

Communication can mitigate disparity between concept and practice because it explicitly solicits information about the critical aspects of a project in terms of who, what, why, when, where, and how of the project. By using this approach, we can avoid taking cooperation for granted. Cooperation must be explicitly pursued through clear communication of the project requirements. Cooperation works only when each cooperating individual inwardly believes in the project and makes a personal commitment to support the project. Ceremonial signing-off on a project is not a guarantee of cooperation.

Rather, subconscious signing-in into the project is what makes a sustainable cooperation. This can only be achieved through communication, extended appropriately and received properly.

20.11   Conclusion and recommendations

This chapter seeks to sensitize decision makers to the diversity of critical issues and factors involved in any government furlough programs, particularly those affecting the acquisition community. For example, the Weapon Systems Acquisition Reform Act (WSARA) and the Defense Acquisition Workforce Improvement Act (DAWIA) represent two of the several initiatives designed to improve the acquisition process. But to realize real and lasting improvements, which have been elusive so far, new practical approaches must be explored and sustained. But when the adverse impact of a furlough program is added on top of the existing challenges, it becomes even more difficult to achieve acquisition excellence or sustain any improvement already achieved. The recommendations are summarized as

1. While changes are essential for organizational improvement, they should be implemented in smaller manageable chunks with respect to implementing furloughs in incremental cost-cutting measures rather than one big furlough period.
2. Focus should be on gradual, incremental improvement rather than the one-fell-swoop drastic implementation of budget cuts.
3. Early and clear communication should be used to clarify the requirements and impacts in order to allay the fear of those affected.
4. The personal needs and welfare of employees should be given priority in the execution of furlough programs.
5. The questions of who, what, why, when, and how of the furlough program should be clearly outlined upfront to minimize ambiguity.

   A final takeaway from this chapter is that 100% of the work during a furlough cannot be done with fewer resources at the original work rate.

References

Altman, M. and Golden, L. 2007. The economics of flexible work scheduling: Theoretical advances and paradoxes. In B. Rubin (ed.), *Research in the Sociology of Work, Vol. 17: Workplace Temporalities*. JAI Press, Oxford, UK, pp. 313–342.

Askenazy, P. 2004. Shorter work time, hours flexibility and labor intensification. *Eastern Economic Journal*, 30(4), 603–614.

Badiru, A. B. 2012. Application of the DEJI model for aerospace product integration. *Journal of Aviation and Aerospace Perspectives (JAAP)*, 2(2), 20–34.

Badiru, A. B. 2014. Adverse impacts of furlough programs on employee work rate and organizational productivity. *Defense ARJ*, 21(2), 595–624.

Badiru, A. B. and Thomas, M. 2013. Quantification of the PICK chart for process improvement decisions. *Journal of Enterprise Transformation*, 3(1), 1–15.

Bailey, D. and Kurland, N. 2002. A review of telework research: Findings, new directions and lessons for the study of modern work. *Journal of Organizational Behavior*, 23(4), 383–400.

Baltes, B., Briggs, T., Wright, J., and Neuman, G. 1999. Flexible and compressed workweek schedules: A meta-analysis of their effects on work-related criteria. *Journal of Applied Psychology*, 84(4), 496–513.

Berg, P., Appelbaum, E., Bailey, T., and Kalleberg, A. 2004. Contesting time: Comparisons of employee control of working time. *Industrial and Labor Relations Review*, 57(3), 531–549.

Bertschek, I. and Kaiser, U. 2004. Productivity effects of organizational change: Microeconometric evidence. *Management Science*, 50(3), 394–404.

Blake, S. 2011. 5-Minute explanation of what an IE does. *Industrial Engineering*, 43(10), 12.

Böheim, R. and Taylor, M. 2004. Actual and preferred working hours. *British Journal of Industrial Relations*, 42(1), 149–166.

Bryson, A. and Forth, J. 2007a. Are There Day-of-the-Week Productivity Effects? Manpower Human Resources Lab Report, Centre for Economic Performance, London School of Economics, July 2007.

Bryson, A. and Forth, J. 2007b. Bad Timing: Are Workers More Productive on Certain Days of the Week? Manpower Human Resources Lab Report, Centre for Economic Performance, London School of Economics, August 2007.

Campolieti, M. and Hyatt, D. E. 2006. Further evidence on the "Monday Effect" in workers' compensation. *Industrial and Labor Relations Review*, 59(3), 438–450.

Carey, B. 2012. F-22 modernization plans progress despite hypoxia issue. AIN Online, June 1, 2012, http://www.ainonline.com/aviation-news/ain-defense-perspective/2012-06-01/f-22-modernization-plans-progress-despite-hypoxia-issue, accessed November 21, 2013.

Chandra, M. 2006. The day-of-the-week effect in conditional correlation. *Review of Quantitative Finance and Accounting*, 27(3), 297–310.

DAU (Defense Acquisition University). 2013. Introduction to Systems Engineering, https://learn.test.dau.mil/CourseWare/801488_10/media/SYS101%20Module%201.pdf, accessed November 23, 2013.

Dawkins, P. and Tulsi, N. 1990. The effects of the compressed workweek: A review of the evidence. *Australian Bulletin of Labour*, 16(2), 104–127.

Golden, L. 2011. The effects of working time on productivity and firm performance: A research synthesis paper. Employment Series Report No. 33, International Labour Office, Geneva.

Hamermesh, D. 1999. The timing of work over time. *Economic Journal*, 109(452), 37–66.

Heisz, A. and LaRochelle-Côté, S. 2006. Work hours instability. *Perspectives on Labour and Income*, 7(12), 17–20.

Hill, R. 2000. New labour force survey questions on working patterns. *Labour Market Trends*, January, pp. 39–50.

Kelliher, C. and Anderson, D. 2010. Doing more with less? Flexible working practices and the intensification of work. *Human Relations*, 63(1), 83–106.

Kronos, Inc. 2004. *Working in America: Absent Workforce Survey Highlights*. Kronos, Chelmsford, MA, http://www.workforceinstitute.org/wp-content/themes/revolution/docs/Working-in-Amer-Survey.pdf, accessed July 17, 2015.

Pettengill, G. N. 1993. An experimental study of the 'Blue Monday' hypothesis. *Journal of Socio-Economics*, 22(3), 241–257.

Pettengill, G. N. 2003. A survey of the Monday effect literature. *Quarterly Journal of Business and Economics*, 42(3–4), 3–27.

Rifkin, S. 2011. Raising questions: How long does it take, how much does it cost, and what will we have when we are done? What we do not know about enterprise transformation. *Journal of Enterprise Transformation*, 1(1), 34–47.

Stone, A. 2013, Under pressure: Tight budgets, aging systems add to logistics challenges. FederalTimes.com, December 9, 2013, http://www.federaltimes.com/article/20131209/ACQUISITION/312090009/Under-pressure-Tight-budgets-aging-systems-add-logistics-challenges.

Taylor, M. P. 2006. Tell me why I don't like Mondays: Investigating day of the week effects on job satisfaction and psychological well-being. *Journal of the Royal Statistical Society: Series A (Statistics in Society)*, 169(1), 127–142.

Weiss, Y. 1996. Synchronization of work schedules. *International Economic Review*, 37(1), 157–179.

chapter twenty-one

Measurement risk analysis

Adedeji B. Badiru

Contents

21.1 Introduction

This chapter addresses risk analysis and estimation techniques relevant for measurement requirements. The specific case example of petroleum reserve estimation presented in the chapter is based on Badiru et al. (2012). In many natural and engineering cases, actual measurements may not be possible or practical. This necessitates resorting to estimations based on sample data. Of course, any analytical estimation will encompass some level of risk. The method used in this chapter consists of a Bayesian technique for parameter estimation. A sampling procedure based on minimizing the mean squared error (MSE) of the posterior Bayesian estimator is presented using the β-density function to model the prior distribution. This Bayesian approach provides several typical representative distributions that are broad enough in scope to provide a satisfactory economic analysis. These distributions reflect general patterns of similar regions, include dry holes as well as several representative class sizes of discoveries, and of course the probabilities associated with each of these classes. Mathematical expressions are provided for the probability estimates for the three-category case, as well as for the general case of k samples. This Bayesian method permits a more detailed economic analysis than is possible by the use of binomial distribution, where wells are simply classified as good or bad.

21.2   Bayesian estimation procedure

This section addresses Bayesian estimation procedure for petroleum discoveries of an exploratory well. Early exploratory efforts in a newly recognized geologic area provide an ideal application for the binomial probability distribution. In using the binomial, no consideration is given to how big or how small a discovery might prove to be. It is taken to be either good or bad, acceptable or unacceptable, dry, or producer, with classification restricted to one or the other of the two groups. However, consider if an exploratory well is grouped into three general classes: (1) $y_i = 0$, zero reserves for a dry well; (2) discovery of y_2 barrels of reserves; and (3) discovery of y_3 barrels of reserves, where y_2 and y_3 are barrels of reserves discovered for two different groups of nondry wells. The probabilities p_i, $I = 1, 2$, and 3 of a well discovering y_i barrels of reserves are not known, and in most cases are assumed. In this chapter, however, we will estimate these probabilities for the three-category case, as well as for the general case of k-category. In addition, we will also provide a procedure for estimating the probabilities of discovering the various total reserves. For the three-category case, the binomial distribution is not adequate because the population of interest is divided into more than two categories. Therefore, a multinominal distribution will be considered for this extended case. The parameter estimation procedure will be based on Bayesian methodology. The sampling procedure will be based on minimizing the MSE of the posterior Bayesian estimator using the β-density function to model the prior distribution (Pore and Dennis, 1980). The estimated probabilities are then used in estimating the total reserves.

21.3   Formulation of the oil and gas discovery problem

Let us consider a certain region where an oil company has grouped the possible outcomes of an exploratory well into three general classes:

1. Discovery of $y_1 = 0$ barrels of reserves (a dry well).
2. Discovery of y_2 barrels of reserves (a nondry well).
3. Discovery of y_3 barrels of reserves (a nondry well), and p_i, $i = 1, 2$, and 3 are the corresponding probabilities of having x_i number of wells discovering y_i barrels of reserves. Let us further represent x_1 = number of wells labeled as discovering $y_1 = 0$ barrels of reserves (dry well); x_2 = number of wells labeled as discovering y_2 barrels of reserves; x_3 = number of wells labeled as discovering y_3 barrels of reserves; p_1 = probability of a well discovering $y_1 = 0$ barrels of reserves; p_2 = probability of a well discovering y_2 barrels of reserves; and p_3 = probability of a well discovering y_3 barrels of reserves. For this case, $p_1 + p_2 + p_3 = 1$, and the conditional distribution of x_1, x_2, and x_3 are given in the next section.

21.4 Computational procedure

The Bayesian procedure considered focuses on the Bayesian technique for the parameter estimation and a sampling procedure based on minimizing the MSE of the posterior Bayesian estimator (Pore and Dennis, 1980). From the above section, the conditional distribution of x_1, x_2, and x_3 can be represented as

$$f(x_1, x_2, x_3 | p_1, p_2, p_3) = \frac{(x_1 + x_2 + x_3)!}{x_1! x_2! x_3!} p_1^{x_1} p_2^{x_2} p_3^{x_3} = A_0 p_1^{x_1} p_2^{x_2} (1 - p_1 - p_2)^{x_3} \tag{21.1}$$

where $p_1 \in (0, 1)$, $x_1 \in (0, 1, 2, \ldots)$, and

$$A_0 = \frac{(x_1 + x_2 + x_3)!}{x_1! x_2! x_3!}$$

This is a multinominal model that is a generalization of the binomial model. If we assume a prior distribution of the form

$$g(p_1, p_2, p_3) = B_0 p_1^{\alpha_1} p_2^{\alpha_2} p_3^{\alpha_3} = B_0 p_1^{\alpha_1} p_2^{\alpha_2} (1 - p_1 - p_2) \alpha_3 \tag{21.2}$$

where α_1, α_2, and $\alpha_3 > -1$; $p_1 \in [0, 1]$, $p_2 \in [0, 1 - p_1]$, and

$$B_0 = \frac{\Gamma[\alpha_1 + \alpha_2 + \alpha_3 + 3]}{\Gamma[\alpha_1 + 1]\Gamma[\alpha_2 + 1]\Gamma[\alpha_3 + 1]}$$

Then, since data are available in discrete form, the posterior conditional probability can be represented as

$$w(p_1, p_2, p_3 | x_1, x_2, x_3) = \frac{g(p_1, p_2, p_3) f(x_1, x_2, x_3 | p_1, p_2, p_3)}{\sum g(p_1, p_2, p_3) f(x_1, x_2, x_3 | p_1, p_2, p_3)} \tag{21.3}$$

and for the continuous case, we have

$$W(p_1, p_2, p_3 | x_1, x_2, x_3) = \frac{g(p_1, p_2, p_3) f(x_1, x_2, x_3 | p_1, p_2, p_3)}{p(x_1, x_2, x_3)} \tag{21.4}$$

where

$$P(x_1, x_2, x_3) = \int_0^1 \int_0^{1-p_2} g(p_1, p_2, p_3) f(x_1, x_2, x_3 | p_1, p_2, p_3) \, dp_1 dp_2$$

$$= A_0 B_0 \int_0^1 \int_0^{1-p_2} p_1^{(\alpha_1+x_1)} p_2^{(\alpha_2+x_2)} (1 - p_1 - p_2)^{(\alpha_3+x_3)} \, dp_1 dp_2$$

$$P(x_1, x_2, x_3) = \frac{A_0 B_0 \Gamma[x_1 + \alpha_1 + 1]\Gamma[x_2 + \alpha_2 + 1]\Gamma[x_3 + \alpha_3 + 1]}{\Gamma[x_1 + x_2 + x_3 + \alpha_1 + \alpha_2 + \alpha_3 + 3]} \quad (21.5)$$

Then, to estimate the probabilities, Pi', $i = 1$, 2, and 3 of a well discovering y_i barrels of reserves, we have

$$\hat{p}_i = E[p_i | x_1, x_2, x_3]$$

which implies that

$$\hat{p}_1 = \int_0^1 \int_0^{1-p_2} p_1 W(p_1, p_2, p_3 | x_1, x_2, x_3) \, dp_1 dp_2$$

$$\hat{p}_1 = \int_0^1 \int_0^{1-p_2} \frac{p_1 g(p_1, p_2, p_3) f(x_1, x_2, x_3 | p_1, p_2, p_3) \, dp_1 dp_2}{p(x_1, x_2, x_3)}$$

$$\hat{p}_1 = \frac{x_1 + \alpha_1 + 1}{x_1 + x_2 + x_3 + \alpha_1 + \alpha_2 + \alpha_3 + 3} \quad (21.6)$$

Similarly

$$\hat{p}_2 = E[p_2 | x_1, x_2, x_3]$$

$$\hat{p}_2 = \frac{x_2 + \alpha_2 + 1}{x_1 + x_2 + x_3 + \alpha_1 + \alpha_2 + \alpha_3 + 3} \quad (21.7)$$

$$\hat{p}_3 = \frac{x_3 + \alpha_3 + 1}{x_1 + x_2 + x_3 + \alpha_1 + \alpha_2 + \alpha_3 + 3} \quad (21.8)$$

$$Q = \sum x_i y_i \quad (21.9)$$

$$E(Q) = \sum W_i(P|X)Q_i \quad (21.10)$$

21.5   The k-category case

The k-category case is a generalization of the three-category case. Pore and Dennis (1980) obtained their k-category pixel expression using Bayesian estimation procedure. Their result is, therefore, extended here to multicategory reservoir systems. Then, the probabilities (P_i), $i = 1$, 2, 3, ..., k of a well discovering y_i barrels of reserves, can be written as

$$\hat{p}_i = \frac{x_i + \alpha_i + 1}{\sum (x_i + \alpha_i + 1)} \quad (21.11)$$

21.5.1 Discussion of results

Table 21.1 is the result obtained by McCray (1975) for various values of x_1, x_2, and x_3 that can appear in a sample. Each line in the table represents one possible sample, and the probability of that particular sample is calculated using Equation 21.1. McCray (1975) assumed $p_1 = 0.5$, $p_2 = 0.3$, and $p_3 = 0.2$ and their corresponding reserves are, respectively, $y_1 = 0$, $y_2 = 15$, and $y_3 = 60$.

On the basis of the assumed probabilities, the Bayesian approach with $\alpha_1 = 0$, $\alpha_2 = 0$, and $\alpha_3 = 0$ provides the total reserves (Q), as well as the expected reserves that correspond to each sample as presented in columns 11 and 12 of Table 21.1. The results from this table are exactly the same as those obtained by McCray (1975) based on the assumed probabilities. The total expected reserve in this case is 49.5 MM barrels. However, since different sampling arrangements should result in different probabilities, the probabilities, p_1, p_2, and p_3 can, therefore, be estimated from the given samples x_1, x_2, and x_3 using Equations 21.11, 21.13, and 21.14 rather than assuming these probabilities. These estimates are based on various values of α's associated with the prior distribution. For the special case of $\alpha_1 = \alpha_2 = \alpha_3 = 0$, the estimated probabilities, as well as the expected reserves, are provided in Table 21.2. These probabilities, which are functions of the sampling arrangements (x_1, x_2, and x_3) provide reserve estimates that do not correspond to McCray's (1975) results. Figure 21.1 shows a histogram plot of the probabilities for the reserve levels.

The total expected reserve in this case is 75 MM barrels. This is only due to the fact that the probabilities used in Table 21.1 are based on assumed probabilities. Table 21.7 is arranged according to the total reserves of each individual sample of three wells for the case of assumed probabilities as well as when probabilities are estimated from the sample. This allows the calculation of the cumulative probability so that it can be interpreted as a probability of not less than certain amounts of discovered crude oil reserves. These results are plotted in Figure 21.1, where one starts with 100% probability of at least 0 (zero) barrels discovered, progresses stepwise to 69.5% chance of discovering at least 45 MM barrels, 40% chance of discovering at least 90 MM barrels, and 11.35% chance of discovering at

Table 21.1 Calculations of expected well reserves (run 1)

	Expected reserves calculation										
	Sample			Probabilities			Joint prob	$f(x\|p)$	$w(p\|x)$	Reserves (MM bbl)	Expected reserves (MM bbl)
OBS	X_1	X_1	X_1	P_1	P_1	P_1					
1	3	0	0	0.5	0.3	0.2	0.250	0.125	0.125	0	0.000
2	2	1	0	0.5	0.3	0.2	0.450	0.225	0.225	15	3.375
3	2	0	1	0.5	0.3	0.2	0.300	0.150	0.150	60	9.000
4	1	2	0	0.5	0.3	0.2	0.270	0.135	0.135	30	4.050
5	1	1	1	0.5	0.3	0.2	0.360	0.180	0.180	75	13.500
6	1	0	2	0.5	0.3	0.2	0.120	0.060	0.060	120	7.200
7	0	3	0	0.5	0.3	0.2	0.054	0.027	0.027	45	1.215
8	0	2	1	0.5	0.3	0.2	0.102	0.054	0.054	90	4.860
9	0	1	2	0.5	0.3	0.2	0.072	0.036	0.036	135	4.860
10	0	0	3	0.5	0.3	0.2	0.016	0.008	0.008	180	1.440
										Total =	49.500

Note: $\alpha_1 = 0$, $\alpha_2 = 0$, and $\alpha_3 = 0$ with assumed probabilities $p_1 = 0.5$, $p_2 = 0.3$, and $p_3 = 0.2$.

Table 21.2 Expected reserves calculations (run 2)

| | Sample | | | Probabilities | | | | | | | Expected |
OBS	X_1	X_1	X_1	P_1	P_1	P_1	Joint prob	$f(x|p)$	$w(p|x)$	Reserves (MM bbl)	reserves (MM bbl)
1	3	0	0	0.66667	0.16667	0.16667	0.593	0.296	0.11348	0	0.000
2	2	1	0	0.50000	0.33333	0.16667	0.500	0.250	0.09574	15	1.4362
3	2	0	1	0.50000	0.16667	0.33333	0.500	0.250	0.09574	60	5.7447
4	1	2	0	0.33333	0.50000	0.16667	0.500	0.250	0.09574	30	2.8723
5	1	1	1	0.33333	0.33333	0.33333	0.444	0.222	0.8511	75	6.3830
6	1	0	2	0.33333	0.16667	0.50000	0.500	0.250	0.09574	120	11.4894
7	0	3	0	0.16667	0.66667	0.16667	0.593	0.296	0.11348	45	5.1064
8	0	2	1	0.16667	0.50000	0.33333	0.500	0.250	0.09574	90	8.6170
9	0	1	2	0.16667	0.33333	0.50000	0.500	0.250	0.09574	135	12.9255
10	0	0	3	0.16667	0.16667	0.66667	0.593	0.296	0.11348	180	20.4255
											Total = 75.0000

Note: $\alpha_1 = 0$, $\alpha_2 = 0$, and $\alpha_3 = 0$ with probabilities calculated from Equations 21.6 through 21.8.

least 180 MM barrels of crude oil reserves. For other values of a (Tables 21.2 through 21.6), different values of the expected reserves can be generated. Table 21.7 shows the probabilities arranged by reserve levels. Table 21.8 contains the total expected reserves for various combinations of α *values*. This result shows that the total expected reserves can vary from a low of 40 MM barrels to a high of 124 MM barrels of reserves.

This approach allows the estimation of the probability of a well discovering certain barrels of reserves from the sample rather than assuming this probability. The results show that the total expected reserve is 49 MM barrels when the probabilities are assumed. However, when the probabilities are estimated from a Bayesian standpoint, the total

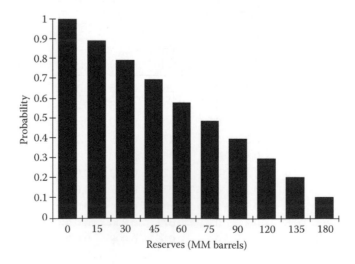

Figure 21.1 Histogram of probabilities versus reserves.

Table 21.3 Expected reserves calculations (run 3)

										Expected
				Expected reserves calculation						
	Sample			Probabilities						
						Joint			Reserves	reserves
OBS	X_1	X_1	X_1	P_1	P_1	P_1	prob	$f(x\|p)$	$w(p\|x)$	(MM bbl)	(MM bbl)
1	3	0	0	0.64286	0.21429	0.14286	0.753	0.265	0.12468	0	0.000
2	2	1	0	0.50000	0.35714	0.14286	0.864	0.267	0.14313	15	2.146
3	2	0	1	0.50000	0.21429	0.28571	0.535	0.214	0.08869	60	5.321
4	1	2	0	0.35714	0.50000	0.14286	0.864	0.267	0.14343	30	4.293
5	1	1	1	0.35714	0.35714	0.28571	0.593	0.218	0.09875	75	7.406
6	1	0	2	0.35714	0.21429	0.42857	0.415	0.196	0.06884	120	8.261
7	0	3	0	0.21429	0.64286	0.14286	0.753	0.265	0.12468	45	5.611
8	0	2	1	0.21429	0.50000	0.28571	0.535	0.214	0.08869	90	7.982
9	0	1	2	0.21429	0.35714	0.42857	0.415	0.196	0.06884	135	9.293
10	0	0	3	0.21429	0.21429	0.57143	0.305	0.186	0.05056	180	9.100
										Total = 59.413	

Note: $\alpha_1 = 0.5$, $\alpha_2 = 0.5$, and $\alpha_3 = 0$.

expected reserves can vary (with different levels of α) from a low of 40 MM barrels of reserves to as high as 124 MM barrels of reserves. This Bayesian approach provides several typical representative distributions that are broad enough in scope to provide a satisfactory economic analysis. These distributions reflect general patterns of similar regions, and include dry holes, as well as several representative class sizes of discoveries, and of course the probabilities associated with each of these classes. In addition, this method permits a more detailed economic analysis than is possible by the use of binomial distribution, where wells are simply classified as good or bad.

Table 21.4 Expected reserves calculations (run 4)

										Expected
				Expected reserves calculation						
	Sample			Probabilities						
						Joint			Reserves	reserves
OBS	X_1	X_1	X_1	P_1	P_1	P_1	prob	$f(x\|p)$	$w(p\|x)$	(MM bbl)	(MM bbl)
1	3	0	0	0.65574	0.18033	0.16393	0.548	0.282	0.10368	0	0.000
2	2	1	0	0.49180	0.34426	0.16393	0.518	0.243	0.09199	15	1.4699
3	2	0	1	0.49180	0.18033	0.32787	0.548	0.254	0.10361	60	5.2490
4	1	2	0	0.32787	0.50820	0.16393	0.548	0.254	0.10361	30	3.1083
5	1	1	1	0.32787	0.34426	0.32787	0.461	0.222	0.08710	75	6.5329
6	1	0	2	0.32787	0.18033	0.49180	0.463	0.238	0.08748	120	10.4980
7	0	3	0	0.16393	0.67213	0.16393	0.674	0.304	0.12736	45	5.7310
8	0	2	1	0.16393	0.50820	0.32787	0.548	0.254	0.10361	90	9.3249
9	0	1	2	0.16393	0.34486	0.49180	0.518	0.249	0.09799	135	13.2291
10	0	0	3	0.16393	0.18033	0.65574	0.548	0.282	0.10368	180	18.6630
										Total = 65.9661	

Note: $\alpha_1 = 0$, $\alpha_2 = 1$, and $\alpha_3 = 0$.

Table 21.5 Expected reserves calculations (run 5)

		Expected reserves calculation
	colspan="3"	Sample	colspan="3"	Probabilities	Joint			Reserves	Expected reserves
OBS	X_1	X_1	X_1	P_1	P_1	P_1	prob	$f(x\|p)$	$w(p\|x)$	(MM bbl)	(MM bbl)
1	3	0	0	0.500	0.250	0.250	0.188	0.125	0.02641	0	0.0000
2	2	1	0	0.375	0.375	0.250	0.356	0.158	0.05014	15	0.7522
3	2	0	1	0.374	0.250	0.375	0.356	0.158	0.05014	60	3.0087
4	1	2	0	0.250	0.500	0.250	0.563	0.188	0.07324	30	2.3772
5	1	1	1	0.250	0.375	0.375	0.712	0.211	0.10029	75	7.5217
6	1	0	2	0.250	0.250	0.500	0.563	0.188	0.07924	120	9.5089
7	0	3	0	0.125	0.625	0.250	0.916	0.244	0.12897	45	5.8038
8	0	2	1	0.125	0.500	0.375	1.266	0.281	0.17829	90	16.0462
9	0	1	2	0.125	0.375	0.500	1.266	0.281	0.17829	135	24.0693
10	0	0	3	0.125	0.250	0.625	0.916	0.244	0.12897	180	23.2150
											Total = 92.3000

Note: $\alpha_1 = 0$, $\alpha_2 = 1$, and $\alpha_3 = 1$.

21.5.2 *Parameter estimation for hyperbolic decline curve*

Here, we present the problem of estimating the nonlinear parameters associated with the hyperbolic decline curve equation. Estimation equations are developed to estimate these nonlinear parameters. Actual field data are used to examine the condition under which the results can be used to predict future oil productions. An approximate linear term is obtained from the nonlinear hyperbolic equation through Taylor's series expansion, and

Table 21.6 Expected reserves calculations (run 6)

		Expected reserves calculation
	colspan="3"	Sample	colspan="3"	Probabilities	Joint			Reserves	Expected reserves
OBS	X_1	X_1	X_1	P_1	P_1	P_1	prob	$f(x\|p)$	$w(p\|x)$	(MM bbl)	(MM bbl)
1	3	0	0	0.500	0.125	0.375	0.211	0.125	0.02572	0	0.0000
2	2	1	0	0.375	0.250	0.375	0.178	0.105	0.02170	15	0.3256
3	2	0	1	0.374	0.125	0.500	0.633	0.211	0.07717	60	4.6302
4	1	2	0	0.250	0.375	0.375	0.178	0.105	0.02170	30	0.6511
5	1	1	1	0.250	0.250	0.500	0.563	0.188	0.06860	75	5.1447
6	1	0	2	0.250	0.125	0.625	1.373	0.293	0.16747	120	20.0965
7	0	3	0	0.125	0.500	0.374	0.211	0.125	0.02572	45	1.1576
8	0	2	1	0.125	0.375	0.500	0.633	0.211	0.07717	90	6.9453
9	0	1	2	0.125	0.250	0.625	1.373	0.293	0.16747	135	22.6085
10	0	0	3	0.125	0.125	0.750	2.848	0.422	0.34727	180	62.5080
											Total = 124.0675

Note: $\alpha_1 = 0$, $\alpha_2 = 0$, and $\alpha_3 = 2$.

Table 21.7 Tabulation of probabilities from three exploratory wells

Reserves (MM bbl)	From Table 21.1 $w(p|x)$	Probability of at least these reserves	From Table 21.2 $w(p|x)$	Probability of at least these reserves
0	0.125	1.00	0.11348	1.0000
10	0.225	0.875	0.09574	0.8865
30	0.135	0.650	0.09574	0.7908
45	0.027	0.515	0.11348	0.6950
60	0.150	0.488	0.09574	0.5815
75	0.180	0.338	0.08511	0.4858
90	0.054	0.158	0.09574	0.4007
120	0.060	0.104	0.09574	0.3049
135	0.036	0.044	0.09574	0.2092
180	0.008	0.008	0.11348	0.1135

the optimum parameter values are determined by employing the method of least squares through an iterative process. The estimated parameters are incorporated into the original hyperbolic decline equation to provide the realistic forecast function. This method does not require any straight-line extrapolation, shifting, correcting, and/or adjusting scales to estimate future oil and gas predictions. The method has been successfully applied to actual oil production data from a West Cameron Block 33 Field in South Louisiana. The results obtained are provided in Figure 21.2.

Table 21.8 Total expected reserves for various combinations of α-values

α_1	α_2	α_3	Total expected reserves
0.0	0.0	0.0	49.50
0.5	0.0	0.0	61.89
0.0	0.5	0.0	69.76
0.0	0.0	0.5	93.36
0.5	0.5	0.0	59.42
0.0	0.5	0.5	86.13
0.5	0.5	0.5	75.00
1.0	0.0	0.0	52.42
0.0	1.0	0.0	65.97
0.0	0.0	1.0	106.62
1.0	1.0	0.0	50.75
0.0	1.0	1.0	92.30
1.0	1.0	1.0	75.00
2.0	0.0	0.0	40.00
0.0	2.0	0.0	60.98
0.0	0.0	2.0	124.06
2.0	2.0	2.0	75.00

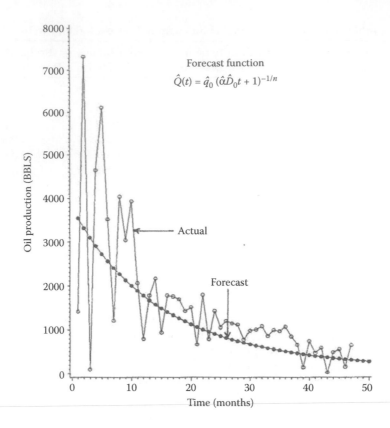

Figure 21.2 Oil production versus time.

21.5.3   Robustness of decline curves

Over the years, the decline curve technique has been extensively used by the oil industry to evaluate future oil and gas predictions, Arps (1945), Gentry (1972), Slider (1968), and Fetkovitch (1980). These predictions are used as the basis for economic analysis to support development, property sale or purchase, industrial loan provisions, and also to determine if a secondary recovery project should be carried out. The graphical solution of the hyperbolic equation is through the use of a log–log paper that sometimes provides a straight line that can be extrapolated for a useful length of time to predict future oil and gas productions. This technique, however, sometimes fails to produce the straight line needed for extrapolation for some oil and gas wells. Furthermore, the graphical method usually involves some manipulation of data, such as shifting, correcting, and/or adjusting scales, which eventually introduce bias into the actual data. To avoid the foregoing graphical problems and to accurately predict future performance of a producing well, a nonlinear least-squares technique is considered. This method does not require any straight-line extrapolation for future predictions.

21.5.4   Mathematical analysis

The general hyperbolic decline equation for oil production rate (q) as a function of time (t) (Arps, 1945), can be represented as

$$q(t) = q_0(1 + mD_0t)^{-1/m} \qquad (21.12)$$

$$0 < m < 1$$

where
    $q(t)$ = oil production at time t
    q_0 = initial oil product
    D_0 = initial decline
    m = decline exponent

Also, the cumulative oil production at time t, $Q(t)$ (Arps, 1945), can be written as

$$Q(t) = \frac{q_0}{(m-1)D_0}[(1 + mD_0t)^{(m-1)/m} - 1] \qquad (21.13)$$

By combining Equations 21.1 and 21.2 and performing some algebraic manipulations (see Arps, 1945), it can be shown that

$$q(t)^{1-m} = q_0^{1-m} + (m-1)D_0q_0^{-m}Q(t) \qquad (21.14)$$

Equation 21.14 shows that the oil production at time t is a nonlinear function of its cumulative oil production. By rewriting Equation 21.14 in terms of cumulative oil production, we have

$$Q(t) = \frac{q_0}{(1-m)D_0} + q(t)^{1-m}\frac{q_0^m}{(m-1)D_0} \qquad (21.15)$$

21.5.5 Statistical analysis

For any given oil well, lease, or property, the oil production at any time t can be observed. The observed production values are always available at discrete equi-spaced time intervals; this will therefore make Equation 21.15 not be satisfied exactly, due to the continuity assumption used in deriving Equation 21.12; hence, it will only define the cumulative production (Q_t) plus the residuals (e_t) as follows (Ayeni, 1989):

$$Q_t = \frac{q_0}{(1-m)D_0} + q_t^{1-m}\frac{q_0^m}{(m-1)D_0} + e_t \qquad (21.16)$$

The general assumption for e_t, Draper and Smith (1981), is that the residuals are assumed to be statistically independent and normally distributed with mean zero and constant variance, s^2, that is, the expected value of e_t, denoted by $E(e_t) = 0$, and the variance of e_t, $Var(e_t) = s^2$. This normality assumption can be checked after the model has been fitted using a residual analysis test or histogram. If this assumption fails due to lack of fit, it may be that there is an outlier or an extreme value in the original data. Then, for this situation, the actual data can first be transformed to stabilize the variability in the data and then use the transformed data in the equations developed in this chapter. The most common transformation method is the log transformation. Other useful transformation techniques are reciprocal, square root, and inverse square root transformations. For more information about outliers and how to handle them, interested readers should see Box et al. (1978).

Now let

$$\alpha_1 = \frac{q_0}{(1-m)D_0}$$

$$\alpha_2 = \frac{q_0^m}{(1-m)D_0}$$ (21.17)

and

$$\alpha_3 = 1 - m$$

By substituting Equation 21.7 into Equation 21.6, we obtain

$$Q = \alpha_1 + \alpha_2 q_1^{\alpha_3} + e_t$$ (21.18)

A close examination of Equation 21.18 shows that it is completely nonlinear in parameters. This is because Equation 21.18 is nonlinear in a_3, which is controlled by the exponent m, just as m controls a_1 and a_2; therefore, both a_1 and a_2 depend on a_3.

21.5.6   *Parameter estimation*

To estimate the parameters in Equation 21.18, we chose to minimize the sum of squares of the residuals given as

$$SS(\alpha) = \sum_{i=1}^{n} (Q_t - \alpha_1 - \alpha_2 q_1^{\alpha_3})^2$$ (21.19)

Since the model is nonlinear in α, the normal equations will be nonlinear. Also, since a_1 and a_2 depend on a_3, an iterative technique will be used to solve the normal equations.

21.5.7   *Optimization technique*

Let us rewrite Equation 21.18 as

$$Q = f(q_t, \alpha) = \alpha_1 + \alpha_2 q_t^{\alpha_3}$$ (21.20)

where

$$f(q_t, \alpha) = \alpha_1 + \alpha_2 q_t^{\alpha_3}$$

Let α_{10}, α_{20}, α_{30} be the initial values for the parameters α_1 and α_3, respectively. If we carry out a Taylor series expansion of $f(q_t, \alpha)$ about the point α_0, where $\alpha_0 = (\alpha_{10}, \alpha_{20},$ and $\alpha_{30})$ and truncate the expansion, we can say that, approximately, when α is close to α_0,

$$f(q_t, \alpha) = f(q_t, \alpha_0) + \sum_{i=1}^{3} \left[\frac{df}{d\alpha_1}(q_t, \alpha) \right]_{\alpha=\alpha_0} (\alpha_1 - \alpha_{10}) \tag{21.21}$$

if we set

$$f_t^0 = f(q_t, \alpha_0)$$

$$b_i^0 = \alpha_1 - \alpha_{10} \tag{21.22}$$

$$Z_{it}^0 = \left[\frac{df(q_t, \alpha)}{d\alpha_1} \right]_{\alpha=\alpha_0}$$

We can see that Equation 21.10 is approximately

$$Q_t = f_t^0 + \sum_{i=1}^{3} b_i^0 Z_{it}^0 + e_t \tag{21.23}$$

$$Q_t - f_t^0 = \sum_{i=1}^{3} b_i^0 Z_{it}^0 + e_t \tag{21.24}$$

which is of linear form. Therefore, we can now estimate the parameters b_i^0, $i = 1, 2,$ and 3 by applying the linear least-squares theory with the assumptions that $E(e_t) = 0$ and $\text{var}(e_t) = s^2$ (Draper and Smith, 1981). This is achieved by minimizing the sum of squares of the residual in Equation 21.17.

If we write

$$Z_0 = \begin{bmatrix} \dfrac{\partial f(q_1, \alpha)}{\partial \alpha_1} & \dfrac{\partial f(q_1, \alpha)}{\partial \alpha_2} & \dfrac{\partial f(q_1, \alpha)}{\partial \alpha_3} \\ \vdots & \vdots & \vdots \\ \dfrac{\partial f(q_n, \alpha)}{\partial \alpha_1} & \dfrac{\partial f(q_n, \alpha)}{\partial \alpha_2} & \dfrac{\partial f(q_n, \alpha)}{\partial \alpha_3} \end{bmatrix} \tag{21.25}$$

$$\beta = \begin{bmatrix} \beta_1^0 \\ \beta_2^0 \\ \beta_3^0 \end{bmatrix} \quad \text{and} \quad Q_0 = \begin{bmatrix} Q_1 - f_1^0 \\ \vdots & \vdots \\ Q_n - f_n^0 \end{bmatrix} = Q - f^0$$

then, the estimate $\hat{\beta}_0 = (\beta_1^0, \beta_2^0, \beta_3^0)$ is given by

$$\hat{\beta} = (Z_0' Z_0)^{-1} Z_0' (Q - f^0) \tag{21.26}$$

The vector $\hat{\beta}_0$ will therefore minimize the sum of squares $SS(\alpha)$ of the residual with respect to β_i^0 for $i = 1, 2$, and 3 where

$$SS(\alpha) = \sum_{t=1}^{n}\left[Q_t - f(q_t, \alpha_0) - \sum_{i=1}^{3} \beta_1^0 Z_{it}^0 \right]^2 \tag{21.27}$$

and $\beta_i^0 = \alpha_i - \alpha_{i0}$.

21.5.8 Iterative procedure

If $b_i^0 = \alpha_1 - \alpha_{10}$, then α_1, $i = 1, 2$, and 3 can be thought of as the revised best estimates of α. We can now place the values α_{i1}, the revised estimates in the same roles as were played in the foregoing by α_{10}, and to through exactly the same procedure as already described, but replacing all zero subscripts by ones. This will lead to another set of revised estimates, α_{i2} and so on. In vector form, extending the previous notation, we can write

$$\alpha_J + 1 = \alpha_J + (Z_J'Z_J)^{-1}Z_J'(Q - f^J) \tag{21.28}$$

where

$$Z_J = Z_{it}^J$$

$$f^J = (f_1^J, f_2^J, ..., f_n^J)$$

$$\alpha^J = (\alpha_{1J}, \alpha_{2J}, \alpha_{3J})$$

The foregoing iterative process is continued until the solution converges, that is, until in successive iterations, $J, J + 1$, such that

$$\left| \frac{\alpha_1, J + 1 - \alpha_{iJ}}{\alpha_{iJ}} \right| < d \quad \text{for } i = 1, 2, 3 \tag{21.29}$$

where d is some prespecified amount, for example, 0.0001. Also, at each stage of the iterative procedure, $SS(\alpha_J)$ can be evaluated to check if a reduction in its value has actually been achieved. For rapid convergence, if $SS(\alpha_{J+1})$ is greater than $SS(\alpha_J)$, the vector b in Equation 21.17 can be amended by having it. But if $SS(\alpha_{J+1})$ is less than $SS(\alpha_J)$, we can double the vector b_J. This halving and/or doubling process is continued until three points between α_J, α_{J+1} are found, which include a local minimum of $SS(\alpha)$ between them. A quadratic interpolation can be used to locate the minimum and the iterative cycle beings again. Figure 21.4. shows a plot of residuals versus normal scores.

After convergence, the optimum parameters estimated can be used to determine the estimates $\hat{m}$, $\hat{q}_0$, $\hat{D}_0$, as follows. From Equation 21.7, we have

$$\hat{m} = 1 - \hat{\alpha}_3$$

$$\hat{q}_0 = (-\hat{\alpha}_1/\hat{\alpha}_2)^{1/\hat{\alpha}_3} \quad 0 < \hat{\alpha}_3 < 1$$

$$\hat{D}_0 = \frac{\hat{q}_0}{\hat{\alpha}_1 \hat{\alpha}_3} \quad \hat{\alpha}_2 < 0 \qquad (21.30)$$

$$\hat{\alpha}_1 > 0$$

where $\hat{\alpha}_1, \hat{\alpha}_2, \hat{\alpha}_3$ are optimum parameters estimated with the minimum residual sum of squares. We can now incorporate Equation 21.20 into Equation 21.1 to give the optimum forecast function that can be used to generate future oil production forecast, as follows:

$$\hat{q}(t) = \hat{q}_0 (1 + \hat{m}_0 \hat{D} t)^{-1/\hat{m}} \qquad (21.31)$$

This technique has been applied to actual oil field data from a West Cameron Block 33 Field in South Louisiana. The results obtained are provided in Figure 21.2. The method described in this chapter can probably be improved if Taylor's series expansion is carried out further; however, this problem is not explored in this chapter.

21.5.9 *Residual analysis test*

A simple residual analysis test is carried out in this section to check the normality assumption used in the development of the estimating equations. A plot of the residuals versus the normal scores shows that the residual is random (Figure 21.3). This is because there is

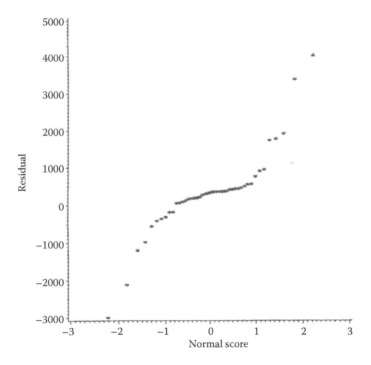

Figure 21.3 Residual versus normal scores.

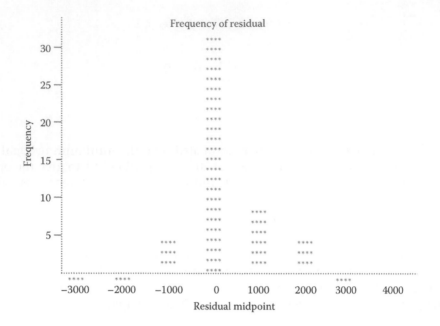

Figure 21.4 Histogram of residuals.

an approximately linear relationship with a correlation coefficient of $r = 0.90$ between the residual and the normal score. In addition, a plot of the histograms of the residual shows that the residual is normally distributed with the mean centered around zero (Figure 21.4). These tests substantiate the normality assumption used to define the residual in the earlier section.

This section has presented a technique for estimating parameters associated with the hyperbolic decline curve equation. The method used provides optimum parameter values with minimum residual sum of squares. The estimated parameters eventually produce the optimum forecast function when incorporated into the original hyperbolic decline equation. This forecast function can be used to generate accurate forecast values needed for estimating future reserves.

21.5.10   Simplified solution to the vector equation

Let $\alpha_0 = (\alpha_{10},\ \alpha_{20},\ \text{and}\ \alpha_{30})^T$ be the initial guess. Then, by using Equation 21.23, set the following:

$$f_1^0 = \alpha_{10} + \alpha_{20}q_t^{\alpha_{30}}$$

$$\beta_1^0 = \alpha_1 + \alpha_{10}$$

$$\beta_2^0 = \alpha_2 + \alpha_{20}$$

$$\beta_3^0 = \alpha_3 + \alpha_{30}$$

$$Z_{1_t}^0 = \left.\frac{\partial f(q_t, \alpha)}{\partial \alpha_1}\right|_{\alpha=\alpha_0} = \frac{\partial}{\partial \alpha_1}(\alpha_1 + \alpha_2 q_t^{\alpha_{30}})$$

$$Z_{2_t}^0 = \left.\frac{\partial f(q_t, \alpha)}{\partial \alpha_2}\right|_{\alpha=\alpha_0} = q_t^{\alpha_{30}}$$

$$Z_{3_t}^0 = \left.\frac{\partial f(q_t, \alpha)}{\partial \alpha_3}\right|_{\alpha=\alpha_0} = \alpha_{20} e^{\alpha_{30} \ln q_t} q_t^{\alpha_{30}} \cdot \ln(q_t)$$

$$Z_{3_t}^0 = \alpha_{20} e^{\alpha_3} \cdot \ln(q_t)$$

where $t = 1, 2, \ldots, n$ observations. Therefore, the $n \times 1$ matrix $Q - f^0$ can be expressed as

$$Q - f^0 = \begin{bmatrix} Q_1 \\ \vdots \\ Q_n \end{bmatrix} - \begin{bmatrix} \alpha_{10} + \alpha_{20} q_1^{\alpha_{30}} \\ \vdots \qquad \vdots \\ \alpha_{10} + \alpha_{20} q_n^{\alpha_{30}} \end{bmatrix} \tag{21.32}$$

Therefore, the $n \times 3$ matrix Z_0 can be written as

$$Z_0 = \begin{bmatrix} 1 & q_1^{\alpha_{30}} & \alpha_{20} q_1^{\alpha_{30}} \ln q_1 \\ \vdots & \vdots & \vdots \\ 1 & q_n^{\alpha_{30}} & \alpha_{20} q_n^{\alpha_{30}} \ln q_n \end{bmatrix} \tag{21.33}$$

Then, the estimate $\hat{\beta}_0$ is given by

$$\hat{\beta}_0 = \begin{bmatrix} n & \sum_{t=1}^{n} q_t^{\alpha_{30}} & \alpha_{20} \sum_{t=1}^{n} q_t^{\alpha_{30}} \ln q_t \\ \sum_{t=1}^{n} q_t^{\alpha_{30}} & \sum_{t=1}^{n} q_t^{2\alpha_{30}} & \alpha_{20} \sum_{t=1}^{n} q_t^{\alpha_{30}} \ln q_t \\ \alpha_{20} \sum_{t=1}^{n} q_t^{\alpha_{30}} \ln q_t & \alpha_{20} \sum_{t=1}^{n} q_t^{2\alpha_{30}} \ln q_t & \alpha_{20}^2 \sum_{t=1}^{n} (q_t^{\alpha_{30}} \ln q_t)^2 \end{bmatrix}^{-1}$$

$$\times \begin{bmatrix} \sum_{t=1}^{n} (Q_t - \alpha_{10} - \alpha_{20} q_t^{\alpha_{30}}) \\ \sum_{t=1}^{n} (Q_t - \alpha_{10} - \alpha_{20} q_t^{\alpha_{30}}) \cdot q_t^{\alpha_{30}} \\ \sum_{t=1}^{n} (Q_t - \alpha_{10} - \alpha_{20} q_t^{\alpha_{30}}) \cdot q_t^{\alpha_{30}} \cdot \ln q_t \end{bmatrix} \tag{21.34}$$

Since $\alpha_{j+1} = \alpha_j + \beta_j$ from Equation 21.28, and by using Equation 21.34, we have

$$
\alpha_{j+1} = \alpha_j + \begin{bmatrix} n & \displaystyle\sum_{t=1}^{n} q_t^{\alpha_3^{(j)}} & \alpha_2^{(j)} \displaystyle\sum_{t=1}^{n} q_t^{\alpha_3^{(j)}} \ln q_t \\[2ex] \displaystyle\sum_{t=1}^{n} q_t^{\alpha_3^{(j)}} & \displaystyle\sum_{t=1}^{n} q_t^{2\alpha_3^{(j)}} & \alpha_2^{(j)} \displaystyle\sum_{t=1}^{n} q_t^{2\alpha_3^{(j)}} \ln q_t \\[2ex] \alpha_{20}^{(j)} \displaystyle\sum_{t=1}^{n} q_t^{\alpha_3^{(j)}} & \alpha_2^{(j)} \displaystyle\sum_{t=1}^{n} q_t^{2\alpha_3^{(j)}} \ln q_t & (\alpha_2^2)^{(j)} \displaystyle\sum_{t=1}^{n} (q_t^{2\alpha_3^{(j)}} \ln q_t)^2 \end{bmatrix}^{-1}
$$

$$
\times \begin{bmatrix} \displaystyle\sum_{t=1}^{n} (Q_t - \alpha_1^{(j)} - \alpha_2^{(j)} q_t^{\alpha_{30}^{(j)}}) \\[2ex] \displaystyle\sum_{t=1}^{n} (Q_t - \alpha_1^{(j)} - \alpha_2^{(j)} q_t^{\alpha_{30}^{(j)}}) \cdot q_t^{\alpha_{30}^{(j)}} \\[2ex] \alpha_2^{(j)} \displaystyle\sum_{t=1}^{n} (Q_t - \alpha_1^{(j)} - \alpha_2^{(j)} q_t^{\alpha_{30}^{(j)}}) \cdot q_t^{\alpha_{30}^{(j)}} \cdot \ln q_t \end{bmatrix}
\tag{21.35}
$$

Equation 21.35 is very easy to program and can be used for the iterative process. The process continues until the solution converges, that is, until successive iterations, j, $j + 1$, such that

$$
\left| \frac{\alpha_i^{(j+1)} - \alpha_i^{(j)}}{\alpha_i^{(j)}} \right| < d \quad i = 1, 2, 3
\tag{21.36}
$$

where d is some prespecified small number.

21.6 *Integrating neural networks and statistics for process control*

This section explores the opportunities available using a neural network (NN) back-propagation technique to model process data. The effects of using different transfer functions (linear, sigmoid, hyperbolic tangent, etc.) as well as multiple hidden nodes are extensively explored for monitoring and controlling manufacturing processes. We emphasize some practical issues useful in developing process control strategies for process optimization. Some actual examples from industrial experiments are presented. These examples are based on statistically designed experiments employing two-level factorial and fractional factorial designs, as well as a central composite response surface design. Experiments from a large system consisting of more than 50 independent variables are also considered in the study. The results obtained from the various NN architectures are compared with those obtained from the statistical linear regression method. Strengths and weaknesses of each method are identified. The conditions under which one method performs better than the other are fully explored. We also investigate the circumstances under which the NN and regression approaches can be integrated with the objective of maximizing the benefits from both methods.

21.6.1   Fundamentals of NN

An NN is a system that is modeled after the human brain and arranged in patterns similar to biological neural nets. The unit analogous to the biological neuron is referred to as a "processing element" (PE). A PE has many input paths representing individual neurons. These artificial neurons receive and sum up information from other neurons or external inputs, perform a transformation on the inputs, and then pass on the transformed information to other neurons or external outputs. The output path of a PE can be connected to input paths of other PEs through connection weights that correspond to the strength of neural connections. The information passed from neuron to neuron can be thought of as a way to activate a response from certain neurons based on the information received.

The most important characteristic of NNs is that they can learn to simulate any behavior and can be used to generate the action necessary to produce a given response. Learning is the process of adapting or modifying the connection weights in response to stimuli being presented at the input buffer and optionally at the output buffer. Back-propagation network is the most common form of NN. This network has an input layer and at least one hidden layer. There is no theoretical limit on the number of hidden layers but typically there will be one or two layers. Each layer is fully connected to the succeeding layer.

21.6.2   The input function

The simplest input function is a simple weighted input

$$I = \sum_{i=1}^{n} w_i x_i$$

A back-propagation element transfers its inputs as

$$X_j = f(I)$$

where f is traditionally the sigmoid function but can be any differentiable function.

21.6.3   Transfer functions

a. Linear transfer function

$$f(z) = Z$$

b. Hyperbolic tangent

$$f(z) = \frac{e^z - e^{-z}}{e^z + e^{-z}}$$

c. Sigmoid transfer function

$$f(z) = \frac{1}{1 + e^{-z}}$$

21.6.4   Statistics and NN predictions

Tables 21.9 through 21.11 provide the actual as well as the predicted values of the statistical regression method with NNs for various transfer functions. These results are respectively based on experiments specifically designed for a 2^3 replicated factorial design with center points; a 2^3 central composite design with center points; as well as a 2^{8-4} fractional factorial design. Plots of the various NN methods are shown in Figures 21.5 through 21.9. Figures 21.10 through 21.12 provide the graphical comparisons of the various methods.

21.6.5   Statistical error analysis

Statistical error analysis is used to compare the performance, as well as the accuracy, of the statistical regression and NN methods. The accuracy of the predicted values relative to the actual value is determined by various statistical methods. The criteria used in this study are average relative error (ARE), forecast root mean square error (FRMSE), minimum error, as well as maximum error. The ARE is the relative deviation of the predicted values from the actual values. The lower the ARE, the more equally distributed is the error between positive and negative values. The forecast mean square error is a measure of the dispersion. A smaller value of FRMSE indicates a better degree of fit.

Tables 21.12 through 21.15 show the results obtained from the error analysis. The results show that statistical linear regression performs better in some cases while the hyperbolic tangent transfer function of the NN performs better than any other transfer function.

Table 21.9 Prediction comparisons for a 2^3 full factorial design

Actual	Statistics prediction	Linear prediction	Hyperbolic tangent	Sigmoid prediction
5.4083	5.8087	6.4648	5.9992	6.1625
13.263	13.484	13.4754	13.4031	13.5137
16.921	16.5432	16.572	16.4019	16.3156
24.57	24.2212	23.5832	24.1862	23.8225
17.094	17.2289	16.7375	17.1906	16.7648
24.272	23.8769	23.7487	23.7327	13.7379
26.667	27.026	26.8453	26.9531	27.1393
32.787	33.1473	33.8565	33.3524	33.3696
19.268	20.167	20.1603	20.2921	20.5331
21.322	20.167	20.1603	2.2921	20.5331
6.1463	5.8087	6.4642	5.9992	6.1625
13.643	13.484	13.4754	13.4031	13.5137
16.103	16.5432	16.572	16.4019	16.3156
23.81	24.2212	23.5832	24.1862	23.8225
17.301	17.2289	16.7375	17.1906	16.7648
23.419	23.8769	23.7487	23.7327	23.7379
27.322	27.026	26.8453	26.9531	27.1393
33.445	33.1473	33.8565	33.3524	33.3696
20.661	20.167	20.1603	20.2921	20.5331
19.92	20.167	20.1603	20.2921	20.5331

Table 21.10 Prediction comparisons for a 2^3 central composite design

Actual	Statistics prediction	Linear prediction	Hyperbolic tangent	Sigmoid prediction
34.2	37.51	30.5013	34.4385	33.4627
12.2	11.25	16.1799	11.0012	11.4857
19.2	14.95	17.7643	19.3777	14.2051
13.85	13.84	3.4429	13.1522	11.9871
42.2	42.71	32.2846	42.1012	40.786
9.8	14.91	17.9633	10.0964	13.2724
12.5	14.3	19.5477	13.1515	14.8784
14.1	11.65	5.2263	12.8587	12.4026
31	30.55	29.3209	30.7354	31.2331
8.3	1.42	6.4067	8.0917	9.7927
41.6	31.01	28.0534	41.1906	36.6313
13.1	16.35	7.6742	13.6192	13.9306
9	11.7	16.9721	10.2448	12.4473
13.2	12.99	17.8638	13.1686	13.0293
18.2	12.99	17.8638	13.1686	13.0293
10.4	12.99	17.8638	13.1686	13.0293
17.3	12.99	17.8638	13.1686	13.0293
8.6	12.99	17.8638	13.1686	13.0293
12.7	12.99	17.8638	13.1686	13.0293
11.6	12.99	17.8638	13.1686	13.0293

Table 21.11 Prediction comparisons for a 2^{8-4} fractional factorial design

Actual	Statistics prediction	NN linear	Linear prediction	Hyperbolic tangent	Sigmoid prediction
2.67	2.67	3.63503	2.66989	2.685	2.73969
2.17	2.17	2.33152	2.16992	2.155	2.11396
1.33	1.33	2.32557	1.3301	1.39	1.71533
1.17	1.17	1.30798	1.71004	1.163	1.18535
4.83	4.86	5.05969	4.82995	4.817	4.80184
2.83	2.83	3.75591	2.82991	2.853	2.90004
5.83	5.83	6.08295	5.83005	5.822	5.83877
4	4	5.06	4.00005	4.038	4.27219
2.17	2.17	−0.1753	2.17012	2.172	1.8063
2.17	2.17	4.36818	2.16999	2.174	2.14609
3.5	3.5	4.20293	3.50008	3.466	3.41812
3.33	3.33	3.058	3.32992	3.33	3.32857
6.67	6.67	6.02276	6.66999	6.666	6.6645
9.67	9.67	7.56678	9.67015	9.664	9.42287
4.67	4.67	4.38513	4.66992	4.67	4.67029
2.5	2.5	3.18801	2.50012	2.449	2.39139

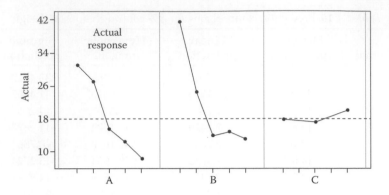

Figure 21.5 Actual effects using Yates's algorithm.

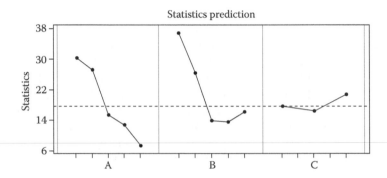

Figure 21.6 Predicted effects using Yates's algorithm: statistics prediction.

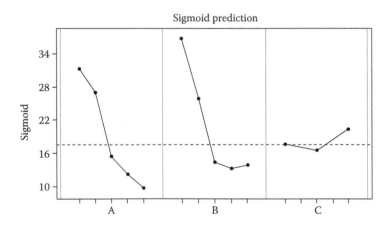

Figure 21.7 Predicted effects using Yates's algorithm: sigmoid prediction function of NNs.

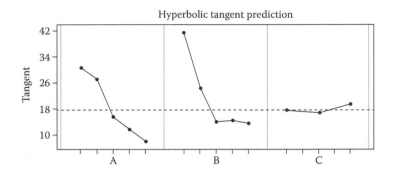

Figure 21.8 Predicted effects using Yates's algorithm: hyperbolic tangent function of NNs.

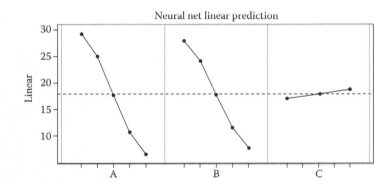

Figure 21.9 Predicted effects using Yates's algorithm: linear prediction NNs.

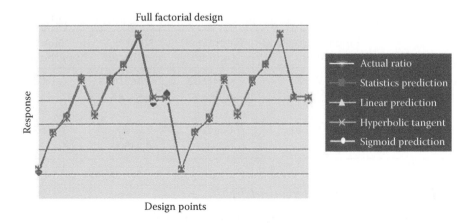

Figure 21.10 Graphical comparison of NNs and statistical method for a three-factor factorial design.

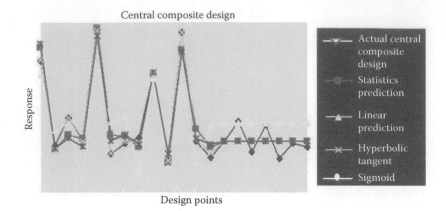

Figure 21.11 Graphical comparison of NNs and statistical method for a three-factor central composite design.

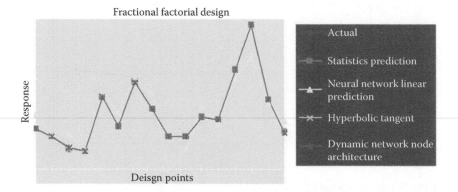

Figure 21.12 Graphical comparison of NNs and statistical method for 2^{8-4} fractional factorial design.

Table 21.12 Central composite design: statistical error comparison of methods

Parameter	Linear regression statistics	Neural net sigmoid	Neural net linear	Neural net hyperbolic tangent
Average relative error	−0.04972	−0.048602	−0.145163	−0.02388
Forecast root mean square error	0.23451	0.224877	0.53887	0.17005
Minimum error	−0.52140	−0.51500	−1.0770	−0.53120
Maximum error	0.28630	0.284100	0.74500	0.27650

21.7   *Integration of statistics and NNs*

In process control, experiments are frequently performed primarily to measure the effects of one or more variables on a response. From these effects, one needs to determine significant variables that can be used to develop control strategies. However, most NN approaches do not provide information about variables that are significant to the response of interest. In addition, the coefficients or weights obtained from NN are difficult to interpret.

Table 21.13 A 2^3 factorial design: statistical error comparison of methods

Parameter	Linear regression statistics	Neural net sigmoid	Neural net linear	Neural net hyperbolic tangent
Average relative error	−0.00125	−0.006015	−0.00812	−0.00371
Forecast root mean square error	0.03059	0.040799	0.053185	0.033937
Minimum error	−0.07403	−0.139450	−0.19520	−0.10926
Maximum error	0.05493	0.037000	0.054500	0.048300

Table 21.14 A 2^{8-4} fractional factorial design: statistical error comparison of methods

Parameter	Linear regression statistics	Neural net sigmoid	Neural net linear	Neural net hyperbolic tangent	DNNA
Average relative error	0.0000	−0.007478	−0.0337	−0.000007	−0.0014
Forecast root mean square error	0.0000	0.090165	0.39863	0.000035	−0.01377
Minimum error	0.0000	−0.289700	−0.7485	−0.000080	−0.04520
Maximum error	0.0000	0.167300	1.08100	0.000040	0.02037

Also, most NN packages are lacking in providing response surface curves needed for process optimization. In this section, we explore the opportunities available in performing Yates's analysis on NN trained data. The results obtained are compared with Yates's analysis of the actual data as well as Yates's analysis from the predicted values of the statistical regression method. The approach used here can be used in identifying significant variables as well as generating response surface curves for process optimization.

Figures 21.5 through 21.9 show the main effects plots from the 2^3 central composite experiment with the actual data, statistical predictions, as well as plots of NN predictions for various NN transfer functions. These effects were obtained using Yates's algorithm. They are the average effects for factor A at five levels (−2, −1, 0, 1, and 2), factor B at five levels (−2, −1, 0, 1, and 2), and factor C at three levels (−1, −1, and 1). Factors A and B are quantitative factors while factor C is a qualitative factor. The results obtained show that NN method provides similar results as compared to the actual Yates's analysis as well as the results obtained under the statistical regression method. These results indicate that

Table 21.15 Statistical error comparison of methods for 50 factors large system

Parameter	Linear regression statistics	Neural net sigmoid	Neural net linear	Neural net hyperbolic tangent	DNNA
Average relative error	−0.01421	0.001216	0.93529	0.017745	−0.019631
Forecast root mean square error	0.13756	0.128879	0.93739	0.101538	0.139777
Minimum error	−1.23053	−1.14555	0.84756	−0.847010	−1.105730
Maximum error	0.41139	0.424750	0.97797	0.355280	0.454540

Yates's analysis can be used on NN trained data to determine the effects of variables as well as identifying statistically significant factors. Figures for interaction plots as well as for other designs provide similar results, but are not provided here due to limited space.

It is recommended that whenever it is necessary to obtain information on significant effects as well as generating response surface curves, Yates's algorithm can be performed on neural nets trained data. This integration of neural nets and statistical methods would enable engineers and statisticians to obtain additional information from NNs approach.

Figures 21.10 through 21.12 provide the graphical comparisons of the various methods. The number of hidden nodes used for each NN approach was equal to the number of input variables. For example, in a three-factor experiment, we used three hidden nodes. Similarly, for the eight-factor fractional factorial experiment, we used eight hidden nodes. We found that too many hidden nodes can cause overtraining that might lead to predictions that look great. However, predictions on a new data set are generally bad.

The above examples confirm that NN approach does work if carefully applied. The results obtained under different types of orthogonally designed experiments show that the statistical linear regression method provides as good results as NNs. Except for the linear transfer function of the NNs that failed in the situation where the quadratic effect is significant, the results of other transfer functions compare favorably with the linear statistical method. The results obtained for the large experiment with 50 factors are consistent with those of smaller factors. The tabulated results for different transfer functions are not provided due to limited manuscript space. NNs trained data can be integrated with statistical methods to obtain additional information about model adequacy, significant factors, as well as generating response surface curves for process optimization.

References

Arps, J.J., 1945, Analysis of decline curves, *Transactions of AIME*, 160, 228–247.

Ayeni, B.J., 1989, Parameter estimation for hyperbolic decline curve, *Published in the Journal of Energy Resources Technology*, 3, 279–283.

Badiru, A.B., Ibidapo-Obe, O., and Ayeni, B.J., 2012, *Industrial Control Systems: Mathematical and Statistical Models and Techniques*, CRC Press/Taylor & Francis, Boca Raton, FL.

Box, G.E.P., Hunter, W.G., and Hunter, J.S., 1978, *Statistics for Experimenters*, John Wiley & Sons, Inc., New York, NY.

Draper, N.K. and Smith, H., 1981, *Applied Regression Analysis*, 2nd edn., John Wiley & Sons, New York, NY.

Fetkovitch, M.J., 1980, Decline curve analysis using type curves, *Journal of Petroleum Technology*, 1067–1077.

Gentry, R.W., 1972, Decline curve analysis, *Journal of Petroleum Technology*, 23(4), 38–41.

McCray, A.W., 1975, *Petroleum Evaluations and Economic Decisions*, Prentice-Hall, Englewood Cliffs, NJ, pp. 448.

Pore, M.D. and Dennis, T.B., 1980, The multicategory case of the sequential Bayesian pixel selection and estimation procedure, *Lockheed Engineering Technical Report*, LEMSCO-14807.

Slider, H.C., 1968, A simplified method of hyperbolic decline curve analysis, *Journal of Petroleum Technology*, 20(3), 38–41.

chapter twenty-two

Data modeling for forecasting

Adedeji B. Badiru

Contents

22.1 Introduction

Forecasting is a kind of predictive "virtual" measurement. The basis for many forecasting applications is a set of historical data derived from a time series event such as energy consumption. In this chapter (based on Badiru, 1981), we present quantitative data modeling of time series for the purpose of generating forecasts. The basic principle of time series analysis is the concept that the sequence of observations is a realization of jointly distributed random variables. Thus, the time series observations $Z_1,\ldots,Z_T$ recorded at discrete and equally spaced time intervals are assumed to be drawn from a probability distribution of the form

$$P_{1,\ldots,T}^{(Z_1,\ldots,Z_T)} \tag{22.1}$$

The ultimate objective is to use the joint distribution to make probability inferences about future observations.

22.2 The concept of stationarity

Simplification in time series analysis is achieved by requiring that the time series be stationary in the sense that the joint distribution be invariant with regard to any time displacement. That is

$$p(Z_t,\ldots,Z_{t+k}) = p(Z_{t+m},\ldots,Z_{t+m+k}) \tag{22.2}$$

where t is any point in time and k and m are any pair of integers. Thus, stationarity implies that the time series process has a constant variance and remains in equilibrium around a constant mean with respect to any time reference. It follows then that we have the following equations:

$$E(Z_t) = E(Z_{t+m}) \tag{22.3}$$

$$V(Z_t) = V(Z_{t+m}) \tag{22.4}$$

$$C(Z_t, Z_{t+1}) = C(Z_t, Z_{t+k+1}) \tag{22.5}$$

In some cases, nonstationarity may be recognized in the graph of the time series. An explosive plot with a lack of tendency for a particular neighborhood is an indication of nonstationarity. In some cases where nonstationarity exists, some form of data transformation may be an aid in achieving stationarity. For most data, the first transformation used is differencing. Differencing implies that a new series is created by taking differences between successive periods of the original series. For example, first regular differences are given by

$$w_t = z_t - z_{t-1} \tag{22.6}$$

In constructing a model for a time series, it is necessary to describe the relationship between a current observation and previous observations. Such relationships are described by the sample autocorrelation function

$$r_j = \frac{\sum_{t=1}^{T-j}(Z_t - \bar{Z})(Z_{t+j} - \bar{Z})}{\sum_{t=1}^{T}(z_t - \bar{Z})^2}, \quad j = 0, 1, 2, \dots \tag{22.7}$$

where T is the number of observations, Z_t is the observation for time t, $\bar{Z}$ is the sample mean of the series, j is the number of periods separating pairs of observations, and r_j is the sample estimate of the theoretical correlation coefficient, p_j. The coefficient of correlation between two variables Y_1 and Y_2 is defined as follows:

$$\rho_{12} = \frac{\sigma_{12}}{\sigma_1 \sigma_2} \tag{22.8}$$

where σ_1 and σ_2 are, respectively, the standard deviations of Y_1 and Y_2, and Y_1 and σ_{12} are the covariances between Y_1 and Y_2. The standard deviations are the positive square roots of the variances defined as

$$\sigma_1^2 = V(Y_1) = E[(Y_1 - \mu_1)^2] \tag{22.9}$$

and

$$\sigma_2^2 = V(Y_2) = E[(Y_2 - \mu_2)^2] \tag{22.10}$$

The covariance σ_{12} is defined as

$$\sigma_{12} = E[(Y_1 - \mu_1)(Y_2 - \mu_2)] \tag{22.11}$$

which will be zero if Y_1 and Y_2 are independent. Thus, when Y_1 and Y_2 are independent, $\rho_{12} = 0$ in Equation 22.8. If Y_1 and Y_2 are positively related, σ_{12} is positive and so is ρ_{12}. If the variables are negatively related, σ_{12} is negative and so is ρ_{12}. The correlation coefficient, ρ_j, is a real number with the property

$$-1.0 \leq \rho_j \leq 1.0 \tag{22.12}$$

The modeling process involves the development of a discrete linear stochastic process in which each observation Z_t may be expressed in the general form

$$Z_t = \mu + u_t + \psi_1 u_{t-1} + \psi_2 u_{t-2} + \cdots \tag{22.13}$$

where μ is the mean of the process, the ψ_1s are fixed parameters which are functions of the autocorrelations, and the time series $(\dots, u_{t-1}, u_t, u_{t+1}, \dots)$ is a sequence of independently and identically distributed random disturbances with mean zero and variance σ_u^2. The expectation of Equation 22.13 is given by

$$E(Z_t) = \mu + E(u_t + \psi_1 u_{t-1} + \psi_2 u_{t-2} + \cdots) = \mu + E(u_t)[1 + \psi_1 + \psi_2 + \cdots] \tag{22.14}$$

Stationarity requires that the expectation in Equation 22.14 be nonexplosive. Thus, from integral calculus, it is required that the infinite sum of the ψ_1s be convergent. That is

$$\sum_{i=0}^{\infty} \psi_i = C \tag{22.15}$$

where $\psi_0 = 1$ and C is some finite number. The theoretical variance of the process, γ_0, can be derived as follows:

$$\begin{aligned}
\gamma_0 &= E[Z_t - E(Z_t)]^2 \\
&= E[\mu + u_t + \psi_1 u_{t-1} + \psi_2 u_{t-2} + \cdots - \mu]^2 \\
&= E[u_t + \psi_1 u_{t-1} + \psi_2 u_{t-2} + \cdots]^2 \\
&= E[u_t^2 + u_t^2 + \psi_2^2 u_{t-2}^2 + \cdots \psi_1 u_t u_{t-1} + \psi_2 u_t u_{t-2} + \cdots] \\
&= E[u_t^2 + \psi_1^2 u_{t-1}^2 + \psi_2^2 u_{t-2}^2 + \cdots] = E[cross - product\ terms] \\
&= E[u_t^2] + \psi_1^2 E[u_{t-1}^2] + \psi_2^2 E[u_{t-2}^2] + \cdots + 0 \\
&= \sigma_u^2 + \psi_1^2 \sigma_u^2 + \psi_2^2 \sigma_u^2 + \cdots \\
&= \sigma_u^2 (1 + \psi_1^2 + \psi_2^2 + \cdots) \\
&= \sigma_u^2 \sum_{i=0}^{\infty} \psi_i^2 \tag{22.16}
\end{aligned}$$

where the expectation of the cross-product terms is zero since the u_ts are independently distributed, and where σ_2^2 stands for the residual variance, which is the variance of the u_ts. Similarly, the theoretical covariance between Z_t and Z_{t+j}, γ_j, can be derived as follows:

$$
\begin{aligned}
\gamma_j &= E\{[Z_t - E(Z_t)][Z_{t+j} - E(Z_{t+j})]\} \\
&= E[(u_t + \psi_1 u_{t-1} + \cdots)(u_{t+j} + \psi_1 u_{t+j-1} + \cdots)] \\
&= E[\psi_j u_{t+j}^2 + \cdots + \psi_1 \psi_{j+1} u_{t+j-1}^2 + \cdots + cross-products] \\
&= \psi_j \sigma_u^2 + \psi_1 \psi_{j+1} \sigma_u^2 + \psi_2 \psi_{j+2} \sigma_u^2 + \cdots \\
&= \sigma_u^2 (\psi_j + \psi_1 \psi_{j+1} + \psi_2 \psi_{j+2} + \cdots) \\
&= \sigma_u^2 \sum_{i=0}^{\infty} \psi_i \psi_{i+j}
\end{aligned}
\tag{22.17}
$$

Sample estimates of the variance and covariances are obtained by

$$
c_j = \frac{1}{T} \sum_{t=1}^{T-j} [(Z_t - \bar{Z})(Z_{t+j} - \bar{Z})], \quad j = 0, 1, 2, \ldots
\tag{22.18}
$$

The theoretical autocorrelations are obtained by dividing each of the autocovariances, γ_j, by γ_0. Thus,

$$
\rho_j = \frac{\gamma_j}{\gamma_0}, \quad j = 0, 1, 2, \ldots
\tag{22.19}
$$

and, the sample autocorrelations given by Equation 22.7 may be obtained by

$$
r_j = \frac{c_j}{c_0}, \quad j = 1, 2, 0, \ldots
\tag{22.20}
$$

22.2.1 Moving-average processes

If it can be safely assumed that $\psi_i = 0$ for $i > q$, where q is some integer, then Equation 22.13 becomes

$$
Z_t = \mu + u_t + \psi_1 u_{t-1} + \cdots + \psi_q u_{t-q}
\tag{22.21}
$$

which is referred to as a moving-average process of order q, or simply MA(q). By notational convention, Equation 22.21 is written as

$$
Z_t = \mu + u_t - \theta_1 u_{t-1} - \cdots - \theta_q u_{t-q}
\tag{22.22}
$$

Any MA process of the form in Equation 22.22 is stationary since the condition $\Sigma_{i=10}^{\infty} \psi_i$ in Equation 22.15 becomes

$$(1 + \psi_1 + \psi_2 + \cdots) = (1 - \theta_1 - \theta_2 - \cdots - \theta_q)$$

$$= 1 - \sum_{i=0}^{q} \theta_1 \tag{22.23}$$

which must always converge since the number of terms in the sum is finite. From Equation 22.16

$$\gamma_0 = \sigma_u^2 \sum_{i=0}^{q} \theta_i^2 \tag{22.24}$$

where $\theta_0 = 1$; and from Equation 22.17

$$\gamma_j = \begin{cases} \sigma_u^2(-\theta_j + \theta_1\theta_{j+1} + \cdots + \theta_{q-j}\theta_q), & j = 1,\ldots,q \\ 0, & j > q \end{cases} \tag{22.25}$$

Thus, from Equations 22.19 through 22.24, and Equation 22.25

$$\rho_j = \begin{cases} \dfrac{-\theta_j + \theta_1\theta_{j+1} + \cdots + \theta_{q-j}\theta_q}{1 + \theta_1^2 + \cdots + \theta_q^2}, & j = 1,\ldots,q \\ 0, & j > q \end{cases} \tag{22.26}$$

22.2.2   *Autoregressive processes*

In Equation 22.13, Z_t is expressed in terms of the current disturbance, u_t, and past disturbances, u_{t-j}. An alternative is to express Z_t in terms of the current disturbance and past observations, Z_{t-1}. This is done by rewriting Equation 22.13 as follows:

$$u_t = Z_t - \mu - \psi_1 u_{t-1} - \psi_2 u_{t-2} - \cdots \tag{22.27}$$

Since Equation 22.27 holds for any time subscript, the following can be written:

$$u_{t-1} = Z_{t-1} - \mu - \psi_1 u_{t-2} - \psi_2 u_{t-3} - \cdots$$
$$u_{t-2} = Z_{t-2} - \mu - \psi_1 u_{t-3} - \psi_2 u_{t-4} - \cdots \tag{22.28}$$
$$u_{t-3} = Z_{t-3} - \mu - \psi_1 u_{t-4} - \psi_2 u_{t-5} - \cdots$$
$$\vdots \qquad \vdots \qquad \vdots \qquad \vdots \qquad \vdots \qquad \vdots$$

By successive substitutions of Equation 22.28 for the u_{t-i}s in Equation 22.27, it is seen that

$$u_t = Z_t - \mu - \psi_1(Z_{t-1} - \mu - \psi_1 u_{t-2} - \cdots) - \psi_2 u_{t-2} - \cdots$$
$$= Z_t - \psi_1 Z_{t-1} - \mu(1 - \psi_1) + (\psi_1^2 - \psi_2)u_{t-2} + \cdots$$
$$\vdots \qquad \vdots \qquad \vdots \qquad \vdots \qquad \vdots \tag{22.29}$$

The continuation of the substitutions in Equation 22.29 yields an expression of the form

$$u_t = Z_t - \pi_1 Z_{t-1} - \pi_2 Z_{t-2} - \cdots - \delta \tag{22.30}$$

where the π_is and δ are fixed parameters resulting from the substitutions and are functions of the ψ_is and μ. From Equation 22.30

$$Z_t = \pi_1 Z_{t-1} + \pi_2 Z_{t-2} + \cdots + \delta + u_t \tag{22.31}$$

if $\pi_i = 0$ for $i > p$, where p is some integer, then Equation 22.31 becomes

$$Z_t = \pi_1 Z_{t-1} + \pi_2 Z_{t-2} + \cdots + \pi_p Z_{t-p} + \delta + u_t \tag{22.32}$$

which is referred to as an autoregressive process of order p, or simply AR(p). By notational convention, Equation 22.32 is written as follows:

$$Z_t = \theta_1 Z_{t-1} + \theta_2 Z_{t-2} + \cdots + \theta_p Z_{t-p} + \delta + u_t \tag{22.33}$$

Thus, AR processes are equivalent to MA processes of infinite order. Stationarity of AR processes is confirmed if the roots of the following characteristic equation lie outside the unit circle in the complex plane.

$$(1 - \theta_1 x - \theta_2 x^2 - \cdots - \theta_p x^p) = 0 \tag{22.34}$$

The variable x is simply an operational algebraic symbol. If stationarity exists, then

$$
\begin{aligned}
E(Z_t) &= \varnothing_1 E(Z_{t-1}) + \varnothing_2 E(Z_{t-2}) + \cdots + \varnothing_p E(Z_{t-p}) + \delta + E(u_t) \\
&= \varnothing_1 E(Z_t) + \varnothing_2 E(Z_t) + \cdots + \varnothing_p E(Z_t) + \delta \\
&= E(Z_t) + (\varnothing_1 + \varnothing_2 + \cdots + \varnothing_p) + \delta
\end{aligned} \tag{22.35}
$$

Therefore

$$E(Z_t) = \frac{\delta}{1 - \varnothing_1 - \varnothing_2 - \cdots - \varnothing_p} \tag{22.36}$$

The theoretical variance and covariances of the AR process may be obtained through the definition

$$\gamma_j = E\{[Z_t - E(Z_t)][Z_{t+j} - E(Z_{t+j})]\} \tag{22.37}$$

Denoting the deviation of the process from its mean by Z_t^d, the following is obtained:

$$
\begin{aligned}
Z_t^d &= Z_t - E(Z_t) \\
&= Z_t - \frac{\delta}{1 - \varnothing_1 - \varnothing_2 - \cdots - \varnothing_p}
\end{aligned} \tag{22.38}
$$

Therefore

$$Z_{t-1}^d = Z_{t-1} - \frac{\delta}{1 - \varnothing_1 - \varnothing_2 - \cdots - \varnothing_p} \tag{22.39}$$

Rewriting Equation 22.39 yields

$$Z_{t-1} = Z_{t-1}^d + \frac{\delta}{1 - \varnothing_1 - \varnothing_2 - \cdots - \varnothing_p} \tag{22.40}$$

Thus

$$Z_{t-k} = Z_{t-k}^d + \frac{\delta}{1 - \varnothing_1 - \varnothing_2 - \cdots - \varnothing_p} \tag{22.41}$$

Substituting Equation 22.33 into Equation 22.38 yields

$$Z_{t-k}^d = \varnothing_1 Z_{t-1} + \varnothing_2 Z_{t-2} + \cdots + \varnothing_p Z_{t-p} + \delta + u_t - \frac{\delta}{1 - \varnothing_1 - \cdots - \varnothing_p} \tag{22.42}$$

Successive substitutions of Equation 22.41 into Equation 22.42 for $k = 1, \ldots, p$ yields

$$Z_t^d = \varnothing_1 Z_{t-1}^d + \varnothing_2 Z_{t-2}^d + \cdots + \varnothing_p Z_{t-p}^d + u_t + \delta + \frac{\varnothing_1 \delta}{1 - \varnothing_1 - \cdots - \varnothing_p}$$

$$+ \frac{\varnothing_2 \delta}{1 - \varnothing_1 - \cdots - \varnothing_p} + \frac{\varnothing_p \delta}{1 - \varnothing_1 - \cdots - \varnothing_p} - \frac{\delta}{1 - \varnothing_1 - \cdots - \varnothing_p}$$

$$= \varnothing_1 Z_{t-1}^d + \varnothing_2 Z_{t-2}^d + \cdots + \varnothing_p Z_{t-p}^d + u_t + \frac{()}{1 - \varnothing_1 - \cdots - \varnothing_p}$$

$$= \varnothing_1 Z_{t-1}^d + \varnothing_2 Z_{t-2}^d + \cdots + \varnothing_p Z_{t-p}^d + u_t \tag{22.43}$$

Thus, the deviations follow the same AR process without a constant term. Definition (Equation 22.37) is now applied as follows:

$$\gamma_j = E[(Z_t^d)(Z_{t+j}^d)] \tag{22.44}$$

Thus

$$\begin{aligned}
\gamma_j &= E[(Z_t^d)(Z_{t+j}^d)] \\
&= E[(\varnothing_1 Z_t^d + \cdots + \varnothing_p Z_{t-p}^d + u_t)(Z_t^d)] \\
&= \varnothing_1 E(Z_t^d Z_{t-1}^d) + \cdots + \varnothing_p E(Z_t^d Z_{t-1}^d) + E(Z_t^d u_t) \\
&= \varnothing_1 \gamma_1 + \cdots + \varnothing_p \gamma_p + E(u_t^2) + E(cross - products) \\
&= \varnothing_1 \gamma_1 + \cdots + \varnothing_p \gamma_p + \sigma_u^2
\end{aligned} \tag{22.45}$$

Similarly

$$
\begin{aligned}
\gamma_j &= E[(Z_t^d)(Z_{t+j}^d)] \\
&= E[(\varnothing_1 Z_{t-1}^d + \cdots + \varnothing_p Z_{t-p}^d + u_t)(Z_{t-1}^d)] \\
&= \varnothing_1 E(Z_{t-1}^d Z_{t-1}^d) + \cdots + \varnothing_p E(Z_{t-1}^d Z_{t-1}^d) + E(Z_{t-1}^d u_t) \\
&= \varnothing_1 \gamma_0 + \cdots + \varnothing_p \gamma_{p-1} + E(cross - products) \\
&= \varnothing_1 \gamma_0 + \cdots + \varnothing_p \gamma_{p-1}
\end{aligned}
\tag{22.46}
$$

The continuation of the above procedure yields the following set of equations referred to as the Yule–Walker equations:

$$
\begin{aligned}
\gamma_0 &= \varnothing_1 \gamma_1 + \varnothing_2 \gamma_2 + \cdots + \varnothing_p \gamma_p + \sigma_u^2 \\
\gamma_1 &= \varnothing_1 \gamma_0 + \varnothing_2 \gamma_1 + \cdots + \varnothing_p \gamma_{p-1} \\
\gamma_2 &= \varnothing_1 \gamma_1 + \varnothing_2 \gamma_0 + \cdots + \varnothing_p \gamma_{p-2} \\
&\vdots \qquad \vdots \qquad \vdots \qquad \vdots \qquad \vdots \\
\gamma_p &= \varnothing_1 \gamma_{p-1} + \varnothing_2 \gamma_{p-2} + \cdots + \varnothing_p \gamma_0
\end{aligned}
\tag{22.47}
$$

and

$$
\gamma_j = \varnothing_1 \gamma_{j-1} + \varnothing_2 \gamma_{j-2} + \ldots + \varnothing_p \gamma_{j-p}, \quad j > p
\tag{22.48}
$$

Given the parameters $\varnothing_1, \varnothing_2, \ldots, \varnothing_p$ and σ_u^2, it is easy to solve the $p + 1$ linear equations in Equation 22.47 for the $p + 1$ unknowns $\gamma_0, \gamma_0, \ldots, \gamma_p$. For $j > p$, γ_j may be computed recursively from Equation 22.48. From Equation 22.19, dividing Equations 22.45 through 22.48 by γ_0, yields the Yule–Walker equations for the autocorrelation function as follows:

$$
\begin{aligned}
\rho_0 &= 1 \\
\rho_1 &= \varnothing_1 \rho_0 + \varnothing_2 \rho_1 + \cdots + \varnothing_p \rho_{p-1} \\
\rho_2 &= \varnothing_1 \rho_1 + \varnothing_2 \rho_0 + \cdots + \varnothing_p \rho_{p-2} \\
&\vdots \qquad \vdots \qquad \vdots \qquad \vdots \qquad \vdots \\
\rho_p &= \varnothing_1 \rho_{p-1} + \varnothing_2 \rho_{p-2} + \cdots + \varnothing_p \rho_0
\end{aligned}
\tag{22.49}
$$

and

$$
\rho_j = \varnothing_1 \rho_{j-1} + \varnothing_2 \rho_{j-2} + \cdots + \varnothing_p \rho_{j-p}, \quad j > p
\tag{22.50}
$$

The principal instruments for model building by the Box–Jenkins methodology (Box and Jenkins, 1976) are the sample autocorrelations, r_j, of the discrete time series. For the model identification procedure, a visual assessment of the plot of r_j against j, called the sample correlogram, is necessary. Figure 22.1 shows a tabulation of the general identification guidelines for sample correlogram.

In building a model, the main tools are the sample autocorrelations, which are only approximations to the theoretical behaviors. Thus, a wide variety of sample correlogram patterns will be encountered in practice. It is the responsibility of the time series analyst

Process type	Correlogram behavior	Pictorial appearance	Model equation
MA	Spikes at lags 1 to q only	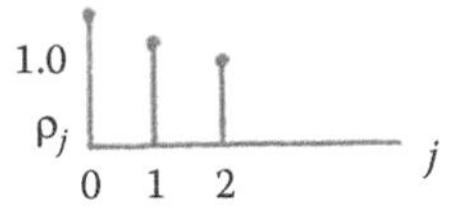	MA(2) $Z_t = u_t - \theta_1 u_{t-1} - \theta_2 u_{t-2}$
AR	Tail off exponentially	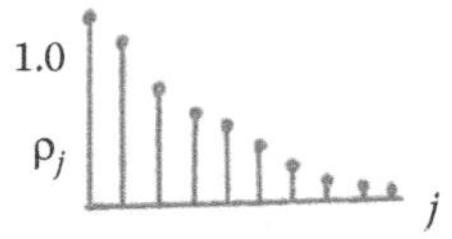	AR(1) $Z_t = \phi_1 Z_{t-1} + u_t$
AR	Damped sine waveform	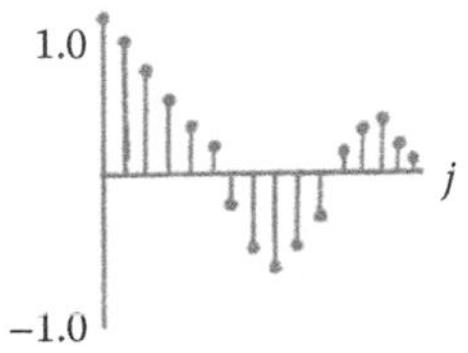	AR(2) $Z_t = \phi_1 Z_{t-1} + \phi_2 Z_{t-2} + u_t$

Figure 22.1 General theoretical characteristics of some time series models.

to choose an appropriate model based on the general guidelines provided by Box and Jenkins (1976). The analyst should be able to determine which behavior pattern is significantly or not significantly different from theoretical behaviors.

22.3   Model formulation

Based upon the guidelines presented earlier, a forecasting model was developed for the monthly electrical energy consumption at a university. The modeling process was performed in three major steps classified as model identification, parameter estimation, and diagnostic checking.

22.3.1   Model identification

The identification procedure involved the computation of the sample autocorrelations, r_j, of the discrete time series data as discussed in the previous section. A visual analysis of the sample correlogram was then used to propose a tentative model according to the Box–Jenkins guidelines. A statistical measure was used to determine which values in a sample correlogram were significantly different from zero. Any point that fell beyond two standard errors was considered to be significant. The significance limits drawn on the sample correlograms were obtained by Bartlett's approximation in 12-lag groups as follows:

$$\mathrm{SE}(r_j) = \frac{1}{\sqrt{T}}\left(1 + 2\sum_{i=1}^{q} r_i^2\right)^{1/2}, \quad j > q \tag{22.51}$$

The hypothesis

$$H_0 : r_j = 0$$
$$H_1 : r_j \neq 0$$ (22.52)

was then tested with the following criterion:

$$\text{If } |r_j| < 2[SE(r_j)], \text{ conclude } H_0$$

$$\text{otherwise, conclude } H_1.$$

Significant r_j values at large lags are an indication of nonstationarity.

Figure 22.2 shows the time series data plot as the corresponding correlogram. The correlogram shows an evidence of nonstationarity and a cyclical correlation in the data. The set of data could, thus, be represented by a mathematical combination of a deterministic periodic function and a stochastic model. Thus

$$Z_t = c_t + \varepsilon_t, \quad t = 0, 1, \ldots, 69$$ (22.53)

where

c_t, the appropriate periodic function, is hypothesized to be given by $\mu + A\cos(\omega t + \alpha)$

ε_t is the stochastic component

μ is a constant estimated by the sample mean of the data series ($\bar{Z} = 1,844,636.41$)

A is the amplitude of c_t

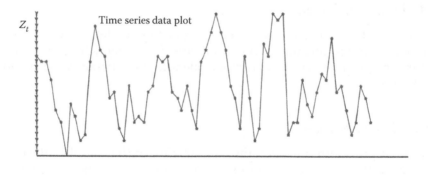

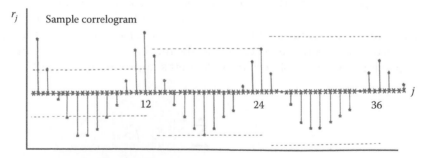

Figure 22.2 Energy consumption time series data plot and correlogram.

ω is the frequency of c_t (since there are 12 periods in one complete cycle of 1 year, ω = 2π/12 rad)
α is the phase angle of c_t

Since ε_t in Equation 22.53 was desired to be as close to zero as possible, it was necessary to minimize

$$\varepsilon_t = Z_t - c_t \tag{22.54}$$

The least-squares method was used to obtain the values of A and α that minimize Equation 22.54 as follows: minimize the sum of squares

$$S(A,\alpha) = \sum_{t=0}^{69}(Z_t - c_t)^2$$

$$= \sum_{t=0}^{69}[Z_t - \bar{Z}_t - A\cos(\varphi t + \alpha)]^2 \tag{22.55}$$

From trigonometric properties, we have the following

$$A\cos(\omega t + \alpha) = A\cos\alpha - A\sin\omega t\sin\alpha \tag{22.56}$$

letting

$$a = A\cos\alpha \tag{22.57}$$

and

$$b = -A\sin\alpha$$

Equation 22.56 becomes

$$A\cos(\omega t + \alpha) = a\cos\omega t + \sin\omega t \tag{22.58}$$

Equation 22.55 becomes

$$S(a,b) = \sum_{t=0}^{69}(Z_t - \bar{Z}_t - a\cos\omega t - b\sin\omega t)^2 \tag{22.59}$$

Now

$$\frac{\partial s}{\partial a} = -2\sum\cos\omega t(Z_t - \bar{Z} - a\cos\omega t - b\sin\omega t) \tag{22.60}$$

and

$$\frac{\partial s}{\partial b} = -2\sum\sin\omega t(Z_t - \bar{Z} - a\cos\omega t - b\sin\omega t) \tag{22.61}$$

Setting the derivatives in Equations 22.60 and 22.61 equal to zero yields two equations in the two unknowns $\hat{a}$ and $\hat{b}$

$$\sum \cos\omega t (Z - \bar{Z} - \hat{a}\cos\omega t - \hat{b}\sin\omega t) = 0 \tag{22.62}$$

and

$$\sum \sin\omega t (Z_t - \bar{Z} - \hat{a}\cos\omega t - \hat{b}\sin\omega t) = 0 \tag{22.63}$$

The expansion of Equations 22.62 and 22.63 yields

$$\sum Z_t \cos\omega t - \bar{Z}\sum \cos\omega t - \hat{a}\sum (\cos\omega t)^2 - \hat{b}\sum \cos\omega t \sin\omega t = 0 \tag{22.64}$$

and

$$\sum Z_t \sin\omega t - \bar{Z}\sum \sin\omega t - \hat{a}\sum \cos\omega t \sin\omega t - \hat{b}\sum (\sin\omega t)^2 = 0 \tag{22.65}$$

Equations 22.64 and 22.65 can now be solved simultaneously to yield $\hat{a}$ and $\hat{b}$ as follows: multiplying Equation 22.64 by $[\Sigma(\sin\omega t)^2]$ yields

$$\sum Z_t \cos\omega t \sum (\sin\omega t)^2 - \bar{Z}\sum \cos\omega t \sum (\sin\omega t)^2 - \hat{a}\sum (\cos\omega t)^2 \sum (\sin\omega t)^2$$
$$- \hat{b}\sum \cos\omega t \sin\omega t \sum (\sin\omega t)^2 = 0 \tag{22.66}$$

Multiplying Equation 22.65 by $[\Sigma\cos\omega t \sin\omega t]$ yields

$$\sum Z_t \sin\omega t \sum \cos\omega t \sin\omega t - \bar{Z}\sum \sin\omega t \sum \cos\omega t \sin\omega t - \hat{a}\left(\sum \cos\omega t \sin\omega t\right)^2$$
$$- \hat{b}\sum \cos\omega t \sin\omega t \sum (\sin\omega t)^2 = 0 \tag{22.67}$$

Subtracting Equation 22.67 from Equation 22.66 yields

$$\sum Z_t \cos\omega t \sum (\sin\omega t)^2 - \bar{Z}\sum \cos\omega t \sum (\sin\omega t)^2$$
$$- \sum Z_t \sin\omega t \sum \cos\omega t \sin\omega t + \bar{Z}\sum \sin\omega t \sum \cos\omega t \sin\omega t$$
$$- \hat{a}\left[\sum (\cos\omega t)^2 \sum (\sin\omega t)^2 \left(\sum \cos\omega t \sin\omega t\right)^2\right] = 0 \tag{22.68}$$

Solving Equation 22.68 for $\hat{a}$ yields

$$\hat{a} = \left\{ \sum Z_t \cos \omega t \sum (\sin \omega t)^2 - \bar{Z} \left[\sum \cos \omega t \sum (\sin \omega t)^2 - \sum \sin \omega t \sum \cos \omega t \sin \omega t \right] \right.$$
$$\left. - \sum Z_t \sin \omega t \sum \cos \omega t \sin \omega t \right\} \div \left[\sum (\cos \omega t)^2 \sum (\sin \omega t)^2 - \left(\sum \cos \omega t \sin \omega t \right)^2 \right]$$

(22.69)

Likewise

$$\hat{b} = \left\{ \sum Z_t \sin \omega t \sum (\cos \omega t)^2 - \bar{Z} \left[\sum \sin \omega t \sum (\cos \omega t)^2 - \sum \cos \omega t \sum \sin \omega t \cos \omega t \right] \right.$$
$$\left. - \sum Z_t \cos \omega t \sum \sin \omega t \cos \omega t \right\} \div \left[\sum (\sin \omega t)^2 \sum (\cos \omega t)^2 - \left(\sum \sin \omega t \cos \omega t \right)^2 \right]$$

(22.70)

The numerical evaluation of Equations 22.69 and 22.70 yielded

$$\hat{a} = 346{,}752.91$$

(22.71)

and

$$\hat{b} = 203{,}918.82$$

(22.72)

and from Equation 22.57

$$\tan \alpha = -\frac{b}{a}$$

(22.73)

Therefore

$$\hat{\alpha} = \arctan \left(-\frac{\hat{b}}{\hat{a}} \right)$$
$$= \arctan(-0.5881)$$
$$= -0.5316 \text{ rad}$$

(22.74)

Also, from Equation 22.57,

$$a^2 + b^2 = A^2(\cos \alpha)^2 + A^2(\sin \alpha)^2$$
$$= A^2(\cos^2 \alpha + \sin^2 \alpha)$$
$$= A^2$$

(22.75)

Therefore

$$\hat{A} = (\hat{a}^2 + \hat{b}^2)^{1/2}$$
$$= 402{,}269.15 \qquad (22.76)$$

Thus

$$c_t = 1{,}884{,}636.41 + 402{,}269.15\cos\left(\frac{\pi}{6}t - 0.5316\right) \qquad (22.77)$$

The model building, thus, involved extracting the cyclic portion of the raw data, and then using the Box–Jenkins approach to develop a stochastic model for the leftover series, ε_t. It was observed that a reduction in variability was effected by fitting the periodic function. The sample variance of the raw data, as obtained from Computer Program 1, is 0.1636474×10^{12} while that of the residual series, ε_t, is 0.821757×10^{11}. Thus, 49.8% of the total variance of the observation is accounted for by the addition of the cyclic term in the model.

Experimentations were done with various plots of ε_t, various differencing of ε_t, and their respective sample autocorrelations. Each differencing tried was computed as follows:

For no differencing

$$\varepsilon_t = (1 - B)^0 (1 - B^{12})^0 \varepsilon_t$$
$$= (1 - B)^0 \varepsilon_t$$
$$= \varepsilon_t \qquad (22.78)$$

where B is the backshift operator. The operator simply shifts the time subscript on a variable backward by one period. Thus

$$B(Z_t) = Z_{t-1} \qquad (22.79)$$

If B is raised to a power, it means B is applied as many times as the power indicates; that is

$$B^3 = BBBZ_t$$
$$= Z_{t-3} \qquad (22.80)$$

For first regular differencing

$$w_t = (1 - B)^1 (1 - B^{12})^0 \varepsilon_t$$
$$= (1 - B)\varepsilon_t$$
$$= \varepsilon_t - B\varepsilon_t$$
$$= \varepsilon_t - \varepsilon_{t-1} \qquad (22.81)$$

For the first seasonal differencing

$$
\begin{aligned}
y_t &= (1-B)^0(1-B^{12})^1\varepsilon_t \\
&= (1-B^{12})\varepsilon_t \\
&= \varepsilon_t - B^{12}\varepsilon_t \\
&= \varepsilon_t - \varepsilon_{t-12}
\end{aligned}
\tag{22.82}
$$

For first regular and first seasonal differencing

$$
\begin{aligned}
s_t &= (1-B)^1(1-B^{12})^1\varepsilon_t \\
&= (1-B)(\varepsilon_t - \varepsilon_{1-12}) \\
&= \varepsilon_t - \varepsilon_{t-1} - \varepsilon_{t-12} + \varepsilon_{t-13} \\
&= y_t - y_{t-1}
\end{aligned}
\tag{22.83}
$$

Four tentative models (one for each differencing) were identified. The parameters, θ_j, included in each model correspond to the significant lags in the sample correlograms. Here, δ is a fixed constant estimated by the process sample mean. An attempt was made to make each model as parsimonious in parameters as possible. The tentative models were as follows:

$$
\varepsilon_t = u_t - \theta_1 u_{t-1} - \theta_{12} u_{t-12} + \delta
\tag{22.84}
$$

$$
w_t = u_t - \theta_1 u_{t-1} - \theta_3 u_{t-3} - \theta_{12} u_{t-12} + \delta
\tag{22.85}
$$

$$
y_t = u_t - \theta_1 u_{t-1} - \theta_7 u_{t-7} - \theta_{12} u_{t-12} + \delta
\tag{22.86}
$$

and

$$
s_t = u_t - \theta_1 u_{t-1} - \theta_5 u_{t-5} - \theta_7 u_{t-7} - \theta_{12} u_{t-12} + \delta
\tag{22.87}
$$

After initial and final parameter estimates were obtained, a computer simulation was carried out for each model. From the simulation outputs, model (22.86) was selected as the overall best based on the residual plots, histogram of residual autocorrelations, residual sum-of-squares, X^2 test, and the closeness of the simulated data, Z_t (*sim*) to the raw data, Z_t. The complete analysis of model (22.86) began with the parameter estimation process.

22.3.2   *Parameter estimation*

The initial estimates for θ_1, θ_7, and θ_{12} were obtained with the aid of Equation 22.26, using the sample autocorrelations. Thus, for regular lags

$$
r_1 = \frac{-\hat{\theta}_1}{1 + \hat{\theta}_1^2 + \hat{\theta}_7^2}
\tag{22.88}
$$

$$r_7 = \frac{-\hat{\theta}_7}{1 + \hat{\theta}_1^2 + \hat{\theta}_7^2}$$ (22.89)

and for the seasonal lags

$$r_{12} = \frac{-\hat{\theta}_{12}}{1 + \hat{\theta}_{12}^2}$$ (22.90)

where θ_j for $j \neq 1, 7,$ and 12 were taken to be zero. From correlogram plots

$$r_1 = r_7 = 0.32$$ (22.91)

which, as a rough estimate, suggested

$$\hat{\theta}_1 = \hat{\theta}_7$$ (22.92)

Therefore, Equation 22.88 became

$$r_1 = \frac{-\hat{\theta}_1}{1 + 2\hat{\theta}_1^2}$$ (22.93)

Substituting r_j in Equations 22.90 and 22.93 yielded

$$0.32 = \frac{-\hat{\theta}_1}{1 + 2\hat{\theta}_1^2}$$ (22.94)

and

$$-0.39 = \frac{-\hat{\theta}_{12}}{1 + \hat{\theta}_{12}^2}$$ (22.95)

The quadratic solutions of Equations 22.94 and 22.85 yielded

$$\hat{\theta}_1 = -0.4491 \quad \text{or} \quad -1.1134$$

and

$$\hat{\theta}_{12} = 0.4798 \quad \text{or} \quad 2.0843$$ (22.96)

Since stationarity requires the absolute value of θ_j to be less than one, the following initial estimates were chosen:

$$\theta_1 = -0.4491$$
$$\theta_7 = -0.4491 \tag{22.97}$$
$$\theta_{12} = 0.4798$$

Under the assumption of normality, the maximum likelihood estimates (MLE) of the parameters were approximated by the least-squares estimates. The likelihood function is defined as

$$L(\emptyset,\theta,\delta,\sigma_u^2|Z)_t = (2\pi\sigma_u^2)^{-T/2}\exp\left[-\frac{1}{2\sigma_u^2}\sum_{t=1}^{T}\hat{u}(\emptyset,\theta,\delta)_t^2\right] \tag{22.98}$$

where $\hat{u}(\emptyset,\theta,\delta)_t$ is the estimate of the residual, u_t, which is a function of the unknown parameters and the given observation sequence, Z. Vectors $\emptyset$ and θ denote the autoregressive and MA parameter vectors, respectively. From model (22.86), $\hat{u}_t$ is given by

$$\hat{u}(\theta,\delta) = y_t + \theta_1 u_{t-1} + \theta_7 u_{t-7} + \theta_{12} u_{t-12} + \delta \tag{22.99}$$

Since only the relative magnitudes of the likelihood function are of interest, it is sufficient to consider the log of the function given by

$$\ln L = \ln(\emptyset,\theta,\delta,\sigma_u^2|Z) = -T\ln\sigma_u - \frac{s(\emptyset,\theta,\delta)}{2\sigma_u^2} \tag{22.100}$$

where

$$S(\emptyset,\theta,\delta) = \sum_{t=1}^{T}\hat{u}(\emptyset,\theta,\delta)_t^2 \tag{22.101}$$

Using Equation 22.101 in Program 3, numerical nonlinear estimation yielded the following estimates which correspond to the minimum sum of squares

$$\hat{\theta}_1 = -0.3145$$
$$\hat{\theta}_7 = -0.2328$$
$$\hat{\theta}_{12} = 0.6385$$
$$\hat{\delta} = \bar{y}$$
$$= 14{,}687.98 \tag{22.102}$$

Thus, model (22.86) became

$$y_t = u_t + 0.3145u_{t-1} + 0.2328u_{t-1} - 0.6385u_{t-12} + 14{,}687.98 \tag{22.103}$$

Substituting Equations 22.54 and 22.82 into Equation 22.103 yielded

$$(Z_t - c_t) - (Z_{t-12} - c_{t-12}) = u_t + 0.3145u_{t-1} - 0.2328u_{t-1} - 0.6385u_{t-12} + 14.687.98 \quad (22.104)$$

Therefore, substituting Equation 22.77 into Equation 22.104 and rearranging yielded

$$Z_t = Z_{t-12} + u_t + 0.3145u_{t-1} + 0.2328u_{t-7} - 0.6385u_{t-12} + 14.687.98 + 402,269.15 \quad (22.105)$$

$$\left\{ \cos\left(\frac{\pi}{6}t - 0.5316\right) - \cos\left[\frac{\pi}{6}(t - 12) - 0.5316\right] \right\}$$

as the fitted model for the monthly electricity consumption.

22.3.3 Diagnostic check

The diagnostic check on the appropriateness of model (22.86) was done with the aid of the estimated residuals, $\hat{u}_t$, from the model. The residual plot does not show any distinct pattern that may invalidate the assumption that the residuals are normally distributed. The histogram of the residual autocorrelations also does not indicate any remarkable departure from normality. Since the ideal situation requires that all the residual autocorrelations be zero, the X^2 test was used as an overall measure of their smallness in groups of 12, 24, and 36 lags. The Q statistic is defined as

$$Q = \sum_{j=1}^{k} r_j^2 \quad (22.106)$$

where T is the sample size of the differenced series, k is the number of lags in the group, and r_j is the residual autocorrelation at lag j. Here, Q is a X^2 distribution with $(k - p - q)$ degrees of freedom, where p and q are, respectively, the number of autoregressive and MA parameters included in the model. The X^2 test was carried out with the following criterion:

$$\text{If } Q > X^2(1 - \alpha; \ k - p - q), \text{ the model is inadequate,}$$
$$\text{otherwise, the model is adequate,} \quad (22.107)$$

where $(1 - \alpha) \times 100\%$ is the level of confidence desired. Using the r_j values from Program 4, the testing was done as follows: for $k = 12$ and $\alpha = 0.25$,

$$Q = 58 \sum_{j=1}^{12} r_j^2$$
$$= 58(0.1194)$$
$$= 6.9252 \quad (22.108)$$

$$X^2(1 - 0.025, 12 - 0 - 3) = X^2(0.975; \ 9) \quad (22.109)$$

For $k = 24$ and $\alpha = 0.25$,

$$Q = 58 \sum_{j=1}^{24} r_j^2$$
$$= 58(0.1852)$$
$$= 10.7400 \tag{22.110}$$

$$X^2(0.975;\ 21) = 34.1700 \tag{22.111}$$

For $k = 36$ and $\alpha = 0.25$,

$$Q = 58 \sum_{j=1}^{36} r_j^2$$
$$= 58(0.3200)$$
$$= 18.5600 \tag{22.112}$$

$$X^2(0.975;\ 33) = 46.9800 \tag{22.113}$$

Since $Q < X^2(1 - \alpha;\ k - p - a)$ for $k = 12$, 24, and 36 at the 97.5% confidence level, it was concluded that the model was adequate. A simultaneous plot of the simulated data, Z_t (*sim*) and the raw data, Z_t, was performed. The close match between the two series in the plot is a further evidence of the adequacy of the fitted model.

22.4   Forecasting

The next step after the favorable diagnostic check of the fitted model involved the development of the forecast functions. From Equation 22.105, the electricity consumption at time $t + 1$ is represented by

$$Z_{t+1} = Z_{t-11} + 0.3145u_t + 0.2328u_{t-6} - 0.6385u_{t-11} + 14.687.98$$
$$+ 402,269.15\left\{\cos\left[\frac{\pi}{6}(t + 1) - 0.5316\right] - \cos\left[\frac{\pi}{6}(t - 11) - 0.5316\right]\right\} \tag{22.114}$$

For simplicity in writing the forecast functions, the cyclic components in Equation 22.114 were represented by $c_{t+\ell}$. Thus, Equation 22.114 could be written as follows:

$$Z_{t+1} = Z_{t-11} + u_{t+1} + 0.3145u_t + 0.2328u_{t-6} - 0.6386u_{t-11} + 14.687.98 + c_{t+1} - c_{t-11} \tag{22.115}$$

22.4.1   One-month-ahead forecasts

The forecast of Z_{t+1} made at time t, is given by

$$\hat{Z}_t(1) = Z_{t-11} + \hat{u}_t(1) + 0.3145u_t + 0.2328u_{t-6} - 0.6385u_{t-11} + 14,687.98 + c_{t+1} - c_{t-11} \tag{22.116}$$

where Z_{t-11} represents the observed electricity consumption at time $t-11$. Since the marginal expectation of the disturbances is zero, the best estimate of u_{t+1}, given by $\hat{u}_t(1)$, is zero. The terms, u_t, u_{t-6}, and u_{t-11}, represent the observed disturbances at times t, $t-6$, and $t-11$, respectively. The terms are defined as

$$u_t = Z_t - \hat{Z}_{t-1}(1)$$
$$u_{t-6} = Z_{t-6} - \hat{Z}_{t-7}(1) \tag{22.117}$$

and

$$u_{t-11} = Z_{t-11} - \hat{Z}_{t-12}(1),$$

where $\hat{Z}_{t-1}(1)$, $\hat{Z}_{t-7}(1)$, and $\hat{Z}_{t-12}(1)$ are the forecasts of Z_t, Z_{t-6}, and Z_{t-11} made, at times $t-1$, $t-7$, and $t-12$, respectively. Thus, the 1-month-ahead forecast function became

$$\hat{Z}_t(1) = Z_{t-11} + 0.3145[Z_t - \hat{Z}_{t-1}(1)] + 0.2328[Z_{t-11} - \hat{Z}_{t-7}(1)]$$
$$- 0.6385[Z_{t-11} - \hat{Z}_{t-12}(1)] + 14{,}687.98 + c_{t+1} - c_{t-11} \tag{22.118}$$

Using Equation 22.118 and the fact that any previously unobserved disturbance could be estimated as being zero while any previously unrecorded consumption could be estimated by the sample average, a sequence of 1-month-ahead forecasts was generated. The forecasts were obtained by recursive computer computations that began at the start of the actual data series. A simultaneous plot of the forecasts and the actual data shows a close match. Since the previous disturbances were not actually zero and the previous consumptions were not actually equal to the average value, the initial forecasts are somehow distorted. However, the errors become smaller the further they are from the first forecast.

22.4.2 Multi-step-ahead forecasts

The functions for ℓ-months-ahead, $(\ell \geq 1)$, forecasts were developed as follows: from Equations 22.105 and 22.115,

$$Z_{t+\ell} = Z_{t+\ell-11} + u_{t+\ell} + 0.3145u_{t+\ell-1} + 0.2328u_{t+\ell-7}$$
$$- 0.6385u_{t+\ell-12} + 14{,}687.98 + c_{t+1} - c_{t-11} \tag{22.119}$$

Thus, for forecast functions for $\ell = 1, 2,\ldots, 12$ at time t, where $\hat{u}_{t+\ell}$ is zero, are given by

$$\ell = 1: \quad \hat{Z}_t(1) = Z_{t-11} + \hat{u}_{t+1} + 0.3145u_t + 0.2328u_{t-6} - 0.6385u_{t-11} + 14{,}687.98 + c_{t+1} - c_{t-11}$$
$$= Z_{t-11} + 0.3145[Z_t - \hat{Z}_{t-1}(1)] + 0.2328[Z_{t-6} - \hat{Z}_{t-7}(1)]$$
$$- 0.6385[Z_{t-11} - \hat{Z}_{t-12}(1)] + 14{,}687.98 + c_{t+1} - c_{t-11} \tag{22.120}$$

$$\ell = 2,\ldots,7: \quad \hat{Z}_t(\ell) = Z_{t+\ell-12} + 0.2328[Z_{t+\ell-7} - \hat{Z}_{t+\ell-8}(1)] - 0.6385[Z_{t+\ell-12} - \hat{Z}_{t+\ell-13}(1)]$$
$$+ 14{,}687.98 + c_{t+\ell} - c_{t=\ell-12} \tag{22.121}$$

$$\ell = 8: \quad \hat{Z}_t(\ell) = Z_{t+\ell-12} + 0.6385[Z_{t+\ell-12} - \hat{Z}_{t+\ell-13}(1)] + 14{,}687.98 + c_{t+\ell} - c_{l=\ell-12}; \quad (22.122)$$

$$\ell = 8,\ldots,12: \quad \hat{Z}_t(\ell) = Z_{t+\ell-12} + 0.6385[Z_{t+\ell-12} - \hat{Z}_{t+\ell-13}(1)] + 14{,}687.98 + c_{t+\ell} - c_{t=\ell-12} \quad (22.123)$$

$$\ell > 12: \quad \hat{Z}_t(\ell) = Z_{t+\ell-12} + 14{,}687.98 + c_{t+\ell} - c_{t=\ell-12} \quad (22.124)$$

From the above forecast function, it is expected that the forecast will become less reliable the further ahead they are made. This is because more of the disturbance terms drop out of the forecast functions as the size of ℓ increases. This condition is evidenced by a plot of the forecast for the next 12 months after the actual data series. It was observed in the plot that the forecasts tend to spread out more as ℓ increases. As a result, it will be more beneficial to update the forecasts about every 6 months.

22.4.3 Confidence limits for forecasts

Since the disturbances are assumed to be normally distributed, 95% confidence limits for the ℓ-months-ahead forecasts were computed as follows:

$$\hat{Z}_t(\ell) - 1.96SD[e_t(\ell)] < Z_{t+\ell} < \hat{Z}_t(\ell) + 1.96SD[e_t(\ell)] \quad (22.125)$$

where "1.96" conveys the fact that approximately two standard deviations cover 95% of the area under the normal curve, $SD[e_t(\ell)]$ is the standard deviation of the forecast error, $e_t(\ell)$, defined as

$$e_t(\ell) - Z_{t+\ell} - \hat{Z}_t(\ell) \quad (22.126)$$

and $\hat{Z}_t(\ell) \pm 1.96SD[e_t(\ell)]$ gives the upper and lower limits for the forecasted consumption. The expression for computing the standard deviation is given by the following

$$SD[e_t(\ell)] = \hat{\sigma}_u(1 + \psi_1^2 + \cdots + \psi_{\ell-1}^2)^{1/2} \quad (22.127)$$

where
$\hat{\sigma}_u = 253{,}798.23$ is the residual standard deviation obtained from Program 4
$\psi_i = i$th MA coefficient derived from Equation 22.105
$\psi_0 = 1$

The ψ_i values were obtained by successive substitutions for Z_{t-j} in Equation 22.105. Thus

$$Z_t = u_t + 0.3145u_{t-1} + 0.2328u_{t-7} - 0.6385u_{t-12} + \cdots + Z_{t-12} \quad (22.128)$$

But

$$Z_{t-12} = u_{t-12} + 0.3145u_{t-13} + 0.2328u_{t-19} + \cdots + Z_{t-24} \quad (22.129)$$

and

$$Z_{t-24} = u_{t-24} + 0.3145u_{t-25} + 0.2328u_{t-31} + \cdots + Z_{t-36}$$
$$\vdots \qquad \vdots \qquad \vdots \qquad \qquad \vdots \qquad \vdots \qquad \vdots \tag{22.130}$$

Therefore

$$Z_t = u_t + 0.3145u_{t-1} + 0.2328u_{t-7} + (1 - 0.6385)u_{t-12} + 0.3145u_{t-13} + \cdots \tag{22.131}$$

Thus

$$\psi_1 = 0.3145$$
$$\psi_1 = \psi_2 = \cdots = \psi_6 = 0$$
$$\psi_7 = 0.2328$$
$$\psi_8 = \psi_9 = \cdots = \psi_{11} = 0$$
$$\psi_{12} = 0.3615$$
$$\vdots \qquad \vdots \tag{22.132}$$

$$SD[e_t(1)] = \hat{\sigma}_u(1)^{1/2}$$
$$= 253{,}798.23$$
$$SD[e_t(2)] = \hat{\sigma}_u[1 + (0.3145)^2]^{1/2}$$
$$= 266{,}053.94$$
$$SD[e_t(\ell)] = 266{,}053.94, \quad \ell = 3,\ldots,7$$
$$SD[e_t(8)] = \hat{\sigma}_u[1 + (0.3145)^2]^{1/2}$$
$$= 272{,}535.59$$
$$SD[e_t(\ell)] = 272{,}535.59, \quad \ell = 9,10,11$$
$$SD[e_t(12)] = \hat{\sigma}_u[1 + (0.3145)^2 + (0.2328)^2 + (0.3615)^2]^{1/2}$$
$$= 287{,}564.52 \tag{22.133}$$

22.4.4   Forecast evaluation and conclusions

The simultaneous plot of the 1-month-ahead forecasts and the actual data shows a very close fit after the initiating forecasts of the first 18 months. The multi-step-ahead forecasts, as mentioned earlier, tend to diverge more as the number of steps ahead increases. Consequently, the best forecasting approach is to update the forecasts as often as conveniently possible. This will allow for the utilization of the most recent consumptions as inputs for subsequent forecasts. As a further check on the efficiency of the forecast functions, subsequently available energy consumption numbers were compared to their forecasts that had been made earlier. The comparison is tabulated below.

It is seen in Table 22.1 that the forecasts reasonably predicted the actual consumptions. Each of the forecasts also falls within the 95% confidence limits. Figure 22.3 shows a graphical comparison of the actual data to the forecasted numbers. As an application of measurement, mathematically based forecasts can be helpful in various management

Table 22.1 Forecast evaluation

Month	Year	Actual (kWh)	Forecast (kWh)
May	1980	1,609,542.00	2,085,127.44
June	1980	2,260,352.00	1,901,127.38
July	1980	2,154,526.00	2,317,098.03
August	1980	2,475,117.00	2,304,690,07

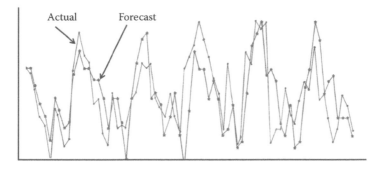

Figure 22.3 Graphical comparison of actual data to forecasted numbers.

decision-making scenarios. The approach presented in this chapter is just one of the many available techniques. With the present availability of powerful computer tools, quantitative analysis of forecasts can be made easily and quickly. The mathematical modeling approach presented in this chapter is expected to serve as a template for researchers and practitioners interested in developing customized forecasting tools.

References

Badiru, A. B. 1981. *Stochastic Modeling of Energy Consumption at Tennessee Technological University*, MS thesis, Tennessee Technological University, Cookeville, TN.

Box, G. E. P. and Jenkins, G. M. 1976. *Time Series Analysis Forecasting and Control*, Holden-Day, San Francisco, CA.

chapter twenty-three

Mathematical measurement of project parameters for multiresource control

Adedeji B. Badiru

Contents

The premise of this chapter is the development of a generic project control measurement tool that incorporates (1) resource characteristics, such as preferences, time-effective capabilities, costs, and availability of project resources, and (2) performance interdependencies among different resource groups, and proceeds to map the most adequate resource units to each newly scheduled project activity for the purpose of better project control. The chapter is based on the work of Milatovic and Badiru (2004). The principal challenge in this measurement model is to make it applicable to realistic project environments, which often involve multifunctional resources whose capabilities or other characteristics may cross activities, as well as within a single activity relative to specific interactions among resources themselves. This measurement methodology is effective wherever the duration, cost, and resource interdependencies exist in a resource-constrained project.

The proposed methodology dynamically executes two alternative procedures: the *activity scheduler* and *resource mapper*. The *activity scheduler* prioritizes and schedules activities based on their attributes and may also attempt to centralize selected resource loading graphs based on activity resource requirements. The *resource mapper* considers resource characteristics, incorporates interdependencies among resource groups or types, and maps the available resource units to newly scheduled activities according to a project manager's prespecified mapping (objective) function.

23.1 Introduction

Project resources, generally limited in quantity, are the most important constraints in scheduling of activities. In cases when resources have prespecified assignments and responsibilities toward one or more activities, their allocation is concurrently performed with the scheduling of applicable activities. In other cases, an activity may only require a certain number of (generic) resource units of particular type(s), which are assigned after the scheduling of the particular activity. These two approaches coarsely represent the dominant paradigms in project scheduling. The objective of this methodology is to propose a new strategy that will shift these paradigms to facilitate a more refined guidance for allocation and assignment of project resources. Accurate measurements are a prerequisite for an effective decision making. There is a need for tools that will provide for more effective resource tracking, control, interaction, and, most importantly, resource–activity mapping.

The main assumption in the methodology is that project environments often involve multicapable resource units with different characteristics. This is especially the case in knowledge intensive settings and industries, which are predominantly staffed with highly trained personnel. The specific characteristics considered were resource preferences, time-effective capabilities, costs, and availability. Each resource unit's characteristics may further vary across project activities, but also within a single activity relative to interaction among resource units. Finally, resource preferences, cost, and time-effective capabilities may also independently vary with time due to additional factors, such as learning, forgetting, weather, type of work, etc. Therefore, although we do not exclude a possibility that an activity duration is independent of resources assigned to it, in this project measurement approach, we assume that it is those resource units assigned to a particular activity that determine how long it will take for the activity to be completed.

The scheduling strategy as illustrated above promotes a more balanced and integrated resource–activity mapping approach. Mapping the most qualified resources to each project activity, and thus preserving the values of resource, is achieved by proper consideration or resource time-effective capabilities and costs. By considering resource preferences and availability, which may be entered in either crisp or fuzzy form, the model enables consideration of personnel's voice and its influence on a project schedule and quality. Furthermore, resource interactive dependencies may also be evaluated for each of the characteristics and their effects incorporated into resource–activity mapping. Finally, by allowing flexible and dynamic modifications of scheduling objectives, the model permits managers or analysts to incorporate some of their tacit knowledge and discretionary input into project schedules.

23.2 Literature review

Literature presents extensive work on scheduling projects by accounting for worker (resource) preferences, qualifications, and skills, as decisive factors to their allocation. Yet, a recent survey of some 400 top contractors in construction showed that 96.2% of them still use basic critical path method (CPM) for project scheduling (Mattila and Abraham, 1998). Roberts (1992) argued that information sources for project planners and schedulers are increasingly nonhuman, and stressed that planners must keep computerized tools for project management and scheduling in line and perspective with human resources used by projects. In other words, the author warns that too much technicality may prompt and mislead the managers into ignoring human aspects of management.

Franz and Miller (1993) considered a problem of scheduling medical residents to rotations and approached it as a large-scale multiperiod staff assignment problem. The objective of the problem was to maximize residents' schedule preferences while meeting hospital's training goals and contractual commitments for staffing assistance.

Gray et al. (1993) discussed the development of an expert system to schedule nurses according to their scheduling preferences. Assuming consistency in nurses' preferences, an expert system was proposed and implemented to produce feasible schedules considering nurses' preferences, but also accounting for overtime needs, desirable staffing levels, patient acuity, etc. Similar problem was also addressed by Yura (1994), where the objective was to satisfy worker's preferences for time off as well as overtime but under due date constraints.

Campbell (1999) further considered allocation of cross-trained resources in multidepartmental service environment. Employers generally value more resource units with various skills and capabilities for performing a greater number of jobs. It is in those cases when managers face challenges of allocating these workers such that the utility of their assignment to a department is maximized. The results of experiments showed that the benefits of cross-training utilization may be significant. In most cases, only a small degree of cross-training captured the most benefits, and tests also showed that beyond a certain amount, the additional cross-training adds little additional benefits.

Finally, many companies face problems of continuous reassignment of people in order to facilitate new projects (Cooprider, 1999). Cooprider suggests a seven-step procedure to help companies consider a wide spectrum of parameters when distributing people within particular projects or disciplines.

Badiru (1993) proposed *critical resource diagramming* (CRD), which is a simple extension to traditional CPM graphs. In other words, criticalities in project activities may also be reflected on resources. Different resource types or units may vary in skills or supply or can be very expensive. This discrimination in resource importance should be accounted for when carrying out their allocation in scheduling activities.

Unlike activity networks, the CRDs use nodes to represent each resource unit. Also, unlike activities, a resource unit may appear more than once in a CRD network, specifying all different tasks for which a particular unit is assigned. Similar to CPM, the same backward and forward computations may be performed to CRDs.

23.3   *Methodology*

The methodology of this chapter represents an analytical extension of CRDs discussed in the preceding chapter. As previously mentioned, the design considerations of the proposed model consist of two distinct procedures: activity scheduling and resource mapping. At each decision instance during a scheduling process, the *activity scheduler* prioritizes and schedules some or all candidate activities, and then the *resource mapper* iteratively assigns the most adequate resource units to each of the newly scheduled activities.

23.4   *Representation of resource interdependencies and multifunctionality*

This study is primarily focused on renewable resources. In addition, resources are not necessarily categorized into types or groups according to their similarities (i.e., into personnel, equipment, space, etc.) but more according to hierarchy of their interdependencies. In other words, we assume that time-effective capabilities, preferences, or even cost of any particular resource unit assigned to work on an activity may be dependent on other resource units also assigned to work on the same activity. Some or all of these other resource units may, in similar fashion, be also dependent on a third group of resources, and so on. Based on the above assumptions, we model competency of project resources in terms of following four resource characteristics: time-effective capabilities, preferences, cost, and availability. Time-effective capability of a resource unit with respect to a particular activity is the amount of time the unit needs to complete its own task if assigned to that particular activity. Preferences are relative numerical weights that indicate personnel's degree of desire to be assigned to an activity, or manager's perception on assigning certain units to particular activities. Similarly, each resource unit may have different costs associated with it relative to which activities it gets assigned. Finally, not all resource units may be available to some or all activities at all times during project execution. Thus, times during which a particular unit is available to some or all activities are also incorporated into the mapping methodology. Each of the characteristics described may vary across different project activities. In addition, some or all of these characteristics (especially time-effective capabilities and preferences) may also vary within a particular activity relative to resource interaction with other resources that are also assigned to work on the same activity.

Those resources whose performance is totally independent of their interaction with other units are grouped together and referred to as the type or group "one" and allocated first to scheduled activities. Resource units whose performance or competency is affected by their interaction with the type or group "one" units are grouped into type or group "two" and assigned (mapped) next. Resource units whose competency or performance is a function of type "two" or both types "one" and "two" are grouped into type "three" and allocated to scheduled activities after the units of the first two types have been assigned to them.

A project manager may consider one, more than one, or all of the four characteristics when performing resource–activity mapping. For example, a manager may wish to keep project costs as low as possible, while at the same time attempting to use resources with the best time-effective capabilities, considering their availability, and even incorporating their voice (in case of humans) or his/her own perception (in cases of human or nonhuman resources) in the form of preferences. This objective may be represented as follows:

$$U_i^{j,k} = f(t_i^{j,k}, c_i^{j,k}, p_i^{j,k}, \alpha_i^{j,k}(t_c))$$

Mapping units of all resource types according to the same mapping function may often be impractical and unrealistic. Cost issues may be of greater importance in mapping some, while inferior to time-effective capabilities of other resource types. To accommodate the need for a resource-specific mapping function as mapping objective, we formulated the mapping function as additive utility function (Keeney and Raiffa, 1993). In such a case, each of its components pertains to a particular resource type and is multiplied by a *Kronecker's delta* function (Bracewell, 1978). Kronecker's delta then detects resource type whose units are currently being mapped and filters out all mapping function components, except the one that pertains to the currently mapped resource type.

As an example, consider again a case where all resource types would be mapped according to their time-effective capabilities, except in the case of resource types "two" and "three" where costs would also be of consideration, and in the case of type "five," resource preferences and availabilities would be considered:

$$U_i^{j,k} = f(t_i^{j,k}) + f_2(c_i^{j,k}) \cdot \delta(j,2) + f_3(c_i^{j,k}) \cdot \delta(j,3) + f_5(p_i^{j,k}, \alpha_i^{j,k}(t_c)) \cdot \delta(j,5)$$

The above example illustrates a case where mapping of resource units is performed according to filtered portions of a manager's objective (mapping) function, which may in turn be dynamically adaptive and varying with project scheduling time. As previously indicated, some resource characteristics may be of greater importance to a manager in the early scheduling stages of a project rather than in the later stages. Such a mapping function may be modeled as follows:

$$U_i^{j,k} = f_g(t_i^{j,k}, c_i^{j,k}, p_i^{j,k}, a_i^{j,k}(t_c)) + \sum_{s \in T} f_s(t_i^{j,k}, c_i^{j,k}, p_i^{j,k}, a_i^{j,k}(t_c)) \cdot w(t_{LO}^s, t_{HI}^s, t_c)$$

where
$f_g \equiv$ component of the mapping function that is common to all resource types
$f_s \equiv$ component of the mapping function that pertains to a *specific* project scheduling interval
$t_{LO}^s, t_{HI}^s \equiv$ specific time interval during which resource mapping must be performed according to a unique function
$T =$ set of above-defined time intervals for a particular project
$w(t_{LO}^s, t_{HI}^s, t_c) \equiv$ window function with a value of one if t_c falls within the interval (t_{LO}^s, t_{HI}^s), and zero otherwise

Finally, it is also possible to map different resource types according to different objectives *and* at different times simultaneously by simply combining the two concepts above. For example, assume again that a manager forms his objective in the early stage of the project based on resources' temporal capabilities, costs, and preferences. Then, at a later stage, the manager wishes to drop the costs and preferences and consider only resource capabilities, with the exception of resource type "three" whose costs should still remain in consideration for mapping. An example of a mapping function that would account for this scenario may be as follows:

$$U_i^{j,k} = f(c_i^{j,k}, p_i^{j,k}, t_i^{j,k}) \cdot w(0, 30, t_c) + (f(t_i^{j,k}) + f(c_i^{j,k}) \cdot \delta(j, 3)) \cdot w(30, 90, t_c)$$

23.5   Modeling of resource characteristics

For resource units whose performance on a particular activity is independent of their interaction with other units, that is, for the *drivers*, $t_i^{j,k}$ is defined as the time, t, it takes k-th unit of resource type j to complete its own task or process when working on activity i. Thus, different resource units, if multicapable, can be expected to perform differently on different activities. Each *dependent* unit, on the other hand, instead of $t_i^{j,k}$, generally has a set of interdependency functions associated with it.

In this measurement approach, we consider two types of interactive dependencies among resources, which due to their simplicity, are expected to be the most commonly used ones: *additive* and *percentual*. *Additive* interaction between a *dependent* and each of its *driver* resource unit indicates the amount of time that the *dependent* will need to complete its own task if assigned to work in conjunction with a particular driver. This is in addition to the time the driver itself needs to spend working on the same activity:

$$(T_i^{j,k})_z (t_i^{j_D, k_D} + \tilde{t}_i^{j,k}) \cdot y_i^{j_D, k_D}$$

where

$\langle j_D, k_D \rangle \in D^{j,k}$, where $D^{j,k}$ is a set of *driver* units (each defined by an indexed pair $\langle j_D, k_D \rangle$) for a particular resource unit $\langle j, k \rangle$.

$(T_i^{j,k})_z \equiv$ z-th interactive time-effective dependency of k-th unit of type j on its *driver* $\langle j_D, k_D \rangle$, $z = 1, \ldots,$ size $(D^{j,k})$. The actual number of these dependencies will depend on a manager's knowledge and familiarity with his/her resources.

$\tilde{t}_i^{j,k} \equiv$ time needed in addition to $t_i^{jD,kD}$ for k-th *dependent* unit of type j to complete its task on activity i if it interacts with its *driver* unit j_D, k_D.

$y_i^{jD,kD} \equiv$ binary (zero–one) variable indicating mapping status of the *driver* unit $\langle j_D, k_D \rangle$. It equals one if the unit $\langle j_D, k_D \rangle$ is assigned to activity i, and zero if the unit $\langle j_D, k_D \rangle$ has been assigned to activity i. Therefore, each $(T_i^{j,k})_z$ will have a nonzero value only if $y_i^{jD,kD}$ is also nonzero (i.e., if the *driver* resource unit $\langle j_D, k_D \rangle$ has been previously assigned to activity i).

The percentual interactive dependency is similarly defined as

$$(T_i^{j,k})_z = t_i^{jD,kD}(1 + \tilde{t}_i^{j,k}\%) \cdot y_i^{jD,kD}$$

where $\tilde{t}_i^{j,k}\%$ is the percentage of time by which $t_i^{jD,kD}$ will be prolonged if the unit k of type j interacts with its *driver* $\langle j_D, k_D \rangle$.

Modeling cost characteristics follows a similar logic used for representation of temporal capabilities and interdependencies. In place of $t_i^{j,k}$, we now define a variable $c_i^{j,k}$, which represents the cost (say, in dollars) of k-th unit of resource type j if it gets assigned to work on activity i. This value of $c_i^{j,k}$ may be invariant regardless of a unit's interaction with other resources, or it may vary relative to interaction among resources, and thus, implying cost interdependencies, which need to be evaluated before any mapping is performed (provided that the cost considerations are, indeed, a part of a manager's utility or objective for mapping).

In cases when a cost of a resource unit for an activity varies depending on its interaction with units of other (lower-indexed) types, we define cost dependencies as

$$(C_i^{j,k})z = \tilde{c}_i^{j,k} \cdot y_i^{jD,kD}$$

where

$y_i^{jD,kD} \equiv$ a binary variable indicating the status of the particular *driver* resource unit $\langle j_D, k_D \rangle$, as defined in the previous section.

$\tilde{c}_i^{j,k} \equiv$ interactive cost of k-th unit of type j on its *driver* $\langle j_D, k_D \rangle$, with respect to activity i.

$(C_i^{j,k})z \equiv$ z-th evaluated interactive cost dependency of k-th unit of type j on its *driver* $\langle j_D, k_D \rangle$, $z = 1, \ldots,$ size $(D^{j,k})$. The values of each $(C_i^{j,k})_z$ equals $\tilde{c}_i^{j,k}$ when $y_i^{jD,kD}$ equals one, and zero otherwise. The actual number of these interactive cost dependencies will again depend on a manager's knowledge and information about available resources.

Given a set of cost dependencies, we compute the overall $c_i^{j,k}$ as a sum of all evaluated $(C_i^{j,k})z$'s as follows:

$$c_i^{j,k} = \sum_{z=1}^{|D^{j,k}|} (C_i^{j,k})_z$$

In many instances, owing to political, environmental, safety or community standards, aesthetics, or other similar nonmonetary reasons, pure monetary factors may not necessarily prevail in decision making. It is those other nonmonetary factors that we wish to capture by introducing preferences in resource mapping to newly scheduled activities. The actual representation of preferences is almost identical to those of the costs:

$$(P_i^{j,k})z = \tilde{p}_i^{j,k} \cdot y_i^{j_D, k_D}$$

where $\tilde{p}_i^{j,k}$ is an interactive preference of k-th unit of type j on its *driver* $<j_D, k_D>$, with respect to activity i. $(P_i^{j,k})z$ is z-th evaluated interactive preference dependency of k-th unit of type j, with respect to activity i. Finally, again identically to modeling costs, $p_i^{j,k}$ is computed as

$$p_i^{j,k} = \sum_{z=1}^{|D^{j,k}|} (P_i^{j,k})_z$$

Having a certain number of resource units of each type available for a project does not necessarily imply that all of the units are available all the time for the project or any of its activities in particular. Owing to transportation, contracts, learning, weather conditions, logistics, or other factors, some units may only have *time preferences* for when they are available to start working on a project activity or the project as a whole. Others may have *strict time intervals* during which they are allowed to start working on a particular activity or the project as a whole. The latter, strictly constrained availability may be easily accommodated by the previously considered *window* function, $w(t_{LO}, t_{III}, t_c)$.

In many cases, especially for humans, resources may have a desired or "ideal" time when to start their work or be available in general. This flexible availability can simply be represented by fuzzifying the specified desired times using the following function:

$$\alpha_i^{j,k}(t_c) = \frac{1}{1 + a(t_c - \tau_i^{j,k})^b}$$

where

$\tau_i^{j,k} \equiv$ desired time for k-th unit of resource type j to start its task on activity i. This desirability may either represent the voice of project personnel (as in the case of preferences), or manager's perception on resource's readiness and availability to take on a given task.

$\alpha_i^{j,k}(t_c) \equiv$ fuzzy membership function indicating a degree of desirability of $<j, k>$-th unit to start working on activity i, at the decision instance t_c.

$a \equiv$ parameter that adjusts for the width of the membership function.

$b \equiv$ parameter that defines the extent of *start time* flexibility.

23.6   *Resource mapper*

At each scheduling time instance, t_c, available resource units are mapped to newly scheduled activities. This is accomplished by solving J number of zero–one linear integer problems (i.e., one for each resource type), where the coefficients of the decision vector correspond to evaluated mapping function for each unit of the currently mapped resource type:

$$\max \sum_{h\in\Omega(t_c)} \sum_{k=1}^{R_j} U_h^{j,k} \cdot y_h^{j,k} \quad \text{for } j = 1,\dots,J$$

where
$\quad y_h^{j,k} \equiv$ binary variable of the decision vector
$\quad \Omega(t_c) \equiv$ set of newly scheduled activities at decision instance t_c

A $y_i^{j,k}$ resulting in a value of one would mean that k-th unit of resource type j is mapped to i-th ($i \in \Omega(t_c)$) newly scheduled activity at t_c. The above objective in each of J number of problems is subjected to four types of constraints, as illustrated below.

I. The first type of constraints ensure that each newly scheduled activity receives its required number of units of each project resource type:

$$\sum_{k=1}^{R_j} y_i^{j,k} = \rho_i^j \quad \text{for } i \in \Omega(t_c) \quad \text{for } j = 1,\dots,J$$

II. The second type of constraints prevent mapping of any resource units to more than one activity at the same time at t_c:

$$\sum_{i\in\Omega(t_c)} y_i^{j,k} \leq 1 \quad \text{for } k = 1,\dots,R_j \quad \text{for } j = 1,\dots,J$$

III. The third type of constraints prevent mapping of those resource units that are currently in use by activities in progress at time t_c:

$$\sum_{k=1}^{R_j} u_{t_c}^{j,k} \cdot y_i^{j,k} = 0 \quad \text{for } i \in \Omega(t_c) \quad \text{for } j = 1,\dots,J$$

IV. The fourth type of constraints ensures that the variables in the decision vector $y_i^{j,k}$ take on binary values:

$$y_i^{j,k} = 0 \quad \text{or} \quad 1 \quad \text{for } k = 1,\dots,R_j, i \in \Omega(t_c), \quad \text{for } j = 1,\dots,J$$

Therefore, in the first of the total of J runs at each decision instance t_c, available units of resource type "one" compete (based on their characteristics and prespecified mapping function) for their assignments to newly scheduled activities. In the second run, resources of type "two" compete for their assignments. Some of their characteristics, however, may vary depending on the "winners" from the first run. Thus, the information from the first run is used to refine the mapping of type or group "two" resources. Furthermore, the information from either or both of the first two runs is then used in tuning the coefficients of the objective function for the third run when resources of type "three" are mapped.

Owing to the nature of linear programming, zeros in the coefficients of the objective do not imply that corresponding variables in the solution will also take the value of zero. In our case, that would mean that although we flagged off a resource unit as unavailable, the solution may still map it to an activity. Thus, we need to strictly enforce the interval (un)availability by adding information into constraints. Thus, we perturbed the third mapping constraint, which was previously set to prohibit mapping of resource units at time t_c, which are in use by activities in progress at that time. The constraint was originally defined as

$$\sum_{k=1}^{R_j} u_{t_c}^{j,k} \cdot y_i^{j,k} = 0 \quad \text{for } i \in \Omega(t_c) \quad \text{for } j = 1,\ldots,J$$

To now further prevent mapping of resource units whose $\alpha_i^{j,k}(t_c)$ equals zero at t_c, we modify the above constraint as follows:

$$\sum_{k=1}^{R_j} (u_{t_c}^{j,k} + (1 - \alpha_i^{j,k}(t_c))) \cdot y_i^{j,k} = 0 \quad \text{for } i \in \Omega(t_c) \quad \text{for } j = 1,\ldots,J$$

This modified constraint now filters out not only those resource units that are engaged in activities in progress at t_c, but also those units that were flagged as unavailable at t_c due to any other reasons.

23.7 Activity scheduler

Traditionally, a project manager estimates duration of each project activity first, and then assigns resources to it. In this study, although we do not exclude a possibility that an activity duration is independent of resources assigned to it, we assume that it is those resource units and their skills or competencies assigned to a particular activity that determine how long it will take for the activity to be completed. Normally, more capable and qualified resource units are likely to complete their tasks faster, and vice versa. Thus, activity duration in this measurement metric is considered a *resource-driven activity attribute*.

At each decision instance t_c (in resource-constrained nonpreemptive scheduling as investigated in this study), activities whose predecessors have been completed enter the set of qualifying activities, $Q(t_c)$. In cases of resource conflicts, we often have to prioritize activities in order to decide which ones to schedule. In this methodology, we prioritize activities based on two (possibly conflicting) objectives:

1. Basic *activity attributes*, such as the *current amount of depleted slack*, number of successors, and initially estimated optimistic activity duration, d_i.
2. Degree of manager's desire to *centralize* (or balance) *the loading* of one or more preselected project resource types.

Amount of depleted slack, $S_i(t_c)$, is defined in this methodology as a measure of how much total slack of an activity from unconstrained CPM computations has been depleted each time the activity is delayed in resource-constrained scheduling due to lack of available resource units. The larger the $S_i(t_c)$ of an activity, the more it has been delayed from its unconstrained schedule, and the greater the probability that it will delay the entire project.

Before resource-constrained scheduling of activities (as well as resource mapping, which is performed concurrently) starts, we perform a single run of CPM computations to determine initial unconstrained *latest finish time, LFT_i,* of each activity. Then, as the resource-constrained activity scheduling starts, at each decision instance t_c, we calculate $S_i(t_c)$ for each candidate activity (from the set $Q(t_c)$) as follows:

$$S_i(t_c) = \frac{t_c + d_i}{LFT_i} = \frac{t_c + d_i}{LST_i + d_i} \quad i \in Q(t_c)$$

$S_i(t_c)$, as a function of time, is always a positive real number. The value of its magnitude is interpreted as follows:

- When $S_i(t_c) < 1$, the activity i still has some slack remaining and it may be safely delayed.
- When $S_i(t_c) = 1$, the activity i has depleted all of its resource-unconstrained slack and any further delay to it will delay its completion as initially computed by conventional unconstrained CPM.
- When $S_i(t_c) > 1$, the activity i has exceeded its slack and its completion will be delayed beyond its unconstrained CPM duration.

Once calculated at each t_c, the current *amount of depleted,* $S_i(t_c)$, is then used in combination with the other two activity attributes for assessing activity priority for scheduling. (These additional attributes are *the number of activity successors,* as well as its *initially estimated duration d_i.)* The number of successors is an important determinant in prioritizing because if an activity with many successors is delayed, chances are that any of its successors will also be delayed, thus eventually prolonging the entire project itself. Therefore, the prioritizing weight, w_p^t, pertaining to basic activity attributes is computed as follows:

$$w_i^p = S_i(t_c) \cdot \left(\frac{\varsigma_i}{\max(\varsigma_i)} \right) \cdot \left(\frac{d_i}{\max(d_i)} \right)$$

where
  $w_i^p \equiv$ activity prioritizing weight that pertains to basic activity attributes
  $\varsigma_i \equiv$ number of activity successors of current candidate activity i
  $\max(\varsigma_i) \equiv$ maximum number of activity successors in project network
  $\max(d_i) \equiv$ maximum of the most optimistic activity durations in a project network

The second objective that may influence activity prioritizing is a manager's desire for a somewhat centralized (i.e., balanced) resource loading graph for one or more resource groups or types. This is generally desirable in cases when a manager does not wish to commit all of the available project funds or resources at the very beginning of the project (Dreger, 1992), or does wish to avoid frequent hiring and firing or project resources (Badiru and Pulat, 1995).

The proposed methodology attempts to balance (centralize) loading of prespecified resources by scheduling those activities whose resource requirements will minimize the increase in loading graph's stair-step size of the early project stages, and then minimize the decrease in the step size in the later stages. A completely balanced resource loading graph contains no depression regions as defined by Konstantinidis (1998), that is, it is a nondecreasing graph up to a certain point at which it becomes nonincreasing.

The activity prioritizing weight that pertains to attempting to centralize resource loading is computed as follows:

$$w_i^r = \sum_{j=1}^{J} \frac{\rho_i^j}{R_j}$$

where

$w_i^r \equiv$ prioritizing weight that incorporates activity resource requirements
$\rho_i^j \equiv$ number of resource type j units required by activity i
$R_j \equiv$ total number of resource type j units required for the project

Note that w_i^p and w_i^r are weights of possibly conflicting objectives in prioritization of candidate activities for scheduling.

To further limit the range of w_i^r between zero and one, we scale it as follows:

$$w_i^r = \frac{w_i^r}{\max(w_i^r)}$$

With the two weights w_i^p and w_i^r defined and computed, we further use them as the coefficients of activity scheduling objective function:

$$\max\left(\sum_{i\in Q(t_c)} w_i^p \cdot x_i\right) + W\left(\sum_{i\in Q(t_c)} (1 - w_i^r \cdot x_i)\right)$$

where

$x_i \equiv$ binary variable whose value becomes one if a candidate activity $i \in Q(t_c)$ is scheduled at t_c, and zero if the activity i is not scheduled at t_c
$W \equiv$ decision maker's supplied weight that conveys the importance of resource centralization (balancing) in project schedule

Note that W is a parameter that allows a manager to further control the influence of w_i^p. Large values of W will place greater emphasis on the importance of resource balancing. However, to again localize the effect of W to the early stages of a project, we dynamically decrease its value at each subsequent decision instance, t_c according to the following formula:

$$W_{\text{new}} = W_{\text{old}}\left(\frac{\sum_{i=1}^{I} d_i - \sum_{i\in H(t_c)} d_i}{\sum_{i=1}^{I} d_i}\right)$$

where

$\sum_{i=1}^{I} d_i \equiv$ the sum of all the most optimistic activity durations (as determined by conventional resource-unconstrained CPM computations) for all activities in project network

$H(t_c) \equiv$ set of activities that have been so far scheduled by the time t_c

Figure 23.1 shows a Gantt chart and resource loading graphs of a sample project with seven activities and two resource types. Clearly, neither of the two resource types is balanced. The same project has been rerun using the above reasoning, and shown in Figure 23.2. The same project has been rerun using the above methodology, and shown in Figure 23.1. Note that the loading of resource type two is now fully balanced. The loading of resource type one still contains depression regions, but to a considerably lesser extent than in Figure 23.1.

With the two weights w_i^p and w_i^r defined and computed, we further use them as the coefficients of activity scheduling objective function:

$$\max\left(\sum_{i \in Q(t_c)} w_i^p \cdot x_i\right) + W\left(\sum_{i \in Q(t_c)} (1 - w_i^r \cdot x_i)\right)$$

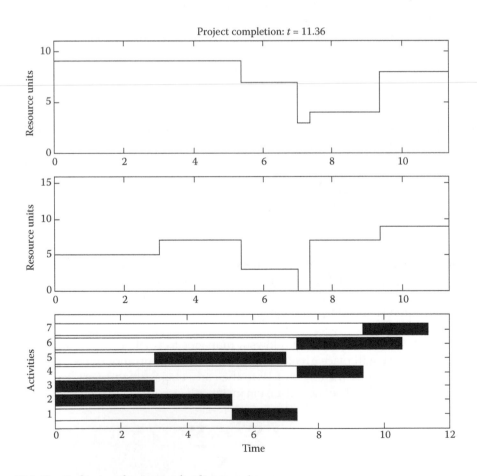

Figure 23.1 Gantt chart and resource loading graphs.

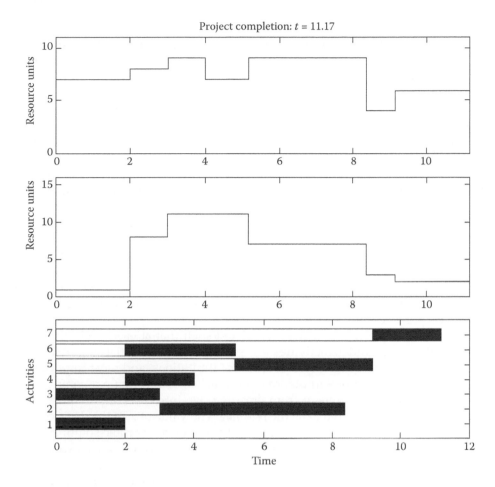

Figure 23.2 Rerun of Gantt chart and resource loading graphs.

where

$x_i \equiv$ binary variable whose value becomes one if a candidate activity $i \in Q(t_c)$ is scheduled at t_c, and zero if the activity i is not scheduled at t_c

$W \equiv$ decision maker's supplied weight that conveys the importance of resource centralization (balancing) in project schedule

Large values of W will place greater emphasis on the importance of resource balancing. However, to again localize the effect of W to the early stages of a project, we dynamically decrease its value at each subsequent decision instance, t_c according to the following formula:

$$W_{\text{new}} = W_{\text{old}} \left(\frac{\sum_{i=1}^{I} d_i - \sum_{i \in H(t_c)} d_i}{\sum_{i=1}^{I} d_i} \right)$$

where

$\sum_{i=1}^{I} d_i \equiv$ the sum of all the most optimistic activity durations (as determined by conventional resource-unconstrained CPM computations) for all activities in project network

$H(t_c) \equiv$ set of activities that have been so far scheduled by the time t_c

Previously, it was proposed that one way of balancing resource loading was to keep minimizing the increase in the stairstep size of the loading graph in the early project stages, and then minimize the decrease in the step size in the later stages. The problem with this reasoning is that a continuous increase in the loading graph in early stages may eventually lead to scheduling infeasibility due to limiting constraints in resource availability. Therefore, an intelligent mechanism is needed that will detect the point when resource constraints become binding and force the scheduling to proceed in a way that will start the decrease in resource loading, as shown in Figure 23.1. In other words, we need to formulate a linear programming model whose constraints will drive the increase in resource stairstep-shaped loading function up to a point when limits in resource availability are reached. At that point, the model must adjust the objective function and modify (relax) the constraints, in order to start minimizing the stairstep decrease of resource loading.

To ensure this, the constraints are formulated such that at each decision instance t_c, maximal number of candidate activities are scheduled, while satisfying activity precedence relations, preventing the excess of resource limitations, and most importantly, flag off the moment when resource limitations are reached. To facilitate a computer implementation and prevent the strategy from crashing, we introduce an auxiliary zero–one variable, $\hat{x}$, in this study referred to as the *peak flag*. The value of $\hat{x}$ in the decision vector is zero as long as current constraints are capable of producing a feasible solution. Once that is impossible, all variables in the decision vector must be forced to zero, except $\hat{x}$, which will then take a value of one and indicate that the peak of resource loading is reached. At that moment, the constraints that force the increase in resource loading are relaxed (eliminated).

The *peak flag* is appended to the previous objective function as follows:

$$\max\left(\sum_{i\in G(t_c)} w_i^p \cdot x_i \right) + W\left(\sum_{i\in G(t_c)} (1 - w_i^r \cdot x_i) \right) - b\hat{x}$$

where

$b \equiv$ arbitrary large positive number (in computer implementation of this study, b was taken as $b = \sum_{i=1}^{I} d_i$)

There are two types of constraints associated with the above objective of scheduling project activities. The first type simply serves to prevent scheduling of activities, which would overuse available resource units:

$$\sum_{i\in Q(t_c)} \rho_i^j \cdot x_i + \left(R_j - \sum_{i\in G(t_c)} \rho_i^j \right)\hat{x} \le \left(R_j - \sum_{i\in G(t_c)} \rho_i^j \right), \quad j = 1,\dots,J$$

where

$x_i \equiv$ candidate activity qualified to be scheduled at t_c

$G(t_c) \equiv$ set of activities that are in progress at time t_c

$(R_j - \Sigma_{i\in G(t_c)}\rho_i^j) \equiv$ difference between the total available units of resource type j (denoted as R_j) and the number of units of the same resource type being currently consumed by the activities in progress during the scheduling instant t_c

The second type of constraints serves to force the gradual increase in the stairstep resource loading graphs. In other words, at each scheduling instant t_c, this group of constraints will attempt to force the model to schedule those candidate activities whose total resource requirements are greater than or equal to the total requirements of the activities that have just finished at t_c. The constraints are formulated as follows:

$$\sum_{i\in Q(t_c)} \rho_i^j x_i + \left(\sum_{i\in F(t_c)} \rho_i^j \right) \hat{x} \geq \left(\sum_{i\in F(t_c)} \rho_i^j \right) \quad j \in D$$

where

$F(t_c) \equiv$ set of activities that have been just completed at t_c

$D \equiv$ set of manager's preselected resource types whose loading graphs are to be centralized (i.e., balanced)

$(\Sigma_{i\in F(t_c)}\rho_i^j) \equiv$ total resource type j requirements by all activities that have been completed at the decision instance t_c

Finally, to ensure an integer zero–one solution, we impose the last type of constraints as follows:

$$x_i = 0 \text{ or } 1, \quad \text{for } i \in Q(t_c)$$

As previously discussed, once $\hat{x}$ becomes unity, we adjust the objective function and modify the constraints that will, from that point on, allow a decrease in resource loading graph(s). Objective function for activity scheduling is modified such that the product $w_i^r \cdot x_i$ is not being subtracted from one any more, while the second type of constraints is eliminated completely:

$$\min\left(-\sum_{i\in Q(t_c)} w_i^t \cdot x_i \right) - W\left(\sum_{i\in Q(t_c)} w_i^r \cdot x_i \right)$$

subject to

$$\sum_{i\in Q(t_c)} \rho_i^j \cdot x_i \leq \left(R_j - \sum_{i\in G(t_c)} \rho_i^j \right) \quad j = 1,\ldots,J$$

$$x_i = 0 \text{ or } 1$$

Since the second type of constraints is eliminated, resource loading function is now allowed to decrease. The first type of constraints still remains in place to prevent any overuse of available resources.

23.8   Model implementation and graphical illustrations

The model as described above has been implemented in a software prototype *project resource mapper* (PROMAP). The output consists of five types of charts. The more traditional ones include project *Gantt chart* (Figure 23.3), and *resource loading graphs* (Figure 23.4) for all resource groups or types involved in a project. More specific graphs include *resource–activity mapping grids* (Figure 23.5), *resource utilization* (Figure 23.6), and *resource cost* (Figure 23.7) bar charts. Based on the imported resource characteristics, their interdependencies, and the form of the objective, the *resource–activity mapping grid* provides a decision support in terms of which units of each specified resource group should be assigned to which particular project activity. Therefore, the *resource–activity grids* are, in effect, the main contributions of this study. *Unit utilization charts* track the resource assignments and provide a relative resource usage of each unit relative to the total project duration. The bottom (darker shaded) bars indicate the total time it takes each unit to complete all of its own project tasks. The upper (lighted shaded) bars indicate the total additional time a unit may be locked in or engaged in an activity by waiting for other units to finish their tasks. In other words, the upper bars indicate the total possible resource idle time during which it cannot be reassigned to other activities because it is blocked waiting for other units to finish their own portions of work. This information is very useful in nonpreemptive

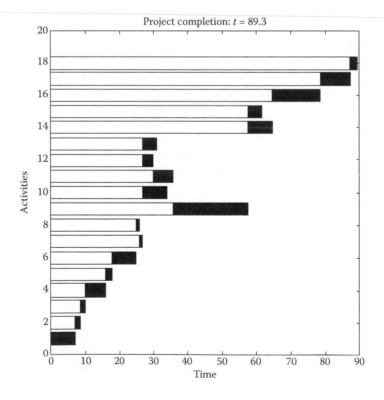

Figure 23.3 Project completion at time 89.3.

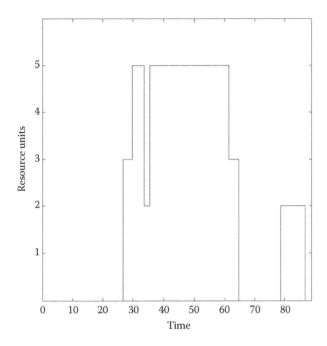

Figure 23.4 Resource type 3 loading graph.

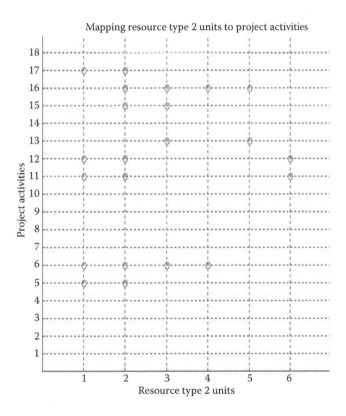

Figure 23.5 Mapping of resource type 2 units to project activities.

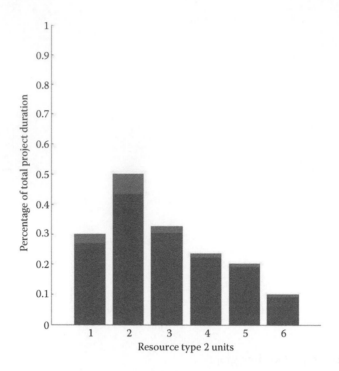

Figure 23.6 Time percentage of resource type 2 units engagement versus total project duration.

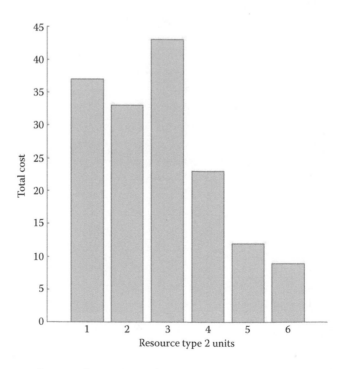

Figure 23.7 Project cost for type 2 resource units.

scheduling as assumed in this study, as well as in contract employment of resources. *Resource cost* charts compare total project resource expenditures for each resource unit.

The model developed in this chapter represents an initial step toward a more comprehensive resource–activity integration in project scheduling and management. It provides for both, effective activity scheduling based on dynamically updated activity attributes, as well as intelligent iterative mapping of resources to each activity based on resource characteristics and preselected shape of project manager's objectives. The model consists of two complementary procedures: an *activity scheduler* and *resource mapper*. The procedures are alternatively being executed throughout the scheduling process at each newly detected decision instance, such that the final output is capable of providing decision support and recommendations with respect to both, scheduling project activities and resource assignments. This approach allows human, social, as well as technical resources to interact and be utilized in value creating ways, while facilitating effective resource tracking and job distribution control.

Notations

$i \equiv$ project activity i, such that $i = 1, \ldots, I$.

$I \equiv$ number of activities in project network.

$t_c \equiv$ decision instance, that is, time moment at which one or more activities qualify to be scheduled since their predecessor activities have been completed.

$PR(i) \equiv$ set of predecessor activities of activity i.

$Q(t_c) \equiv$ set of activities qualifying to be scheduled at t_c, that is, $Q(t_c) = \{i \mid PR(i) = \varnothing\}$.

$j \equiv$ resource type j, $j = 1, \ldots, J$.

$J \equiv$ number of resource types involved in the project.

$R_j \equiv$ number of units of resource type j available for the project.

$<j,k> \equiv$ notation for k-th unit of type j.

$\rho_i^j \equiv$ number of resource units type j required by activity i.

$u_{t_c}^{j,k} \equiv$ a binary variable with a value of one if k-th unit of type j is engaged in one of the project activities that are in progress at the decision instance t_c, and zero otherwise. All $u_{t_c}^{j,k}$'s are initially set to zero.

$t_i^{j,k} \equiv$ time-effective executive capability of k-th unit of resource type j if assigned to work on activity i.

$p_i^{j,k} \equiv$ preference of k-th unit of resource type j to work on activity i.

$c_i^{j,k} \equiv$ estimated cost of k-th unit of resource type j if assigned to work on activity i.

$\alpha_i^{j,k}(t_c) \equiv$ desired start time or interval availability of k-th unit of type j to work on activity i at the decision instance t_c. In many cases this parameter is invariant across activities, and the subscript i may often be dropped.

References

Badiru, A. B. 1993. Activity resource assignments using critical resource diagramming. *Project Management Journal*, 14(3), 15–21.

Badiru, A. B. and P. S. Pulat. 1995. *Comprehensive Project Management: Integrating Optimization Models, Management Principles, and Computers*. Prentice Hall, NJ, pp. 162–209.

Bracewell, R. N. 1978. *The Fourier Transform and Its Applications*. McGraw-Hill, Inc., New York, p. 97.

Campbell, G. M. 1999. Cross-utilization of workers whose capabilities differ. *Management Science*, 45(5), 722–732.

Cooprider, C. 1999. Solving a skill allocation problem. *Production and Inventory Management Journal*, Third Quarter, 34(3), 1–6.

Dreger, J. B. 1992. *Project Management: Effective Scheduling*. Van Nostrand Reinhold, New York, pp. 202–231.

Franz, L. S. and J. L. Miller. 1993. Scheduling medical residents to rotations: Solving the large scale multiperiod staff assignment problem. *Operations Research*, 41(2), 269–279.

Gray, J. J., D. McIntire and H. J. Doller. 1993. Preferences for specific work schedulers: Foundation for an expert-system scheduling program. *Computers in Nursing*, 11(3), 115–121.

Keeney, R. L. and H. Raiffa. 1993. *Decisions with Multiple Objectives: Preferences and Value Tradeoffs*. Cambridge University Press, Cambridge, New York.

Konstantinidis, P. D. 1998. A model to optimize project resource allocation by construction of a balanced histogram. *European Journal of Operational Research*, 104, 559–571.

Mattila, K. G. and D. M. Abraham. 1998. Resource leveling of linear schedules using integer linear programming. *Journal of Construction Engineering and Management*, 124(3), 232–244.

Milatovic, M. and A. B. Badiru. 2004. Applied mathematics modeling of intelligent mapping and scheduling of interdependent and multifunctional project resources. *Applied Mathematics and Computation*, 149(3), 703–721.

Roberts, S. M. 1992. Human skills–keys to effectiveness. *Cost Engineering*, 34(2), 17–19.

Yura, K. 1994. Production scheduling to satisfy worker's preferences for days off and overtime under due-date constraints. *International Journal of Production Economics*, 33, 265–270.

chapter twenty-four

Measurement and control of imprecision in engineering design

Ronald E. Giachetti

Contents

24.1 Introduction

Engineering design is characterized by a high level of imprecision, vague parameters, and ill-defined relationships. In design, imprecision reduction must occur to arrive at a final product specification. Few design systems exist for adequately representing design imprecision and formally reducing it to precise values. Fuzzy set theory has considerable potential for addressing the imprecision in design. However, it lacks a formal methodology for system development and operation. One repercussion is that imprecision reduction is, at present, implemented in a relatively ad hoc manner. The main contribution of this chapter is to introduce a methodology called *precision convergence* introduced by Giachetti et al. (1997) for making the transition from imprecise goals and requirements to the precise specifications needed to manufacture a product. A hierarchy of fuzzy constraint networks is presented along with a methodology for creating transitional links between different levels of the hierarchy. The solution methodology is illustrated with an example within which an imprecision reduction of 98% is achieved in only three stages of the design process. The imprecision is measured using the *coefficient of imprecision*, a new metric introduced to quantify imprecision.

24.1.1   Embodiment of design

The design process encompasses all of the activities which are preformed to arrive at a final product specification. It is during the design process when an estimated 60–85% of the product's costs is determined (Zangwill, 1992; Eversheim et al., 1995). This led to increased interest in the activities which comprise the design process. Rapid product development cycles require design methodologies to efficiently explore the design space to build products with reduced cost, improved functionality, and improved quality to compete in global markets.

One of the phases of design which is least supported is the transition from early design phases to the final design stages (Abeln, 1990). This is the transition from vague and imprecise specifications to precise and exact values and is a major activity of the design process. Many systems only attempt to provide design support in the domain of well-defined variables and specifications in which all values used during design must be known with certainty. This restriction to certainty limits the utility of these systems to the later stages of the design process. Although routinely preformed, a lack of formal mechanism for quantitatively representing the transition has hindered the development of tools to support the early phases of design.

A promising technology for representing the imprecision in design is fuzzy set theory. A few researchers have begun to explore this possibility. Zimmermann and Sebastian (1994) believe fuzzy representation of imprecision will play an increasingly important role in design systems. Standard mathematical functions used in design can be applied to imprecise quantities through use of the extension principle (Zadeh, 1965). The extension principle determines the possibility function of $f(\tilde{Q}_1, \tilde{Q}_2)$ to encompass all possible values that could result from combining the ranges of the arguments $\tilde{Q}_1$ and $\tilde{Q}_2$ (Dubois and Prade, 1993). The spreads, which are representative of the imprecision, *expand* in fuzzy calculations. Operations on imprecise quantities will result in equally imprecise or more imprecise results. The imprecision of the results cannot decrease. The problem with supporting the entire design process using fuzzy set theory representation is that the operators used to compute fuzzy numbers *increase* the imprecision while a necessary condition to design an artifact which is manufacturable is to *decrease* the imprecision. Although some work has been done on an extension principle defined for "optimistic" functions, where the imprecision does not necessarily increase during calculations, it appears to have limited practical applicability (Dubois and Prade, 1988).

This problem of increased imprecision from operations on fuzzy sets can be illustrated using interval arithmetic. The relationship between fuzzy sets and intervals is well documented (Dubois and Prade, 1988; Klir and Yuan, 1995). Operations on fuzzy sets are equivalent to performing interval arithmetic at each α-level. Fuzzy multiplication is related to interval multiplication where interval multiplication is defined as (Moore, 1966)

$$[a,b] * [c,d] = [\min(ac, ad, bc, bd), \max(ac, ad, bc, bd)] \tag{24.1}$$

The product will always have greater imprecision (magnitude of interval, Kim et al., 1995) than the input values. This is referred to as a "pessimistic" calculation by Dubois and Prade (1988, p. 43). For a specific example of this problem, let us compute the volume of a cube with imprecisely defined edges. Each edge is defined by the triangular fuzzy number 2, 3, 4 and the volume is the fuzzy product

$$2,3,4 \otimes 2,3,4 \otimes 2,3,4 = 8,27,64 \tag{24.2}$$

Intuitively, many engineers can accept the concept of a dimension with a tolerance 3 ± 1 (i.e., a triangular fuzzy number 2,3,4) yet find it unacceptable that the tolerance for the volume is 27^{+37}_{-19} (i.e., a triangular fuzzy number 8,27,64). We can see that the result is much more imprecise than the input values and that extended arithmetic can produce such a large spread of possible values that the results become meaningless. Consequently, incorporating a representation for imprecision into existing design systems and utilizing existing design methodologies is inadequate for operating with imprecise quantities, and it becomes imperative to develop methodologies for manipulating fuzzy numbers to obtain reasonable results.

In this chapter, we present a solution to this problem using a new approach, termed *precision convergence*, as we move from conceptual design to detailed design. To quantify the imprecision reduction, a new metric is also introduced—the *coefficient of imprecision*. The new approach will be discussed and then described through an example—the design of a cellular telephone. However, prior to this discussion, brief discussions in current design theory, fuzzy set theory, and constraint networks will be presented.

24.2   The engineering design process

There are many definitions of design which can be found in the literature. Suh (1990) defines design as the *"continuous interplay between what we want to achieve and how we want to achieve it."* Design is a mapping from the functional domain into the physical realm (Tomiyama, 1990), it is a decision-making process (Bradley and Agogino, 1993), it is a search process (Gero, 1990), and it is the formulation and satisfaction of constraints (Serrano and Goddard, 1987). Design exhibits all of these traits and conforms to these definitions to varying degrees. The research in engineering design is focused on finding suitable theories of design hopefully leading to methods to arrive at better designs (Dixon, 1988).

In studying design, researchers have found it convenient to classify the different types of design which are described as a method of design (N.N, 1993). There are several definitions of design types (Ehrlenspiel, 1985; Hintzen et al., 1989; Eversheim, 1990; Pahl and Beitz, 1993). In general, four types of design can be differentiated: original design, adaptive design, variant design, and design by fixed principles. An analysis by Hintzen et al. (1989) shows that about 5% of design activities are original design, about 55% are adaptive design, about 20% are variant design, and the other 20% is design by fixed principles.

Regardless of the type of design problem encountered and how they are classified, it is generally acknowledged that the design process consists of stages of progressively finer detailed designs (Ehrlenspiel, 1983; Hubka, 1984; Koller, 1985; Seifert, 1989; Rodenacker, 1991; Pahl and Beitz, 1993; Roth, 1994). Finger and Dixon (1989) provide a good review of the many and varied perspectives on representing the engineering design process. However, they note that their review is missing the large body of research published in German. We concur and for our purposes use the classification of Pahl and Beitz (1988), one of the few works available in translation.

Pahl and Beitz (1988) list four design stages: clarification of task, conceptual design, embodiment design, and detail design. Let us describe the four design stages in more detail. *Clarification of task* is a problem formulation activity in which the functional requirements are specified. *Conceptual design* is synthesis of an abstract structure that can be a solution to the design problem. *Embodiment design* is the development of an abstract concept into a preliminary scaled engineering drawing. *Detailed design* involves the specification of attribute values to the design parameters. Sufficient detail must be added to each stage for evaluation and analysis. Clearly, design is an iterative process that moves an imprecise concept to a precise definition of the product and how it is to be manufactured. Despite the different breakdown and terms used, there are similarities in all descriptions of the design stages and the design process in general.

24.2.1 Imprecision in design

Imprecision is most notable in the early phases of the design process and has been defined as the choice between alternatives (Antonsson and Otto, 1995). Attempts have been made to address the issue of imprecision and inconsistency in design employing intervals (Davis, 1987; Navinchandra and Rinderle, 1990; Kim et al., 1995). However, these were found unsatisfactory for the general simultaneous engineering problem. Other approaches to representing imprecision in design include using utility theory, implicit representations using optimization methods, matrix methods such as quality function deployment, probability methods, and necessity methods. These methods have all had limited success in solving design problems with imprecision. Antonsson and Otto (1995) provide an extensive review of these approaches for interested readers.

Reusch (1993) examined the general problem of imprecision and inconsistency in design and concluded that the problems are well suited to be solved using fuzzy technology. There are two aspects of imprecision when modeled with fuzzy sets—a preference view and a plausibility view. Imprecision can be defined as the preference a designer has for a particular value but will accept other values to a lesser degree. This interpretation of using fuzzy sets to model preference was put forward by Dubois (1987) and demonstrated in the domain of mechanical engineering by Wood and Antonsson (1989). Imprecision can also be the plausibility of a value under a given possibility distribution. There exists important conceptual differences between the two based on whether the parameter is controllable or not (Dubois and Prade, 1993).

Young et al. (1995) classify the different sources of imprecision found in engineering design as: relationship imprecision, data imprecision, linguistic imprecision, and inconsistency imprecision. *Relationship imprecision* is the ambiguity that exists between the design parameters. *Data imprecision* is when a parameter's value is not explicitly known. *Linguistic imprecision* arises from the qualitative descriptions of goals, constraints, and preferences made by humans. *Inconsistency imprecision* arises from the inherent conflicting objectives among various areas in a product's life cycle. Regardless of the different sources of imprecision we believe it can be modeled using fuzzy set theory.

24.3 Fuzzy set theory

Fuzzy set theory is a generalization of classical set theory. In normal set theory, an object is either a member of a set or not a member of the set. There are only two states. This is referred to as a *crisp set*. Fuzzy sets contain elements to a certain degree. Thus, it is possible to represent an object which has partial membership in a set. The membership value of element x in a fuzzy set is represented by $\mu(x)$ and is normalized such that the membership degree of element x is in [0, 1]. All elements x such that $\mu(x) \geq \alpha$ define an "α-level" which is a bounded interval.

Operations on fuzzy sets are performed across these α-levels, either by discretizing μ into a set of α-levels with $\alpha \in [0, 1]$ or by treating μ as a continuum, and then applying the extension principle. This forms the basis of fuzzy set theory (Zadeh, 1965). Since the boundaries of inclusion in a set are *fuzzy* and not definite, we are able to directly represent ambiguity or imprecision in our models.

Fuzzy sets can represent linguistic terms and imprecise quantities. Linguistic terms are used to model the imprecision in natural language statements such as "tall" or "inexpensive." A fuzzy quantity is a set defined on $\mathbb{R}$, the set of real numbers. It represents information such as, "about 5 in." Thus, 4.9 in. would be a member of this set. 4.5 in. may also be a number of this set but to a lesser degree. Fuzzy numbers have a membership function that is normal, piecewise continuous, and convex. The value with membership of 1 is called the modal value. If all three of these properties hold for the membership function then this is a LR fuzzy number (Zimmermann, 1985; Dubois and Prade, 1988). Standard mathematical operators can be used with fuzzy numbers through application of the extension principle, and specific implementations of these operations are fuzzy operators (Zadeh, 1965; Dubois and Prade, 1988).

24.4   Constraint satisfaction

The representation of a constraint satisfaction problem, defined as a constraint network problem, can be defined as follows (adapted from Dechter and Pearl, 1988):

24.4.1   Fuzzy constraint network problem

A fuzzy constraint network problem has a set of n fuzzy variables, $\tilde{X} = \{\tilde{X}_1, \tilde{X}_2, \ldots, \tilde{X}_n\}$, and a set of m constraints, $C = \{C_1, C_2, \ldots, C_m\}$. A fuzzy variable $\tilde{X}_1$ has its domain, Ω_i, which defines the set of values that the variable can have. A constraint C_i is a *k-ary relation* on

$$\tilde{X}_i = \{\tilde{X}_{i1}, \tilde{X}_{i2}, \ldots, \tilde{X}_{ik}\} \subseteq \tilde{X}, \tag{24.3}$$

that is, $C_i(\tilde{X}_{i1}, \tilde{X}_{i2}, \ldots, \tilde{X}_{ik})$, and is a subset of the Cartesian product $\Omega_{i1} \times \Omega_{i2} \times \ldots \times \Omega_{ik}$.

In this formulation each constraint is satisfied to a degree $\mu_{C_i} \in [0,1]$, depending on the instantiation of the variables. This is the membership value of the constraint. A solution of the network is defined as an assignment of values to all the variables such that the constraints are satisfied. The constraints are satisfied when $\mu_{C_i} \geq \alpha_S$ where α_S is the *system truth threshold*. It is a level of satisfaction a solution must fulfill within the entire network to be accepted by the designer. This value is set *a priori* by the user (Young et al., 1996).

The fuzzy constraint processing system used to implement the precision convergence method described in this chapter is FuzCon. In Young et al. (1996) the system and its operator set is described along with a brief review of prior fuzzy constraint processing work. FuzCon is part of a series of constraint processing systems which include SPARK (Young et al., 1992), SATURN (Fohn et al., 1993), and JUPITER (Liau et al., 1996). The approach taken in these systems is to view each constraint as a logic sentence with an associated truth value. Refer to Greef et al. (1995) for an elaboration of the equivalency of a logic-based system and the constraint satisfaction network problem defined above. These are interactive constraint processing systems which aid the user by propagating user assignments through the network in an attempt to satisfy all the constraints using direct calculations. This approach supports a feature-rich representation schema and supports omni-directional constraint propagation. Finding a solution that satisfies the constraints is supported by a subsystem that identifies the source of a constraint violation and assists in correcting the problem. This type of system is crucial to successful modeling design because

1. Representing design information requires a wide variety of data types that constrain the design through complex relationships beyond the equality relation.
2. Design is an iterative process that draws upon the knowledge of the user requiring that propagation not be restricted to a single direction nor a prescribed starting point.

24.5 Related work

Other researchers have also explored the use of fuzzy set theory in design. Many of these systems are for conducting design evaluation (Knosala and Pedrycz, 1987; Müller and Thärigen, 1994). Fuzzy set theory has also been applied to specific design problems such as design for assembly (Jackson et al., 1993). However, none of these systems profess to support the entire design process. They are all targeted to a specific type of design problem or to a known and specific phase of the design process.

Fuzzy constraint processing has been studied by Kim and Cho (1994) who extended the definitions for numeric- and interval-based constraint networks to fuzzy constraint networks. Ruttkay (1994) discusses different methods for determining the joint satisfaction of constraints. Fargier (1994) and Dubois et al. (1995) have used fuzzy constraint networks to model flexible temporal constraints in job-shop scheduling. Bowen et al. (1992) discuss the link between fuzzy constraint satisfaction and semantics in classical logic. Of these methods the constraint model of Dubois et al. (1995) has the greatest similarities to the model used here.

Recent work by Chen and Otto (1995) used fuzzy set theory to solve a variational system of constraints for CAD (computer-aided design). However, the approach is limited to numeric variables and equality constraints and can only be solved when the number of variables equals the number of constraints. Wood and Antonsson (1989) developed the *Method of Imprecision* (MoI) which uses fuzzy set theory to model the preferences of a designer. Instead of direct calculations the MoI technique allows a designer to rank the coupling between design parameters and performance parameters. Those with the strongest coupling become candidates for modification in an attempt to reduce the imprecision of the output. In this sense, it is related to Taguchi's method (Otto and Antonsson, 1993a) and to utility theory (Otto and Antonsson, 1993b).

The computations in these two approaches are performed using a discretized solution algorithm developed by Dong and Wong (1987) and modified by Wood et al. (1992). The algorithm has computational complexity of the order $O(M.2^{N-1}k)$, where N is the number of imprecise parameters, M is the number of α-levels into which the membership function is divided, and k is the number of multiplications and divisions in the function $f(\tilde{d})$, containing all the calculations.

The algorithm was initially limited to performing calculations on a single expression but has been extended to perform calculations for a system of equations (Chen and Otto, 1995). Clearly, in addition to representational restrictions, these approaches are computationally bounded when applied to even small-sized problems and indicate a new approach is needed to support modeling imprecision in design using fuzzy set theory. Giachetti and Young (1997) and Giachetti (1996) have developed a parametric representation which facilitates efficient computation of fuzzy arithmetic on triangular and trapezoidal fuzzy numbers. It has computational complexity of order $O(k)$ where k is the number of arithmetic operations in the equations. Additionally, the fuzzy constraint system developed by Young et al. (1996) is rich in features with the ability to represent linguistic variables and terms, fuzzy variables and numbers, and crisp variables and numbers in complex relations that include not only crisp equality, but also fuzzy equality relations, fuzzy and crisp inequality relations, and fuzzy and crisp logical relations. These developments set the stage for the development of the *precision convergence* methodology presented in this chapter.

24.6   Hierarchical model of design

To be successful, a design must have an overall monotonically decreasing trend with regard to imprecision. This is shown graphically in Figure 24.1. Different stages of the design process require different knowledge representation methods. These are identified along the abscissa of the figure. The different stages through which a design passes can be modeled by decomposing the design process into a hierarchy of constraint networks. This hierarchy matches the movement of decisions from product conception to product manufacture. The hierarchy proceeds from higher level, abstract models of the product to lower level, detailed models of the product. A constraint network is a node in the hierarchy. The top node of the hierarchy is the root node. A constraint network can be linked from above and from below to other networks in the hierarchy. The approach of decomposing the design problem into a hierarchy of constraint networks conforms to the formal hierarchical model presented in O'Grady et al. (1994).

Figure 24.2 shows a hierarchy of fuzzy constraint networks. The constraints are represented as circles and the variables are represented as solid-line links. The dashed-line links connect different levels of the hierarchy. They represent a variable's value in a higher level of the hierarchy becoming a constraint in a lower level of the hierarchy. These dashed-line links are *transition links* through which precision convergence is realized. In the example presented in Section 24.7, the *clarification of task* constraint network is the root node of the hierarchy. The next lower level network is the *conceptual design* network. The lowest level network in Figure 24.2 is the *embodiment design* network.

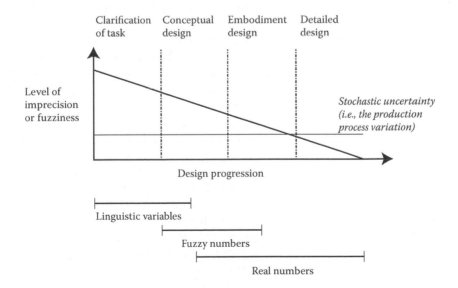

Figure 24.1 Design stages versus imprecision level and the type of variables to use.

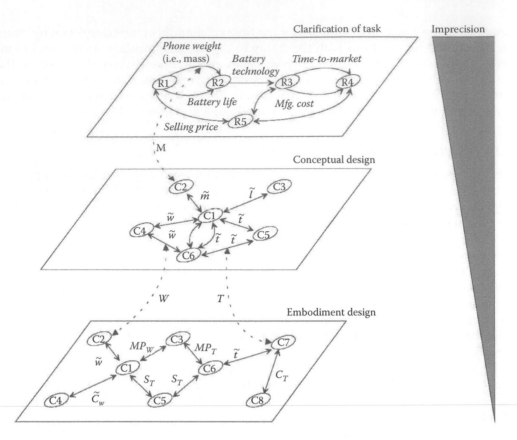

Figure 24.2 Design hierarchy for the cellular phone example.

A *transition link* connects a variable's value in one level of the hierarchy to a constraint in a lower level of the hierarchy. As an example of linking between constraint networks in different levels of the design hierarchy, let us examine transition link T connecting the *conceptual design* network to the *embodiment design* network. In the *conceptual design* network, the fuzzy variable $\tilde{t}$ represents the phone thickness. Through constraint propagation a solution is found to the *conceptual design* network that satisfies its constraints. As part of this solution, $\tilde{t}$ is assigned a value. We now move down the design hierarchy to the next stage in the design process—*embodiment design*.

The *embodiment design* stage for the cellular phone is modeled by the *embodiment design* constraint network. Via transition link T, the value assigned to $\tilde{t}$ as part of the solution to the *conceptual design* network defines constraint C7 in the *embodiment design* network. A solution to the *embodiment design* constraint network must satisfy the relationship defined by constraint C7. Since operations on fuzzy numbers only produce values with increased imprecision, the fuzzy sets selected for use in this lower-level constraint network must have sufficiently reduced imprecision such that their combined effect will not exceed the imprecision of T within constraint C7. This bounds the solution in the *embodiment design* phase and has the effect of forcing the designer to work backwards from imprecise quantities to more precise quantities. Successive applications of this principle—higher levels set requirements for lower levels—reduces the overall imprecision as the design progresses through the design stages. We call the result of this principle *precision convergence*.

24.7   Precision convergence using fuzzy constraint satisfaction: An example

In this example taken from Young et al. (1995), we design a cellular telephone in which we balance phone weight, battery life, battery technology, phone selling price, manufacturing cost, and time-to-market. The design parameters are

- Phone weight—ranges between 0.2 and 1.0 kg
- Battery life—ranges between 2 and 10 h
- Time-to-market—ranges between 9 and 24 months
- Selling prices—ranges between $50 and $200
- Manufacturing cost—ranges between $20 and $90
- Battery technology is one of the following: new technology, nickel–cadmium type 1, nickel–cadmium type 2, alkaline.

Although these variables are stated as quantitative ranges, values for them at the *clarification of task* stage are discussed in terms such as *undesirable, less desirable, more or less satisfactory, more desirable,* and *very desirable.* Chen and Hwang (1992) examined this type of representation and concluded that typically descriptive terms are rarely less than three, rarely more than 10, and typically five. Figure 24.3 restates the problem variables using descriptive terms and shows how each variable's term is mapped into a common set of terms.

Figure 24.4 shows an example mapping for phone weight from its quantitative range onto five triangular fuzzy numbers whose combined range is normalized between zero and one. The mapping onto a normalized range allows us to transform the different data types into a common *fuzzy linguistic* representation so we can more easily compute with them. They are now each a fuzzy linguistic variable whose possible values are one of the descriptive terms in Figure 24.3. As an example, the variable *Phone weight* can now only be assigned one of the five descriptive terms (high weight, above average weight, average weight, below average weight, low weight). The solution to the problem will be a descriptive term for each variable selected from the possible descriptive terms listed in Figure 24.3. It is then mapped back into its quantitative range to identify a more restrictive range (see Figure 24.4). In this way, a solution reduces the imprecision for each variable.

Continuing the problem statement, the variables are related to each other through the following types of relationships.

Marketing requirements: The relationship of phone weight and battery life to selling price.
Production requirements: The relationship of time-to-market to manufacturing cost.
Cost requirements: The relationship of selling price to manufacturing cost.
1*st technology restrictions*: The relationship of *battery technology to phone weight and to battery life.*
2*nd technology restrictions*: The relationship of battery technology to time-to-market and manufacturing cost.

These requirements constitute design restrictions and can be represented as a fuzzy constraint network in which the requirements are constraints interconnected through shared variables. The fuzzy constraint network for this problem is the root node shown in Figure 24.2. The marketing constraint is shown in Figure 24.5. In the marketing constraint we see that one of four conditions must be met to satisfy the constraint.

Battery technology (not a fuzzy number)	One of the following:	{new technology, nickel-cadmium type 1, nickel-cadmium type 2, alkaline}			

Variable	Satisfaction range				
	Undesirable *but not unacceptable*	**Less desirable**	**More or less satisfactory**	**More desirable**	**Very desirable**
Phone weight	High weight	Above average weight	Average weight	Below average weight	Low weight
Battery life	Short life	Below average life	Average life	Above average life	Long life
Time-to-market	Long	Longer than average	Average	Short than average	Short
Selling price	High	Higher than average	Average	Lower than average	Low
Manufacturing cost	High	Higher than average	Average	Lower than average	Low

Figure 24.3 Typical terms used to describe the range for each variable for the clarification of task constraint network and their mapping onto a common term set.

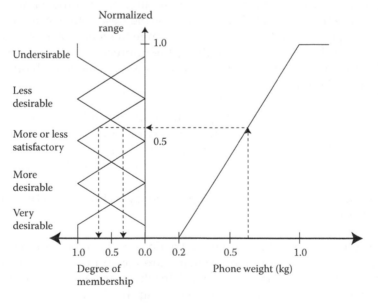

Figure 24.4 An example mapping for the phone weight to a normalized range of linguistic terms.

Phone weight *is* low weight and **Battery life** *is* long life and **Selling price** *is* high
or

Phone weight *is* low weight and **Battery life** *is* above average life and
Selling price *is* average
or

Phone weight *is* below average weight and **Battery life** *is* long life and
Selling price *is* average
or

Phone weight *is* below average weight and **Battery life** *is* above average life and
Selling price *is* lower than average

Figure 24.5 The marketing constraint expressed using fuzzy linguistic variables, descriptive terms, and linguistic operators.

Examining the first condition in the constraint we have the following

Phone weight is low weight
Battery life is long life
Selling price is high

This condition expresses marketing's view that customers will pay more for a light weight phone that lasts a long time. In the constraint, *"is"* is a fuzzy linguistic operator that compares a linguistic variable to a descriptive term and is an example of modeling *relationship imprecision*. *Phone weight, Battery life,* and *Selling price* are linguistic variables, and "low weight," "long life," and "high" are descriptive terms from Figure 24.3. These are examples of modeling *linguistic imprecision*. The remaining constraints have similar relationships based on the concerns of the production, marketing, management, and technological departments within the company. Using fuzzy linguistic variables, descriptive terms, and linguistic operators allows us to build a model using imprecise descriptions typically employed in requirement specifications while supporting those descriptions with a well-defined mathematical basis and processing approach. In this way, the highest level constraint network of Figure 24.2 is a *fuzzy constraint network* and addresses the "clarification of task phase" in Pahl and Beitz's (1988) model. The fuzzy constraint network is described in Young et al. (1995) and solved using fuzzy constraint satisfaction techniques. It was shown to have a solution only when a compromise could be found between two conflicting constraints—the 2nd technology restrictions and the production requirements. Conflicting constraints are common in design and are an example of *inconsistency imprecision*. The ability to find compromise solutions through a sound mathematical basis is a recently identified strength of fuzzy constraint satisfaction (Lang and Schiex, 1993; Martin-Clouaire, 1993; Young et al., 1995).

The solution to the clarification-of-task is as follows (Young et al., 1995):

Phone weight: Below average (0.28–0.6 kg)
Battery life: Above average (6–9.2 h)
Battery technology: Nickel–cadmium type 2
Time-to-market: Average (13.5–19.5 months)
Selling price: Lower than average ($65 to $125).

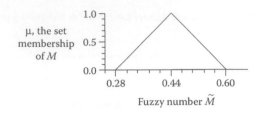

Figure 24.6 Triangular fuzzy set for the phone's weight (i.e., mass), $\tilde{M} = \langle 0.28,\ 0.44,\ 0.60 \rangle$.

We can see in the solution all the variables have a reduced range and are more precise than when we began. This is the objective of strategic planning and is not unexpected. However, by using fuzzy technology we now have a formal basis for what has previously been an ad hoc procedure. This approach, in and of itself, is useful. However, the issue is how to vertically integrate this solution with the next stage to continue the progression towards more precise results. To show how this might be accomplished, let us focus upon one of the parameters from the solution of the strategic planning problem—the phone weight. In the solution, the phone's weight is determined to be "below average" which maps onto a reduced range of acceptable weights defined by the interval 0.28–0.6 kg which is represented by the fuzzy number $\langle 0.28, 0.44, 0.60 \rangle$. $\langle 0.28, 0.44, 0.60 \rangle$ is a triple describing the triangular fuzzy number shown in Figure 24.6 and is an example of *data imprecision*. In the next lower level of the design hierarchy the phone's weight becomes the upper bound for the phone's mass in constraint C7 of the *conceptual design* constraint network. A solution to the *conceptual design* phase must satisfy constraint C7.

24.7.1 Conceptual design network

The second network in the design hierarchy represents the *conceptual design* phase. The solution to the *clarification of task* constraint network becomes a requirement for conceptual design. To specify the requirement we state it as a fuzzy constraint in which the mass of the cellular phone must be less than or equal to the triangular fuzzy number $\tilde{M}$ (represented by the triple $\langle 0.28, 0.44, 0.60 \rangle$). Thus, $\tilde{M}$ is an upper bound on the mass of the phone and receives its value via transition link M shown connecting these two constrain networks in Figure 24.2. Table 24.1 lists the constraints for this level of the example problem for conceptual design. Table 24.2 lists the constraints for the embodiment design. The *conceptual design* constraint network is the second level of Figure 24.2. The objective at this second level is to determine the phone's dimensions (length, width, and thickness) subject to the constraints.

The designer has a concept of a thin, slender phone and selects values appropriately. The values selected for width and thickness are $\tilde{w} = 5, 7, 9$ cm and $\tilde{t} = 1.5, 2, 2.5$ cm. The mass is calculated in constraint C1 automatically by constraint propagation and is $\tilde{m} = 254, 882, 2295$ g. This result is automatically checked by constraint propagation against constraint C2 using possibility theory (Dubois and Prade, 1988)

$$\mu_{C2} = \text{Poss}(\tilde{m}| < M) = 0.57 \tag{24.4}$$

Since $\mu_{C2} \geq \alpha_S$, where α_S is the system threshold for determining acceptability, the relationship defined in constraint C2 is satisfied. The problem solution is an approximate size

Table 24.1 Fuzzy constraints for the conceptual design constraint network

Constraint number	Fuzzy constraint	Description
C1	$\tilde{m} = \tilde{\rho} \otimes \tilde{l} \otimes \tilde{w} \otimes \tilde{t}$	Mass calculation
C2	$\tilde{m} \, \tilde{<} \, M$	Upper bound of mass (from the solution of *clarification task* network)
C3	$\tilde{l} = \widetilde{1.5} \otimes \tilde{P}$	Length calculation, where length is about 1.5 times a typical pocket depth
C4	$4 \le \tilde{w} \le 12$	Typical pocket width
C5	$\tilde{t} \le 3$	Typical thickness
C6	$2\tilde{t} \oplus \tilde{w} \, \tilde{<} \, W$	A cellular phone will fit into a pocket if the sum of its thickness and width are approximately less than a typical pocket width

Fuzzy variables:
$\tilde{w}$—Width (cm)
$\tilde{t}$—Thickness (cm)
PN—Part number of mouthpiece
MP_{WT}—Mouthpiece weight (g)
MP_{W}—Mouthpiece width (cm)
MP_{L}—Mouthpiece length (cm)
MP_{T}—Mouthpiece thickness (cm)
T—Upper bound on thickness—1.5, 2, 2.5 cm (receives its value via transition link T from the *conceptual design* network)
W—Upper bound on width—5, 7, 9 cm (receives its value via transition link W from the *conceptual design* network)
$\tilde{S}_{T}$—Shell thickness of plastic case (cm)
$\tilde{c}_{W}$—Clearance between mouthpiece and shell (cm)
$\tilde{c}_{T}$—Clearance between mouthpiece and shell in thickness direction (cm)

for the cellular phone that satisfies the requirements from the *clarification of task* phase. In finding a solution to the *conceptual design* constraint network, the constraint propagation process infers fuzzy values to assign to the fuzzy variables such that the application of fuzzy multiplication in constraint C1 produces a value that does not violate the upper bound for the phone's mass in constraint C2. The solution is produced by satisfying constraint C2 through successive applications of fuzzy arithmetic operators. Because the application of a fuzzy arithmetic operator always results in an increase in imprecision, a side effect of using constraint processing technology is to effectively solve the problem backwards so that the fuzzy values assigned to variables $\tilde{l}$, $\tilde{w}$, and $\tilde{t}$ (representing length, weight, and thickness) are more precise than the fuzzy value for the phone's mass. In the next lower level of the design hierarchy, the phone's width and thickness become upper bounds on thickness and width which the embodiment design must satisfy.

24.7.2   Embodiment design of the cellular phone mouthpiece

The solution to the *conceptual design* constraint network sets the fuzzy values for the dimensions of the phone. Its solution was determined by satisfying constraints which modeled the intended use of the phone. In the *embodiment design* phase, the designer selects components that will become part of the phone as manufactured. The selection of these

Table 24.2 Fuzzy constraints for the embodiment design constraint network

Constraint number	Fuzzy constraint	Description
C1	$\tilde{w} = 2\tilde{S}_T \oplus MP_w \oplus 2\tilde{c}_w$	Width calculation
C2	$\tilde{w} \tilde{<} W$	Upper bound of width (from the solution of *clarification design* network via transition link W)
C3	$MP(PN, MP_L, MP_T,$ $MP_W, MP_{WT})$	Relation that defines the structure of the mouthpiece inventory data
C4	$\tilde{c}_w \geq 0.25$	Clearance should be larger than 0.25 cm for wires
C5	$0.2 \leq \tilde{S}_T \leq 0.5$ cm	Shell thickness should be in this range to meet strength and weight constraints
C6	$\tilde{t} = MP_T \oplus 2\tilde{S}_T \oplus 2\tilde{c}_T$	Thickness calculation
C7	$\tilde{t} \tilde{<} \tilde{T}$	Upper bound of thickness (from the solution of the *conceptual design* network via transition link T)
C8	$0.2 \leq \tilde{c}_T \leq 0.5$	Clearance must be large enough not to interfere with lead protrusion and small enough not to muffle sound

Embodiment design variables:
$\tilde{w}$—Width (cm)
$\tilde{t}$—Thickness (cm)
PN—Part number of mouthpiece
MP_{WT}—Mouthpiece weight (g)
MP_W—Mouthpiece width (cm)
MP_L—Mouthpiece length (cm)
MP_T—Mouthpiece thickness (cm)
T—Upper bound on thickness—1.5, 2, 2.5 cm (receives its value via transition link T from the *conceptual design* network)
W—Upper bound on width—5, 7, 9 cm (receives its value via transition link W from the *conceptual design* network)
$\tilde{S}_T$—Shell thickness of plastic case (cm)
$\tilde{c}_W$—Clearance between mouthpiece and shell (cm)
$\tilde{c}_T$—Clearance between mouthpiece and shell in thickness direction (cm)

components as well as the relationship among them and other variables must further reduce the imprecision involved in the design problem to correspond as closely as possible to the specifications of standard components. The standard component's specifications are the lower limit on the level of imprecision and represent the production process variation shown in Figure 24.1. When this level of imprecision is achieved, the precision convergence process can terminate.

In our example, we concentrate on a subcomponent of the cellular phone, the mouthpiece design and the selection of a particular mouthpiece from those available in inventory. Table 24.2 shows the constraints for the *embodiment design* fuzzy constraint network. Table 24.3 shows the inventory data on which the relation constraint, C3, operates. This constraint network is the bottom node in the design hierarchy of Figure 24.2. The values assigned to variables in solving the *conceptual design* constraint network become constraints in the *embodiment design* constraint network via transition links W and T. These transition links become constraints C2 and C7 in the *embodiment design* network.

The designer selects the fuzzy set 0.20, 0.25, 0.30 as the fuzzy value for the shell thickness, $\tilde{S}_T$. The designer also selects the fuzzy set 0.15, 0.2, 0.25 as the fuzzy value for the

Table 24.3 Mouthpiece inventory data used by constraint C3

Part number (PN)	MP_L (cm)	MP_W (cm)	MP_T (cm)	MP_{WT} (g)
MP-01	4	4	1.25	100
MP-02	5	5	1.25	125
MP-03	6	6	1.5	150
MP-04	6.5	6.5	1.75	150
MP-05	7	7	1.75	200

clearance variable $\tilde{c}_T$, between the mouthpiece and the phone's outer shell in the thickness direction. The designer selects part numbers from the mouthpiece inventory data table that are automatically propagated to constraints C1 and C6, inferring values for $\tilde{w}$ and $\tilde{t}$. Constraint propagation automatically checks requirement constraints, C2 and C7. A partial ordering can be produced based on μ_{C2} and μ_{C7}, which shows how well the mouthpieces satisfy these constraints. The designer is free to choose any mouthpiece which satisfies the constraints where $\mu \geq \alpha_S$. The partially ordered set of mouthpieces provides the designer with knowledge as to which mouthpiece best satisfies the preference set in prior design stages. As an example, mouthpiece MP-01 satisfies constraints C2 and C7 at fuzzy set membership values of $\mu_{C2} = 1.0$ and $\mu_{C7} = 0.75$ with an overall width of $\tilde{w} = 4.7, 4.9, 5.1$ and $\tilde{t} = 1.95, 2.15, 2.35$. Alternatively, the designer can search for a system truth threshold level, α_S, at which the constraint network is satisfied. The α_S value is indicative of the solution's quality. As α_S approaches one the solution quality is better and as α_S approaches zero the solution quality is worse.

24.7.3   Detailed design

The detailed design stage could be accomplished using a parametric CAD system. The physical layout of the product has been determined, and the parameters have converged to within the capabilities of the production process so the nominal dimensions and tolerances are defined by the fuzzy numbers in the solution set. The procedure followed in the previous stages can be continued where the solution values from embodiment design become constraints in detailed design. Using a parametric CAD system, these constraints can be represented, thus ensuring the continuation of the original design objectives into the detailed design stage.

24.8   A metric for the reduction of design imprecision

Common measures of fuzziness are defined as the lack of distinction between a fuzzy set and its complement. The less a set differs from its complement the fuzzier it is (Klir and Yuan, 1995). Wood and Antonsson (1989) showed that these measures are inadequate for quantifying design imprecision and developed a new metric to measure the design imprecision of a fuzzy set. This new metric, called the gamma function, quantifies the spread of the membership function about the mode value and is given by the expression

$$D(\tilde{C}) = \sum_{i=1}^{|X|} (e^{\alpha_{\tilde{C}}(x_i)} - 1) \tag{24.5}$$

However, this metric is not adequate to compare imprecise quantities of different units and scale because it only accounts for the absolute spread of the fuzzy set. It does not account for the relative difference in magnitude different imprecise quantities may have. A similar problem exists in statistics in comparing the variability of samples drawn from populations that differ in scale or units. In such cases, the standard deviations are scaled by dividing each by its own mean to produce the coefficient of variation as a relative measure of variability. In an analogous manner, we can compare the relative imprecision of disparate fuzzy sets by modifying the gamma function to account for the relative scale of the fuzzy set. We do this by scaling expression (1) by the mode of the fuzzy set, denoted by b. We call this the *coefficient of imprecision* and define it as

$$c(\tilde{A}) = \left| \frac{1}{b} \int_a^c (e^{\alpha_{\tilde{A}}(x)} - 1) dx \right| \qquad (24.6)$$

A crisp number has a measure of fuzziness equal to 0. The higher the coefficient of imprecision, the more imprecise the underlying membership function. When applied to crisp intervals, the *mode* parameter is determined by $= (1/2)(c - a)$. This measure of fuzziness appears to correctly rank a mixture of triangular fuzzy numbers and crisp intervals regardless of the mode value b.

The coefficients of variation for the design parameters in the solutions to the example constraint networks are computed and compiled in Table 24.4.

In the *clarification of task* constraint network the original range for the mass was reduced to $\tilde{m} = 0.28, 0.44, 0.60$ with $c(\tilde{m}) = 0.522$. The *conceptual design* constraint network reduced this mass into three separate variables with lower coefficients of imprecision as listed in Table 24.4. All three values show a reduction in imprecision from the imprecision in the mass. The hierarchy of constraint networks reduced the imprecision from 2.29 to a low value of 0.287 for a 87% total reduction in imprecision from the *clarification of task* stage to the *conceptual design* stage. Further precision convergence is realized by making each of the three product dimensions a constraint in a lower level network. An example of *embodiment design* is presented using variables for thickness and width. This network demonstrates a method for reducing the number of feasible alternatives and leads to the selection of a discrete part from a component database. The new thickness and width

Table 24.4 Coefficient of imprecision for design parameters in the example problem

Design parameter and its corresponding fuzzy set	Design stage in which the parameter appears	Coefficient of imprecision, $c(\tilde{A})$
Phone weight$_{Clarification\ of\ task}$ ∈ [0.2, 1.0]	*Clarification of task*	2.29
$\tilde{m}_{Clarification\ of\ task}$ = ⟨0.28, 0.44, 0.60⟩	*Clarification of task*	0.522
$\tilde{l}_{Clarification\ of\ task}$ = ⟨16.9, 21, 25.5⟩	*Clarification of task*	0.287
$\tilde{w}_{Conceptual\ design}$ = ⟨5, 7, 9⟩	*Clarification of task*	0.410
$\tilde{t}_{Conceptual\ design}$ = ⟨1.5, 2, 2, 5⟩	*Conceptual design*	0.359
$\tilde{w}_{Embodiment\ design}$ = ⟨4.7, 4.9, 5.1⟩	*Embodiment design*	0.059
$\tilde{t}_{Embodiment\ design}$ = ⟨1.95, 2.15, 2.35⟩	*Embodiment design*	0.032

are $\langle 1.95, 2.15, 2.35 \rangle$ and $\langle 4.7, 4.9, 5.1 \rangle$. This reduces the imprecision further, and when measured as a percentage of the initial mass, a 98% reduction in imprecision is realized as the design progressed from the *clarification of task* stage to the *embodiment design* stage. The solution to the *embodiment design* constraint network reduces the imprecision to a magnitude approaching the tolerances of the manufacturing processes. This signals the crossover from a fuzzy analysis of imprecision to the uncertainty from the stochastic variation that is an inherent property of the manufacturing processes. This termination is shown in Figure 24.1 where the imprecision line intersects the uncertainty line.

24.9   Conclusions

Imprecision is an inherent characteristic of engineering design. All designs begin with goals and requirements that are descriptive statements of what functions a product should perform. These imprecise product specifications are transformed by the design process to the precise specifications necessary for manufacturing. A methodology was outlined, whereby a concurrent engineering design problem could be structured so that the effect of constraint propagation would be a reduction in design imprecision. A total reduction of 98% was achieved in the example problem. Using a hierarchy of fuzzy constraint networks, we decomposed the design process according to levels of decision making appropriate to the chronological design stages. The hierarchy models information at different levels of abstraction and corresponds to actual design processes. It facilitates team-based design since conceptual design decisions are usually made by different people than embodiment design decisions. This precision convergence approach is quantifiable. We can determine when the imprecision in a design has been sufficiently reduced to the point that it matches the imprecision inherent in the production process. This gives us a termination metric for the design iteration cycle and also a means to map inherent production process variation to its impact on the design. This, in essence, provides a higher fidelity for the measurement and control of imprecision in engineering processes.

References

Aheln, O. 1990. *CAD-Systeme der 90er Jahre -Vision und Realität (CAD systems 90s -Vision and reality)*, VDI-Berichte Nr. 861.1, Düsseldorf.

Antonsson, E.K. and Otto, K.N. 1995. Imprecision in engineering design, *ASME Journal of Mechanical Design*, 117, 25–32.

Bowen, J., Lai, R., and Bahler, D. 1992. Fuzzy semantics and fuzzy constraint networks, *Proceedings of the IEEE International Conference on Fuzzy Systems*, San Diego, CA, Vol. 3, pp. 1009–1016.

Bradley, S.R. and Agogino A.M. 1993. Open problems in uncertain reasoning in design, *EUFIT '93—First European Congress on Fuzzy and Intelligent Technologies*, Aachen, September 7–10, pp. 387–392.

Chen, J.E. and Otto, K.N. 1995. A tool for imprecise calculation in engineering systems, *Proceedings of the 4th International IEEE Conference on Fuzzy Systems*, Yokohama, Japan, pp. 389–394.

Chen, S.-J. and Hwang, C.-L. 1992. *Fuzzy Multiple Attribute Decision Making*, Springer-Verlag, Berlin.

Davis, E. 1987. Constraint propagation with interval labels, *Artificial Intelligence*, 32, 281–331.

Dechter, R. and Pearl, J. 1988. Network-based heuristics for constraint-satisfaction problems, *Artificial Intelligence*, 34, 1–38.

Dixon, J.R. 1988. On research methodology towards a scientific theory of engineering design, *Artificial Intelligence for Engineering Design Analysis and Manufacturing (AIEDAM)*, 1(3), 145–157.

Dong, W.M. and Wong, F.S. 1987. Fuzzy weighted averages and implementation of the extension principle, *Fuzzy Sets and Systems*, 24(2), 183–199.

Dubois, D. 1987. An application of fuzzy arithmetic to the optimization of industrial machining processes, *Mathematical Modelling*, 9, 461–475.

Dubois, D., Fargier, H., and Prade, H. 1995. Fuzzy constraints in job-shop scheduling, *Journal of Intelligent Manufacturing*, 6, 215–234.

Dubois, D. and Prade, H. 1988. *Possibility Theory*, Plenum Press, New York.

Dubois, D. and Prade, H. 1993. Fuzzy numbers: An overview, in: Dubois, D., Prade, H., and Yager, R.R. (Eds.), *Readings in Fuzzy Sets for Intelligent Systems*, Kaufmann Publishers, San Mateo, CA.

Ehrlenspiel, K. 1983. Ein Denkmodell des Konstruktionsprozesses (A conceptual model of the design process), *Proceedings of the ICED 83*, Bd.1, Heurista, Zürich.

Ehrlenspiel, K. 1985. *Konstengüstig Konstruieren, Kostenwissen, Kosteneinflüsse, Kostensenkung (Cost knowledge, cost factors, cost reduction)*, Springer-Verlag, Berlin, .

Eversheim, W. 1990. *Konstruktion*, Organisation in der Produktionstechnik, Bd. 2, 2. neubearb. und erweiterte Auflage (Design, organization of production technology, new and updated version), VDI-Verlag, Düsseldorf.

Eversheim, W., Bochtler, W., and Laufenberg, L. 1995. *Simultaneous Engineering—von der Strategie zur Realisierung, Erfahrungen aus der Industrie für die Industrie (Simultaneous engineering of the strategy to implementation, experiences from the industry for the industry)*, Springer-Verlag, Heidelberg.

Fargier, H. 1994. *Problèmes de Satisfaction de Constraintes Flexibles Application à L'ordinnancement de Production (Constraint satisfaction problems flexible application to production management)*, PhD dissertation, l'Université Paul Sabatien de Toulouse.

Finger, S. and Dixion, J.R. 1989. A review of research in mechanical engineering design, Part I: Descriptive, prescriptive, and computer-based models of design processes, *Research in Engineering Design*, 1, 51–67.

Fohn, S., Greef, A., Young, R.E., and O'Grady, P. 1993. A constraint-system shell to support concurrent engineering approaches to design, *Artificial Intelligence in Engineering*, 9(1), 1–17.

Gero, J.S. 1990. Design prototypes: A knowledge representation schema for design, *Proceedings of the 2nd IJCAI Conference*, London, pp. 621–640.

Giachetti, R.E. 1996. *The mathematics of triangular fuzzy numbers to support a model of imprecision in design*, PhD dissertation, Industrial Engineering, North Carolina State University.

Giachetti, R.E. and Young, R.E. 1997. Analysis of the error in the standard approximation used for multiplication of triangular and trapezoidal fuzzy numbers and the development of a new approximation, *Fuzzy Sets and Systems*, 91, 1–13.

Giachetti, R.E., Young, R.E., Roggatz, A., Eversheim, W., and Perrone, G. 1997, A methodology for the reduction of imprecision in the engineering process, *European Journal of Operational Research*, 100, 277–292.

Greef, A.R., Fohn, S.M., Young, R.E., and O'Grady, P.J. 1995. Implementation of a logic-based support system for concurrent engineering, *Data and Knowledge Engineering*, 15(1), 31–61.

Hintzen, H., Laufenberg, H., Matek, W., Muhs, D., and Wittel, H. 1989. *Konstruieren und Gestalten (Construction and Design)*, Vieweg-Verlag, Braunschweig, Wiesbaden, 1989.

Hubka, V. 1984. *Theorie de Konstrukionsprozesse-Grundlage einer wissenschaftlichen Konstruktionslehre (Theory of the Design Process—Foundation of a knowledge-based design)*, 2, Auflage, Springer-Verlag, Berlin.

Jackson, S.D., Sutton, J.C., and Zorowski, C.F. 1993. Design for assembly using fuzzy sets, DE-Vol. 52, *Design for Manufacturability*, ASME, Dearborn, MI, pp. 117–122.

Kim, K. and Cho, P.D. 1994. A fuzzy-based approach to numeric constraint networks, *Proceedings of the 3rd IEEE Conference on Fuzzy Systems*, Orlando, FL, June 26–29, pp. 1878–1882.

Kim, K., Cormier, D., Young, R.E., and O'Grady, P. 1995. A system for design and concurrent engineering under imprecision, *Journal of Intelligent Manufacturing*, 6(1), 11–27.

Klir, G. and Yuan, B. 1995. *Fuzzy Sets and Fuzzy Logic: Theory and Applications*, Prentice-Hall, New Jersey, pp. 254–258.

Knosala, R. and Pedrycz, W. 1987. Multicriteria design problem—Fuzzy set approach, *International Conference on Engineering Design—ICED 87*, Boston, MA, Vol. 1, pp. 715–722.

Koller, R. 1985. *Konstruktionslehre für den Maschinenbau (Design for Mechanical Systems)*, 2, Auflage, Springer-Verlag, Berlin.

Lang, H.F.-J. and Schiex, T. 1993. Selecting preferred solutions in fuzzy constraint satisfaction problems, *Proceedings of EUFIT'93—First European Congress on Fuzzy and Intelligent Technologies*, Aachen, Germany, pp. 1128–1134.

Liau, J., Young R. E., and O'Grady P. 1996. An interactive constraint modeling system for concurrent engineering, *Engineering Design and Automation*, 1(2), 105–119.

Martin-Clouaire, R. 1993. CSP techniques with fuzzy linear constraints: Practical issues, *Proceedings of EUFIT'93- First European Congress on Fuzzy and Intelligent Technologies*, Aachen, Germany, pp. 1117–1123.

Moore, R.E. 1966. *Interval Analysis*, Prentice-Hall, Englewood Cliffs, N.J.

Müller, K. and Thärigen, M. 1994. Applications of fuzzy hierarchies and fuzzy MADM methods to innovative system design, *1994 IEEE International Conference on Fuzzy Systems*, Orlando, FL, Vol. 1, pp. 364–367.

N.N. 1993. *Methodik zum Entwickeln und Konstruieren technischer Systeme und Produkte, VDI Richtlinie 2221* (Approach to the development and design of technical systems and products), 2, Auflage, VDI-Verlag, Düsseldorf.

Navinchandra, D. and Rinderle, J. 1990. Interval approaches for concurrent evaluation of design constraints, *Proceedings of Concurrent Product and Process Design*, ASME Publication DE-21, San Francisco, CA.

O'Grady, P.J., Kim, Y., and Young, R.E. 1994. A hierarchical approach to concurrent engineering systems, *International Journal of Computer Integrated Manufacturing*, 7(3), 152–162.

Otto, K.N. and Antonsson, E.K. 1993a. Extension to the Taguchi method of product design, *Journal of Mechanical Design*, 115, 5–13.

Otto, K.N. and Antonsson, E.K. 1993b. The method of imprecision compared to utility theory for design selection problems, DE-Vol. 53, *5th International Conference on Design Theory and Methodology*, ASME, Albuquerque, New Mexico, September 19–22, pp. 167–173.

Pahl, G. and Beitz, W. 1988. *Engineering Design*, The Design Council, Springer-Verlag, London.

Pahl, G. and Beitz, W. 1993. *Konstruktionslehre: Methoden und Anwendung*, 3. neu überarbeitete und erweiterte (Engineering Design: Methods and Application, newly revised and advanced edition), Auflage, Springer-Verlag, Berlin, Heidelberg, New York.

Reusch, B. 1993. Potentiale der Fuzzy-Technologie in Nordhein-Westfalen, *Studie der Fuzzy-Initiative NRW, Ministerium für Wirtschaft, Mittelstand und Technologie des Landes Nordhein-Westfalen* (Potentials of Fuzzy Technology in North Rhine-Westphalia, Study of Fuzzy Initiative NRW, Ministry of Economic Affairs and Technology of the State North Rhine-Westphalia), Düsseldorf.

Rodenacker, W.G. 1991. *Methodisches Konstruieren (Systematic Design)*, 4, Auflage, Springer-Verlag, Berlin.

Roth, K. 1994. *Konstruieren mit Konstruktionskatalogen: Systematisierung und zweckmäßige Aufbereitung techn. Sachverhalte für das methodische Konstruieren (Designing with design catalogs: systematization and expedient processing techn. Issues for the methodological construct)*, Springer-Verlag, Berlin, Heidelberg, New York.

Ruttkay, Z. 1994. Fuzzy constraint satisfaction, *Proceedings of the 3rd IEEE Conference on Fuzzy Systems*, Orlando Fl., June 26–29, pp. 1263–1268.

Seifert, H. 1989. Grundzüge des methodischen Vorgehens bei der Synthese von Maschinen Des Stoff-, Energie-, und Signalumsatzes (Basic features of the methodological approach in the synthesis of machines material, energy, and signal conversion), Ruhr-Universität Bochum, Institut für Konstruk-tionstechnik.

Serrano, D. and Goddard, D. 1987. Constraint management in conceptual design, In: Sriram, D. and Adey, R.A. (Eds.), *Knowledge Based Expert Systems in Engineering: Planning and Design*, Computational Mechanics Publications, New York, NY.

Suh, N.P. 1990. *The Principles of Design*, Oxford University Press, New York.

Tomiyama, T. 1990. *Intelligent CAD Systems*, Eurographics, Switzerland.

Wood, K.L. and Antonsson, E.K. 1989. Computations with imprecise parameters in preliminary engineering design: Background and theory, *Transactions of the ASME*, 111, 616–625.

Wood, K.L., Otto, K.N., and Antonsson, E.K. 1992. Engineering design calculations with fuzzy parameters, *Fuzzy Sets and Systems*, 52, 1–20.

Young, R.E., Giachetti, R.E., and Ress, D.A. 1996. Fuzzy constraint satisfaction in design and manufacturing, in *6th International IEEE Conference on Fuzzy Systems (FUZZ-IEEE'96)*, September, New Orleans.

Young, R.E., Greef, A., and O'Grady, P. 1992. SPARK: An artificial intelligence constraint network system for concurrent engineering, *International Journal of Production Research*, 30(7), 1715–1735.

Young, R.E., Perrone, G., Eversheim, W., and Roggatz, A. 1995. Fuzzy constraint satisfaction for simultaneous engineering, *Annals of the German Academic Society for Production Engineering*, II(2), 181–184.

Zadeh, L.A. 1965. Fuzzy sets, *Information and Control*, 8(3), 338–353.

Zangwill, W.I. 1992. Concurrent engineering: Concepts and implementation, *IEEE Management Review*, 20(4), 40–52.

Zimmermann, H.-J. 1985. *Fuzzy Set Theory and Its Applications*, Kluwer, Boston.

Zimmermann, H.-J. and Sebastian H.-J. 1994. Fuzzy design—Integration of fuzzy theory with knowledge-based system-design, *1994 IEEE International Conference on Fuzzy Systems*, Orlando, FL, Vol. 1, pp. 352–357.

chapter twenty-five

Fuzzy measurements in systems modeling

Adedeji B. Badiru

Contents

This chapter discusses scientific uncertainty, fuzziness, and application of stochastic models in systems decision making. It highlights the importance of decision support systems (DSS) in providing a robust platform for effective actions in organization and business enterprises. Scientists now recognize the importance of studying scientific phenomenon having complex interactions among their components. These components include not only electrical or mechanical parts, but also "soft science" (human behavior, etc.) and how information is used in models. Most real-world data for studying models are uncertain (Ibidapo-Obe, 2006). Uncertainty exists when facts, state, or outcome of an event cannot be determined with probability of 1 (in a scale of 0 to 1). If uncertainty is not accounted for in model synthesis and analysis, deductions from such models become at best uncertain. The "lacuna" in understanding the concept of uncertainty and developmental policy formulation/implementation are not only due to the nonacceptability of its existence in policy foci, but also due to the radically different expectations and modes of operation that scientists and policy makers use. It is therefore necessary to understand these differences and provide better methods to incorporate uncertainty into policy making and developmental strategies (Figure 25.1).

Scientists treat uncertainty as a given, a characteristic of all data and information (as processed data). Over the years, however, sophisticated methods to measure and

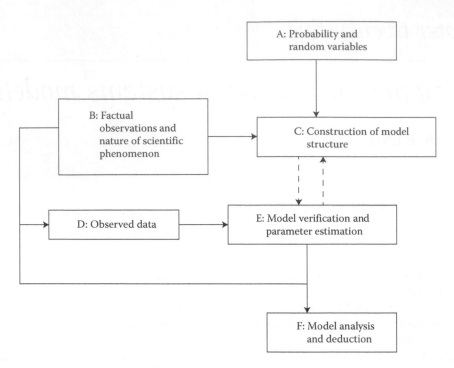

Figure 25.1 Basic cycle of probabilistic modeling and analysis.

communicate uncertainty, arising from various causes, have been developed. In general, more uncertainty has been uncovered than *absolute precision*. Scientific inquiry can only set boundaries on the limits of knowledge. It can define the *edges of the envelope of known possibilities*, but often the envelope is very large and the probabilities of the content (the known possibilities) occurring can be a complete mystery. For instance, scientists can describe the range of uncertainty about global warming and toxic chemicals and perhaps about the relative probabilities of different outcomes, but in most important cases they cannot say which of the possible outcomes will occur at any particular time with any degree of accuracy. Current approaches to policy making, however, try to avoid uncertainty and gravitate to the edges of the scientific envelope. The reasons for this bias are clear. Policy makers want to make unambiguous, defensible decisions, which are to be codified into laws and regulations.

Although legislative language is often open to interpretation, regulations are much easier to write and enforce if they are stated in absolutely certain terms. *Science defines the envelope while the policy process gravitates to an edge*—usually the edge that best advances the policy maker's political agenda! But to use science rationally to make policy, the whole envelope and all of its contents must be considered.

25.1   *Decision support systems*

A decision is a judgment, choice, or resolution on a subject made after due consideration of alternatives or options. It involves setting the basic objectives, optimizing the resources, determining the main line of strategy, planning and coordinating the means to achieve them, managing relationships, and keeping things relevant in the operating environment. A competitive situation exists when conflicting interests must be resolved.

A good decision must be *auditable*; that is, it must be the best decision at the time it was taken and must have been based on facts or assumptions as well as technical, sociopolitical, cultural, and other considerations.

DSS are computer-based tools from problem formulation through solution, simulation implementation, archiving, and reporting. These tools include operations research software tools, statistical and simulation programs, expert systems, and project management utilities. The DSS scheme is as follows:

- Problem statement (modeling—dry testing)
- Data information requirements (collection and evaluation procedure)
- Performance measure (alternatives-perceived worth/utility)
- Decision model (logical framework for guiding project decision)
- Decision making (real-world situation, sensitivity analysis)
- Decision implementation (schedule and control)

25.2   Uncertainty

In science, information can be, for example, objective, subjective, dubious, incomplete, fragmentary, imprecise, fluctuating, linguistic, data-based, or expert-specified. In each particular case, this information must be analyzed and classified to be eligible for quantification. The choice of an appropriate uncertainty model primarily depends on the characteristics of the available information. In other words, the underlying reality with the sources of the uncertainty dictates the model.

A form of uncertainty stems from the *variability* of the data. In equal and or comparable situations, each datum in question may not show identical states. This kind of uncertainty is typically found in data taken from plants and animals and reflects the rich variability of nature. Another kind of uncertainty is the impossibility of observing or measuring to a certain level of precision. This kind of precision depends not only on the power of the sensors applied, but also on the environment, including the observer. This type of uncertainty can also be termed as uncertainty due to *partial ignorance* or *imprecision*. Finally, uncertainty is introduced by using a natural or professional language to describe the observation as a datum. This *vagueness* is a peculiar property of humans and uses the special structure of human thinking. Vagueness becomes more transparent in a case, such as when dealing with grades, shades, or nuances expressed verbally and represented by marks or some natural numbers. Typical phrases used in such vague descriptions include "in many cases"; "frequently"; "small"; "high"; "possibly"; "probably"; etc. All these kinds of uncertainty (uncertainty due to variability, imprecision, and vagueness) can also occur in combinations.

One approach is to investigate uncertainty by use of *sensitivity analysis* whereby the given data are subjected to small variations to see how these variations will influence the conclusions drawn from the data. The problem with sensitivity analysis, however, has to do with its "point-oriented" approach as to where and in what dimensions the variations are to be fixed. Another approach is in the use of *interval mathematics*, in which each datum is replaced by a set of "possible" surrounding data on the real line. The problem with interval mathematics is the difficulty of specifying sharp boundaries for the data sets, for example, the ends of the intervals. A third approach for tackling uncertainty is the *stochastic* approach. This involves realization of each datum as a random variable in time, that is, the datum is assumed to be chosen from a hypothetical population according to some fixed probability law. This approach works well with modeling of variability and small imprecision. Finally, uncertainty can be taken into account using notions and tools from *fuzzy set theory*. In this approach, each datum is represented by a fuzzy set over a suitable universe. *The main idea of fuzzy set is the allowance of membership, to a grade, for every element of a specified set.* With this notion, uncertainty can be modeled mathematically more adequately and subtly using only the common notion of membership of an element to a set. Fuzzy set models both imprecision and vagueness.

However, in typical real-world systems and decision-making processes, virtually all the three types of uncertainty (variability, imprecision, and vagueness) manifest. Since, stochasticity captures variability and small imprecision well and fuzzy set captures imprecision and vagueness in data description well, then for a comprehensive treatment of uncertainty in data, it is advisable to exploit a combined effect of stochastic and fuzzy types of uncertainty. Simply put, *randomness* caters for objective data while *fuzziness* caters for subjective data.

25.3   Fuzziness

Fuzzy logic (and reasoning) is a scientific methodology for handling uncertainty and imprecision. Unlike in conventional (crisp) sets, the members of fuzzy sets are permitted varying degrees of membership. An element can belong to different fuzzy sets with varying membership grade in each set. The main advantage of fuzzy sets is that it allows classification and gradation to be expressed in a more natural language; this modeling concept is a useful technique when reasoning in uncertain circumstances or with inexact information, which is typical of human situations. Fuzzy models are constructed based on expert knowledge rather than on pure mathematical knowledge; *therefore, they are both quantitative and qualitative, but are considered to be more qualitative than quantitative*. Therefore, a fuzzy expert system is a computer-based decision tool that manipulates imprecise inputs based on the knowledge of an expert in that domain.

A fuzzy logic controller (FLC) makes control decisions by its well-known fuzzy IF–THEN rules. In the antecedence of the fuzzy rules (i.e., *the IF part*), the control space is partitioned into small regions with respect to different input conditions. Membership function (MF) is used to fuzzify each of the input variables. For continuity of the fuzzy space, the regions are usually overlapped by their neighbors. By manipulating all the input values in the fuzzy rule base, an output will be given in the consequent (i.e., *the THEN part*). FLCs can be classified into two major categories: the Mamdani (M)-type FLC that uses fuzzy numbers to make decisions and the Takagi–Sugeno (TS)-type FLC that generates control actions by linear functions of the input variables.

25.4   Fuzzy set specifications

In classical set theory, the membership of elements in relation to a set is assessed in binary terms according to a crisp condition. An element either belongs or does not belong to the set; the boundary of the set is crisp. As a further development of classical set theory, fuzzy set theory permits the gradual assessment of the membership of elements in relation to a set; this is described with the aid of an MF.

If X represents a fundamental set and x is the elements of this fundamental set, to be assessed according to an (lexical or informal) uncertain proposition and assigned to a subset A of X, the set

$$\tilde{A} = \{(x, \mu_A(X)) \mid x \in X\} \tag{25.1}$$

is referred to as the uncertain set or fuzzy set on X (Figure 25.2).

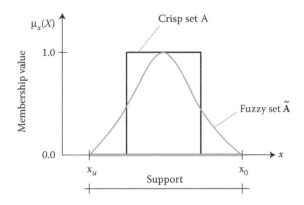

Figure 25.2 Fuzzy set.

$\mu_A(x)$ is the MF of the fuzzy set and may be continuous or discrete with elements assessed by membership values.

The uncertainty model fuzziness lends itself to describing imprecise, subjective, linguistic, and expert-specified information. It is capable of representing dubious, incomplete, and fragmentary information and can additionally incorporate objective, fluctuating, and data-based information in the fuzziness description. Requirements regarding special properties of the information do not generally exist. With respect to the regularity of information within the uncertainty, the uncertainty model fuzziness is less rigorous in comparison with probabilistic models. It specifies lower information content and thus possesses the advantage of requiring less information for adequate uncertainty quantification.

Primarily, fuzzification (Figure 25.3) is a subjective assessment that depends on the available information. In this context, four types of information are distinguished to formulate guidelines for fuzzification. If the information consists of various types, different fuzzification methods may be combined.

25.5   Information type I: Sample of very small size

The MF is specified on the basis of existing data comprising elements of a sample. The assessment criterion for the elements x is directly related to numerical values derived from X. An initial draft for an MF may be generated with the aid of simple interpolation algorithms applied to the objective information, for example, represented by a histogram. This is subsequently adapted, corrected, or modified by means of subjective aspects.

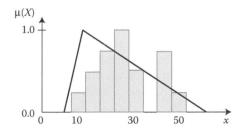

Figure 25.3 Fuzzification of information from a very small sample.

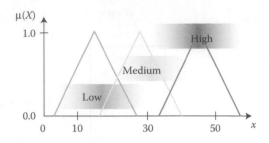

Figure 25.4 Fuzzification of information from a linguistic assessment.

25.6 Information type II: Linguistic assessment

The assessment criterion for the elements x of X may be expressed using linguistic variables and associated terms, such as "low" or "high" as shown in Figure 25.4. As numerical values are required for a fuzzy analysis, it is necessary to transform the linguistic variables to a numerical scale. By combining the terms of a linguistic variable with modifiers, such as "very" or "reasonably," a wide spectrum is available for the purpose of assessment.

25.7 Information type III: Single uncertain measured value

If only a single numerical value from X is available as an uncertain result of measurement, m, the assessment criterion for the elements x may be derived from the uncertainty of the measurement, which is quantified on the assigned numerical scale. The uncertainty of the measurement is obtained as a "gray zone" comprising more or less trustworthy values. This can be induced, for example, by the imprecision of a measurement device or by a not clearly specifiable measuring point.

The experimenter evaluates the uncertain observation for different membership levels. For the level $\mu_A(x) = 1$, a single measurement or a measurement interval is specified in such a way that the observation may be considered to be "as crisp as possible." For the level of the support, $\mu_A(x) = 0$, a measurement interval is determined that contains all possible measurements within the scope of the observation. An assessment of the uncertain measurements for intermediate levels is left up to the experimenter. The MF is generated by interpolation or by connecting the determined points $(x, \mu_A(x))$. Figure 25.5 shows an example.

25.8 Information type IV: Knowledge based on experience

The specification of an MF generally requires the consideration of opinions of experts or expert groups, of experience gained from comparable problems, and of additional information where necessary. Also, possible errors in measurement and other inaccuracies attached to the fuzzification process may be accounted for. These subjective aspects generally supplement the initial draft of an MF. If neither reliable data nor linguistic assessments are available, fuzzification depends entirely on estimates by experts.

As an example, consider a single measurement carried out under dubious conditions, which only yields some plausible value range. In those cases, a crisp set may initially be specified as a kernel set of the fuzzy set. The boundary regions of this crisp kernel set

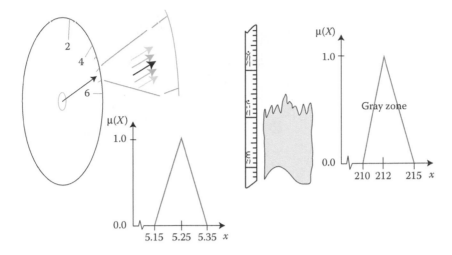

Figure 25.5 Fuzzification of a single uncertain measurement due to imprecision of the measuring device or imprecise measuring point.

are finally "smeared" by assigned membership values $\mu_A(x) < 1$ to elements close to the boundary and leading the branches of $\mu_A(x)$ beyond the boundaries of the crisp kernel set monotonically to $\mu_A(x) = 0$. By this means, elements that do not belong to the crisp kernel set, but are located "in the neighborhood" of the latter, are also assessed with membership values of $\mu_A(x) > 0$. This approach may be extended by selecting several crisp kernel sets for different membership levels (α-level sets) and by specifying the $\mu_A(x)$ in level increments.

25.9   Stochastic–fuzzy models

Fuzzy randomness simultaneously describes objective and subjective information as a fuzzy set of possible probabilistic models over some range of imprecision. This generalized uncertainty model contains fuzziness and randomness as special cases.

Objective uncertainty in the form of observed/measured data is modeled as *randomness*, whereas subjective uncertainty (see Figure 25.6), for example, due to a lack of trustworthiness or imprecision of measurement results, of distribution parameters, of environmental conditions, or of the data sources is described as *fuzziness*. The fuzzy–random model then combines but does not mix objectivity and subjectivity; these are separately visible at any time. It may be understood as an imprecise probabilistic model, which allows for simultaneously considering all possible probability models that are relevant to describing the problem.

The uncertainty model fuzzy randomness is particularly suitable for adequately quantifying uncertainty that comprises only some (incomplete, fragmentary) objective, data-based, randomly fluctuating information, which can simultaneously be dubious or imprecise and may additionally be amended by subjective, linguistic, expert-specified evaluations.

This generalized uncertainty model is capable of describing the whole range of uncertain information reaching from the special case of *fuzziness* to the special case of *randomness*. That is, it represents a viable model if the available information is too rich in content

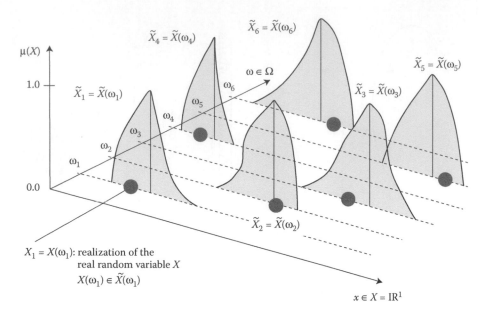

Figure 25.6 Model of a fuzzy–random variable.

to be quantified as fuzziness without a loss in information but, on the other hand, cannot be quantified as randomness due to imprecision, subjectivity, and nonsatisfied requirements. *This is probably the most common case in applied science.*

Uncertainty quantification with fuzzy randomness represents an imprecise probabilistic modeling, which incorporates imprecise data as well as uncertain or imprecise subjective assessments in terms of probability. The quantification procedure is a *combination of established methods from mathematical statistics for specifying the random part and of fuzzification methods for describing the fuzzy part of the uncertainty.*

25.10   Development model

Development is the vital summation of all efforts made by man to increase the quality of life while sustainability is the continued successful upholding and enhancement of this quality of life by replenishing the necessary ingredient/resources such as human labor and ecological resources. Development occurs when the intrinsic aspect is applied through technology to generate the physical aspect. Technology confirms the existence of the intrinsic aspects and creates the physical aspect to manifest development.

Considering development as a hierarchical (Figure 25.7) multivariate nonlinear function

$$D = (C, T, I, K, F, H) \tag{25.2}$$

where C is capital, T is technology, I is information technology, K is knowledge, F is food, and H is health; clearly, C, T, I, K, F, and H are not independent variables.

A compact analysis of the behavior of D can be facilitated by the following *dimensionality reduction.*

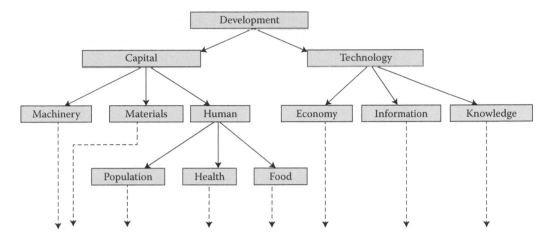

Figure 25.7 Hierarchical representation of macro- and micro-development variables.

Capital can be defined as an inventory of infrastructures machinery m, that is, plant, equipment, etc.; materials m_2; and human resources h.

$$C = C(m_1, m_2, h) \tag{25.3}$$

Since human resources, h, itself can be viewed as a function of population, health, and food, the functional representation for h alone becomes

$$h = h(P, H, F) \tag{25.4}$$

Thus, capital C can be represented as a compound functional representation as

$$C = C(m_1, m_2, h(P, H, F)) \tag{25.5}$$

Following a similar argument, since the technology variable T depends on the average performance measure of all productive processes in the economy, E, the level of development and utilization of information technology, I, the average experience and intelligence in the society, that is, knowledge, K, T becomes

$$T = T(E, I, K) \tag{25.6}$$

Thus, the development nonlinear equation can be written as

$$D = D(C, T) = D(C(m_1, m_2, h(P, H, F)), T(E, I, K)) \tag{25.7}$$

This is obviously functionalizing development at a macro level. At the micro level are myriads of variables on which each of the above variables depends. The decision-making organelle in optimizing for development definitely consists of optimization at the macro- and micro-level functions.

Uncertainty of different kinds influences decision making and hence development. In optimizing functions of development, the effects of uncertainty cannot be ignored or

Table 25.1 A typical uncertain description of variables influencing food production

Food-related variable	Description	Class of uncertainty
Mechanization	Linguistic description, for example, "high" level of mechanization	Fuzzy
Population	A range of numbers, random over time and space	Fuzzy–stochastic
Incentives	A range of numbers	Fuzzy
Soil conditions	A random variable subject to variation over time	Stochastic
Climatic conditions	Quantitative and variable over time and space	Fuzzy–stochastic
Yield	Quantitative description	Fuzzy–stochastic

rationalized or else the results of such decisions will at best be out of relevance to affect economic growth. At the micro level, where the numerous variables that influence decision making at the macro level are, cognizance must be taken of uncertainty in quantifying development variables. Inherent in the quantification are elements of uncertainty in terms of variability, imprecision, vagueness of description, and randomness. As amply explained in earlier sections, such quantities are better realized as combinations of fuzzy and random variables.

For example, food security (see Table 25.1) is influenced by a myriad of variables, including level of mechanization of agriculture, population, incentives, soil conditions, climatic conditions, etc. These variables may not be measured precisely. Some may be fuzzy, some stochastic, and others manifesting a combination of stochasticity and fuzziness in their quantification. In effect, quantification of *food* (*F*) in the development model is a *fuzzy–stochastic* variable.

The other variables in the above development can also be viewed as fuzzy/stochastic variables, thus the *developmental model* above can be viewed as a *fuzzy–stochastic developmental model*.

$$D = D\left(\widetilde{\overline{C}}, \widetilde{\overline{T}}\right) = D\left(C\left(\widetilde{\overline{m}}_1, \widetilde{\overline{m}}_2, h\left(\widetilde{\overline{P}}, \widetilde{\overline{H}}, \widetilde{\overline{F}}\right)\right) T\left(\widetilde{\overline{E}}, \widetilde{\overline{I}}, \widetilde{\overline{K}}\right)\right) \quad (25.8)$$

where the symbols _ and ~ represent fuzzification and randomization, respectively, of the various variables.

25.10.1 Optimization of the fuzzy–stochastic development model

The realization of the relational effects of the variables of development are in hierarchies, even at the macro level. A good strategy for optimizing the function is to use the concept of multilevel optimization model.

Multilevel-optimizing models are employed to solve decentralized planning decision problems in which decisions are made at different hierarchical decision levels in a top-to-down fashion. Essentially, the features of such multilevel planning organizations are that

- Interactive decision-making units exist within a predominantly hierarchical structure.
- Execution of decisions is sequential, from top to lower level.

- Each unit independently maximizes its own net benefit but is affected by the actions of other units through externalities.
- The external effect of a decision maker's (DM's) problem can be reflected in both the objective function and the feasible decision space.

The mode of execution of such decision problem is that

- The upper-level DM sets his goal and accepts the independent decisions at the lower levels of the organization.
- The upper-level DM modifies it within the framework of the overall benefit of the organization.
- The upper-level DM's action further constrain the lower-level decision space and may or may not be acceptable at that level. If it is not acceptable, the upper-level DM can still make a consensus that the constraints are relaxed further.
- This process is carried out until a satisfactory solution to all levels and units of decision making is arrived at.

In solving the fuzzy–stochastic development model, each level of the macro-level model is taken as a decision level optimizing variables of its concern.

The complete fuzzy–stochastic multilevel formulation of the fuzzy–stochastic development model is, therefore

$$\underset{\tilde{C},\tilde{T}}{\text{Max}}\ \tilde{\tilde{D}} \qquad (25.9)$$

where $\tilde{\tilde{C}}, \tilde{\tilde{T}}$ are obtainable from

$$\underset{V_3}{\text{Max}}\ f(V_2) \qquad (25.10)$$

with $V_2 = \{\tilde{\tilde{m_1}}, \tilde{\tilde{m_2}}, \tilde{\tilde{h}}, \tilde{\tilde{E}}, \tilde{\tilde{I}}, \tilde{\tilde{K}}\}$, which again is obtainable from

$$\underset{V_3}{\text{Max}}\ f(V_2) \qquad (25.11)$$

where $V_2 = \{\tilde{\tilde{X}}, \tilde{\tilde{P}}, \tilde{\tilde{H}}, \tilde{\tilde{F}}\}$ and $\tilde{\tilde{X}}$ = relevant variables at the micro level of the development model.

Subject to:

$$\tilde{\tilde{C}}, \tilde{\tilde{T}}, v_2, v_3, \tilde{\tilde{X}} \in S$$

where S is the constrained space of development variables (constrained by the limitations in realizing the various variables).

Realistically, this type of model may be difficult to solve for large spaces such as the national development pursuit but may be solved with smaller decision spaces. The worst-solution scenario will be those that are not amenable to analytical solutions for which there are many heuristics to be coupled with simulation-optimization techniques.

25.11   Urban transit systems under uncertainty

A relevant work (Ibidapo-Obe and Ogunwolu, 2004) on DSS under uncertainty is the investigation and characterization of combinations of effects of fuzzy and stochastic forms of uncertainty in urban transit time-scheduling. The results of the study vindicate the necessity for taking both comprehensive combinations of fuzzy and stochastic uncertainties as well as multistakeholders' objective interests into account in urban transit time scheduling. It shows that transit time-scheduling performance objectives are better enhanced under fuzzy–stochastic and stochastic–fuzzy uncertainties and with the multistakeholders' objective formulations than with lopsided single stakeholder's interests under uncertainty. Figure 25.8 shows the modeling framework, which covers the following:

- Mathematical modeling
- Algebraic, fuzzy, and stochastic methods
- Multiobjective genetic algorithm
- Algorithmic computer
- Simulation technique
- Mixed integer models
- Max–min techniques

25.12   Water resources management under uncertainty

An ongoing research on water resources management also suggests the need for incorporating comprehensive scientific uncertainty in developmental models (Figure 25.9). This is using a three-level hierarchical system model for the purpose of optimal timing and sequencing of project development for water supply, the optimal allocation of land, water, and funds for crop growth, and optimal timing for irrigation. Major functions and inputs in the hierarchical fuzzy–stochastic dynamic programming developmental model are to be realized as fuzzy, stochastic, and or fuzzy–stochastic inputs of the model. Particularly, the demand and supply of water over a time horizon are taken as fuzzy–stochastic.

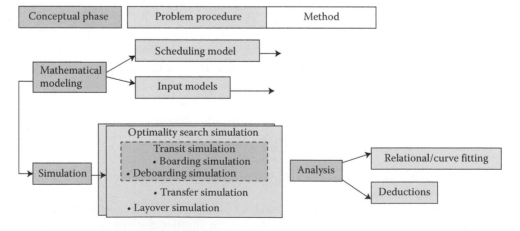

Figure 25.8 Conceptual, procedural, and methodic frames of the transit time-scheduling research study.

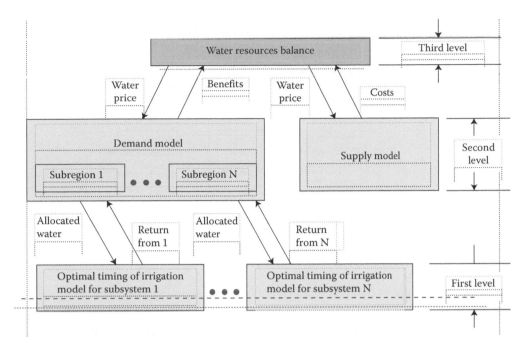

Figure 25.9 Hypothetical generic multilevel dynamic programming structure for planning and management of agricultural/water system.

25.13 Energy planning and management under uncertainty

One of the immediate future thrusts is on *energy planning and management modeling under uncertainty*. Energy information systems are organized bodies of energy data often indicating energy sources and uses with information on historical as well as current supply–demand patterns, which are naturally bugged with uncertainties (see Figure 25.10). These information systems draw upon energy surveys of various kinds as well as on other sources of information such as the national census (another potential source of uncertainty in evaluation over time and space), information on energy resources and conversion technologies, as well as consumption patterns (which are also better realized considering inherent uncertainties). A typical structure of national energy planning system is shown below.

25.14 University admissions process in Nigeria: Post-UME test selection decision problem

There are many issues related to the Nigerian education system (universal basic education [UBE], the 3–3 component of the 6–3–3–4 system, funding, higher education institution admissions, consistency of policies, etc.).

In spite of the opening up of the higher educational institution (HEI) space to more states, private, open, and transnational institutions, the ratio of available space to the number of eligible candidates is 87:1000 (University of Lagos, 2006). It is therefore imperative that the selection mechanism be open and rational. The realization problem for rational access is further compounded with the introduction of post-University matriculation

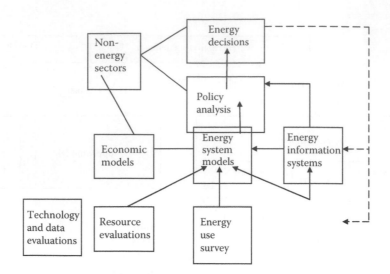

Figure 25.10 A structure for national energy planning.

examination (UME) test. With two assessment scores per candidate given the department/faculty/university of choice quota, the question is how can we best optimize a candidate's opportunity to be admitted? That is

- How do we establish that a particular candidate actually sat for the Joint Admissions and Matriculations Board (JAMB) examination?
- Is the JAMB examination score enough to measure competence?
- Is it sufficient for the various universities to select based on their own examinations?
- To what extent will the university post-UME test help judge candidates' suitability?
- Could there be a rational middle-of-the-road approach in determining a candidate's admission to the university based on the two regimes of scores he holds?
- To what confidence level can a university assert that it has selected rationally?
- Answers to these and many more questions and issues on this subject matter are concerns of wide variability, imprecision, and vagueness. The objective is to formulate fuzzy–stochastic decision support models for the resolution of this matter.
- The 2010 University of Lagos admission process reflected the underlying uncertainty in rationally giving candidates to its programs based on the two regimes of scores:
  - *Stochastic aspect*: A candidate for admission must reach a particular scholastic level at the JAMB UME examination. UME is a highly stochastic scheme with randomness in performance over time and space. The various performance levels of all candidates for a particular course/faculty/university may be assumed as normally distributed.
  - *Fuzzy aspect*: Candidates for admission need another scholastic assessment based on the score of the post-UME screening exercise conducted by the university. This is fuzzy based on a reference course faculty/university mean score(s). The post-UME screening score may be assumed as having membership grade distribution, which is equivalent to the fuzzy set (represented by the course/faculty/university mean score (x_i) to 100/400).

In other words, candidates must satisfy post-UME screening (fuzzy) criterion before being considered for admission based on the UME scores (stochastic).

Mathematically expressed,

$$\text{Max}\, A = A\left(Pu_s, \frac{U_s}{q_o} \right) \tag{25.12}$$

subject to

$q_o = q_o$ (merit, catchment, ELDS)

where

A = probability of admission at the university of first choice and first course
U_s = UME score
q_0 = course admission quota
Pu_s = post-UME screening score
merit = absolute performance (independent of state of origin of candidate)
catchment = contingent states of location of institution
ELDS = educationally less disadvantaged states in the federation

25.15   Conclusions

This chapter outlined the scientific uncertainty in relation to optimal DSS management for development. The two principal inherent forms of uncertainty (in data acquisition, realization, and processing), fuzziness and stochasticity, as well as their combinations are introduced. Possible effects on developmental issues and the necessity for DSS, which incorporate combined forms of this uncertainty, are presented.

In particular, the major focus of this submission is the proposal to deal with scientific uncertainty inherent in developmental concerns:

- First, uncertainty should be accepted as a basic component of developmental decision making at all levels and thus the quest is to correctly quantify uncertainty at both macro and micro levels of development.
- Second, expert and scientific DSS, which enhance correct evaluation, analysis, and synthesis of uncertainty inherent in data management, should be utilized at each level of developmental planning and execution.

References

Ibidapo-Obe, O. 2006. Modeling, identification/estimation in stochastic systems. In Badiru, A.B. (ed.). *Handbook of Industrial and Systems Engineering*. Vol. 14, pp.1–16. CRC Press, Taylor & Francis Group, New York.

Ibidapo-Obe, O. and Ogunwolu, L. 2004. Simulation and analysis of urban transit system under uncertainty—A fuzzy-stochastic approach. *Proceedings of the Practice and Theory of Automated Timetabling, PATAT*, Pittsburgh, PA.

chapter twenty-six

Using metrics to manage contractor performance

R. Marshall Engelbeck

Contents

26.1 Introduction

Acquisition transformation has brought numerous changes to the way the Department of Defense (DoD) purchases goods and services, as well as to the way it manages the contractor's performance after contract award. One of the most significant changes has been the emphasis on the use of commercial goods and services coupled with the use of performance-based contracts to satisfy mission-critical needs.

As the performance of contractors has become more critical to the mission, there has been a significant reduction in the number of acquisition personnel available to perform what traditionally has been referred to as "contract administration tasks." In addition, with the increase in the amount of services contracts, there has been a change in the skills needed to manage contracts after award.

Contract administration is currently undergoing a paradigm shift because of changes to what the DoD buys and to how purchases are being made. The move to minimize the importance of detailed specifications and to rely on commercial processes requires more management than administrative skills. The bottom line is that new methods and skills must be explored and developed to more effectively perform what traditionally has been referred to as "contract administration."

26.2 Foundation and framework for how the federal government should do business every day

The idea of integrating performance into the federal budget process is not new. Since the Hoover Commission proposed "performance budgeting" in 1949, there has been continued interest on the part of the Legislative Branch in measuring the results achieved in the past and linking them to future budgets. The Planning, Programming, and Budgeting System (PPBS) introduced in the 1960s, and the Zero-Based Budgeting System employed during the Carter Administration are just two examples of structures that tried to correlate program objectives both to results and to the means of achieving program goals (Cavanagh et al., 1999).

Another example of a governmental attempt to determine how effective an agency is in managing its programs is the Government Performance and Results Act (GPRA) of 1993 and the GPRA Modernization Act of 2010. GPRA requires departments and agencies to have long-range strategic plans, develop measurable outcomes, and report results achieved annually for each program in their budgets. Previously, managers concentrated on administrating the processes and reporting how well they were functioning rather than on being concerned if the output satisfied the goals upon which the program was based (Kessler and Kelly, 2000).

The bottom line for most government agencies is their mission: What they want to achieve. Under GPRA, what the agency wants to achieve is expressed in its strategic plans. These plans also include performance indicators to be used to measure the outputs and evaluate the outcomes of each program. The primary goals of GPRA are to ensure strategic planning drives budgeting and that resource decisions reflect specific strategic priorities, not just any "firestorm" that occurs (Whittaker, 2003).

The key to strategic success under GPRA is an understanding of the elements of its framework and interrelationships.

- The congressional budget includes a strategic performance plan that defines what major programs the agency intends to accomplish.
- Strategic performance plans are designed to establish a "bottom line" to which each government program can be held accountable.
- Activities are actions and tasks performed within the organization to produce outputs (goods or services).
- Performance indicators are particular values or characteristics that are used to measure outputs and/or outcomes. They include factors such as timeliness, quality, product integrity, customer satisfaction, and quantity.
- Output is the goods or services produced by a program.
- Outcomes are the things that occur as a result of output when the customer receives and uses the goods or services produced (Kessler and Kelly, 2000) (see Figure 26.1).

The National Partnership for Reinventing Government (National Performance Review), which was established to make government run more efficiently and effectively, continued

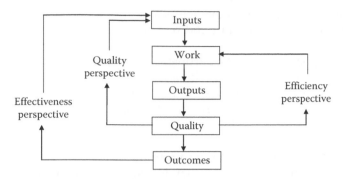

Figure 26.1 The GPRA performance accountability framework. (Adapted from Martin, L.L. *Making Performance-Based Contracting Perform: What the Federal Government Can Learn from State and Local Governments, New Ways to Manage Series*, Arlington, VA: IBM Endowment for the Business of Government, November 2002.)

the GPRA theme. This was followed by three other pieces of legislation that required strategic plans to describe how each government agency will deliver quality results and then measure its performance in terms of output and outcomes. These pieces of legislation were the Federal Acquisition Streamlining Act of 1994, the Government Management Reform Act of 1994, and the Information Technology Management Reform Act of 1996 (Clinger–Cohen Act).

Some agencies and departments have also adapted Kaplan and Norton's balanced scorecard approach and are using balanced performance measures to evaluate performance of their organizations as a whole. As the name implies, the balanced scorecard does not concentrate on just measuring financial performance. It is multidimensional and includes the customer's perspective, internal processes' perspective, and learning and growth as measurements of the financial perspective in order to obtain some balance of these categories in relation to mission achievement. This approach allows managers to consider whether improvements in one area are being achieved at the expense of another. Examples of the use of the balanced scorecard by public organizations include Naval Undersea Warfare Center, FAA Logistics Center, Internal Revenue Service, and the Department of Transportation (Whittaker, 2003).

26.3   Federal Acquisition Regulations guide to acquisition of goods and services

The Federal Acquisition Regulations (FAR) provide the foundation to guide acquisition of goods and services procured by the Executive Branch. The FAR includes a strategic planning perspective in the form of a "primary vision" for the federal acquisition system. The vision statement dictates: The primary goal of the acquisition system is to "deliver on a timely basis a best value product or service to the customer, while maintaining the public trust and fulfilling public policy objectives" (see Figure 26.2).

The FAR goes on to enumerate four major critical success factors by which the process should be evaluated. These are

- Customer satisfaction of cost, quality, and timeliness
- Minimization of administrative cost
- Maintenance of the public trust
- Fulfillment of public policy objectives (Federal Acquisition Regulation (FAR) 1.102, 2015)

Each of these critical success factors lends itself to the establishment of performance indicators, which measure the output and outcome as they are presented in the GPRA framework.

For defense weapons systems, Department of Defense 5000 (DoD 5000) series directives govern all acquisition programs. Important to the management of each defense program is the Acquisition Program Baseline (APB). The APB serves as a critical success indicator because the baseline also represents desired results in measurable terms. Every Program Manager (PM) is required to

> Establish program goals for a minimum number of cost, schedule and performance parameters that describe the program over its life cycle. Approved program baseline parameters shall serve as control objectives. PM shall identify deviation from approved program baseline parameters and exit criteria.

> **Defense Acquisition System**
> *2003*

In addition to ABP data, DoD Instruction 5000.2 contains references to 50 pieces of information required by the statutes or regulations that may be applicable during product development milestones and phases. This regulatory program information is to be tailored to fit the particular conditions of each individual program. This program includes a Contractor Cost Data Report, which is required for noncommercial contracts not procured under competitive procedures that include information on major contracts and subcontracts for Acquisition Category I Programs over $50 million and high-risk or high-technical interest contracts between $7 and $50 M (DoD Instruction 5000.2, 2003).

26.4   Changing what we buy and the way we buy it

In FY 2002, the Federal Government continued its long-standing record as the largest buyer of goods and services in the world. In that year, contract actions exceeded $250.2 billion. DoD contracts represented 66% of that total—for example, $164.7 billion.

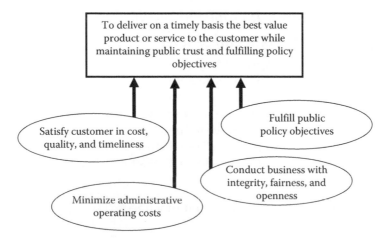

Figure 26.2 Acquisition vision and performance standards.

In recent years, there have been significant increases in the quantity of commercial items purchased by the DoD. Primary reasons for this were (a) the fact that more and more innovative products are available commercially-off-the-shelf (COTS) that would satisfy government requirements; (b) Federal Acquisition Streamlining Act (FASA) has simplified procedures for procurement of low-cost, low-risk commercial supplies and services; and (c) military specifications and standards are replaced by performance standards or commercial standards when possible. Consequently, FAR policy now requires the contracting officer of each acquisition to initially look to the commercial market as a source for goods and services (FAR 10.002, 2015).

The dollar value of purchases for services now exceeds what is spent for supplies and equipment. In 2003, in particular, the General Accounting Office (GAO) stated that during the last 5 years, DoD spent more to procure services than it did for supplies and equipment. In fact, approximately 50% of DoD's spending in 2002 was for services. Between 1999 and 2002, spending for services increased approximately 19% to $93 billion (GAO, 2003). During this same period, the number of contractors grew by over 727,000, while the civil service workforce decreased (Peckenpaugh, 2003).

Part of this trend has been due to the fact that over the past few years, policy has emphasized consideration of private sector firms as alternative service providers of tasks that can be classified as being inherently commercial. The Federal Activities Inventory Reform Act (FAIRA) calls for the head of each agency to decide which activity would be subject to competition under the provisions of OMB Circular A-76 (Federal Acquisition Council, 2004).

Two examples of activities that have been contracted out to the private sector have been facility maintenance and personnel support services at military installations and bases. Information technology is another area considered to be commercial in nature and, therefore, subject to competitive sourcing.

There have been other changes to what the federal government buys and how it purchases those goods and services. Beginning in 1995, acquisition managers have been encouraged to think more creatively and look for innovative ways to procure supplies and services. This paradigm shift was propelled by two significant pieces of legislation, which have been discussed previously, FASA and the FARA. Both urged acquisition managers to look for alternative ways of doing business that were cheaper, better, and faster—as long as it made good business sense. The only constraint was that the FAR, Executive Order, or agency regulations could not prohibit the particular course of action (FAR 1.102d, 2015).

Examples of some of the more significant changes are listed below. Each was brought about by a policy that tolerated thinking outside-the-box (see Table 26.1).

1. *Use of government purchase cards*: Government-wide policy has made use of the purchase cards as the preferred method of procuring commercial items below the micropurchase threshold of $2500 ($2000 for construction). In FY 2003, purchase card usage history showed there were 10.7 million purchase card transactions totaling $7.2 billion in sales; these sales thereby eliminated a significant workload, which previously had been performed by contracting personnel and the management of relatively small-value purchases.
2. *Use of e-commerce*: The DoD has further facilitated the purchase of consumable supplies by establishing the DoD EMALL. The DoD EMALL's objective is to be the single electronic point of entry for purchasers to locate and acquire off-the-shelf, finished goods and services from the commercial market place as well as from government sources. It creates a paperless environment, thereby simplifying the contracting job.

Table 26.1 Results of thinking out-of-the-box

Purchase card used for micro purchases
Purchases <$2500
Reduced role of contracting officer
Use of e-commerce
Single point of entry to commercial market
Expanded use of federal supply schedules
Using previously established contracts
Strategic purchasing by commodities
Leverage buying power
Use of performance-based contracting methods
50% of services contracts to be PBSC

3. *Expanded use of federal supply schedules*: There has been a significant increase in the use of indefinite-delivery, indefinite-quantity (IDIQ) contracts. Use of a federal supply schedule permits contracting officers to initiate individual orders for supplies or services from commercial firms using IDIQ contracts previously established by the General Services Administration (GSA). This system was not previously permitted because the source of contracting authority was outside the GSA. Under this method, however, the ordering agency is responsible for determining if contract performance is acceptable and for taking the necessary action if it is not. This method, formerly known as the Federal Supply Schedule Program, accounted for contracts which totaled 7 billion dollars in FY 2002 alone. It has not only simplified the contracting officers' jobs by eliminating the need to participate in the source-selection process, but it has also greatly reduced procurement lead times.

4. *Strategic purchasing of commodities*: Use of cross-functional sourcing teams' leverage in terms of buying power by strategic source centralized purchasing: The services are now looking to optimize their buying power by creating "commodity councils." The objective is to develop centralized purchasing strategies for different commodities. IBM, Daimler/Chrysler, and Cessna have successfully used it. Adoption of this concept means the service will enter into enterprise-wide larger volume contracts with fewer suppliers rather than having each user contract for its own needs. This method of acquisition gives the buyer the advantages of leverage. It also eliminates a significant amount of administrative workload at multiple installations (Burt, 2003).

5. *Use of performance-based contracting methods*: Performance-based contracts are structured around terms that describe measurable output, quality, or outcomes in the form of measurable standards for the work to be performed, rather than broad or precise statements of work regarding the way services are to be performed. This method of procurement emphasizes telling the provider what is wanted rather than how. It also encourages contractor rewards for exceptional performance or fee reduction when performance is not satisfactory. It was DoD policy that 50% of all service contracts be performance-based by 2005. Experience has shown that although additional work is required with this method, initially describing tasks to be performed and quantifying desired outputs and outcomes greatly reduces the contracting officers' involvement after contract award.

As government acquires more and more services via contract, the lead-time from acquisition decision to mission impact has become less and less. Regarding a weapon system, years can pass before the mission impact of a poor decision made during design, production, or provisioning of spare parts can be discovered. In contrast, failure to perform a mission-critical service can have an almost instantaneous negative impact.

This reality requires a reexamination of the relationships between the buyer and seller. Policies initiated during acquisition reform have made it possible to move away from an arms-length buyer–seller relationship in favor of greater communication and even inclusion of the contractor as a member of the acquisition team (FAR 1.102(c), 2015).

26.5   Contractor oversight after contract award

Reform of federal procurement policy has resulted in increased flexibility for contracting officers, program managers, and other acquisition personnel. Changing times call for innovative ways to achieve results. As noted above, this innovation occurred primarily in the methods by which supplies and services are purchased. In relation to contracts, the impact of acquisition reform on the function of contract administration has not been fully recognized (Lawther, 2002).

After contract award, oversight of the contractor's performance has traditionally been referred to as "contract administration." The objectives of contract administration is to gain assurance that a quality product will be delivered to the ultimate user in accordance with cost and schedule projections, as well as to retain the confidence of the tax payer in proving his/her tax dollars are being spent wisely.

Oversight includes the employment of government personnel for tasks such as monitoring contractor performance, inspecting and accepting delivered products or services, modifying the contract, resolving technical and delivery problems, reviewing contractor's processes, auditing contractor systems, and applying rewards and penalties when needed (FAR 42.302, 2015).

Professor Steve Kelman, former head of the Office of Federal Procurement Policy (OFPP) and now a Professor of Public Management at John F. Kennedy School of Government, believes contracting can no longer be considered primarily an administrative function. His thesis is supported by the fact that a high percentage of an agency's budget is spent on contracts for supplies and services. He recommended contracting management be considered a core competence within each federal organization. He further believed critical responsibilities that accompany the job include performance measurement and management of the contractor (Kelman, 2002).

The nature and extent of contract administration varies from contract to contract. It can vary from a minimum level of involvement in the contractor's activities to a maximum level of almost dictating the methods by which supplies and services are purchased. This is in contrast to the impact of total engagement (Cibinic and Nash, 1986). The depth of oversight has traditionally taken one of the following approaches (see Figure 26.3).

- Minimum oversight is associated with small-scale jobs, projects where the management and control is essentially left to the discretion of the supplier in the purchase of commercial items. The contractor is most often permitted to follow his own production, assembly, testing, and quality process/procedures. For commercial items, inspection and testing by the government is completed in a manner consistent with accepted industry practices.
- The intermediate level of oversight occurs when the procurement is large in scale, or the supplies and equipment being procured are unique to the government. The buyer sometimes imposes extensive data collection and status reporting on the seller.

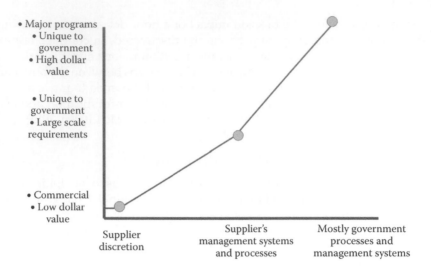

Figure 26.3 Level of contractor oversight by government.

However, in this situation, the seller's managing systems and techniques generally provide the buyer with the necessary performance-status information. When government unique standards are required, the contractor can also be required to follow government production, assembly and testing processes as a term and condition of the contract.

- The most intensive level involves large projects, research and development efforts, or the acquisition of major systems. In this case, the supplier's internal management systems and processes may not be fully acceptable. Also in this case, the buyer often imposes process specifications or standards of performance upon the supplier. Often, the contractor is required to follow government production, assembly, and testing processes as a term and condition of the contract.

An important responsibility of contracting personnel after contract award is annual evaluation of a contractor's performance information on designated contracts, as well as at the time of completion of the contract. The FAR then requires the source-selection authority to reconsider this record of past performance for future source-selection purposes. Knowledge of a contractor's record of contract performance is seen as a way to mitigate the risk of selecting a contractor with a poor track record.

Contractor performance reports are collected and maintained in a web-enabled library developed by Naval Sea Logistics Center for the DoD, commonly referred to as CPARS (Contractor Performance Assessment Reporting System). The type of information reported includes

- Contractor's record of conforming to contract requirements and standards of good workmanship
- Contractor's record of forecasting and controlling cost
- Contractor's adherence to contract schedules, including the administrative aspects of performance contractor's history of reasonable and cooperative behavior and commitment to customer satisfaction
- Contractor's business-like concern for the interests of the customer

The study concludes that both parties would jointly benefit by managing these indicators concurrently throughout the life of the contract, rather than reporting on them after the fact.

26.6   Cost of oversight

The premium paid by the government for the oversight methods and procedures used by the buyer to assure the timely delivery of the needed supplies, services, and the confidence of the taxpayer have not gone unnoticed. Between 1986 and 1992, at least six studies were made of the regulatory/oversight cost premium being passed on to the government buyer by contractors. Rough order of magnitude (ROM) estimates of the added cost premium cited in these reports ranged from 5% to 50% in potential cost reductions. The potential cost avoidance cited by all of these studies no doubt had an influence on FASA and other DoD acquisition policies before and after the Acquisition Streamlining Act in 1994. Another assessment of the DoD regulation and oversight burden, published by RAND in 2001, estimated the oversight burden to range from 1% to 6% (Lorell and Graser, 2001).

However, there has been recent criticism regarding how well the DoD oversees contractor performance. In response to this criticism, Professor Kelman pointed out that oversight is not without cost and that fixating on traditional oversight methods while ignoring other critical contract management tasks (such as establishing performance metrics, inciting contractor technical experts and government subject-matter experts to share information, and motivating both government and contractor personnel to achieve results) is not the answer (Kelman, 2004).

26.7   What is performance measurement?

Performance may be defined as "The execution or accomplishment of work, acts, feats, etc. [...] The act of performing. [It is] the manner in which, or efficiency with which something reacts or fulfills its intended purpose" (*Random House Webster's College Dictionary*, 1994).

Professors Michael Lebas and Kenneth Euske stated that performance is continuous and has a cause and an effect. Performance exists only if outcome and results can be described or measured so as to be accurately communicated; only in this way can someone decide to do something within the shared model of causal relationships. Performance does not have the same meaning if the evaluator is inside rather than outside the organization. The use of the term "performance" should be reserved for all processes that lead to a potential or future sequence of outcomes and results (Euske and Lebas, 2002).

The Government Accounting Standards Board views performance as a multidimensional concept consisting of outputs, quality, and results (Martin, 2002). *Random House Webster's College Dictionary* (1994) defines measurement as "the act of measuring [...] a dimension, extent, range, etc. or ascertained by measuring." In its simplest terms, performance measurement is the comparison of actual levels of performance to preestablished target levels of performance (PBM SIG, 2001). (In the terminology of GPRA, this is comparing output to preestablished outcome objectives.)

Traditionally, performance measurement has been a key element of what management gurus have referred to as the "control function." The objectives of performance measurement are

> Assessing progress toward achieving predetermined goals, including information on the efficiency with which resources are transformed

into goods and services (outputs), quality of those outputs (how well they are delivered to clients and the extent to which clients are satisfied) and outcomes (the results of a program activity compared to its intended purpose) and the effectiveness of government operations in terms of their specific contributions to program objectives.

Whittaker
2003

Kessler and Kelly (2000) recommended that performance measures must be placed in any plan immediately after the statements of goals. This strategic location forces the planner to think critically to define applicable performance measures and tie them to each goal early in the planning process. The authors went further by stating, in order to be effective, a performance measure must include the following elements:

- *Performance measures*: Characteristics relative to the overall strategy, such as cost, cycle time, quality, and quantity.
- *Performance indicators*: Unit of measure as dollars, hours, errors per 1000, number of reports each period that can be used to measure output and outcome.
- *Baseline*: An initial starting measurement.
- *Performance target*: Desired level of performance expressed as a tangible, measurable objective against which achievement can be compared relative to a baseline measure.
- *Tolerances*: Acceptable ranges of variance from performance targets.

As anyone in public or private industry can attest, performance measurement does not stop when variances from planned performance have been identified. Leading-edge organizations seek to create an efficient and effective performance-management system. Whittaker (2003) dictated that managers must also

Use performance measurement information to effect positive change in organizational culture, systems, and processes by helping to set agreed-upon performance goals, allocating and prioritizing resources, informing managers to either confirm or change current policy or program directions to meet those goals, and sharing results of performance in pursuing those goals.

These actions are part of the performance management system.

Professor Marshall Meyer of the University of Pennsylvania's Wharton School provided a schematic of how performance measures are used (see Figure 26.4). He contended that performance measurement fulfills seven purposes. In addition to serving as a control metric, which he referred to as "Look back," performance metrics are also used to "Compensate," "Motivate," and "Look ahead" to assess the future. The use of performance measures becomes even more significant as an organization increases in size. In larger organizations (the area within the triangle), metrics are reported up in the chain of command ("Roll[ed] up"), so management can determine if the desired results have been achieved and can adjust goals or establish future goals accordingly. These are then "Roll[ed] back" down the organization in the form of financial and nonfinancial objectives. Another value performance measurement this model provides is that it can also be used to "Compare" to other organizations across functional boundaries horizontally or to compare to other business units.

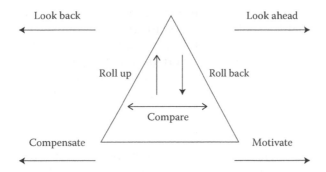

Figure 26.4 The seven purposes of performance measures. (Adapted from Meyer, M.W. *Rethinking Performance Measurement: Beyond the Balanced Scorecard.* Cambridge, UK: Cambridge University Press, 2002.)

26.8 Some examples of how DoD now uses performance measurement to manage contracts after award

26.8.1 Award fee type of pricing arrangements

This type of contract provides a monetary base fee, plus an additional fee that is awarded periodically based on a subjective evaluation by the acquisition team of how well the contractor satisfied predetermined evaluation factors. Contractor performance in areas such as responsiveness, quality, timeliness, management of risk, and cost contract are examples of possible evaluation factors. It is important to note, however, that the acquisition team can change its evaluation criteria at designated milestones during contract performance.

A good example of how performance measurement can be used to evaluate contractor effectiveness can be found in the contractor-operated Arnold Engineering Development Center (AEDC) USAF. AEDC, whose mission is to support new aerospace systems by conducting R&D testing and to evaluate for the DoD and other government and commercial agencies, has been using competitively awarded cost-plus-award fee-type pricing arrangements for over 25 years.

Bowling (2004) felt the award fee feature has enabled the Air Force to use the profit motive to incentivize contractors to continually improve productivity and quality while controlling cost. On a certain contract, in over eight award fee evaluation periods, AEDC tracked an average of 283 separate performance metrics. It should be noted that the importance the government placed on performance factors to be measured did not remain static. Between evaluation periods, on average, 22% of the performance goals were adjusted; 28% of the metrics from the previous period were deleted; and an average of 18% new performance measures were added.

An advantage of this type of contract is that it improves communication between the two parties. A disadvantage to both parties is the fact that there is a substantial administrative burden associated with the evaluation and review of the award fee evaluation information.

26.8.2 Program baseline parameters

DoD 5000 series directives, which provide management principles and mandatory policies and procedures for managing all acquisition programs, require every program manager to establish goals for the minimum number of cost, schedule, and performance parameters that describe the program over its life cycle (DoD Directive 5000.1, 2003).

These goals represent the key performance parameters that serve as the program baseline that is essentially a contract between the program manager, Defense Acquisition Executive, and Service Acquisition Executive. The approved program baseline parameters then serve as control objectives in that the program manager must report on the status of approved baseline parameters to the acquisition executives; the status of those baselines is also key to a program's migration throughout the system's development.

26.8.3   Contracts for performance-based services

The contract is the common language between the buyer and seller and, therefore, it should lend itself to quantifiable measures of performance. A guide to best practices for writing Performance-Based Service Contracts (P-BSC) was published by the OFPP. This document applies the GPRA concept in its instructions on the preparation of performance-based work statements (P-BWS). The guide recommends describing the requirement in measurable performance standards. These standards should include elements such as "what, when, where, how many, and how well" the work is to be performed. This requires defining the required work in quantifiable terms with an acceptable error rate, which is referred to as an allowable quality level (AQL) (OFPP, 1998).

26.8.4   Aircraft service contracts

In describing the value of performance measures in the management of aircraft and facility maintenance service contracts, Ms. Cheryl Smith of Headquarters Air Mobility Command (AMC/LGCO) stated (personal communication, 2004): "performance measures quantitatively tell us something important about the contractor's products, services and the processes that produce them, that is, how well the contractor is doing, is the contractor meeting its goals, are customers satisfied, are the contractor's processes in statistical control, and if improvements are necessary, where are they needed?"

AMC best practices include: (a) relying on commercial-quality standards, (b) having the contractor propose metrics and a quality plan, (c) only measuring critical outcomes because excessive metrics can be costly, and (d) including language in the contract permitting the negotiation of changes to the measures (Cheryl Smith, personal communication, 2004).

26.8.5   Contractor performance evaluation reporting

As noted earlier, contractor performance assessments reports are required and maintained in the CPARS database. Dollar thresholds by business sectors for reporting positive as well as negative performance information on DoD contracts are given in Table 26.2.

26.9   What can be learned from private industry about measuring contractor performance?

26.9.1   Measuring supplier performance

From literature research, one can conclude that private industry collects supplier-performance data primarily for strategic sourcing purposes and to determine how well the purchasing department as a whole is performing. The primary purpose is to use

Table 26.2   CPARS dollar thresholds by business sectors

Business sector	Dollar threshold
Systems	$5,000,000
Ship repair/overhaul	$500,000
Services	$1,000,000
Information technology	$1,000,000
Operations support	$5,000,000
Fuels	$100,000

metrics at a macro level to judge performance of the procurement organization at the macro level.

Typical is a comment by Terry Sueltman, Corporate Vice President of Purchasing and Logistics, Sonoco Products. He believed purchasing and supply-chain management measures, to be effective, "must be more strategic, at a higher level to have a broad business impact and should lead change. Instead of talking of the price per part we use, talk about the impact on production cost" (Sueltman, 2001).

The Center for Advanced Purchasing (CAPS) publishes a list of "Standard" Purchasing Performance Measures. The significance placed on the effectiveness of purchasing organizations in a macro sense can be judged by looking at the standards. Examples of the 21 CAPS standards are purchasing operating expense dollars as a percent of sales dollars, percent change in total purchase dollars influenced/assisted by purchasing and the number of active suppliers accounting for 80% of purchase dollars, percent of total purchase dollars spent through reverse auctions, and percent of purchase dollars spent through strategic alliances (CAPS, 2001).

26.9.2   What are the most important supplier characteristics?

In February 2002, *The Journal of Supply Chain Management* published the results of a random sampling of the over 2000 members of the Institute for Supply Management (ISM) members on formal supplier evaluation programs. An analysis of the 299 responses was published as a report titled "Formal Supplier Evaluation Programs and the Factors Used to Channel Evaluations." A summary of the responses showed the following:

- Less than half of the responding firms had formal supplier evaluation processes in place. (Formal supplier evaluations were aggregate evaluations of vendor performance rather than measuring ongoing outputs of a contract as part of contractor surveillance.)
- For those who had a formal rating system:
  - Two-thirds ranked supplier performance relative to one another.
  - Over half evaluated suppliers on a routine basis.
  - Twenty-one percent of the suppliers were asked to perform self-evaluations.
  - Less than one-third of the responses indicated they had a system for weighing categories of evaluation.
  - There were over 30 different supplier characteristics (Simpson, 2002). (Table 26.3 illustrates the top ten categories.)

Table 26.3 Top ten categories

Category (%)	Relative importance
Quality and process control (24.9)	1
Continuous improvement (9.2)	2[a]
Facility environment (8.2)	2[a]
Customer relationship (8.2)	2[a]
Delivery (8.1)	2[a]
Inventory and warehousing (7.0)	2[a]
Ordering (5.8)	2[a]
Financial condition (5.5)	2[a]
Certifications—ISO (3.6)	3
Price (3.6)	3

[a] No statistical difference among categories.

From this survey, one can conclude that, for the most part, commercial industry uses performance measurement primarily to evaluate the effectiveness and efficiency of its purchasing department. The lack of a formal supplier evaluation program, like the DoD CPARS program, by less than half of those surveyed may be the result of the fact private industry is not as subject to bid-protest rules as the federal government is.

26.9.3 How do commercial firms measure services?

An analysis published by RAND on a study of "fourteen innovative commercial customers and world-class providers of facilities management services" provided insight into how these firms used performance metrics to execute sourcing activities and to help manage their relationships and promote continuous improvement (Baldwin et al., 2000). The report concluded the following:

- Customer and provider jointly choose metrics.
  - The customer tended to track output-oriented metrics.
  - The service provider may also track process-oriented metrics.
- Both qualitative and quantitative measures were used.
- Both parties refined their set of metrics as the relationship evolved.
- Benchmarking studies were used to set feasible goals.
- Both formal and informal reviews were held to communicate performance.
- Both the generation of metrics and tracking of performance are costly.
- Most common characteristics of metrics used are
  - Finance cost and customer satisfaction.
  - Qualitative metrics on performance of specific services.
- Special interests topics were also tracked.

26.10 Conclusions

Performance measurement can be used to manage contractor's performance after award. The GPRA requires managers in the Executive Branch to develop strategic plans and to use performance indicators to record output and evaluate the outcome of each program. The Act establishes a management framework that includes

- Development of strategic performance program plans with well-defined desired outcomes
- Development of performance indicators for desired outcomes
- Definitions of outputs to track and measure results
- Comparison of outputs to desired outcomes
- Documentation of report results

This framework readily lends itself to the measurement of contractor performance and appears to be an appropriate way to manage contracts in the public sector.

Currently, an integral part of the contract administration (management) process is contractor-performance-evaluation reporting. However, such documentation is seen primarily as a way to mitigate the risk of selecting a contractor with a poor performance track record in the future. Examples of data that must be reported on a contractor's performance include (a) contractor's record of conforming to contract requirements and standards of good workmanship, (b) contractor's record of forecasting and controlling cost, (c) contractor's adherence to contract schedules, including the administrative aspects of performance, (d) contractor's history of reasonable and cooperative behavior and commitment to customer satisfaction, and (e) contractor's business-like concern for the interests of the customer. It appears the reported data could easily be tailored to fit the need of a performance-measurement process.

Kaplan and Norton's (1993) Balanced Scorecard concept enables managers to visualize the breadth of an operation as well as to link measurements to strategy. The balanced scorecard methodology could be adapted to measure contractor performance in a manner that would enable the contract strategy, as well as critical success factors and desired outcomes, to be included in a contractor-performance measurement process. The result would be to add joint accountability for results to the buyer–seller relationship.

It is also believed that both parties would jointly benefit from the enhancement of communications—communications that would no doubt result in the use of these indicators throughout the life of the contract rather than as a reporting mechanism after the fact.

References

Baldwin, L.H., F. Camm, and N. Moore. *Strategic Sourcing: Measuring and Managing Performance*, DB-287-AF, Santa Monica, CA: RAND, 2000.

Bowling, T. IV. *Senior Contracting Official*, Arnold Engineering Development Center/PK, Tullahoma, TN, February 2004.

Burt, D., D. Dobler, and S.L. Starling. *World Class Supply Management SM: The Key to Supply Chain Management*, Seventh Edition, New York: McGraw-Hill, 2003.

Cavanagh, J.J. et al. The balanced scorecard for managing procurement performance, *Contract Management*, February 1999, p. 13.

Center for Advanced Purchasing Studies (CAPS), *CAPS "Standard" Purchasing Performance Measures: Supplier Performance Metrics That Really Make a Difference*, Supplier Selection and Management Report, September 2001.

Cibinic, J., Jr. and R.C. Nash, Jr. *Administration of Government Contracts*, Second Edition, Washington, D.C.: George Washington University, 1986.

Defense Acquisition System, DoD Directive 5000.1, 4.3.4 Discipline, May 12, 2003.

Euske, K. and M. Lebas. A conceptual and operational delineation of performance, in *Business Performance Measurement: Theory and Practice*, Edited by A.D. Neely, Cambridge, UK: Cambridge University Press, 2002, pp. 125–140.

Federal Acquisition Council, *Manager's Guide to Competitive Sourcing*, Second Edition, February 20, 2004.

Federal Acquisition Regulation (FAR) 1.102, http://www.acquisition.gov/far/, accessed February 24, 2015.

Federal Acquisition Regulation (FAR) 1.102 (c), http://www.acquisition.gov/far/, accessed February 24, 2015.

Federal Acquisition Regulation (FAR) 1.102 (d), http://www.acquisition.gov/far/, accessed February 24, 2015.

Federal Acquisition Regulation (FAR) 10.002, http://www.acquisition.gov/far/, accessed February 24, 2015.

Federal Acquisition Regulation (FAR) 42.302 list 6, http://www.acquisition.gov/far/, accessed February 24, 2015.

Government Accountability Office (GAO), *Contract Management: High-Level Attention Needed to Transform DoD Services Acquisition*, GAO Report GAO-03-935, September 2003.

Kaplan, R.S. and D.P. Norton. Putting the balanced scorecard to work, *Harvard Business Review*, September–October 1993, pp. 134–147.

Kelman, S. *Remaking Federal Procurement: Visions of Governance in the 21st Century*, Working Paper No. 3, May 2002.

Kelman, S. Oversight is not the only way to judge a procurement system, *Federal Times*, May 31, 2004, p. A29.

Kessler, T.G. and P. Kelly. *The Business of Government: Strategy, Implementation and Results, Management Concepts*, Vienna, VA: Management Concepts, 2000.

Lawther, W.C. *Contracting in the 21st Century: A Partnership Model*, Arlington, VA: Pricewaterhouse Coopers Endowment for the Business of Government, 2002.

Lorell, M. and J.C. Graser. *An Overview of Acquisition reform Savings Estimates, (MR-1329-AF) Project Air Force*, Santa Monica, CA: RAND, 2001.

Martin, L.L. *Making Performance-Based Contracting Perform: What the Federal Government Can Learn from State and Local Governments, New Ways to Manage Series*, Arlington, VA: IBM Endowment for the Business of Government, November 2002.

Meyer, M.W. *Rethinking Performance Measurement: Beyond the Balanced Scorecard*. Cambridge, UK: Cambridge University Press, 2002.

Office of Federal Procurement Policy (OFPP), Office of Management and Budget (OMB), Executive Office of the Government. *A Guide to Best Practices for Performance-Based Service Contracting*, Final Edition, October 1998.

Peckenpaugh, J. Contractor workforce grows as civil service shrinks, *Government Executive Magazine*, 35(13), 5, September 4, 2003.

Performance-Based Management Special Interest Group (PBM SIG), Establishing and maintaining a performance-based management program, *The Performance-Based Management Handbook*, Vol. 1, September 2001, http://www.orau.gov/pbm, accessed February 24, 2015.

Random House Webster's College Dictionary, New York: Random House, 1994.

Simpson, P.M., J.A. Siguaw, and S.C. White. Measuring the performance of suppliers: An analysis of evaluation processes, *Journal of Supply Chain Management: A Global Review of Purchasing and Supply Chain*, 38(1), 29–41, February 2002.

Statutory, Regulatory, and Contract Reporting Information and Milestone Requirements, Enclosure 3, DoD Instruction 5000.2, May 12, 2003.

Sueltman, T. Supplier performance metrics that really make a difference, *Supplier Selection and Management Report*, September 2001: 11.

Whittaker, J.B. *President's Management Agenda, A Balanced Scorecard Approach*, Vienna, VA: Management Concepts, 2003.

chapter twenty-seven

Low-clutter method for bistatic RCS measurement

Peter J. Collins

Contents

27.1   Introduction

Bistatic radar cross-section (RCS) measurements have seen a recent resurgence as interest in passive coherent location (PCL) systems and other survivable detection schemes has increased. Low RCS bistatic measurements, in particular, provide a unique challenge to the RCS metrologist as noise, clutter, and interfering signals become significant relative to the desired signal. Traditional methods for supporting the bistatic antenna/receiver often provide a significant source of moving clutter that interacts strongly with the target. This chapter presents a new low clutter method based on recent advances in string-based target support systems. A simple experiment is described, and the resulting fixed angle bistatic measurement results are qualitatively analyzed, clearly demonstrating the benefits of the proposed methodology.

Bistatic radars and, consequently, bistatic RCS measurements have existed from the advent of radar nearly 80 years ago. During that period, interest in bistatic RCS metrology has ebbed and flowed in an approximate 15-year cycle depending on the perceived advantages for specific applications over the ubiquitous monostatic techniques (Willis, 1995). Recently, interest has turned to bistatic radar again as a possible means to provide a survivable counterstealth capability (Griffiths, 2003). Low RCS measurements provide a unique challenge to the RCS metrologist as many of the contaminants such as noise, clutter, and interfering signals, safely ignored for relatively high RCS targets, become significant sources of measurement uncertainty. The issue is compounded for bistatic measurements, as the clutter becomes a function of the bistatic angle and often the receiver structure itself.

This chapter presents a new mitigation strategy aimed at solving this problem by reducing the clutter associated with the bistatic antenna support structure. The concept leverages the lessons learned from recent string-based target support designs (Buterbaugh

and Mentzer, 2002). Key among these is the fact the stability of the inverted Stewart platform far exceeds the traditional three and four string designs, providing a very stable, low
interaction support system for RCS measurements (Buterbaugh et al., 2007). These same
properties are exactly what are needed to reduce clutter introduced by the antenna support structure in a bistatic measurement.

The following section develops this idea in the context of an indoor bistatic measurement. After briefly discussing the concepts related to bistatic RCS measurements,
the chapter describes the method whereby a string-based inverted Stewart platform is
used to suspend the bistatic receiver/antenna assembly of a bistatic measurement radar.
This approach is compared to the traditional method of supporting the assembly with an
absorber-coated tower. Next, an experiment is described with the intent of demonstrating
the benefits of the proposed design in reducing the measurement uncertainty. The measurement results for a 6"-square metal plate are presented showing a significant improvement in measurement quality.

27.2   *Bistatic RCS and bistatic measurements*

Bistatic RCS is defined by the IEEE Std 686-2008 as one of the cases of the more general
RCS definition whereby energy is "reflected or scattered in any direction other than the
incident direction or opposite of the incident direction." According to this standard, RCS is
"a measure of the reflective strength of a radar target, usually represented by the symbol σ
and measured in square meters" (IEEE, 2008a,b). The reflective strength is not only a function of physical size, but of shape, frequency, polarization, aspect angle, and target material
properties. Figure 27.1 (Currie, 1989) depicts a typical bistatic measurement configuration.

Notice for any target position, there are associated lines of constant phase delay where
scattering from the measurement environment can be introduced through antenna side

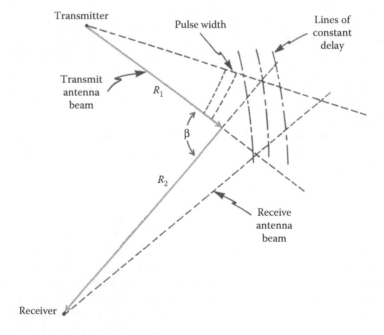

Figure 27.1 Bistatic measurement geometry with bistatic angle β.

lobes or target bistatic illumination to provide a contamination source. Equation 27.1 gives the received power for the case of distributed clutter characterized by an RCS per unit area σ^c as

$$P_r = \frac{P_t \lambda^2}{(4\pi)^3} \int_A \frac{G_t G_r \sigma^c}{R_1^2 R_2^2} dA \tag{27.1}$$

where P_r is the received clutter power, P_t is the transmitted power, G_t and G_r are the transmit and receive antenna gains, λ is the wavelength, and R_1 and R_2 are the ranges from the transmitter and receiver to the target (Currie, 1989). If there is a relatively large clutter source such as an antenna support structure in the clutter ellipse containing the desired target, Equation 27.1 says it will be integrated into the clutter power competing with the desired target return.

Common clutter suppression techniques such as range gating and coherent background subtraction may be ineffective if the target has strong bistatic scattering lobes that interact with these background structures. Since the exact interaction geometry is target dependent and, in the case of a variable bistatic angle measurement, angle dependent, it is difficult to design an absorber treatment that is equally effective for all measurement geometries. The only alternative is to create a noninteracting support mechanism for the bistatic receiving channel. Section 27.3 presents just such a minimally interacting support mechanism and describes an experiment to demonstrate its advantage over traditional bistatic channel designs.

27.3 Concept demonstration

The basic concept proposed is simply to replace the ridged antenna support structure commonly used in bistatic measurement ranges with a low scatter string support system. The scattering characteristics of high tensile strength dielectric strings have been shown sufficient for low RCS measurements while being able to support significant target loads (Buterbaugh et al., 2007). Until recently, however, the inherent instability of a suspension type support system limited their use as compared to ogival pylons and foam columns. Buterbaugh and Mentzer (2002) have overcome this limitation in their inverted Stewart platform design, which we adapt here.

One of the challenges in moving from target support to supporting an active bistatic antenna receiver is the need to get the received signal back to the radar without reintroducing scattering structures such as radiofrequency (RF) cables into the range. The author's approach is to use a wide-band RF-to-optical modem to convert the received RF signal to an optical signal that can be sent to the receiver via a relatively small, low scattering fiber optic cable. Thus, the only significant scattering structure in the range is the actual receive antenna and a small box housing the low-noise amplifier (LNA), modem, and battery power supply (see Figure 27.2). Also shown in the figure is the ceiling mounted optical position measurement system used to control the position and orientation of the receive antenna.

In order to demonstrate the viability of this new method, the author directed a comparative experiment whereby the same target was measured in the RCS measurement range at the Air Force Institute of Technology (AFIT), using the traditional bistatic arm and the proposed string antenna support system. The experiment was designed specifically to keep all variables the same except the clutter introduced by the bistatic receiver

Figure 27.2 Battery-powered bistatic-receiving channel with a quad-ridge, wideband antenna.

support system. A description of the experiment and qualitative analysis of representative results follows.

27.3.1 AFIT's RCS measurement facility

The AFIT RCS measurement range is an anechoic chamber with a tapered design approximately 14′ × 16′ at the throat, 24′ × 26′ at the back wall, and 45′ in length. The facility is designed primarily for academic research and thus is relatively flexible, being able to be configured as a far-field or compact range with either ogive pylon, foam column, or string target support systems. When configured for bistatic measurements, the bistatic arm/string support system can be used in conjunction with the latter target support mechanisms to provide either swept or fixed angle bistatic measurements. The range uses a Lintek 4000, parametric network analyzer-(PNA)-based pulsed continuous wave (CW) radar with frequency coverage from 2 to 18 gigahertz (GHz). For the subject demonstration, Flamm & Russell FR6400 low-side-lobe-diagonal horns were used in the transmit channel to minimize clutter illumination. The penalty was the reduction in the frequency range to 6–18 GHz due to antenna cutoff.

27.3.2 Experiment design

As the main purpose of the demonstration is to compare the impact of bistatic clutter on an RCS measurement for the two support systems, the author chose a target with a fairly wide dynamic range and relatively large bistatic signature. The 6″- square metal plate served the purpose for the size and frequency range available in AFIT's range. The author used horizontal ($\varphi\varphi$) polarization for the resulting low RCS near the edge-on illumination angles. For the sake of brevity, only a single fixed angle bistatic measurement is reported here. A more extensive examination of both fixed and variable angle bistatic measurements of a variety of interacting targets is presented in a companion journal paper currently in the submission process.

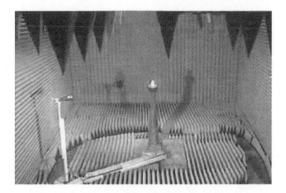

Figure 27.3 AFIT's RCS measurement range configured with a quad-ridge, wideband horn mounted on the bistatic arm.

For both the traditional bistatic arm measurement and the string support measurement, the standard RCS calibration procedure as expressed in Equation 27.2 is performed. Specifically, the calibrated bistatic RCS is defined by

$$\sigma = \frac{\sigma_{tar} - \sigma_{bkg}}{\sigma_{cal} - \sigma_{cbk}} \cdot \sigma_{thr} \qquad (27.2)$$

where subscript *tar* denotes the target measurement, *bkg* denotes target background, *cal* denotes calibration object, *cbk* denotes calibration background, and *thr* denotes the theoretical calibration object bistatic RCS. The calibration was verified using the procedure recommended in Kent (2001) resulting in measurement uncertainties of ±0.75 decibel relative to 1 square meter (dBsm).

In the two measurements presented, the bistatic receiver was positioned 20° from the transmitter approximately 8' from the target. The target was mounted on a low RCS ogival pylon and rotated 360° about an axis in the plane of the plate and perpendicular to the plane formed by the transmit and receive legs (see Figure 27.3).

The 6" square plate (not shown in the figure) is mounted on the short foam column atop the low RCS ogival pylon. The transmit antenna orientation is the same as the camera except it is in the same plane as the bistatic antenna and target height.

27.3.3 Comparative data analysis

We began the analysis by examining the global RCS plots shown in Figures 27.4 and 27.5. As presented, the global RCS represents the frequency dependent bistatic scattering as a function of the plate's angle with respect to the illumination direction. The inner radius corresponds to a frequency of 6 GHz and the outer 18 GHz. From both figures, we see a peak RCS response at 170° and 350°. This is consistent with the 20° fixed bistatic angle and relative position of the receive antenna as the plate rotated. We also note the peak values are of equal magnitude in dBsm, indicating a relatively high signal to clutter level.

Turning our attention to the lower RCS sectors of the plots, we see a distinct difference in scattering behavior. While the plate measured with the string support system shows the distinctive lobing structure associated with near grazing incidence angles, the traditional bistatic arm exhibits less lobing and a higher overall RCS level indicative of an interfering

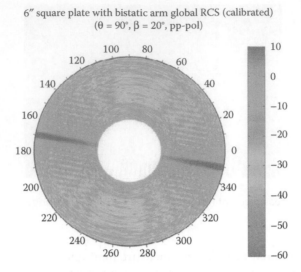

6" square plate with bistatic arm global RCS (calibrated)
(θ = 90°, β = 20°, pp-pol)

Figure 27.4 Global RCS for 6" square plate with a fixed bistatic angle (β) of 20° where the receiver is supported by the ridged support tower.

source. We also note a few discrete frequencies with what appears to be radio frequency interference (RFI) in the bistatic arm measurement of Figure 27.4. Clearly, there is a higher clutter level in the bistatic arm measurements. To recognize the source of this clutter better, we next present the associated global range plots generated by concatenating a series of down range images at each measurement angle.

Figures 27.6 and 27.7 tell a consistent story in that the bistatic arm measurements contain a significantly higher clutter level with a notable peak at approximately 20" up range from the plate. For the most part, the clutter seems to be independent of plate angle

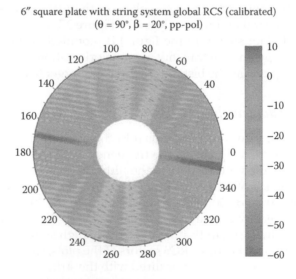

6" square plate with string system global RCS (calibrated)
(θ = 90°, β = 20°, pp-pol)

Figure 27.5 Global RCS for 6" square plate with a fixed bistatic angle (β) of 20° where the receiver is supported by low-scatter dielectric strings.

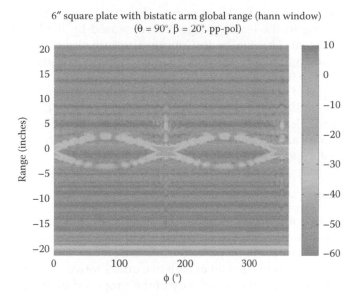

Figure 27.6 Global range for 6″ square plate with a fixed bistatic angle (β) of 20° where the receiver is supported by the ridged support tower.

indicating a direct clutter source rather than a target/background interaction term. Both plots show the distinctive pattern produced by the edge scattering from the plate as it rotates. The weaker scattering from the leading edge (relative to the illumination direction) is consistent with horizontal polarization scattering. Both plots also show the nulls in the trailing edge scatter as the plate comes into alignment with either the transmit antenna

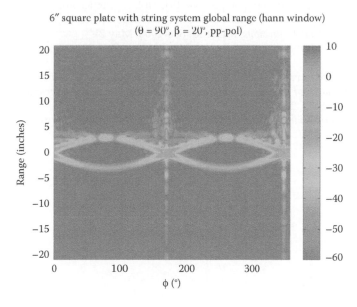

Figure 27.7 Global range for 6″ square plate with a fixed bistatic angle (β) of 20° where the receiver is supported by low-scatter dielectric strings.

or the receive antenna separated by 20°. However, the string support system measurement has none of the nulling in the leading edge return caused by the bistatic clutter in the bistatic arm measurement. This is why the lobing structure around plate edge-on illumination angles is missing in the corresponding global RCS plot (Figure 27.4). The two-point interference between the leading and trailing edges is destroyed.

Clearly, the proposed string antenna support system is a significantly less interfering alternative to the traditional mechanical towers. The inverted Stewart platform design also gives equal mechanical stability with six degrees of freedom for antenna orientation. Perhaps the most significant disadvantage of the proposed methodology is limited battery life. Our current prototype system requires a recharge every 1.5 h.

27.4   Conclusions

A novel low clutter antenna support method for bistatic RCS measurements is proposed with significant benefits over traditional mechanical means. We present the results of a small experiment designed to compare the influence of antenna support clutter on a bistatic RCS measurement. While not an exhaustive quantitative examination, the qualitative results clearly demonstrate the superiority of the proposed method.

References

Buterbaugh, A., B. M. Kent, C. Mentzer, M. Scott, and W. Forster. 2007. Demonstration of an inverted Stewart platform target-suspension system using lightweight, high-tensile strings, *IEEE Antennas and Propagation Magazine*, 49(5), 185–190.

Buterbaugh, A. and C. Mentzer. 2002. Critical technologies for performing RCS target measurements using a string support system, *Proceedings of the Antenna Measurement Techniques Association*, Cleveland, OH, pp. 23–28, November, 2002.

Currie, N. C. 1989. *Radar Reflectivity Measurement: Techniques and Applications*, 1st ed. Norwood, MA: Artech House.

Griffiths, H. D. 2003. From a different perspective: Principles, practice and potential of bistatic radar, *Proceedings of the International Radar Conference*, Adelaide, Australia, pp. 1–7.

IEEE. 2008a. *Standard Radar Definitions*, IEEE Std. 686-2008, May 21.

IEEE. 2008b. *Recommended Practice for Radar Cross-Section Test Procedures*, IEEE Std. 1502–2007, September 7.

Kent, B. M. 2001. Comparative measurements of precision radar cross section (RCS) calibration targets, *IEEE International Symposium of the Antennas and Propagation Society*, Boston, MA, Vol. 4, pp. 412–415, July 8–13, 2001.

Willis, N. J. 1995. *Bistatic Radar*, 2nd ed. Silver Spring, MD: Technology Service Corporation.

Appendix A: Measurement references and conversion factors

Natural constants

Speed of light	$2.997{,}925 \times 10^{10}$ cm/s
	983.6×10^{6} ft/s
	186,284 miles/s
Speed of sound	340.3 m/s
	1,116 ft/s
Gravity (acceleration)	9.80665 m/s²
	32.174 ft/s²
	386.089 in./s²

Numbers and prefixes

Prefix	Symbol	Number
yotta (10^{24})	Y	1 000 000 000 000 000 000 000 000
zetta (10^{21})	Z	1 000 000 000 000 000 000 000
exa (10^{18})	E	1 000 000 000 000 000 000
peta (10^{15})	P	1 000 000 000 000 000
tera (10^{12})	T	1 000 000 000 000
giga (10^{9})	G	1 000 000 000
mega (10^{6})	M	1 000 000
kilo (10^{3})	k	1 000
hecto (10^{2})	h	1 00
deca (10^{1})	D	1 0
deci (10^{-1})	d	0.1
centi (10^{-2})	c	0.01
milli (10^{-3})	m	0.001
micro (10^{-6})	μ	0.000 001
nano (10^{-9})	n	0.000 000 001
pico (10^{-12})	p	0.000 000 000 001
femto (10^{-15})	f	0.000 000 000 000 001
atto (10^{-18})	a	0.000 000 000 000 000 001
zepto (10^{-21})	z	0.000 000 000 000 000 000 001
yocto (10^{-24})	y	0.000 000 000 000 000 000 000 001

Common abbreviations

Abbreviation	Description	Abbreviation	Description
Å	Angstrom unit of length	liq	Liquid
Abs	Absolute	L	Liter and lambert(s)
Amb	Ambient	LP-gas	Liquefied petroleum gas
Amp	Ampere	log	Logarithm (common)
app mol wt	Apparent molecular weight	ln	Logarithm (natural)
Atm	Atmospheric	m, M	Meter
at wt	Atomic weight	mA	Milliampere
bé	Degrees Baumé	MAC	Maximum allowable concentration
bp	Boiling point	max	Maximum
bbl	Barrel	mp	Melting point
Btu	British thermal unit	μ	Micron
Btuh	Btu per hour	mks system	Meter–kilogram–second system
c	Cycles per second (Hz)	mph	Mile per hour
cal	Calorie	mg	Milligram
cfh	Cubic feet per hour	mi	Milliliter
cfm	Cubic feet per minute	mm	Millimeter
cfs	Cubic feet per second	mm (Hg)	mm of mercury
cg	Centigram	mμ	Millimicron
cm	Centimeter	mppcf	Million particles per cubic feet
cgs system	Centimeter–gram–second system	mr	Milliren
conc	Concentrated, concentration	mR	1/1000 Roentgen
cc, cm³	Cubic centimeter	min	Minute or minimum
cu ft, ft³	Cubic foot	mol wt, MW	Molecular weight
cu in.	Cubic inch	N	Newton
°or deg	Degree	OD	Outside diameter
C	Degree Centigrade, degree Celsius	oz	Ounce
F	Degree Fahrenheit	ppb	Parts per billion
K	Degree Kelvin	pphm	Parts per hundred million
R	Degree Reaumur, degree Rankine	ppm	Parts per million
dB	Decibel	lb	Pound
ET	Effective temperature	psf	Pounds per square foot
ft	Foot	psi	Pounds per square inch
ft-c	Foot-candle	psia	Pounds per square inch absolute
ft lb	Foot pound	psig	Pounds per square inch gage
fpm	Feet per minute	Rem	Roentgen equivalent man (manual)
fps	Feet per second	rpm	Revolution per minute
fps system	Foot–pound–second system	s	Second
fp	Freezing point	sp gr	Specific gravity

(Continued)

Common abbreviations (*Continued*)

Abbreviation	Description	Abbreviation	Description
gal	Gallon	sp ht	Specific heat
gr	Grain	sp wt	Specific weight
gm	Gram	sq	Square
gpm	Gallons per minute	scf	Standard cubic foot
Hz	Hertz (cycles per second)	STP	Standard temperature and pressure
hp	Horsepower	temp	Temperature
h	Hour	TLV	Threshold limit value
ID	Inside diameter	v	Volt
in.	Inch	W	Watt
kcal	Kilocalorie	wt	Weight
kg	Kilogram	x	Times
km	Kilometer	Z	Zulu time (coordinated universal time [UTC])

Greek alphabets and symbols

Capital	Lower case	Greek name	Pronunciation	English
A	α	Alpha	al-fah	a
B	β	Beta	bay-tah	b
Γ	γ	Gamma	gam-ah	g
Δ	δ	Delta	del-tah	d
E	ε	Epsilon	ep-si-lon	e
Z	ζ	Zeta	zat-tah	z
H	η	Eta	ay-tah	h
Θ	θ	Theta	thay-tah	th
I	ι	Iota	eye-oh-tah	i
K	κ	Kappa	cap-ah	k
Λ	λ	Lambda	lamb-da	l
M	μ	Mu	mew	m
N	ν	Nu	new	n
Ξ	ξ	Xi	sah-eye	x
O	o	Omicron	oh-mi-cron	o
Π	π	Pi	pie	p
P	ρ	Rho	roe	r
Σ	σ	Sigma	sig-mah	s
T	τ	Tau	tah-hoe	t
Υ	υ	Upsilon	oop-si-lon	u
Φ	φ	Phi	fah-eye	ph
X	χ	Chi	kigh	ch
Ψ	ψ	Psi	sigh	ps
Ω	ω	Omega	oh-mega	o

Roman numerals

1	I	14	XIV	27	XXVII	150	CL
2	II	15	XV	28	XXVIII	200	CC
3	III	16	XVI	29	XXIX	300	CCC
4	IV	17	XVII	30	XXX	400	CD
5	V	18	XVIII	31	XXXI	500	D
6	VI	19	XIX	40	XL	600	DC
7	VII	20	XX	50	L	700	DCC
8	VIII	21	XXI	60	LX	800	DCCC
9	IX	22	XXII	70	LXX	900	CM
10	X	23	XXIII	80	LXXX	1000	M
11	XI	24	XXIV	90	XC	1600	MDC
12	XII	25	XXV	100	C	1700	MDCC
13	XIII	26	XXVI	101	CI	1900	MCM

Note: C (*centum*, Latin for 100); Derivatives: "centurion"; "century"; "cent."

Signs and symbols

+	Plus, addition, positive
−	Minus, subtraction, negative
±	Plus or minus, positive or negative
∓	Minus or plus, negative or positive
÷, /, —	Division
×, ·, ( )( ),	Multiplication
( )[ ]	Collection
=	Is equal to
≠	Is not equal to
≡	Is identical to
≅	Equals approximately, congruent
>	Greater than
≯	Not greater than
≧	Greater than or equal to
<	Less than
≮	Not less than
≦	Less than or equal to
::	Proportional to
:	Ratio
~	Similar to
∝	Varies as, proportional to
→	Approaches
∞	Infinity
∴	Therefore
√	Square root
$\sqrt[n]{}$	*n*th Root

(*Continued*)

Signs and symbols (*Continued*)

a^n	nth Power of a
log, $\log_{10}$	Common logarithm
$1n$, $\log_e$	Natural logarithm
e or $\in$	Base of natural logs, 2.718
Π	Pi, 3.1416
$\angle$	Angle
$\perp$	Perpendicular to
$\parallel$	Parallel to
N	Any number
$\lvert n \rvert$	Absolute value of n
$\bar{n}$	Average value of n
a^{-n}	Reciprocal of nth power of a, of a, or $\left\{\dfrac{1}{a^n}\right\}$
n^0	n degrees (angle)
n'	n minutes, n feet
n''	n seconds, n inches
$f(x)$	Function of x
Δx	Increment of x
dx	Differential of x
$\Sigma\Omega$	Summation of
sin	Sine
cos	Cosine
tan	Tangent
■	End of proof

Common derived SI units

Quantity	Unit	Symbol	Formula
Frequency (of a periodic phenomenon)	Hertz	Hz	$1/s$
Force	Newton	N	$kg/m/s^2$
Pressure, stress	Pascal	Pa	N/m^2
Energy, work, and quantity of heat	Joule	J	N/m
Power, radiant flux	Watt	W	J/s
Electric charge, quantity of electricity	Coulomb	C	A/s
Electric potential, potential difference, and electromotive force	Volt	V	W/A
Electric capacitance	Farad	F	C/V
Electric resistance	Ohm	Ω	V/A
Conductance	Siemens	S	A/V
Magnetic flux	Weber	Wb	V/s
Magnetic flux density	Tesla	T	Wb/m^2

(*Continued*)

Common derived SI units (*Continued*)

Quantity	Unit	Symbol	Formula
Inductance	Henry	H	Wb/A
Celsius temperature	Degree Celsius	°C	K
Fahrenheit temperature	Degree Fahrenheit	°F	F
Luminous flux	Lumen	lm	cd/sr
Illuminance	Lux	lx	lm/m^2
Activity (of a radionuclide)	Becquerel	Bq	1/s
Absorbed dose	Gray	Gy	J/kg
Dose equivalent	Sievert	Sv	J/kg

Table summary of SI units

Units	Symbol	Relation	Units	Symbol	Relation
Meter	m	Length	Degree Celsius	°C	Temperature
Hectare	ha	Area	Kelvin	K	Thermodynamic temperature
Tonne	t	Mass	Pascal	Pa	Pressure, stress
Kilogram	kg	Mass	Joule	J	Energy, work
Nautical mile	M	Distance (navigation)	Newton	N	Force
Knot	kn	Speed (navigation)	Watt	W	Power, radiant flux
Liter	L	Volume or capacity	Ampere	A	Electric current
Second	s	Time	Volt	V	Electric potential
Hertz	Hz	Frequency	Ohm	Ω	Electric resistance
Candela	cd	Luminous intensity	Coulomb	C	Electric charge

Length equivalents

English system			Metric system	
1 foot (ft)	= 12 inches (in) 1′ = 12″	mm	Millimeter	0.001 m
1 yard (yd)	= 3 feet	cm	Centimeter	0.01 m
1 mile (mi)	= 1760 yards	dm	Decimeter	0.1 m
1 sq. foot	= 144 sq. inches	m	Meter	1 m
1 sq. yard	= 9 sq. feet	dam	Decameter	10 m
1 acre	= 4840 sq. yards = 43560 ft^2	hm	Hectometer	100 m
1 sq. mile	= 640 acres	km	Kilometer	1000 m

Area conversion factors

Multiply	By	To obtain
Acres	43,560	Sq. feet
	4,047	Sq. meters
	4,840	Sq. yards
	0.405	Hectare
Sq. centimeter	0.155	Sq. inches
Sq. feet	144	Sq. inches

(*Continued*)

Area conversion factors (*Continued*)

Multiply	By	To obtain
	0.09290	Sq. meters
	0.1111	Sq. yards
Sq. inches	645.16	Sq. millimeters
Sq. kilometers	0.3861	Sq. miles
Sq. meters	10.764	Sq. feet
	1.196	Sq. yards
Sq. miles	640	Acres
	2.590	Sq. kilometers

Volume conversion factors

Multiply	By	To obtain
Acre-foot	1233.5	Cubic meters
Cubic centimeter	0.06102	Cubic inches
Cubic feet	1728	Cubic inches
	7.480	Gallons (US)
	0.02832	Cubic meters
	0.03704	Cubic yards
Liter	1.057	Liquid quarts
	0.908	Dry quarts
	61.024	Cubic inches
Gallons (US)	231	Cubic inches
	3.7854	Liters
	4	Quarts
	0.833	British gallons
	128	U.S. fluid ounces
Quarts (US)	0.9463	Liters

Energy, heat, and power conversion factors

Multiply	By	To obtain
BTU	1055.9	Joules
	0.2520	kg-calories
Watt-hour	3600	Joules
	3.409	BTU
HP (electric)	746	Watts
BTU/second	1055.9	Watts
Watt-second	1.00	Joules

Mass conversion factors

Multiply	By	To obtain
Carat	0.200	Cubic grams
Grams	0.03527	Ounces
Kilograms	2.2046	Pounds
Ounces	28.350	Grams
Pound	16	Ounces
	453.6	Grams
Stone (UK)	6.35	Kilograms
	14	Pounds
Ton (net)	907.2	Kilograms
	2000	Pounds
	0.893	Gross ton
	0.907	Metric ton
Ton (gross)	2240	Pounds
	1.12	Net tons
	1.016	Metric tons
Tonne (metric)	2204.623	Pounds
	0.984	Gross pounds
	1000	Kilograms

Pressure conversion factors

Multiply	By	To obtain
Atmospheres	1.01325	Bars
	33.90	Feet of water
	29.92	Inches of mercury
	760.0	mm of mercury
Bar	75.01	cm of mercury
	14.50	Pounds/sq. inch
Dyne/sq. cm	0.1	N/sq. meter
Newtons/sq. cm	1.450	Pound/sq. inch
Pounds/sq. inch	0.06805	Atmospheres
	2.036	Inches of mercury
	27.708	Inches of water
	68.948	Millibars
	51.72	mm of mercury

Length conversion factors

Multiply	By	To obtain
Angstrom	10^{-10}	Meters
Feet	0.30480	Meters
	12	Inches
Inches	25.40	Millimeters
	0.02540	Meters
	0.08333	Feet
Kilometers	3280.8	Feet
	0.6214	Miles
	1094	Yards
Meters	39.370	Inches
	3.2808	Feet
	1.094	Yards
Miles	5280	Feet
	1.6093	Kilometers
	0.8694	Nautical miles
Millimeters	0.03937	Inches
Nautical miles	6076	Feet
	1.852	Kilometers
Yards	0.9144	Meters
	3	Feet
	36	Inches

Physical science measurements

$D = \dfrac{m}{V}\left(\dfrac{\text{g}}{\text{cm}^3} = \dfrac{\text{kg}}{\text{m}^3}\right)$

D = density
m = mass
V = volume

$P = \dfrac{W}{t}$

P = power W (watts)
W = work (J)
t = time (s)

$d = v \cdot t$

d = distance (m)
v = velocity (m/s)
t = time (s)

$KE = \dfrac{1}{2} \cdot m \cdot v^2$

KE = kinetic energy
m = mass (kg)
v = velocity (m/s)

$a = \dfrac{vf - vi}{t}$

a = acceleration (m/s^2)
vf = final velocity (m/s)
vi = initial velocity (m/s)
t = time (s)

$F = \dfrac{k \cdot Q_1 \cdot Q_2}{d^2}$

F = electrical force
k = Coulomb's constant
$\left(k = 9 \times 10^9 \dfrac{\text{N} \cdot \text{m}^2}{\text{c}^2}\right)$

Q_1, Q_2 = electrical charges
d = separation distance

$d = vi \cdot t + \dfrac{1}{2} \cdot a \cdot t^2$

d = distance (m)
vi = initial velocity (m/s)
t = time (s)
a = acceleration (m/s^2)

$F = m \cdot a$

F = net force N (newtons)
m = mass (kg)
a = acceleration (m/s^2)

$V = \dfrac{W}{Q}$

V = electrical potential difference V (volts)
W = work done (J)
Q = electric charge moving (C)

(Continued)

Physical science measurements (*Continued*)

$$Fg = \frac{G \cdot m_1 \cdot m_2}{d^2}$$

Fg = force of gravity (N)
G = universal gravitational constant

$$\left(G = 6.67 \times 10^{-11} \frac{N \cdot m^2}{kg^2} \right)$$

m_1, m_2 = masses of the two objects (kg)
d = separation distance (m)

$$I = \frac{Q}{t}$$

I = electric current ampères
Q = electric charge flowing (C)
t = time (s)

$$W = V \cdot I \cdot t$$

W = electrical energy (J)
V = voltage (V)
I = current (A)
t = time (s)

$$p = m \cdot v$$

p = momentum (kg/m/s)
m = mass
v = velocity

$$P = V \cdot I$$

P = power (W)
V = voltage (V)
I = current (A)

$$W = F \cdot d$$

W = work J (joules)
F = force (N)
d = distance (m)

$$H = c \cdot m \cdot \Delta T$$

H = heat energy (J)
m = mass (kg)
ΔT = change in temperature (°C)
c = specific heat J/kg (°C)

Kitchen measurement conversions

A pinch	Less than 1/8 teaspoon (tsp.)
3 teaspoons	1 tablespoon (tbsp.)
1 tablespoon	3 tsp = 1/16 cup = 1/2 oz
2 tablespoons	1/8 cup
4 tablespoons	1/4 cup
16 tablespoons	1 cup
5 tbsp. + 1 tsp.	1/3 cup
4 ounces (oz)	1/2 cup
8 ounces (oz)	1 cup
16 ounces (oz)	1 pound (lb)
1 ounce	2 tbsp. = 6 tsp. = 1/8 cup
1 cup	1/2 pint = 8 oz = 16 tbsp. = 48 tsp.
1 pint	2 cups = 16 oz = 1/2 quart
2 pint	1 quart
4 cups	1 quart
1 quart	2 pints = 4 cups = 1/4 gallon
4 quarts	1 gallon
8 quarts	1 peck (apples, pears, etc.)
1 jigger	1 1/2 fl ounces (fl oz)
1 jigger	3 tablespoons
1 gallon	4 quarts = 8 pints = 16 cups
1 pound	16 oz = 2 cups = 453 g

Appendix B: Measurement equations and formulas

B.1 Quadratic equation

$$ax^2 + bx + c = 0$$

$$x = \frac{-b \pm \sqrt{b^2 - 4ac}}{2a}$$

Overall mean:

$$\bar{x} = \frac{n_1\bar{x}_1 + n_2\bar{x}_2 + n_3\bar{x}_3 + \cdots + n_k\bar{x}_k}{n_1 + n_2 + n_3 + \cdots + n_k} = \frac{\sum n\bar{x}}{\sum n}$$

Permutations:

$$P(n, m) = \frac{n!}{(n - m)!}, \quad (n \geq m)$$

Combinations:

$$C(n, m) = \frac{n!}{m!(n - m)!}, \quad (n \geq m)$$

Failure rate:

$$q = 1 - p = \frac{n - s}{n}$$

Probability:

$$P(X \leq x) = F(x) = \int_{-\infty}^{x} f(x)\, dx$$

Expected value:

$$\mu = \sum (xf(x))$$

Variance:

$$\sigma^2 = \sum (x - \mu)^2 f(x) \quad \text{or} \quad \sigma^2 = \int_{-\infty}^{\infty} (x - \mu)^2 f(x)dx$$

Binomial distribution:

$$f(x) = {}^n c_x p^x (1 - p)^{n-x}$$

Poisson distribution:

$$f(x) = \frac{(np)^x e^{-np}}{x!}$$

Normal distribution:

$$f(x) = \frac{1}{\sigma\sqrt{2\pi}} e^{\frac{-(x-\mu)^2}{2\sigma^2}}$$

Cumulative distribution function:

$$F(x) = P(X \leq x) = \frac{1}{\sigma\sqrt{2\pi}} \int_{-\infty}^{x} e^{\frac{-(x-\mu)^2}{2\sigma^2}} dx$$

Population mean:

$$\mu_{\bar{x}} = \mu$$

Standard error of the mean:

$$\sigma_{\bar{x}} = \frac{\sigma}{\sqrt{n}}$$

Set:

$$A = \{a_1, a_2, a_3, \ldots\ldots\ldots\}$$

Empty set:

$$\Phi = \{\}$$

Subset:

$$A \subseteq B$$

Set union:

$$A \cup b = \{x \mid x \in A \quad \text{or} \quad x \in B \quad \text{or} \quad \text{both}\}$$

Set operations:

$$A \cup B = B \cup A$$

$$A \cap B = B \cap A$$

$$A \cup (B \cup C) = (A \cup B) \cup C$$

$$A \cap (B \cap C) = (A \cap B) \cap C$$

$$A \cup (B \cap C) = (A \cup B) \cap (A \cup C)$$

$$A \cap (B \cap C) = (A \cap B) \cup (A \cap C)$$

$$(A \cup B)' = A' \cap B'$$

$$(A \cap B)' = A' \cup B'$$

$$n(A \cup B) = n(A) + n(B)$$

$$n(A \cup B) = n(A) + n(B) - n(A \cap B)$$

Arithmetic progression:

$$S = a + (a + d) + (a + 2d) + \cdots + [a + (n - 1)d]$$

Geometric progression:

$$S = a + ar + ar^2 + \cdots + ar^{n-1} = a\left(\frac{1 - r^n}{1 - r}\right)$$

Greek alphabet:

Name	Letter	Capital	Pronunciation (as in)	English
Alpha	α	A	al-fah (hat)	a, A
Beta	β	B	bay-tah (ball)	b, B
Gamma	γ	Γ	gam-ah (gift)	g, G
Delta	δ	Δ	del-tah (den)	d, D
Epsilon	ε	E	ep-si-lon (met)	e, E

(Continued)

Name	Letter	Capital	Pronunciation (as in)	English
Zeta	ζ	Z	zay-tah (zoo)	z, Z
Eta	η	H	ay-tay, ay-tah (they)	e, E
Theta	θ	Θ	thay-tah (thing)	Th
Iota	ι	I	eye-o-tah (kit)	i, I
Kappa	κ	K	cap-ah (kitchen)	k, K
Lambda	λ	Λ	lamb-dah (lamb)	l, L
Mu	μ	M	mew (mother)	m, M
Nu	ν	N	new (nice)	n, N
Xi	ξ	Ξ	zzEee, zee-eye (taxi)	x, X
Omicron	o	O	om-ah-cron (pot)	o, O
Pi	π	Π	pie (pie)	p, P
Rho	ρ	P	row (row)	r, R
Sigma	σ, ς	Σ	sig-ma (sigma)	s, S
Tau	τ	T	tawh (tau)	t, T
Upsilon	υ	Y	oop-si-lon (put)	U, U
Phi	φ	Φ	figh, fie, fah-ee (phone)	Ph
Chi	χ	X	kigh (kah-i)	Ch
Psi	ψ	Ψ	sigh (sigh)	Ps
Omega	ω	Ω	o-may-gah (bone)	O

Products and factors:

$$(x + y)^2 = x^2 + 2xy + y^2$$

$$(x - y)^2 = x^2 - 2xy + y^2$$

$$(x + y)^3 = x^3 + 3x^2y + 3xy^2 + y^3$$

$$(x - y)^3 = x^3 - 3x^2y + 3xy^2 - y^3$$

$$(x + y)^4 = x^4 + 4x^3y + 6x^2y^2 + 4xy^3 + y^4$$

$$(x - y)^4 = x^4 - 4x^3y + 6x^2y^2 - 4xy^3 + y^4$$

$$(x + y)^5 = x^5 + 5x^4y + 10x^3y^2 + 10x^2y^3 + 5xy^4 + y^5$$

$$(x - y)^5 = x^5 - 5x^4y + 10x^3y^2 - 10x^2y^3 + 5xy^4 - y^5$$

$$(x + y)^6 = x^6 + 6x^5y + 15x^4y^2 + 20x^3y^3 + 15x^2y^4 + 6xy^5 + y^6$$

$$(x - y)^6 = x^6 - 6x^5y + 15x^4y^2 - 20x^3y^3 + 15x^2y^4 - 6xy^5 + y^6$$

$$x^2 - y^2 = (x - y)(x + y)$$

$$x^3 - y^3 = (x - y)(x^2 + xy + y^2)$$

$$x^3 + y^3 = (x + y)(x^2 - xy + y^2)$$

$$x^4 - y^4 = (x - y)(x + y)(x^2 + y^2)$$

$$x^5 - y^5 = (x - y)(x^4 + x^3y + x^2y^2 + xy^3 + y^4)$$

$$x^5 + y^5 = (x + y)(x^4 - x^3y + x^2y^2 - xy^3 + y^4)$$

$$x^6 - y^6 = (x - y)(x + y)(x^2 + xy + y^2)(x^2 - xy + y^2)$$

$$x^4 + x^2y^2 + y^4 = (x^2 + xy + y^2)(x^2 - xy + y^2)$$

$$x^4 + 4y^4 = (x^2 + 2xy + 2y^2)(x^2 - 2xy + 2y^2)$$

Powers and roots:

$$a^x \cdot a^y = a^{(x+y)}$$

$$a^0 = 1[\text{if } a \neq 0]$$

$$(ab)^x = a^x b^x$$

$$\frac{a^x}{a^y} = a^{(x-y)}$$

$$a^{-x} = \frac{1}{a^x}$$

$$\left(\frac{a}{b}\right)^x = \frac{a^x}{b^x}$$

$$(a^x)^y = a^{xy}$$

$$a^{\frac{1}{x}} = \sqrt[x]{a}$$

$$\sqrt[x]{ab} = \sqrt[x]{a}\sqrt[x]{b}$$

$$\sqrt[x]{\sqrt[y]{a}} = \sqrt[xy]{a}$$

$$a^{\frac{x}{y}} = \sqrt[y]{a^x}$$

$$\sqrt[x]{\frac{a}{b}} = \frac{\sqrt[x]{a}}{\sqrt[x]{b}}$$

Proportion:
  If $a/b = c/d$, then

$$ad = bc$$

$$\frac{a+b}{b} = \frac{c+d}{d}$$

$$\frac{a-b}{b} = \frac{c-d}{d}$$

$$\frac{a-b}{a+b} = \frac{c-d}{c+d}$$

Sum of arithmetic progression to n terms:

$$a + (a+d) + (a+2d) + \cdots + (a + (n-1)d) = na + \frac{1}{2}n(n-1)d = \frac{n}{2}(a+l)$$

last term in series $= l = a + (n-1)d$

Sum of geometric progression to n terms:

$$s_n = a + ar + ar^2 + \cdots + ar^{n-1} = \frac{a(1-r^n)}{1-r}$$

$$\lim_{n \to \infty} s_n = a!(1-r) \quad (-1 < r < 1)$$

Arithmetic mean:

$$A = \frac{a_1 + a_2 + \cdots + a_n}{n}$$

Geometric mean:

$$G = (a_1 a_2 \ldots a_n)^{1/n}, \quad (a_k > 0, \ k = 1, 2, \ldots, n)$$

Harmonic mean:

$$\frac{1}{H} = \frac{1}{n}\left(\frac{1}{a_1} + \frac{1}{a_2} + \cdots + \frac{1}{a_n}\right), \quad (a_k > 0, \ k = 1, 2, \ldots, n)$$

Circle:

$$x^2 + y^2 = r^2$$

Cassinian curves:

$$x^2 + y^2 = 2ax$$

$$r = 2a\cos\theta$$

$$x^2 + y^2 = ax + by$$

$$r = a\cos\theta + b\sin\theta$$

Cotangent curve:

$$y = \cot x$$

Cubical parabola:

$$y = ax^3, \quad a > 0$$

$$r^2 = \frac{1}{a}\sec^2\theta\tan\theta, \quad a > 0$$

Cosecant curve:

$$y = \csc x$$

Cosine curve:

$$y = \cos x$$

Ellipse:

$$x^2/a^2 + y^2/b^2 = 1$$

$$\begin{cases} x = a\cos\phi \\ y = b\sin\phi \end{cases}$$

Parabola:

$$y = x^2$$

Cubical parabola:

$$y = x^3$$

Tangent curve:

$$y = \tan x$$

Ellipsoid:

$$\frac{x^2}{a^2} + \frac{y^2}{b^2} + \frac{z^2}{c^2} = 1$$

Elliptic cone:

$$\frac{x^2}{a^2} + \frac{y^2}{b^2} - \frac{z^2}{c^2} = 0$$

Elliptic cylinder:

$$\frac{x^2}{a^2} + \frac{y^2}{b^2} = 1$$

Hyperboloid of one sheet:

$$\frac{x^2}{a^2} + \frac{y^2}{b^2} - \frac{z^2}{c^2} = 1$$

Elliptic paraboloid:

$$\frac{x^2}{a^2} + \frac{y^2}{b^2} = cz$$

Hyperboloid of two sheets:

$$\frac{z^2}{c^2} - \frac{x^2}{a^2} - \frac{y^2}{b^2} = 1$$

Hyperbolic paraboloid:

$$\frac{x^2}{a^2} - \frac{y^2}{b^2} = cz$$

Sphere:

$$x^2 + y^2 + z^2 = a^2$$

Surface area of cylinder:

$$2\pi rh + 2\pi r^2$$

Volume of cylinder:

$$\pi r^2 h$$

Surface area of a cone:

$$\pi r^2 + \pi r s$$

Volume of a cone:

$$\frac{\pi r^2 h}{3}$$

Volume of a pyramid:

$$\frac{Bh}{3}, \quad (B = \text{area of base})$$

Distance d between two points:

$$P_1(x_1, y_1, z_1) \quad \text{and} \quad P_2(x_2, y_2, z_2)$$

$$d = \sqrt{(x_2 - x_1)^2 + (y_2 - y_1)^2 + (z_2 - z_1)^2}$$

Angle ϕ between two lines with direction cosines l_1, m_1, n_1 and l_2, m_2, n_2

$$\cos \phi = l_1 l_2 + m_1 m_2 + n_1 n_2$$

Plane:

$$Ax + By + Cz + D = 0, \quad A, B, C, D \text{ are constants}$$

Equation of plane passing through points:

$$(x_1, y_1, z_1), \quad (x_2, y_2, z_2), \quad (x_3, y_3, z_3)$$

$$\begin{vmatrix} x - x_1 & y - y_1 & z - z_1 \\ x_2 - x_1 & y_2 - y_1 & z_2 - z_1 \\ x_3 - x_1 & y_3 - y_1 & z_3 - z_1 \end{vmatrix} = 0$$

or

$$\begin{vmatrix} y_2 - y_1 & z_2 - z_1 \\ y_3 - y_1 & z_3 - z_1 \end{vmatrix}(x - x_1) + \begin{vmatrix} z_2 - z_1 & x_2 - x_1 \\ z_3 - z_1 & x_3 - x_1 \end{vmatrix}(y - y_1) + \begin{vmatrix} x_2 - x_1 & y_2 - y_1 \\ x_3 - x_1 & y_3 - y_1 \end{vmatrix}(z - z_1) = 0$$

Plane equation in intercept form:

$x/a + y/b + z/c = 1,$ a,b,c are the intercepts on the x,y,z axes respectively.

Equations of line through (x_0,y_0,z_0) and perpendicular to plane:

$$Ax + By + Cz + D = 0$$

$$\frac{x - x_0}{A} = \frac{y - y_0}{B} = \frac{z - z_0}{C}$$

or $x = x_0 + At,$ $y = y_0 + Bt,$ $z = z_0 + Ct$

Distance from point (x,y,z) to plane:

$$Ax + By + D = 0$$

$Ax_0 + By_0 + Cz_0 + D/ \pm\sqrt{A^2 + B^2 + C^2}$, where the sign is chosen so that the distance is nonnegative.

Cylindrical coordinates (r, θ, z):

$$\begin{cases} x = r\cos\theta \\ y = r\sin\theta \\ z = z \end{cases} \text{ or } \begin{cases} r = \sqrt{x^2 + y^2} \\ \theta = \tan^{-1}(y/x) \\ z = z \end{cases}$$

Spherical coordinates (r, θ, ϕ):

$$\begin{cases} x = r\cos\theta\cos\phi \\ y = r\sin\theta\sin\phi \\ z = r\cos\theta \end{cases} \text{ or } \begin{cases} r = \sqrt{x^2 + y^2 + z^2} \\ \phi = \tan^{-1}(y/x) \\ \theta = \cos^{-1}\left(z/\sqrt{x^2 + y^2 + z^2}\right) \end{cases}$$

Equation of sphere in rectangular coordinates:

$(x - x_0)^2 + (y - y_0)^2 + (z - z_0)^2 = R^2$, center $= (x_0, y_0, z_0)$ and radius $= R$

Equation of sphere in cylindrical coordinates:

$r^2 - 2r_0 r \cos(\theta - \theta_0) + r_0^2 + (z - z_0)^2 = R^2$, center $= (r_0, \theta_0, z_0)$ and radius $= R$

If the center is at the origin, the equation is $r^2 + z^2 = R^2$

Equation of sphere in spherical coordinates:

$r^2 + r_0^2 - 2r_0 r \sin\theta \sin\theta_0 \cos(\phi - \phi_0) = R^2$, center $= (r_0, \theta_0, \phi_0)$ and radius $= R$

If the center is at the origin, the equation is $r = R$

Logarithmic identities:

$$\ln(z_1 z_2) = \ln z_1 + \ln z_2$$

$$\ln \frac{z_1}{z_2} = \ln z_1 - \ln z_2$$

$$\ln z^n = n \ln z \quad (n \text{ integer})$$

Special logarithmic values:

$$\ln 1 = 0$$

$$\ln 0 = -\infty$$

$$\ln(-1) = \pi i$$

$$\ln(\pm i) = \pm \frac{1}{2} \pi i$$

$$\ln e = 1, e \text{ is the real number such that } \int_1^e dt/t = 1$$

$$e = \lim_{n \to \infty} \left(1 + \frac{1}{n}\right)^n = 2.7182818284\ldots$$

Logarithms to general base:

$$\log_a z = \ln z / \ln a$$

$$\log_a z = \frac{\log_b z}{\log_b a}$$

$$\log_a b = \frac{1}{\log_b a}$$

$$\log_e z = \ln z$$

$$\log_{10} z = \ln z / \ln 10 = \log_{10} e \ln z = (.4342944819\ldots)\ln z$$

$$\ln z = \ln 10 \log_{10} z = (2.3025850929\ldots)\log_{10} z$$

Series expansions:

$$\ln(1 + z) = z - \frac{1}{2}z^2 + \frac{1}{3}z^3 - \cdots \quad (|z| \leq 1 \quad \text{and} \quad z \neq -1)$$

$$\ln z = \left(\frac{z-1}{z}\right) + \frac{1}{2}\left(\frac{z-1}{z}\right)^2 + \frac{1}{3}\left(\frac{z-1}{z}\right)^3 + \cdots \quad \left(\Re z \geq \frac{1}{2}\right)$$

$$\ln z = (z-1) - \frac{1}{2}(z-1)^2 + \frac{1}{3}(z-1)^3 - \cdots \quad (|z-1| \leq 1, \ z \neq 0)$$

$$\ln z = 2\left[\left(\frac{z-1}{z+1}\right) + \frac{1}{3}\left(\frac{z-1}{z+1}\right)^3 + \frac{1}{5}\left(\frac{z-1}{z+1}\right)^5 + \cdots\right] \quad (\Re z \geq 0, \ z \neq 0)$$

$$\ln\left(\frac{z+1}{z-1}\right) = 2\left(\frac{1}{z} + \frac{1}{3z^3} + \frac{1}{5z^5} + \cdots\right) \quad (|z| \geq 1, \ z \neq \pm 1)$$

$$\ln(z + a) = \ln a + 2\left[\left(\frac{z}{2a+z}\right) + \frac{1}{3}\left(\frac{z}{2a+z}\right)^3 + \frac{1}{5}\left(\frac{z}{2a+z}\right)^5 + \cdots\right] \quad (a > 0, \ \Re z \geq -a \neq z)$$

Limiting values:

$$\lim_{x \to \infty} x^{-\alpha} \ln x = 0 \quad (\alpha \text{ constant}, \Re\alpha > 0)$$

$$\lim_{x \to 0} x^{\alpha} \ln x = 0 \quad (\alpha \text{ constant}, \Re\alpha > 0)$$

$$\lim_{m \to \infty}\left(\sum_{k=1}^{m} \frac{1}{k} - \ln m\right) = \gamma(\text{Euler's constant}) = .57721\,56649\ldots$$

Inequalities:

$$\frac{x}{1+x} < \ln(1+x) < x \quad (x > -1, \quad x \neq 0)$$

$$x < -\ln(1-x) < \frac{x}{1+x} \quad (x < 1, \quad x \neq 0)$$

$$|\ln(1-x)| < \frac{3x}{2} \quad (0 < x \leq .5828)$$

$$\ln x \leq x - 1 \quad (x > 0)$$

$$\ln x \le n(x^{1/n} - 1) \text{ for any positive } n \quad (x > 0)$$

$$|\ln(1 - z)| \le -\ln(1 - |z|) \quad (|z| < 1)$$

Continued fractions:

$$\ln(1 + z) = \frac{z}{1} + \frac{z}{2} + \frac{z}{3} + \frac{4z}{4} + \frac{4z}{5} + \frac{9z}{6} \cdots \quad (z \text{ in the plane cut from } -1 \text{ to } -\infty)$$

$$\ln\left(\frac{1 + z}{1 - z}\right) = \frac{2z}{1} - \frac{z^2}{3} - \frac{4z^2}{5} - \frac{9z^2}{7} - \cdots$$

Slopes:

Equation of a straight line: $y - y_1 = m(x - x_1)$

$$m = \text{slope} = \frac{\text{rise}}{\text{run}} = \frac{\Delta y}{\Delta x} = \frac{y_2 - y_1}{x_2 - x_1}$$

or

$$y = mx + b, \quad m = \text{slope}, \quad b = \text{y-intercept}$$

Trigonometric identities:

$$\tan 0 = \frac{\sin \theta}{\cos \theta}$$

$$\sin^2 \theta + \cos^2 \theta = 1$$

$$1 + \tan^2 \theta = \sec^2 \theta$$

$$1 + \cot^2 \theta = \csc^2 \theta$$

$$\cos^2 \theta - \sin^2 \theta = \cos 2\theta$$

$$\sin 45° = \frac{1}{\sqrt{2}}$$

$$\cos 45° = \frac{1}{\sqrt{2}}$$

$$\tan 45° = 1$$

$$\sin(A + B) = \sin A \cos B + \cos A \sin B$$

$$\sin (A - B) = \sin A \cos B - \cos A \sin B$$

$$\cos (A + B) = \cos A \cos B - \sin A \sin B$$

$$\cos(A - B) = \cos A \cos B + \sin A \sin B$$

$$\tan(A + B) = \frac{\tan A + \tan B}{1 - \tan A \tan B}$$

$$\tan(A - B) = \frac{\tan A - \tan B}{1 + \tan A \tan B}$$

$$\sin \theta = \frac{y}{r} (\text{opposite/hypotenuse}) = 1/\csc \theta$$

$$\cos \theta = \frac{x}{r} (\text{adjacent/hypotenuse}) = 1/\sec \theta$$

$$\tan \theta = \frac{y}{x} (\text{opposite/adjacent}) = 1/\cot \theta$$

$$\sin 30° = \frac{1}{2} \quad \sin 60° = \frac{\sqrt{3}}{2}$$

$$\cos 30° = \frac{\sqrt{3}}{2} \quad \cos 60° = \frac{1}{2}$$

$$\tan 30° = \frac{1}{\sqrt{3}} \quad \tan 60° = \sqrt{3}$$

Sine law:

$$\frac{a}{\sin A} = \frac{b}{\sin B} = \frac{c}{\sin C}$$

Cosine law:

$$a^2 = b^2 + c^2 - 2bc \cos A$$

$$b^2 = a^2 + c^2 - 2ac \cos B$$

$$c^2 = a^2 + b^2 - 2ab \cos C$$

$$\theta = 1 \text{ radian}$$

$$2\pi \text{ radians} = 360°$$

Algebraic expansions:

$$a(b + c) = ab + ac$$

$$(a + b)^2 = a^2 + 2ab + b^2$$

$$(a - b)^2 = a^2 - 2ab + b^2$$

$$(a + b)(c + d) = ac + ad + bc + bd$$

$$(a + b)^3 = a^3 + 3a^2b + 3ab^2 + b^3$$

$$(a - b)^3 = a^3 - 3a^2b + 3ab^2 - b^3$$

Algebraic factoring:

$$a^2 - b^2 = (a + b)(a - b)$$

$$a^2 + 2ab + b^2 = (a + b)^2$$

$$a^3 + b^3 = (a + b)(a^2 - ab + b^2)$$

$$a^3b - ab = ab(a + 1)(a - 1)$$

$$a^2 - 2ab + b^2 = (a - b)^2$$

$$a^3 - b^3 = (a - b)(a^2 + ab + b^2)$$

Mechanical advantage:
  Ratio of force of resistance to force of effort:

$$MA = \frac{F_R}{F_E}$$

F_R = *force of resistance*
F_E = *force of effort*
Mechanical advantage formula for levers:

$$F_R \cdot L_R = F_E \cdot L_E$$

Mechanical advantage formula for axles:

$$MA_{\text{wheel and axle}} = \frac{r_E}{r_R}$$

$$F_R \cdot r_{R^.} = F_E \cdot r_E$$

r_R = radius of resistance wheel
r_E = radius of effort wheel (m)

Mechanical advantage formula for pulleys:

$$MA_{\text{pulley}} = \frac{F_R}{F_E} = \frac{nT}{T} = n$$

T = tension in each supporting strand
n = number of strands holding the resistance

Mechanical advantage formula for inclined plane:

$$MA_{\text{inclined plane}} = \frac{F_R}{F_E} = \frac{l}{h}$$

l = length of plane
h = height of plane (m)

Mechanical advantage formula for wedges:

$$MA = \frac{s}{T}$$

s = length of either slope
T = thickness of the longer end

Mechanical advantage formula for screws:

$$MA_{\text{screw}} = \frac{F_R}{F_E} = \frac{U_E}{h}$$

$$F_R \cdot h = F_E \cdot U_E$$

h = pitch of screw
U_E = circumference of the handle of the screw
Distance (S), speed (V), and time (t) formulas

$$S = Vt$$

Acceleration formulas:
  Distance: $s = v_0 t + at^2/2$
  Speed: $v = v_0 + at$
  s = distance, v = speed, t = time, v_0 = initial speed, a = acceleration

Rotational motion

Distance: $s = I\varphi$
Velocity: $v = I\omega$
Tangential acceleration: $a_t = r \cdot \alpha$
Centripetal acceleration: $a_n = \omega^2 r = v^2/r$
$\hat{\varphi}$ = angle determined by s and r (rad)
ω = angular velocity
α = angular acceleration
a_t = tangential acceleration
a_n = centripetal acceleration

Pendulum:

$$T = 2\pi\sqrt{\frac{l}{g}}$$

T = period (s)
l = length of pendulum (m)
$g = 9.81$ (m/s^2) or (32.2 ft/s^2)

Vertical project:

a. Initial speed: $v_0 > 0\,(upward); v_0 < 0\,(downward)$
b. Distance: $h = v_0 t - gt^2/2 = (v_0 + v)t/2; h_{max} = v_0^2/2g$
c. Time: $t = (v_0 - v)/g = 2h/(v_0 + v); t_{hmax} = v_0/g$

v = velocity; h = distance; and g = acceleration due to gravity.

Angled projections:

$$\text{Upward}\,(\alpha > 0); \quad \text{downward}\,(\alpha < 0)$$

a. Distance: $s = v_0 \cdot t\cos\alpha$

b. Altitude: $h = v_0 t\sin\alpha - \dfrac{g \cdot t^2}{2} = s\tan\alpha - \dfrac{g \cdot s^2}{2v_0^2\cos\alpha}$

c. $h_{max} = (v_0^2\sin^2\alpha)/2g$

d. Velocity: $v = \sqrt{v_0^2 - 2gh} = \sqrt{v_0^2 + g^2 t^2 - 2gv_0 t\sin\alpha}$

e. Time: $t_{hmax} = (v_0\sin\alpha)/g; \quad t_{s1} = (2v_0\sin\alpha)/g$

Horizontal projection: $(\alpha = 0)$:

a. Distance: $s = v_0 t = v_0\sqrt{2h/g}$
b. Altitude: $h = -gt^2/2$
c. Trajectory velocity: $v = \sqrt{v_0^2 + g^2 t^2}$

v_0 = initial velocity
v = trajectory velocity
s = distance
h = height

Sliding motion on inclined plane:

1. If excluding friction $\mu = 0$, then
   a.   Velocity: $v = at = 2s/t = \sqrt{2as}$
   b.   Distance: $s = at^2/2 = vt/2 = v^2/2a$
   c.   Acceleration: $a = g\sin\alpha$
2. If including friction ($\mu > 0$), then
   a.   Velocity: $v = at = 2s/t = \sqrt{2as}$
   b.   Distance: $s = at^2/2 = vt/2 = v^2/2a$
   c.   Accelerations: $s = at^2/2 = vt/2 = v^2/2a$

   μ = coefficient of sliding friction
   g = acceleration due to gravity
   $g = 9.81$ (m/s^2)
   v_0 = initial velocity (m/s)
   v = trajectory velocity (m/s)
   s = distance (m)
   a = acceleration (m/s^2)
   α = inclined angle

Rolling motion on inclined plane:

1. If excluding friction ($\mu = 0$), then
   a.   Velocity:

$$v = at = \frac{2s}{t} = \sqrt{2as}$$

   b.   Acceleration:

$$a = \frac{gr^2}{I^2 + k^2}\sin\alpha$$

   c.   Distance:

$$s = \frac{at^2}{2} = \frac{vt}{2} = \frac{v^2}{2a}$$

   d.   Tilting angle:

$$\tan\alpha = \mu_0 \frac{r^2 + k^2}{k^2}$$

2. If including friction ($\mu > 0$), then
   a. Distance:

$$s = \frac{at^2}{2} = \frac{vt}{2} = \frac{v^2}{2a}$$

   b. Velocity:

$$v = at = \frac{2s}{t} = \sqrt{2as}$$

   c. Accelerations:

$$a = gr^2 \frac{\sin\alpha - (f/r)\cos\alpha}{I^2 + k^2}$$

   d. Tilting angle:

$$\tan\alpha_{min} = \frac{f}{r}; \quad \tan\alpha_{max} = \mu_0 \frac{r^2 + k^2 - fr}{k^2}$$

The value of k can be the calculated by the formulas below:

Ball	Solid cylinder	Pipe with low wall thickness
$k^2 = 2r^2/5$	$k^2 = r^2/2$	$k^2 = r_i^2 + r_0^2/2 \approx r^2$

s = distance (m)
v = velocity (m/s)
a = acceleration (m/s²)
α = tilting angle (°)
f = lever arm of rolling resistance (m)
k = radius of gyration (m)
μ_0 = coefficient of static friction
g = acceleration due to gravity (m/s²)

Dynamics

Newton's first law of motion:
Newton's first law (law of inertia): An object that is in motion continues in motion with the same velocity at constant speed and in a straight line, and an object at rest continues at rest unless an unbalanced (outside) force acts upon it.

Newton's second law:
The second law of motion (law of accelerations): The total force acting on an object equals the mass of the object times its acceleration.

$F = ma$
F = total force

m = mass
a = acceleration

Newton's third law:
The third law of motion (law of action and reaction): For every force applied by object A to object B (action), there is a force exerted by object B on object A (the reaction) which has the same magnitude but is opposite in direction.

$$F_B = -F_A$$

F_B = force of action
F_A = force of reaction

Momentum of force:
The momentum can be defined as mass in motion. Momentum is a vector quantity, for which direction is important

$$p = mv$$

Impulse of force:
The impulse of a force is equal to the change in momentum that the force causes in an object:

$$I = Ft$$

p = momentum
m = mass of object
v = velocity of object
I = impulse of force
F = force
t = time

Conservation of momentum:
In the absence of external forces, the total momentum of the system is constant. If two objects of mass m_1 and mass m_2, having velocity v_1 and v_2, collide and then separate with velocity v_1' and v_2', the equation for the conservation of momentum is:

$$m_1v_1 + m_2v_2 = m_1v_1 + m_2v_2$$

Friction:

$$F_f = \mu F_n$$

F_f = frictional force (N)
F_n = normal force (N)
μ = coefficient of friction (μ = tan α)

Gravity:
Gravity is a force that attracts bodies of matter toward each other. Gravity is the attraction between any two objects that have mass.

$$F = \Gamma \frac{m_A m_B}{r^2}$$

m_A, m_B = mass of objects A and B
F = magnitude of attractive force between objects A and B
r = distance between object A and B
Γ = gravitational constant, $\Gamma = 6.67 \times 10^{-11} \, \text{Nm}^2/\text{kg}^2$

Gravitational force:

$$F_G = g \frac{R_e^2 m}{(R_e + h)^2}$$

On the earth surface, $h = 0$; so

$$F_G = mg$$

F_G = force of gravity
R_e = radius of the Earth ($R_e = 6.37 \times 10^6 \text{m}$)
m = mass
g = acceleration due to gravity
$g = 9.81 \, (\text{m/s}^2)$ or $g = 32.2 \, (\text{ft/s}^2)$

The acceleration of a falling body is independent of the mass of the object. The weight F_w on an object is actually the force of gravity on that object:

$$F_w = mg$$

Centrifugal force:

$$F_c = \frac{mv^2}{r} = m\omega^2 r$$

Centripetal force:

$$F_{cp} = -F_c = \frac{mv^2}{r}$$

F_c = centrifugal force (N)
F_{cp} = centripetal force (N)
m = mass of the body (kg)
v = velocity of the body (m/s)

r = radius of curvature of the path of the body (m)
ω = angular velocity (s^{-1})

Torque:

$$T = F \cdot l$$

T = torque (Nm or 1b ft)
F = applied force (N or 1b)
l = length of torque arm (m or ft)

Work:
  Work is the product of a force in the direction of the motion and the displacement.
  a. Work done by a constant force

$$W = F_s \cdot s = F \cdot s \cdot \cos\alpha$$

W = work (Nm = J)
F_s = component of force along the direction of movement (N)
s = distance the system is displaced (m)

  b. Work done by a variable force
      If the force is not constant along the path of the object, then

$$W = \int_{s_i}^{s_f} F_s(s) \cdot ds = \int_{s_i}^{s_f} F(s)\cos\alpha \cdot ds$$

$F_s(s)$ = component of the force function along the direction of movement
$F(s)$ = function of the magnitude of the force vector along the displacement curve
s_i = initial location of the body
s_f = final location of the body
α = angle between the displacement and the force

Energy:
  Energy is defined as the ability to do work.

$$TME_i + W_{ext} = TME_f$$

TME_i = initial amount of total mechanical energy (J)
W_{ext} = work done by external forces (J)
TME_f = final amount of total mechanical energy (J)

  a. Kinetic energy
      Kinetic energy is the energy due to motion.

$$E_k = \frac{1}{2}mv^2$$

m = mass of moving object (kg)
v = velocity of moving object (m/s)

b. Potential energy
Potential energy is the stored energy of a body and is due to its position.

$$E_{pg} = m \cdot g \cdot h$$

E_{pg} = gravitational potential energy (J)
m = mass of object (kg)
h = height above reference level (m)
g = acceleration due to gravity (m/s²)

Conservation of energy:
In any isolated system, energy can be transformed from one kind to another, but the total amount of energy is constant (conserved):

$$E = E_k + E_p + E_e + \cdots = \text{constant}$$

Conservation of mechanical energy:

$$E_k + E_p = \text{constant}$$

Power:
Power is the rate at which work is done, or the rate at which energy is transformed from one form to another.

$$P = \frac{W}{t}$$

P = power (W)
W = work (J)
t = time (s)

Since the expression for work is $W = F \cdot s$, the expression for power can be rewritten as

$$P = F \cdot v$$

s = displacement (m)
v = speed (m/s)

Project management notations:
ES: Earliest starting time
EC: Earliest completion time
LS: Latest starting time
LC: Latest completion time
t: Activity duration
T_p: Project duration
n: Number of activities in the project network

Earliest start time for activity i:

$$ES(i) = \text{Max}\{EC(j)\}$$
$$j \in P\{i\}$$

$P\{i\}$ = {set of immediate predecessors of activity i}.

Earliest completion time of activity i:

$$EC(i) = ES(i) + t_i$$

Earliest completion time of a project:

$$EC(\text{Project}) = EC(n), n \text{ is the last node.}$$

Latest completion time of a project:

$$LC(\text{Project}) = EC(\text{Project}), \text{ if no external deadline is specified}$$

LC time of a project:
if a desired deadline, T_p, is specified:

$$LC(\text{Project}) = T_p$$

LC time for activity i:

$$LC(i) = \text{Min}\{LS(j)\}$$
$$j \in S\{i\}$$

where $S\{i\}$ = set of immediate successors of activity i.

Latest start time for activity i:

$$LS(i) = LC(i) - t_i$$

Total slack:

$$TS(i) = LC(i) - EC(i)$$

$$TS(i) = LS(i) - ES(i)$$

Free slack:

$$FS(i) = \text{Min}\{ES(j)\} - EC(i)\}$$
$$j \in S(i)$$

Interfering slack:

$$IS(i) = TS(i) - FS(i)$$

Independent float:

$$IF(i) = \text{Max}\{0,(\text{Min ES}_j - \text{Max LC}_k - t_i)\}$$
$$j \in S\{i\} \quad \text{and} \quad k \in P\{i\}$$

ES_j = earliest starting time of a succeeding activity from the set of successors of activity i
LC_k = latest completion time of a preceding activity from the set of predecessors of activity i.

PERT formulas:

$$t_e = \frac{a + 4m + b}{6}$$

$$s^2 = \frac{(b - a)^2}{36}$$

a = optimistic time estimate
b = pessimistic time estimate
m = most likely time estimate
t_e = expected time
s^2 = variance of expected time

CPM path criticality computation:

$$\lambda = \frac{\alpha_2 - \beta}{\alpha_2 - \alpha_1}(100\%)$$

α_1 the minimum total slack in the CPM network
α_2 the maximum total slack in the CPM network
β total slack for the path whose criticality is to be calculated

Task weight:
Work content of a project is expressed in terms of days.

$$\text{Task weight} = \frac{\text{Work-days for activity}}{\text{Total work-days for project}}$$

Expected % completion:

$$\text{Expected \% completion} = \frac{\text{Work-days for activity}}{\text{Work-days for planned}}$$

Expected relative % completion:

$$\text{Expected relative \% completion} = (\text{expected \% completion}) \cdot (\text{task weight})$$

Actual relative % completion:

$$\text{Actual relative \% completion} = (\text{actual \% completion}) \cdot (\text{task weight})$$

Planned project % completion:

$$\text{Planned project \% completion} = \frac{\text{Work-days completed on project}}{\text{Total work-days planned}}$$

Project-tracking index:

$$\text{Project-tracking index} = \frac{\text{Actual relative \% completion}}{\text{Expected relative \% completion}} - 1$$

Appendix C: Slides of statistics for measurement

Statistics, whether in qualitative or quantitative form, is the foundation for accuracy and precision of measurements. This appendix presents a collection of useful statistical definitions, explanations, interpretations, illustrations, examples, formulations, formulas, and equations.
Using the tools and techniques presented in this appendix, readers can

- Identify and quantify the sources of measurement variability.
- Assess the effect of the measurement system variability on process variability.
- Discover opportunities for measurement system and total process variability improvement.

Statistics
- **A discipline of study dealing with the collection, analysis, interpretation, and presentation of data.**

Descriptive Statistics
- **Measure to organize and summarize information**
- **Uses graph, charts and tables**
- **Reduces information to a manageable size and put it into focus**

Inferential Statistics

- **Definitions**
  - Population – complete collection of items under consideration in a statistical study
  - Sample – a group of items selected for analysis

- **Reaching conclusions about a population based upon information in a sample**

Basic Definitions

- **Variable – characteristic of interest**
- **Observation – the value of a variable for one particular element in a population or sample**
- **Data Set – the observations of a variable for the elements of a sample**

Variables

- **Quantitative**
  - Discrete – a variable whose values are countable
  - Continuous – a variable that can assume any numerical value over an interval
- **Qualitative**
  - A variable whose value is nonnumeric

Levels of Measurement

- **Nominal**
  - Mutually exclusive categorical data
- **Ordinal**
  - Mutually exclusive categorical data
  - Data can be arranged in some order
- **Interval**
  - Data that can be arranged in some order
  - The differences in data values are meaningful
  - The value of zero is arbitrarily chosen
- **Ratio**
  - Data that can be ranked
  - All arithmetic operations can be performed
  - There is an absolute zero

Measurement Characteristics

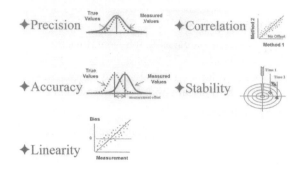

♦Precision

♦Correlation

♦Accuracy

♦Stability

♦Linearity

Frequency Distributions

- **Qualitative Data**
  - Frequency - the number of elements that belong to each category
  - Frequency distribution – lists all categories of data and their frequency
  - Relative frequency (percent) – frequency divided by sum of all elements

Frequency Distributions

- **Qualitative Data**

A	B	B	AB	O
O	O	B	AB	B
B	B	O	A	O
A	O	O	O	AB
AB	A	O	B	A

Category	Tally	Frequency	Relative Frequency	Percent
A	〤	5	0.20	20%
B	〤 //	7	0.28	28%
O	〤 ////	9	0.36	36%
AB	////	4	0.16	16%

Frequency Distributions

- **Qualitative Data Displays**
  - Bar chart – frequencies

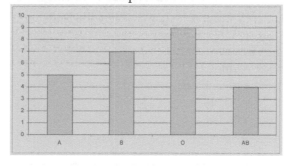

Frequency Distributions

- **Qualitative Data Displays**
  - Pie chart – relative frequencies

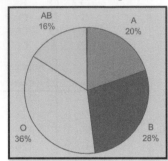

Frequency Distributions

- **Quantitative Data**
  - Similar to qualitative data, but categories are replaced by values or ranges of values, called classes
  - Determining classes
    - Limits
    - Boundaries
    - Widths
    - Marks

Frequency Distributions

• Quantitative Data

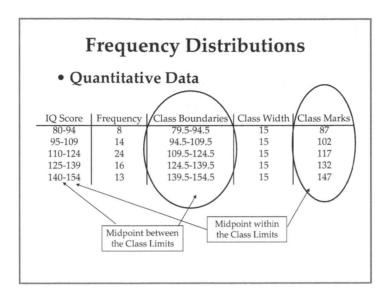

IQ Score	Frequency	Class Boundaries	Class Width	Class Marks
80-94	8	79.5-94.5	15	87
95-109	14	94.5-109.5	15	102
110-124	24	109.5-124.5	15	117
125-139	16	124.5-139.5	15	132
140-154	13	139.5-154.5	15	147

Midpoint between the Class Limits

Midpoint within the Class Limits

Frequency Distributions

• Quantitative Data Displays
• Histograms – frequencies

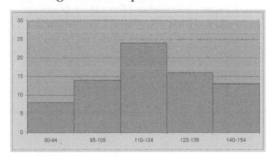

Frequency Distributions

- **Histogram Guidelines**
  - Between 5 and 15 classes
  - Each data value belongs to one and only one class
  - When possible, all classes should be of equal width

Frequency Distributions

- **Quantitative Data**
  - Cumulative Frequencies

IQ Score	Frequency	Cumulative Frequency
80-94	8	8
95-109	14	22
110-124	24	46
125-139	16	62
140-154	13	75

Frequency Distributions

- **Quantitative Data Displays**
  - Ogives – cumulative frequencies

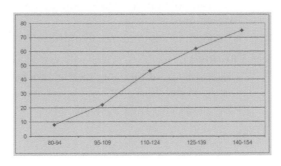

Frequency Distributions

- **Quantitative Data Displays**
  - Stem and Leaf Displays – individual data

```
4  9
5  6
6  1455889
7  12337999
8  2356667788
9  1123
```

Measures

- **Statistic**
  - Measure obtained by using the data values of samples
  - Denoted using the English alphabet
- **Parameter**
  - Measure found by using all the data values in the population
  - Denoted using the Greek alphabet

Measures of Central Tendency

- **Mean – the arithmetic average**

Sample
statistic
$$\overline{x} = \frac{\sum x}{n}$$

Population
parameter
$$\mu = \frac{\sum x}{N}$$

Measures of Central Tendency

- **Mean**
  - The number of calls in one day, collected for 30 randomly selected days

Example:

| 25 | 46 | 34 | 45 | 37 | 36 | 40 | 30 | 29 | 37 | 44 | 56 | 50 | 47 | 23 |
| 40 | 30 | 27 | 38 | 47 | 58 | 22 | 29 | 56 | 40 | 46 | 38 | 19 | 49 | 50 |

Sample statistic

$$\bar{x} = \frac{\sum x}{n} = \frac{1168}{30} = 38.9$$

Measures of Central Tendency

- **Mean**
  - Total number of calls made in one non-leap year is 14,950
  - Mean number per day?

Population parameter

$$\mu = \frac{\sum x}{N} = \frac{14,950}{365} = 41.0$$

Measures of Central Tendency

- **Median – the value that divides the data set in half**

Example:

19 22 23 25 27 29 29 30 30 34 36 37 37 38 (38)
(40) 40 40 44 45 46 46 47 47 49 50 50 56 56 58

$$\text{median} = \frac{38 + 40}{2} = 39$$

Measures of Central Tendency

- **Mode – the value that occurs most often**
  - A data set can have more than one mode
  - A mode need not exist

Example:

19 22 23 25 27 29 29 30 30 34 36 37 37 38 38
(40 40 40) 44 45 46 46 47 47 49 50 50 56 56 58

$$\text{mode} = 40$$

Measures of Dispersion

- **Range – the maximum value minus the minimum value**

Example:

(19) 22 23 25 27 29 29 30 30 34 36 37 37 38 38
40 40 40 44 45 46 46 47 47 49 50 50 56 56 (58)

$$\text{range} = 58 - 19 = 39$$

Measures of Dispersion

- **Variance – indicates the spread of the data about the mean**

Sample statistic

$$s^2 = \frac{\sum (x - \bar{x})^2}{n - 1}$$

Population parameter

$$\sigma^2 = \frac{\sum (x - \mu)^2}{N}$$

Measures of Dispersion

- **Variance**
  - Times to complete a task (min): 5, 10, 15, 3, 7

$$\bar{x} = \frac{\sum x}{n} = \frac{40}{5} = 8$$

x	$x - \bar{x}$	$(x - \bar{x})^2$
5	-3	9
10	2	4
15	7	49
3	-5	25
7	-1	1
		$\sum(x-\bar{x})^2 = 88$

Sample statistic

$$s^2 = \frac{\sum(x-\bar{x})^2}{n-1} = \frac{88}{5-1} = 22 \text{ min}^2$$

Measures of Dispersion

- **Variance – shortcut formulas**

Sample statistic

$$s^2 = \frac{\sum x^2 - \frac{\left(\sum x\right)^2}{n}}{n-1}$$

Population parameter

$$\sigma^2 = \frac{\sum x^2 - \frac{\left(\sum x\right)^2}{N}}{N}$$

Measures of Dispersion

- **Variance**

$$\sum x^2 = 25 + 100 + 225 + 9 + 49 = 408$$

$$\left(\sum x\right)^2 = (5 + 10 + 15 + 3 + 7)^2 = 1600$$

Sample statistic

$$s^2 = \frac{\sum x^2 - \dfrac{\left(\sum x\right)^2}{n}}{n-1} = \frac{408 - \dfrac{1600}{5}}{4} = \frac{408 - 320}{4} = 22$$

Measures of Dispersion

- **Standard Deviation – the square root of variance**
  - Typically easier to comprehend because it is in the same units as the value being measured

Sample statistic $\quad s = \sqrt{s^2}$ Population parameter $\quad \sigma = \sqrt{\sigma^2}$

Sample statistic $\quad s = \sqrt{s^2} = \sqrt{22} = 4.7$ min

Grouped Data

- **Mean for grouped data**
  - x = the class mark
  - f = the class frequency

  $$\bar{x} = \frac{\sum xf}{n}$$

- **Median for grouped data**
  - The value that divides the data into two equal parts
  - Assume data in each class are uniformly spread across the class

Grouped Data

- **Mean for grouped data**

IQ Score	Frequency	Class Boundaries	Class Width	Class Marks
80-94	8	79.5-94.5	15	87
95-109	14	94.5-109.5	15	102
110-124	24	109.5-124.5	15	117
125-139	16	124.5-139.5	15	132
140-154	13	139.5-154.5	15	147

$$\bar{x} = \frac{\sum xf}{n} = \frac{87 \cdot 8 + 102 \cdot 14 + 117 \cdot 24 + 132 \cdot 16 + 147 \cdot 13}{8 + 14 + 24 + 16 + 13} = \frac{8955}{75} = 119.4$$

- **Variance for grouped data**

$$\sigma^2 = \frac{\sum x^2 f - \frac{(\sum xf)^2}{n}}{n - 1}$$

Grouped Data

- **Median for grouped data**

IQ Score	Frequency	Cumulative Frequency
80-94	8	8
95-109	14	22
110-124	24	46
125-139	16	62
140-154	13	75

- **Median = 118.75**

Grouped Data

- **Modal class for grouped data**
  - Class with maximum frequency

IQ Score	Frequency	Cumulative Frequency
80-94	8	8
95-109	14	22
110-124	24	46
125-139	16	62
140-154	13	75

- **Modal class = 110-124 class**

Grouped Data

- **Range for grouped data**
  - Upper boundary for largest class minus lower boundary for smallest class

IQ Score	Frequency	Cumulative Frequency
80-94	8	8
95-109	14	22
110-124	24	46
125-139	16	62
140-154	13	75

- **Range = 154-80 = 74**

Chebyshev's Theorem

- **The fraction of _any_ data set that lies within 'k' standard deviations (k>1) is at least:** $1 - \dfrac{1}{k^2}$

- **If k=2, at least 75% of the data set will fall between $\overline{x} - 2s$ and $\overline{x} + 2s$**

Chebyshev's Theorem tells us that regardless of the distribution, the probability that a randomly selected value lies in the interval $\mu \pm k\sigma$

is at least $1 - \dfrac{1}{k^2}$ (or $(1 - \dfrac{1}{k^2})100\%$, in percentage terms)

Empirical Rule

- ## Applies to a bell-shaped distribution

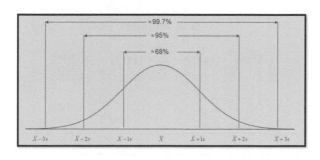

Coefficient of Variation

- ## Measure of relative variation

Sample
statistic

$$CV = \frac{s}{\overline{x}} \cdot 100\%$$

Population
parameter

$$CV = \frac{\sigma}{\mu} \cdot 100\%$$

Coefficient of Variation

Example: Time required to complete a task

Sample statistic

$$CV = \frac{s}{\overline{x}} \cdot 100\% = \frac{4.7}{8} \cdot 100\% = 58.75\%$$

z-scores

- **The number of standard deviations that a given observation, x, is below or above the mean**

  Sample statistic $\qquad z = \dfrac{x - \overline{x}}{s}$

  Population parameter $\qquad z = \dfrac{x - \mu}{\sigma}$

z-scores

- **Two counties:**
  - A: mean = $27,500; sd = $4500
    - Deputy earns $30,000
  - B: mean = $24,250; sd = $2750
    - Deputy earns $28,500
  - Which deputy has a higher salary relative to their county?

  Sample statistic $$z = \frac{x - \bar{x}}{s}$$

z-scores

- **Which deputy has a higher salary relative to their county?**

  - A: $z = \dfrac{x - \mu}{\sigma} = \dfrac{30,000 - 27,500}{4500} = 0.56$

  - B: $z = \dfrac{x - \mu}{\sigma} = \dfrac{28,500 - 24,250}{2750} = 1.55$

Measures of Position

- **Percentiles**
  - Percentile of x is found by dividing the number of observations less than x by the total number of observations and multiplying by 100%
  - To find the pth percentile, compute index i and round up to identify that position within the data set

$$i = \frac{p \cdot n}{100}$$

Measures of Position

- **Percentiles**

Example:

19	22	23	25	27	29	29	30	30	34	36	37	37	38	38
40	40	40	44	45	46	46	47	47	49	50	50	56	56	58

- **A score of 36 is the 33rd percentile or P_{33}**

$$\frac{10}{30} \cdot 100\% = 33\%$$

Measures of Position

• **Percentiles**

Example:

19 22 ⟨23⟩⟨25⟩ 27 29 29 30 30 34 36 37 37 38 38
40 40 40 44 45 46 46 47 47 49 50 50 56 56 58

• **The 60th percentile score, or P_{60} is 42**

$$i = \frac{60 \cdot 30}{100} = 18$$

Measures of Position

• **Deciles**
  • $D_1 = P_{10}, D_2 = P_{20}, D_3 = P_{30}$, etc.
• **Quartiles**
  • $Q_1 = P_{25}, Q_2 = P_{50}, Q_3 = P_{75}$
• **Interquartile Range**
  • $Q_3 - Q_1$
• **Median**
  • $P_{50} = D_5 = Q_2$

Box and Whisker Plot

- Graphical display of the quartiles
- Box contains the middle 50% of the data
- Whiskers indicate the smallest and largest values
- Median is identified with a line in the box

Box and Whisker Plot

Example:

19	22	23	25	27	29	29	30	30	34	36	37	37	38	38
40	40	40	44	45	46	46	47	47	49	50	50	56	56	58

- $Q_1 = 30$
- $Q_2 = 39$
- $Q_3 = 47$

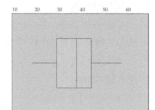

Experiment, Outcomes, and Sample Space

- **Experiment**
  - Any operation or procedure whose outcomes cannot be predicated with certainty
- **The set of all possible *Outcomes* for an experiment is called the *Sample Space* for the experiment**

- **Example: games of chance: coin toss, roll of a die, choosing a card from a full deck**

Tree Diagram

- **Tree Diagram**
  - Each outcome of an experiment is represented by a branch of a tree

- **Example: 3 tosses of a coin**

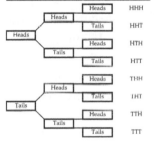

Counting

- **Counting Rule**
  - In a multi-step experiment, the total number of possible outcomes is the product of the number of possible outcomes for each step

- **Example: 3 tosses of a coin**

$$2 \times 2 \times 2 = 8$$

An Event:
Simple and Compound

- **An *Event* is a subset of the sample space consisting of at least one outcome**
- **Simple event**
  - Exactly one outcome
  - Example: Roll of the number 1 on a die
- **Compound event**
  - More than one outcome
  - Example: Roll of an odd number on a die

Probability

- **The measure of the likelihood of the occurrence of some event**
  - Notation: the probability of an event E occurring is denoted as $P(E)$
  - Read *"P* of *E"* or *"probability of event E"*
- **Properties**
  - $P(E)$ is a real number between 0 and 1
    $$0 \leq P(E) \leq 1$$
  - The sum of probabilities for all simple events of an experiment must equal 1
    $$P(E_1) + P(E_2) + \ldots + P(E_n) = 1$$

Classical Probability

- **All outcomes of an experiment are equally likely**
- **For an experiment with n outcomes**
  - The probability of any simple event
    $$P(E) = \frac{1}{n}$$
  - The probability of an event consisting of k outcomes
    $$P(E) = \frac{k}{n}$$

Relative Frequency Probability

- For experiments not having equally likely outcomes
- If an experiment is performed 'n' times, and if event E occurs 'f' times

$$P(E) = \frac{f}{n}$$

Subjective Probability

- Neither classical nor relative frequency definitions of probability are applicable
- Utilizes intuition, experience, and collective wisdom to assign a degree of belief that an event will occur

Joint, Marginal, and Conditional Probabilities

- **Joint probability**
  - The probability of multiple events occurring together
- **Marginal probability**
  - The probability of an event, regardless of the other events
- **Conditional probability**
  - The probability of an event, given that another event has already occurred

Joint, Marginal, and Conditional Probabilities

- **Example:**

Convert the table to probabilities

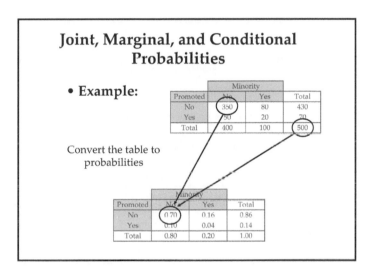

Promoted	Minority		
	No	Yes	Total
No	350	80	430
Yes	50	20	70
Total	400	100	500

Promoted	Minority		
	No	Yes	Total
No	0.70	0.16	0.86
Yes	0.10	0.04	0.14
Total	0.80	0.20	1.00

• **Example:**

Promoted	Minority		Total
	No	Yes	
No	350	80	430
Yes	50	20	70
Total	400	100	500

Probability that a police officer is not a minority and did not get promoted = 0.70

Joint probabilities

A: minority
B: promoted

Promoted	Minority		Total
	No	Yes	
No	0.70	0.16	0.86
Yes	0.10	0.04	0.14
Total	0.80	0.20	1.00

$$P(\text{not } A \text{ and } B) = 0.70$$

• **Example:**

Promoted	Minority		Total
	No	Yes	
No	350	80	430
Yes	50	20	70
Total	400	100	500

Probability that a police officer is a minority = 0.20

Marginal probabilities

A: minority
B: promoted

Promoted	Minority		Total
	No	Yes	
No	0.70	0.16	0.86
Yes	0.10	0.04	0.14
Total	0.80	0.20	1.00

$$P(A) = 0.20$$

• **Example:**

Promoted	Minority		Total
	No	Yes	
No	350	80	430
Yes	50	20	70
Total	400	100	500

Probability that a police officer that is not a minority is promoted = 0.10/0.80 = 0.125

Conditional probability

A: minority
B: promoted

Promoted	Minority		Total
	No	Yes	
No	0.70	0.16	0.86
Yes	0.10	0.04	0.14
Total	0.80	0.20	1.00

$$P(A|B) = \frac{P(A \text{ and } B)}{P(B)}$$

$$P(B|\text{not } A) = 0.125$$

Mutually Exclusive Events

• **Two events are said to be mutually exclusive if they do not have any outcomes in common**
  • They cannot both occur together
  • Their joint probability equals zero
  $$P(A \text{ and } B) = 0$$

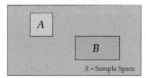

Dependent Events

- Two events are said to be *Dependent Events* if the knowledge that one of the events has occurred influences the probability that the other will occur
- Example:
  - Being diabetic and having a family history of diabetes

Independent Events

- If the occurrence of one event does not influence the probability of occurrence of another event, these events are said to be *Independent Events*

$$P(A|B) = P(A)$$

- Example:
  - Having 10 letters in your last name and being a sociology major
  - Drawing a club from a standard deck and drawing a face card from a standard deck
- **Not always obvious – do the math**

Complementary Events

- **An event that consists of all other outcomes in the sample space not in the original event**
  - If we have an event A, its complement would be denoted as A^c
  - Read "not A" or "complement of A"
- **Since an event and its complement account for all the outcomes of an experiment, then:**

$$P(A) + P(A^c) = 1$$

65

Complementary Events

- **Complementary events are always mutually exclusive**
- **Mutually exclusive events are not always complementary**

Intersection of Events

- The intersection of events refers to all outcomes which are common to both events
- Denoted as either *A* and *B* or *A* ∩ *B*
- Read as "*A* intersect *B*"

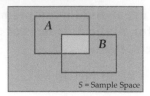

S = Sample Space

Intersection of Events

- The probability of the intersection of two events is given by the *multiplication rule*
- Derived from the formula for conditional probabilities

$$P(A \text{ and } B) = P(A)P(B|A)$$

- If A and B are independent

$$P(A \text{ and } B) = P(A)P(B)$$

Union of Events

- The union of two events consists of all those outcomes that belong to either or both events
- Denoted as either A or B or $A \cup B$
- Read as "A union B"

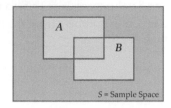

S = Sample Space

Union of Events

- The probability of the union of two events is given by the *addition rule*

$$P(A \text{ or } B) = P(A) + P(B) - P(A \text{ and } B)$$

- If A and B are mutually exclusive

$$P(A \text{ or } B) = P(A) + P(B)$$

Permutations and Combinations

- **Involve the selection of a subset of items from a larger group of items**
- **Examples**
  - Selecting 5 volunteers from this class
  - Selecting 6 lotto numbers from the numbers 1 through 45

Permutations and Combinations

- **Combinations – the order of their selection does not matter**
- **Denoted by**

$$C_n^N \text{ or } {}_N C_n \text{ or } C(N,n) \text{ or } \binom{N}{n}$$

- **The number of combinations possible when selecting *n* from *N* items**

$$C_n^N = \frac{N!}{n!(N-n)!}$$

Permutations and Combinations

- **Permutations – the order of their selection does matter**
- **Denoted by**

$$P_n^N \text{ or } {}_NP_n \text{ or } P(N,n) \text{ or } (N)_n$$

- **The number of permutations possible when selecting n from N items**

$$P_n^N = \frac{N!}{(N-n)!}$$

Permutations and Combinations

- **Example: Selecting the Final Four**

$$C_4^{64} = \frac{64!}{4!(64-4)!} = \frac{64 \cdot 63 \cdot 62 \cdot 61 \cdot 60!}{4 \cdot 3 \cdot 2 \cdot 1 \cdot 60!} = 635,376$$

- **Example: Selecting the Top 4 Teams in Order**

$$P_4^{64} = \frac{64!}{(64-4)!} = \frac{64 \cdot 63 \cdot 62 \cdot 61 \cdot 60!}{60!} = 15,249,024$$

Random Variable

- **Associates a numerical value with each outcome of an experiment**
- **Mathematical definition: a real-valued function defined on a sample space, represented by a letter such as X or Y**

Random Variable

- **Example**
  - Experiment: Flipping a coin twice
    - Random variable X = the number of tails
    - Random variable Y = the number of heads minus the number of tails

Outcome	Value of X	Value of Y
HH	0	2
HT	1	0
TH	1	0
TT	2	–2

  - Experiment: Blood test
    - Fasting blood sugar, hemoglobin, triglycerides

Discrete R.V.

- **A discrete random variable assumes a countable number of values**
- **Usually associated with counting**
- **Example**
  - 100 individuals get flu shots
    - R.V. X = the number who experience a reaction
    - The variable X may assume 101 different values, from 0 to 100
  - R.V. W = the number of individuals who receive a flu shot before the first person has a reaction
    - The variable W can assume the values 1, 2, 3, …
      (a countable infinite number of values)

Probability Distribution of a Discrete R.V.

- **A list or table of the distinct numerical values of X and the probabilities associated with those values**

Probability Distribution of a Discrete Random Variable

• *X* = the sum of the roll of a pair of dice

Outcome		Value of *x*	Outcome		Value of *x*	Outcome		Value of *x*
1	1	2	3	1	4	5	1	6
1	2	3	3	2	5	5	2	7
1	3	4	3	3	6	5	3	8
1	4	5	3	4	7	5	4	9
1	5	6	3	5	8	5	5	10
1	6	7	3	6	9	5	6	11
2	1	3	4	1	5	6	1	7
2	2	4	4	2	6	6	2	8
2	3	5	4	3	7	6	3	9
2	4	6	4	4	8	6	4	10
2	5	7	4	5	9	6	5	11
2	6	8	4	6	10	6	6	12

Probability Distribution of a Discrete Random Variable

• Probability Distribution

x	2	3	4	5	6	7	8	9	10	11	12
P(x)	1/36	2/36	3/36	4/36	5/36	6/36	5/36	4/36	3/36	2/36	1/36
P(x)	0.028	0.056	0.082	0.111	0.139	0.167	0.139	0.111	0.082	0.056	0.028

$$P(4 \leq X \leq 7) = P(4) + P(5) + P(6) + P(7)$$

$$= \frac{3}{36} + \frac{4}{36} + \frac{5}{36} + \frac{6}{36} = \frac{18}{36} = 0.5$$

Discrete Random Variable

- **Mean** $\mu = \sum xP(x)$

  - Also called "expected value" or "population mean"
- **Example**

x	2	3	4	5	6	7	8	9	10	11	12
$P(x)$	1/36	2/36	3/36	4/36	5/36	6/36	5/36	4/36	3/36	2/36	1/36
$xP(x)$	2/36	6/36	12/36	20/36	30/36	42/36	40/36	36/36	30/36	22/36	12/36

$$\mu = \sum xP(x) = \frac{252}{36} = 7$$

Discrete R.V.

- **Variance**
$$\sigma^2 = \sum (x - \mu)^2 P(x)$$

$$\sigma^2 = \left[\sum x^2 P(x)\right] - \mu^2$$

- **Example**

x	2	3	4	5	6	7	8	9	10	11	12
$P(x)$	1/36	2/36	3/36	4/36	5/36	6/36	5/36	4/36	3/36	2/36	1/36
$x^2P(x)$	4/36	18/36	48/36	100/36	180/36	294/36	320/36	324/36	300/36	242/36	144/36

$$\sigma^2 = \left[\sum x^2 P(x)\right] - \mu^2 = 54.833 - 49 = 5.833$$

$$\sigma = \sqrt{\sigma^2} = \sqrt{5.833} = 2.415$$

Binomial Random Variable

- **A discrete random variable that is defined when the conditions of a *binomial experiment* are satisfied**

- **Conditions of a binomial experiment**
    - There are n identical trials
    - Each trial has only two possible outcomes
    - The probabilities of the two outcomes remain constant for each trial
    - The trials are independent

Binomial Random Variable

- **The two outcomes: *success* and *failure***
- **The probability of success = p**
- **The probability of failure = q**

$$p + q = 1$$

Binomial Random Variable

- When the conditions of the binomial experiment are satisfied, the binomial random variable X is defined to equal the number of successes to occur in the n trials
- The binomial random variable may assume any one of the whole numbers from 0 to n

Binomial Probability Formula

- To compute the probabilities for the binomial random variables

- $P(x)$ = the probability of x successes in n trials

$$P(x) = \binom{n}{x} p^x q^{(n-x)} = \frac{n!}{x!(n-x)!} p^x q^{(n-x)}$$

Binomial Probability Formula

- **Example: a die rolled 3 times**
  - X = the # of times 6 turns up
    - X can assume values of 0, 1, 2, or 3
  - $p = 1/6$ and $q = 5/6$

$$P(0) = \frac{3!}{0!(3-0)!}\left(\frac{1}{6}\right)^0\left(\frac{5}{6}\right)^{(3-0)} = 0.578$$

$$P(1) = \frac{3!}{1!(3-1)!}\left(\frac{1}{6}\right)^1\left(\frac{5}{6}\right)^{(3-1)} = 0.348$$

$$P(2) = \frac{3!}{2!(3-2)!}\left(\frac{1}{6}\right)^2\left(\frac{5}{6}\right)^{(3-2)} = 0.070$$

$$P(3) = \frac{3!}{3!(3-3)!}\left(\frac{1}{6}\right)^3\left(\frac{5}{6}\right)^{(3-3)} = 0.005$$

Binomial Random Variable

- **Mean** $\mu = np$

- **Variance** $\sigma^2 = npq$

- **Example: a die rolled 3 times**

$$\mu = np = 3\left(\frac{1}{6}\right) = 0.5$$

$$\sigma^2 = npq = 3\left(\frac{1}{6}\right)\left(\frac{5}{6}\right) = 0.4166$$

Other Discrete Random Variable

- **Poisson Random Variable**
  - When the number of trials is large or potentially infinite
  - Outcomes occur randomly and independently
  - Ex: number of 911 calls to arrive in an hour
- **Hypergeometric Random Variable**
  - There is not independence from trial to trial
  - Predetermined number of success and failures
  - Ex: number of women chosen if 5 volunteers are selected from the class

Continuous Random Variable

- **A continuous random variable is capable of assuming all the values in an interval or several intervals**
- **It is not possible to list all the values and their probabilities**
- **Examples**
  - Time between failures of an aircraft engine
  - Time to repair an aircraft engine
  - Number of hours spent reading statistics

Continuous Random Variable

- **The probability distribution is represented as the area under the probability density function (*pdf*) curve**
  - The graph of a *pdf* is never below the *x* axis
  - The total area under the *pdf* curve = 1
- **The probability associated with a single value is always equal to 0**

Uniform Probability Distribution

- **Conditions**
  - Minimum value that is fixed
  - Maximum value that is fixed
  - All values between the minimum and maximum occur with equal likelihood

$$f(x) = \begin{cases} \dfrac{1}{b-a} & \text{if } a \leq x \leq b, \\ 0 & \text{otherwise.} \end{cases}$$

Uniform Probability Distribution

- **Example**

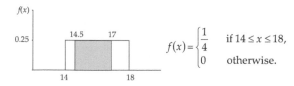

$$f(x) = \begin{cases} \dfrac{1}{4} & \text{if } 14 \le x \le 18, \\ 0 & \text{otherwise.} \end{cases}$$

$$P(14.5 \le X \le 17) = (17 - 14.5) \cdot 0.25 = 0.625$$

Uniform Probability Distribution

- **Mean**
$$\mu = \frac{a+b}{2}$$

- Example:
$$\mu = \frac{a+b}{2} = \frac{14+18}{2} = 16$$

- **Variance**
$$\sigma^2 - \frac{(b-a)^2}{12}$$

- Example:
$$\sigma^2 = \frac{(18-14)^2}{12} = \frac{16}{12} = 1.33$$

Normal Probability Distribution

- **Most important and most widely used of the continuous distributions**

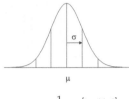

$$f(x) = \frac{1}{\sqrt{2\pi}\sigma} e^{-[(x-\mu)^2/2\sigma^2]}$$

Normal Probability Distribution

- **Properties**
  - Symmetric about μ and the area under the curve on each side of the mean is 0.5
  - Tails extend indefinitely
  - Values of μ and σ define the curve
  - Highest point of the curve occurs at μ
  - Mean = median = mode
  - Larger σ causes a wider and flatter curve

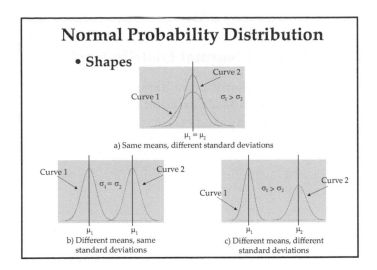

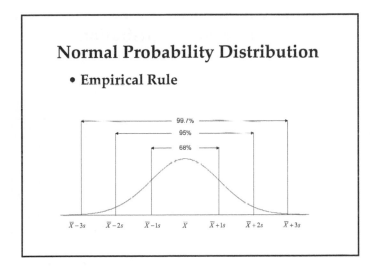

Standard Normal Distribution

- Normal distribution having mean of 0 and a standard deviation of 1
- Denoted by the letter Z

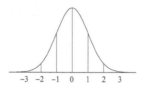

−3 −2 −1 0 1 2 3

Standard Normal Distribution

- Table

Z	0	0.01	0.02	0.03	0.04	0.05	0.06	0.07	0.08	0.09
0	0.5	0.504	0.508	0.512	0.516	0.5199	0.5239	0.5279	0.5319	0.5359
0.1	0.5398	0.5438	0.5478	0.5517	0.5557	0.5596	0.5636	0.5675	0.5714	0.5753
0.2	0.5793	0.5832	0.5871	0.591	0.5948	0.5987	0.6026	0.6064	0.6103	0.6141
0.3	0.6179	0.6217	0.6255	0.6293	0.6331	0.6368	0.6406	0.6443	0.648	0.6517
0.4	0.6554	0.6591	0.6628	0.6664	0.67	0.6736	0.6772	0.6808	0.6844	0.6879
0.5	0.6915	0.695	0.6985	0.7019	0.7054	0.7088	0.7123	0.7157	0.719	0.7224
0.6	0.7257	0.7291	0.7324	0.7357	0.7389	0.7422	0.7454	0.7486	0.7517	0.7549
0.7	0.758	0.7611	0.7642	0.7673	0.7704	0.7734	0.7764	0.7794	0.7823	0.7852
0.8	0.7881	0.791	0.7939	0.7967	0.7995	0.8023	0.8051	0.8078	0.8106	0.8133
0.9	0.8159	0.8186	0.8212	0.8238	0.8264	0.8289	0.8315	0.834	0.8365	0.8389
1	0.8413	0.8438	0.8461	0.8485	0.8508	0.8531	0.8554	0.8577	0.8599	0.8621
1.1	0.8643	0.8665	0.8686	0.8708	0.8729	0.8749	0.877	0.879	0.881	0.883
1.2	0.8849	0.8869	0.8888	0.8907	0.8925	0.8944	0.8962	0.898	0.8997	0.9015
1.3	0.9032	0.9049	0.9066	0.9082	0.9099	0.9115	0.9131	0.9147	0.9162	0.9177
1.4	0.9192	0.9207	0.9222	0.9236	0.9251	0.9265	0.9279	0.9292	0.9306	0.9319
1.5	0.9332	0.9345	0.9357	0.937	0.9382	0.9394	0.9406	0.9418	0.9429	0.9441
1.6	0.9452	0.9463	0.9474	0.9484	0.9495	**0.9505**	0.9515	0.9525	0.9535	0.9545
1.7	0.9554	0.9564	0.9573	0.9582	0.9591	0.9599	0.9608	0.9616	0.9625	0.9633
1.8	0.9641	0.9649	0.9656	0.9664	0.9671	0.9678	0.9686	0.9693	0.9699	0.9706
1.9	0.9713	0.9719	0.9726	0.9732	0.9738	0.9744	**0.975**	0.9756	0.9761	0.9767
2	0.9772	0.9778	0.9783	0.9788	0.9793	0.9798	0.9803	0.9808	0.9812	0.9817
2.1	0.9821	0.9826	0.983	0.9834	0.9838	0.9842	0.9846	0.985	0.9854	0.9857
2.2	0.9861	0.9864	0.9868	0.9871	0.9875	0.9878	0.9881	0.9884	0.9887	0.989
2.3	0.9893	0.9896	0.9898	0.9901	0.9904	0.9906	0.9909	0.9911	0.9913	0.9916
2.4	0.9918	0.992	0.9922	0.9925	0.9927	0.9929	0.9931	0.9932	0.9934	0.9936
2.5	0.9938	0.994	0.9941	0.9943	0.9945	0.9946	0.9948	0.9949	0.9951	0.9952
2.6	0.9953	0.9955	0.9956	0.9957	0.9959	0.996	0.9961	0.9962	0.9963	0.9964
2.7	0.9965	0.9966	0.9967	0.9968	0.9969	0.997	0.9971	0.9972	0.9973	0.9974
2.8	0.9974	0.9975	0.9976	0.9977	0.9977	0.9978	0.9979	0.9979	0.998	0.9981
2.9	0.9981	0.9982	0.9982	0.9983	0.9984	0.9984	0.9985	0.9985	0.9986	0.9986
3	0.9987	0.9987	0.9987	0.9988	0.9988	0.9989	0.9989	0.9989	0.999	0.999

Standard Normal Distribution

- Area under the curve denotes probability
- Finding the area under the curve between two values equates to the probability that a randomly occurring event will fall between those values

Calculating Area Under the Curve

- Find $P(x < 1.11)$ = the area under the curve to the left of $z = 1.11$
  - Look up 1.11 on the z-table
  - Solution: 0.3665 + 0.5 = 0.8665 $z = 1.11$

- Find $P(0 < x < 2.34)$
  - Look up 2.34 on the z-table
  - Solution: 0.4904

$z = 0$ $z = 2.34$

Standardizing a Normal Distribution

- *z*-score
- **The number of standard deviations that a data value falls above or below the mean**

$$z = \frac{X - \mu}{\sigma}$$

Application: Normal Distribution

- **The mean number of hours an American worker spends on the computer is 3.1 hours per work day with standard deviation is 0.5 hours.**
- **Find the percentage of workers who spend less than 3.5 hours on the computer.**

$$z = \frac{X - \mu}{\sigma} = \frac{3.5 - 3.1}{0.5} = 0.80$$

- Find the *z*-value corresponding to 3.5
- Find $P(x < 0.80)$ on the *z*-table
- Solution: 0.2881 + 0.5 = 0.7881 or 78.81%

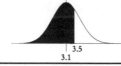

3.5
3.1

Normal Approximations to the Binomial Distribution

- When p is approximately 0.5, and as n increases, the shape of the binomial distribution becomes similar to the normal distribution
- Appropriate when

$$np \geq 5$$
$$nq \geq 5$$

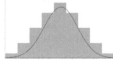

Sampling

- **Census**
  - Measuring every item of a population
- **Sample**
  - Measuring a subset of the population
  - Selecting a sample of size n from a population of size N

Replacement?

- **With replacement**
  - An item is returned to the population prior to the next selection
- **Without replacement**
  - An item is removed from the population upon selection

- **Replacement impacts the number of possible outcomes for your sample**

Types of Sampling

- **Simple Random Sampling**
  - Each item has the same chance of being selected
- **Systematic Random Sampling**
  - Randomly select the first item
  - Select every kth item thereafter
- **Cluster Sampling**
  - Divide items into clusters
  - Randomly choose clusters to sample
- **Stratified Sampling**
  - Divide items into strata – groups with commonality
    - Age, annual income, region of the country
  - Randomly sample each strata

The Sample Mean

- **The parameter of interest is usually the population mean, μ**
- **From a selected sample, only the sample mean is determined, $\bar{x}$**

- **The sampling dilemma**
  - The sample mean can assume several different values
  - The population mean is constant

Sampling Distribution of the Sampling Mean

- **The set of all possible values of the sample mean along with the probabilities of occurrence of the possible values**

Example

- # of African-American-owned businesses

City	#, in thousands
A: New York	42
B: Washington D.C.	39
C: Los Angeles	36
D: Chicago	33
E: Atlanta	30

x	30	33	36	39	42
$P(x)$	0.2	0.2	0.2	0.2	0.2

$$\mu = \sum xP(x) = 30 \cdot 0.2 + 33 \cdot 0.2 + 36 \cdot 0.2 + 39 \cdot 0.2 + 42 \cdot 0.2 = 36$$

$$\sigma^2 = \sum x^2 P(x) - \mu^2 = 900 \cdot 0.2 + 1089 \cdot 0.2 + 1296 \cdot 0.2 + 1521 \cdot 0.2 + 1764 \cdot 0.2 - 1296$$

$$= 1314 - 1296 = 18$$

$$\sigma = \sqrt{\sigma^2} = \sqrt{18} = 4.24$$

Example

- Select three cities at random

$$C_3^5 = \frac{5!}{3! \cdot 2!} = 10$$

Samples	# of businesses in the samples	Sample Mean
A, B, C	42, 39, 36	39
A, B, D	42, 39, 33	38
A, B, E	42, 39, 30	37
A, C, D	42, 36, 33	37
A, C, E	42, 36, 30	36
A, D, E	42, 33, 30	35
B, C, D	39, 36, 33	36
B, C, E	39, 36, 30	35
B, D, E	39, 33, 30	34
C, D, E	36, 33, 30	33

$\bar{x}$	33	34	35	36	37	38	39
$P(\bar{x})$	0.1	0.1	0.2	0.2	0.2	0.1	0.1

Sampling Error

- **Absolute difference between the sample mean and population mean**

$$\text{sampling error} = |\overline{x} - \mu|$$

Sample Mean	Sampling Error	Probability
33	3	0.1
34	2	0.1
35	1	0.2
36	0	0.2
37	1	0.2
38	2	0.1
39	3	0.1

The Sample Mean

$\overline{x}$	33	34	35	36	37	38	39
$P(\overline{x})$	0.1	0.1	0.2	0.2	0.2	0.1	0.1

- **Mean**

$$\mu_{\overline{x}} = \sum xP(x) = 33 \times .1 + 34 \times .1 + 35 \times .2 + 36 \times .2$$
$$+ 37 \times .2 + 38 \times .1 + 39 \times .1 = 36$$

- **Variance**

$$\sigma_{\overline{x}}^2 = \sum x^2 P(x) - \mu^2 = 1089 \times .1 + 1156 \times .1 + 1225 \times .2 +$$
$$1296 \times .2 + 1369 \times .2 + 1444 \times .1 + 1521 \times .1 \quad 1296 = 3$$

$$\sigma = \sqrt{\sigma^2} = \sqrt{3} = 1.73$$

Sample vs. Population?

- **Mean** $\quad\quad\quad \mu_{\bar{x}} = \mu$

  $$\mu_{\bar{x}} = \mu = 36$$

- **Variance** $\quad \sigma_{\bar{x}}^2 = \dfrac{\sigma^2}{n}\left(\dfrac{N-n}{N-1}\right)$

$$\sqrt{\dfrac{N-n}{N-1}}$$

The finite population correction factor

$$3 = \sigma_{\bar{x}}^2 = \frac{\sigma^2}{n} \cdot \frac{N-n}{N-1} = \frac{18}{3} \cdot \frac{5-3}{5-1} = 3$$

- **Do not use the finite population correction factor when *n* is less than 5% of *N***

$$\sigma_{\bar{x}}^2 = \frac{\sigma^2}{n}$$

The Central Limit Theorem

- **The shape of the sampling distribution of the sample mean is normal (bell-shaped)**
  - Regardless of the shape of the distribution of the population
  - Sampling from a large population
  - Sample size is 30 or more

$$\mu_{\bar{x}} = \mu \quad\quad\quad \sigma_{\bar{x}}^2 = \frac{\sigma^2}{n}$$

Application Example

- **Government Report**
  - Mean = $500, St. Dev. = $75
- **Study reports**
  - 40 cities sampled
  - Sample mean = $465

- **Does this make sense?**

Application Example

- **If the Government Report is correct**
  - The probability of finding a sample that is $35 below the national average is given by

$$P(\overline{X} < \$465)$$

  - By the central limit theorem (sample size >30), our sample mean has a normal distribution, and

$$\mu_x - \overline{X} = \$500$$

$$\sigma_{\bar{x}} = \frac{\sigma}{\sqrt{n}} = \frac{\$75}{\sqrt{40}} = \$11.86$$

Application Example

- **The area under the curve left of $465 for the sample mean is the same as the area to the left of the standard score**

$$z = \frac{\$465 - \$500}{\$11.86} = -2.95$$

$$P(\overline{X} < \$465) = P(z < -2.95) = 0.0016$$

- **Conclusion: results suggest that we have a highly unusual sample, or the government claim is incorrect**

Estimation

- **The assignment of a numerical value, or range of values, to a population parameter**
- **Point Estimate**
  - The value of a sample statistic assigned to an unknown population parameter
- **Interval Estimate**
  - An interval believed to contain the population parameter

Confidence Interval

- **An interval of numbers obtained from**
  - The point estimate of the parameter
  - The standard error of the sample statistic and its sampling distribution
  - A percentage that specifies how confident we are that the value of the parameter lies in the interval.
    - The percentage is called the confidence level

Confidence Interval for Large Samples

- **Derivation**
  - Step 1: Central Limit Theorem

  $$z = \frac{\bar{x} - \mu}{\sigma_{\bar{x}}}$$

  - Step 2: 95% of the area under the curve is between $z = -1.96$ and $z = 1.96$

  $$P\left(-1.96 < \frac{\bar{x} - \mu}{\sigma_{\bar{x}}} < 1.96\right) = 0.95$$

  - Step 3: The inequality is solved for μ

The 95% confidence interval	$\bar{x} - 1.96\sigma_{\bar{x}} < \mu < \bar{x} + 1.96\sigma_{\bar{x}}$

Confidence Levels

Confidence Level	Z Value
80	1.28
90	1.65
95	1.96
99	2.58

Confidence Level

- **What is the confidence interval?**
  - We are 95% confident that the population mean lies within that interval

- **What does it mean?**
  - Sample the population in 100 separate samples
  - Create the 100 confidence intervals at the 95% confidence level
  - 95 of them will contain the population mean, but 5 will not

Estimated standard error

- **If no historical data exists concerning mean or standard deviation, we would use the estimated standard error**

$$s_{\bar{x}} = \frac{s}{\sqrt{n}}$$

where, s = the sample standard deviation

Student's t-distribution

- **When sample size is less than 30 and the estimated standard error is used**
  - The confidence interval will generally be incorrect – too narrow
- **In this case, we would use the Student's t-distribution in place of the normal distribution as a correction**

Student's *t*-distribution

- Student's *t*-distribution is a family of distributions
- Similar to the normal distribution, but its standard deviation is greater than 1, determined by degrees of freedom
- Generally speaking
  - Lower maximum height
  - Thicker tails

Confidence Interval for Small Samples

- Use the estimated standard error and the Student's *t*-distribution

$$\bar{x} - ts_{\bar{x}} < \mu < \bar{x} + ts_{\bar{x}}$$

where, t = Student's *t*-distribution value determined by the confidence level chosen and the degrees of freedom available $(n-1)$

What is a Hypothesis?

A hypothesis is a supposition or proposed explanation made on the basis of limited evidence as a starting point for further investigation.
- It is a tentative statement about the relationship between two or more variables.
- It is an educated prediction that can be tested.

Hypotheses

- Suppose a company determines its historical mean of sales per order is $155 with a standard deviation of $50
- This company hypothesizes that sales per order from this year are different than the historical mean of $155

$$\text{Is } \mu \neq \$155?$$

- What evidence can we use to support this conclusion?

Random Variable

- Sales per order is a random variable
- Mean sales per order for a given year is also a random variable
- Essentially, we are trying to find out if the random variable of mean sales per order calculated for this year is likely to have come from the same distribution as the historical distribution of mean sales

Hypotheses

- **Alternate hypothesis**
  - Research hypothesis
  - That which we wish to prove

$$H_a : \mu_0 \neq \$155$$

- **Null hypothesis**
  - That which will be concluded if there is not enough evidence to reject it

$$H_0 : \mu_0 = \$155$$

Testing Our Hypothesis

- **To test our hypothesis, a sample of 100 orders for the current year are selected**
- **From the Central Limit Theorem, we know that**

$$\bar{x} = \$155$$

$$\sigma_{\bar{x}} = \frac{\$50}{\sqrt{100}} = \$5$$

To Reject or Not to Reject?

- **What would we consider to be different enough from our mean to say that they weren't from the same distribution?**
  - 1 standard deviation different?
  - 2 standard deviations different?

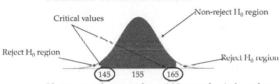

- If our sample mean from 100 samples is less than 145 or greater than 165, we will conclude that it is not likely to have come from this distribution.

Type I and Type II Errors

- α – probability of making a Type I error
- β – probability of making a Type II error

Conclusion	H_0 True	H_0 False
Do not rejct H_0	Correct conclusion	Type II error
Reject H_0	Type I error	Correct conclusion

- **We call α the level of significance of the hypothesis test**

Types of Hypotheses

- **Two-tailed**

$$H_a : \mu_0 \neq \$155$$
$$H_0 : \mu_0 = \$155$$

- **One-tailed**

$$H_a : \mu_0 > \$155 \qquad H_a : \mu_0 < \$155$$
$$H_0 : \mu_0 \leq \$155 \qquad H_0 : \mu_0 \geq \$155$$

Level of Significance

- **Two-tailed significance, α = 0.05 (95% CL)**

Area = 0.025   Area = 0.025

- **One-tailed significance, α = 0.05 (95% CL)**

Area = 0.05

Testing Our Hypothesis

- **Large sample: Z-test**
- **z-statistic**

$$z = \frac{\overline{x} - \mu_0}{\sigma_{\overline{x}}}$$

- **Compare our z-statistic with our Standard normal distribution for a given level of confidence based on one- or two-tailed hypothesis**

Testing Our Hypothesis

- Small sample: t-test
- t-statistic

$$t = \frac{\overline{x} - \mu_0}{s_{\overline{x}}}$$

- Compare our t-statistic with our Student's t-distribution for a given level of confidence and degrees of freedom based on one- or two-tailed hypothesis

Example: Large Sample

- Mean age = 32.5; St. Dev. = 5.5
- Has the mean average changed this year?
- Hypothesis: $H_0 : \mu_0 = 32.5$

$$H_a : \mu_0 \neq 32.5$$

- Let's test our hypothesis at the 95% CL

Reject region	Non-reject Region	Reject Region
−1.96	1.96	

Example: Large Sample

- **A sample of 50 policyholders was taken**
  - Mean age (50 policyholders) = 34.4
- **Standard error**

$$\sigma_{\bar{x}} = \frac{5.5}{\sqrt{50}} = 0.778$$

Therefore, we reject the Null Hypothesis and conclude that mean age has changed.

- **Test statistic**

$$z = \frac{\bar{x} - \mu_0}{s_{\bar{x}}} = \frac{34.4 - 32.5}{0.778} = 2.44$$

Reject region	Non-reject Region	Reject Region

            −1.96   1.96   **X**

p-value

- **Our rejection region is determined by our choice of confidence level**
- **At what confidence level would my rejection decision change?**
  - That confidence level is called the *p*-value
  - The *p*-value equals the area of the rejection region determined by the test statistic

Area = ?   Area = ?

              −2.44   2.44

Two samples

- **Are they different?**
- **Different enough that we can say with certainty that they are different?**

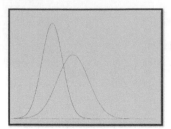

Notation

μ_1 and μ_2 Means of populations 1 and 2

σ_1 and σ_2 Standard deviation of populations 1 and 2

n_1 and n_2 Sample size taken from populations 1 and 2

$\bar{x}_1$ and $\bar{x}_2$ Sample means from populations 1 and 2

s_1 and s_2 Sample standard deviations from populations 1 and 2

Two Sample Hypothesis

- **Two-tailed**

$$H_0 : \mu_1 - \mu_2 = D_0$$
$$H_a : \mu_1 - \mu_2 \neq D_0$$

- **One-tailed**

$$H_0 : \mu_1 - \mu_2 = D_0 \qquad H_0 : \mu_1 - \mu_2 = D_0$$
$$H_a : \mu_1 - \mu_2 < D_0 \qquad H_a : \mu_1 - \mu_2 > D_0$$

Test Statistic

$$z^* = \frac{\bar{x}_1 - \bar{x}_2 - D_0}{\sigma_{\bar{x}_1 - \bar{x}_2}}$$

where,

$$\sigma_{\bar{x}_1 - \bar{x}_2} = \sqrt{\frac{\sigma_1^2}{n_1} + \frac{\sigma_2^2}{n_2}}$$

Or, if population information is unknown

$$S_{\bar{x}_1 - \bar{x}_2} = \sqrt{\frac{s_1^2}{n_1} + \frac{s_2^2}{n_2}}$$

Example

- **Option 1:**
  - Mean = 3.5 days; st. dev. = 1.5 days
  - Sample Size = 48
- **Option 2:**
  - Mean = 8.0 days; st. dev. = 2.0 days
  - Sample Size = 55
- **Hypothesis** $H_0 : \mu_1 - \mu_2 = 0$
  $$H_a : \mu_1 - \mu_2 < 0$$

Example

- **Estimate the standard error**

$$S_{\bar{x}_1 - \bar{x}_2} = \sqrt{\frac{s_1^2}{n_1} + \frac{s_2^2}{n_2}} = \sqrt{\frac{(1.5)^2}{48} + \frac{(2.0)^2}{55}} = 0.346$$

- **Compute the test statistic**

$$z^* = \frac{\bar{x}_1 - \bar{x}_2 - D_0}{\sigma_{\bar{x}_1 - \bar{x}_2}} = \frac{3.5 - 8.0 - 0}{0.346} = -13.0$$

- **Compare to the rejection region**
  - Reject the null hypothesis
  - Conclude that Option 1 < Option 2

Confidence Interval

- 95% confidence interval

$$\left(\overline{x}_1 - \overline{x}_2\right) - 1.96 \cdot \sigma_{\overline{x}_1 - \overline{x}_2} < \mu_1 - \mu_2 < \left(\overline{x}_1 - \overline{x}_2\right) + 1.96 \cdot \sigma_{\overline{x}_1 - \overline{x}_2}$$

Determined by the
normal distribution pdf
corresponding to the
confidence level desired

- How do we interpret this confidence interval for two populations?

Two Sample Test Variations

- Small samples or large samples
- Equal or unequal variances
- Known or unknown variances
- Independent or dependent samples
- Paired or unpaired observations

Chi-square [χ^2] Distribution

- **As degrees of freedom become larger, the χ^2 curves look more like normal curves**

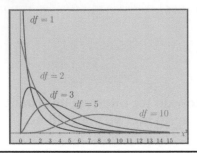

Goodness of Fit Tests

- **If each element of a population is assigned to one and only one of k categories or classes**
- **Multinomial probability distribution**
- **Example: dieting**
    - 85% most likely go off a diet on a weekend
    - 10% most likely go off a diet on a weekday
    - 5% don't know $p_1 = 0.85$
                              $p_2 = 0.10$
                              $p_3 = 0.05$

Goodness of Fit Hypothesis

$$H_0 : p_1 = 0.85, p_2 = 0.10, p_3 = 0.05$$
are correct

$$H_a : p_1 = 0.85, p_2 = 0.10, p_3 = 0.05$$
are not correct

Observed vs. Expected Frequencies

- **Observed frequencies: from the sample**
- **Expected frequencies: from the null hypothesis**
- **Example: diet (n = 200)**
  - Observed frequencies
    - Weekend cheaters: 160
    - Weekday cheaters: 22
    - Don't know: 18
  - Expected frequencies, $n * p$
    - Weekend cheaters: 200 * 0.85 = 170
    - Weekday cheaters: 200 * 0.10 = 20
    - Don't know: 200 * 0.05 = 10

Goodness of Fit Test Statistic

- **Test statistic** $\chi^2 = \sum_k \dfrac{(o-e)^2}{e}$
  - *k*-1 degrees of freedom

- **Always a one-tailed test**
  - If observed values are close to expected, the test statistic is close to zero
  - If observed values are not close to expected, the test statistic will be large

Example

- **Compute the test statistic**

Category	o	e	o-e	$(o\text{-}e)^2$	$(o\text{-}e)^2/e$
1	160	170	−10	100	0.588
2	22	20	2	4	0.200
3	18	10	8	64	6.400
Sum	200	200	0		7.188

- **Determine the 95% critical value from the χ^2 distribution [Excel: chiinv(0.05,2)]**
$$\chi^2_{0.05,2} = 5.991$$

- **Compare to the rejection region**
$$7.188 > 5.991$$
  - Reject the null hypothesis: the probabilities are incorrect

ANOVA

- **Used to compare means of several populations**

Design 1	Design 2	Design 3
10	24	17
10	22	17
8	24	15
10	24	19
12	26	17
mean = 10	mean = 24	mean = 17

F-Distribution

- **The ratio of two χ^2 distributions**
- **Two parameters**
  - Degrees of freedom of the numerator
  - Degrees of freedom of the denominator

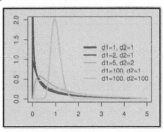

Logic of a One-way ANOVA

Design 1	Design 2	Design 3
10	24	17
10	22	17
8	24	15
10	24	19
12	26	17
mean = 10	mean = 24	mean = 17

KEYBOARD FACTOR

Treatments ($k = 3$)

$\overline{x} = 17$

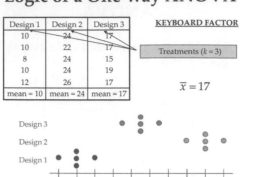

Logic of a One-way ANOVA

Design 1	Design 2	Design 3
10	34	29
12	14	17
5	24	5
1	19	10
22	29	24
mean = 10	mean = 24	mean = 17

KEYBOARD FACTOR

$\overline{x} = 17$

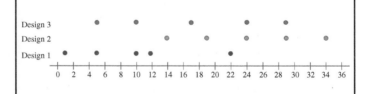

One-way ANOVA Hypothesis

$$H_0 : \mu_1 = \mu_2 = ... = \mu_k$$

$$H_a : \text{All means are not equal}$$

ANOVA Test Statistic

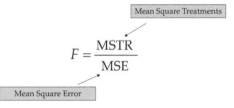

$$F = \frac{\text{MSTR}}{\text{MSE}}$$

Mean Square Treatments

Mean Square Error

Mean Square Treatments

$$\text{MSTR} = \frac{\text{SSTR}}{k-1}$$

Sum of Squares Treatments

$$\text{SSTR} = \sum_k n_k (\overline{x}_k - \overline{x})^2$$

$$\text{SSTR} = \sum_k n_k (\overline{x}_k - \overline{x})^2$$
$$= n_1 (\overline{x}_1 - \overline{x})^2 + n_2 (\overline{x}_2 - \overline{x})^2 + n_3 (\overline{x}_3 - \overline{x})^2$$
$$= 5(10-17)^2 + 5(24-17)^2 + 5(17-17)^2$$
$$= 490$$

$$\text{MSTR} = \frac{\text{SSTR}}{k-1} = \frac{490}{2} = 245$$

Mean Square Error

$$\text{MSE} = \frac{\text{SSE}}{n-k}$$

Sum of Squares Error

$$\text{SSE} = \sum_k (n_k - 1) s_k^2$$

$$\text{SSTR} = \sum_k (n_k - 1) s_k^2$$

$$= (n_1 - 1) s_1^2 + (n_2 - 1) s_2^2 + (n_3 - 1) s_3^2$$

$$= (5 - 1)2 + (5 - 1)2 + (5 - 1)2$$

$$= 24$$

$$\text{MSE} = \frac{\text{SSE}}{n-k} = \frac{24}{15-3} = \frac{24}{12} = 2$$

Test Statistic

$$F^* = \frac{\text{MSTR}}{\text{MSE}} = \frac{245}{2} = 122.5$$

$$\begin{cases} df_1 = k - 1 = 3 - 1 = 2 \\ df_2 = n - k = 15 - 3 = 12 \end{cases}$$

$$F_{0.01}(2,12) = 6.93$$

Since $F^* > F_{0.01}(2,12)$

Reject the null hypothesis

Shortcut Formulas

| Sum of Squares Total |

$$SST = SSTR + SSE$$

$$SST = \sum x^2 - \frac{\left(\sum x\right)^2}{n}$$

$$SSTR = \sum \frac{T_i^2}{n_i} - \frac{\left(\sum x\right)^2}{n}$$

T_i = sum of values for the ith treatment

$$SSE = SST - SSTR$$

One-way ANOVA Table

• Standard ANOVA Table

Source	df	SS	MS = SS/df	F-Statistic
Treatment	$k-1$	SSTR	MSTR	F^*
Error	$n-k$	SSE	MSE	
Total	$n-1$	SST		

• From the Example

Source	df	SS	MS = SS/df	F-Statistic
Treatment	2	490	245	122.5
Error	12	24	2	
Total	14	514		

Two-way ANOVA

- Used to test two factors and their interaction
- In addition to the Keyboard Factor, we also want to test a Seating Factor

Seating Design	Keyboard Design		
	1	2	3
1	10, 12	20, 22	16, 18
2	14, 16	25, 27	20, 22
3	8, 8	18, 16	14, 16

Two-way Data Table

Seating Design	Keyboard Design			Row Mean
	1	2	3	
1	11	21	17	16.33
2	15	26	21	20.67
3	8	17	15	13.33
Column Mean	11.33	21.33	17.67	

Main Effects Plots

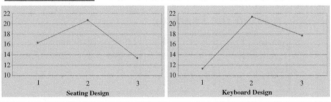

Two-way ANOVA Hypotheses

- **Interaction**

H_0 : There is no interaction between Factors A and B

H_a : Factors A and B interact

- **Factor A** $H_0 : \mu_{A1} = \mu_{A2} = \mu_{A3}$

  H_a : Factor A means differ

- **Factor B** $H_0 : \mu_{B1} = \mu_{B2} = \mu_{B3}$

  H_a : Factor B means differ

Two-way ANOVA Table

Source	df	SS	MS = SS/df	F-Statistic
Factor A	$a - 1$	SSA	MSA	F_A = MSA/MSE
Factor B	$b - 1$	SSB	MSB	F_B = MSB/MSE
Interaction	$(a-1)(b-1)$	SSAB	MSAB	F_{AB} = MSAB/MSE
Error	$n - ab$	SSE	MSE	
Total	$n - 1$	SST		

Correlation

- **Strength of linear association between two variables**
  - ρ = population correlation
  - r = sample correlation

Properties of the Correlation Coefficient

- **Always between –1 and +1**

- **Magnitude of *r* indicates strength of linear relationship**
  - Negative *r* – inverse relationship
  - Positive *r* – direct relationship

- **Value of *r* close to –1 or +1**
  - Strong linear relationship

- **Value of *r* close to 0**
  - Weak linear relationship

Pearson Correlation Coefficient

$$r = \frac{S_{xy}}{\sqrt{S_{xx}S_{yy}}}$$

where,

$$S_{xx} = \sum x^2 - \frac{\left(\sum x\right)^2}{n} \qquad S_{yy} = \sum y^2 - \frac{\left(\sum y\right)^2}{n}$$

$$S_{xy} = \sum xy - \frac{\left(\sum x\right)\left(\sum y\right)}{n}$$

Example

Years	Days off	Years	Days off
x	y	x	y
2	10	16	20
2	14	20	20
2	12	20	22
4	11	20	21
4	15	24	23
8	14	24	24
8	16	24	22
10	14	30	27
10	18	30	25
16	18	30	26

• Correlation Coefficient

$$r = \frac{S_{xy}}{\sqrt{S_{xx}S_{yy}}} = \frac{945.6}{\sqrt{1891.2 \cdot 506.8}} = 0.966$$

• Correlation Matrix

r	Years	Days off
Years	1	
Days off	0.966	1

Testing Correlation

- **Hypothesis:**

$$H_0 : \rho = 0$$
$$H_a : \rho \neq 0, \rho > 0, \text{or } \rho < 0$$

- **Test statistic:**

$$t^* = r\sqrt{\frac{n-2}{1-r^2}}$$

Correlation is not Causality

- **Causality – the existence of one characteristic causes another characteristic to also exist**
- **Causality requires three components**
  - The two variables are meaningfully related (correlation)
  - The causal variable precedes or occurs contemporaneously with the resulting variable
  - Causality is reasonable – no plausible alternative explanations exist

Linear Regression

- Correlation tells us the strength of a linear relationship
- Correlation does not quantify the linear relationship

Linear Regression

- A linear regression model assumes some dependent variable, y, is linearly related to an independent variable, x

$$y = \beta_0 + \beta_1 x + \varepsilon$$

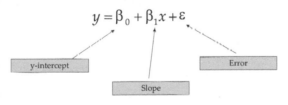

y-intercept   Error

Slope

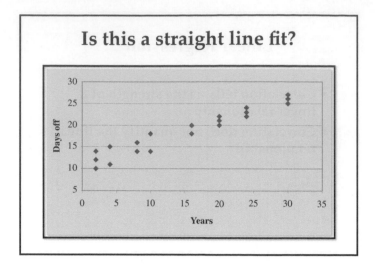

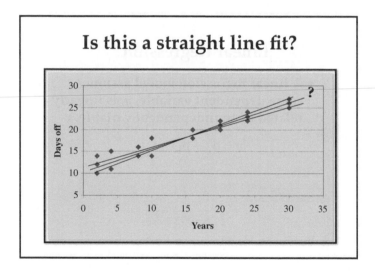

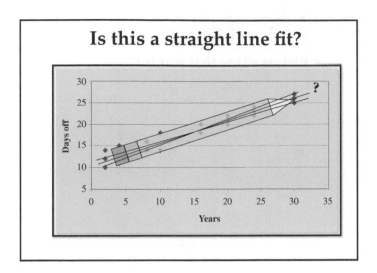

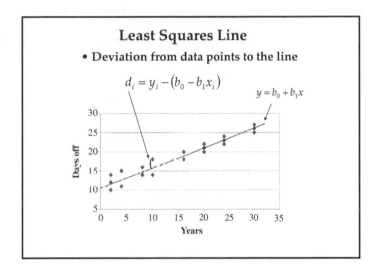

Least Squares Line

• **Pick the line that minimizes**

$$D = \sum d_i^2$$

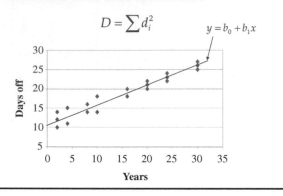

$y = b_0 + b_1 x$

Least Squares Line

$$y = b_0 + b_1 x$$

$$b_1 = \frac{S_{xy}}{S_{xx}} \qquad\qquad b_0 = \overline{y} - b_1 \overline{x}$$

• **This line has several names**
  • The least squares line
  • The line of best fit
  • The estimated regression line
  • The prediction line

Example

$$b_1 = \frac{s_{xy}}{s_{xx}} = \frac{945.6}{1891.2} = 0.5$$

$$b_0 = \overline{y} - b_1\overline{x} = 18.6 - 0.5 \cdot 15.2 = 11.0$$

$$y = 11.0 + 0.5x$$

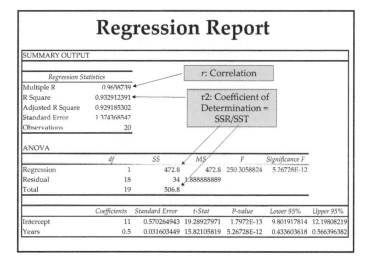

Regression Report

SUMMARY OUTPUT

Regression Statistics	
Multiple R	0.9658739
R Square	0.932912391
Adjusted R Square	0.929185302
Standard Error	1.374368542
Observations	20

r: Correlation

r2: Coefficient of Determination = SSR/SST

ANOVA

	df	SS	MS	F	Significance F
Regression	1	472.8	472.8	250.3058824	5.26728E-12
Residual	18	34	1.888888889		
Total	19	506.8			

	Coefficients	Standard Error	t-Stat	P-value	Lower 95%	Upper 95%
Intercept	11	0.570264943	19.28927971	1.7972E-13	9.801917814	12.19808219
Years	0.5	0.031603449	15.82105819	5.26728E-12	0.433603618	0.566396382

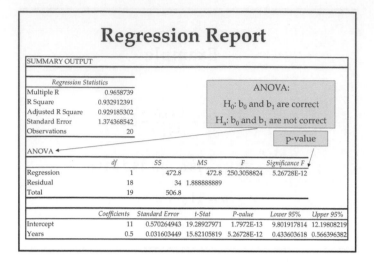

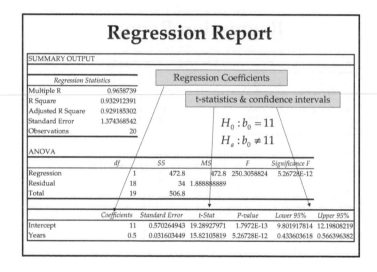

Precision to Tolerance

$$\hat{\sigma}^2_{Gage} = \hat{\sigma}^2_{Measurement\ Error} = \hat{\sigma}^2_{Repeatability} + \hat{\sigma}^2_{Reproducibility}$$

$$\frac{P}{T} = \frac{5.15\ \hat{\sigma}_{Meas.\ Error}}{USL - LSL} \le 0.1$$

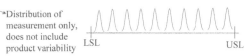

→Distribution of
measurement only,
does not include
product variability   LSL   USL

Precision to Variability

$$\hat{\sigma}^2_{Process} = \hat{\sigma}^2_{Product} + \hat{\sigma}^2_{Measurement\ Error}$$

$$\hat{\sigma}^2_{Measurement\ Error} = \hat{\sigma}^2_{\substack{Repeatability \\ Gage}} = \hat{\sigma}^2_{\substack{Reproducibility \\ Operator}}$$

$$\frac{Precision}{Process} = \frac{\hat{\sigma}_{Measurement\ Error}}{\hat{\sigma}_{Process}} \le 0.1$$

Process Variation

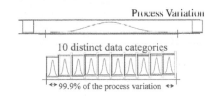

10 distinct data categories

↔ 99.9% of the process variation ↔

Controlling Repeatability

- Note: If you want to decrease your gage error take advantage of the standard error square root of the sample.
- The signal averaging technique uses:

$$\frac{1}{\sqrt{n}}$$

Distribution of Individuals

- n = the number of repeat measures taken on the same part
- the measurement = the average of "n" readings
- Example: a gage error of 50% can be cut in half if your point estimate is an average of 4 repeat measurements

$$\frac{1}{\sqrt{4}} = 1/2$$

Distribution of Means

- This technique should be used as a short term approach to perform a study, but you must fix the gage.

Accuracy vs. Precision

Averages

Measurement System Bias -Determined through "Calibration Study"

Accuracy

$$\mu_{total} = \mu_{product} + \mu_{measurement}$$

Variability

Measurement System Variability - Determined through "R&R Study"

Precision

$$\sigma^2_{total} = \sigma^2_{product} + \sigma^2_{measurement}$$

Accuracy vs. Precision
(Example)

- Suppose we have a reference material with a 'true' hardness of 5.0.
- Method 1 gives the following readings:
  3.8, 4.4, 4.2, 4.0
- Method 2 gives the following readings:
  6.5, 4.0, 3.2, 6.3

- Which method is more accurate?
- Which method is more precise?
- Which method do you prefer? Why?

Accuracy vs. Precision

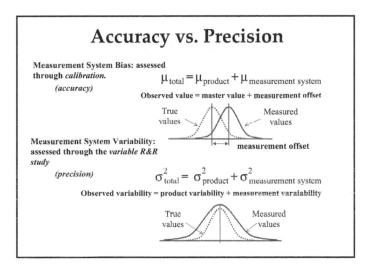

Measurement System Bias: assessed through *calibration.*
(accuracy)

$$\mu_{total} = \mu_{product} + \mu_{measurement\ system}$$

Observed value = master value + measurement offset

Measurement System Variability: assessed through the *variable R&R study*
(precision)

$$\sigma^2_{total} = \sigma^2_{product} + \sigma^2_{measurement\ system}$$

Observed variability = product variability + measurement varaiability

Measure of Accuracy: Bias

$$\mu_{total} = \mu_{product} + \mu_{measurement\ system}$$

Bias: The difference between the observed average value of measurements and the true value. The true value is an accepted, traceable reference standard (e.g., NIST).

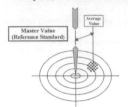

The accuracy of the measurement system is determined by conducting a bias study.

Examples of Bias

Bias: Average of measurements are different by a fixed amount. Effects include:

– Operator bias - different operators get detectably different averages for the same measurements of the same part

– Machine bias - different machines or tooling get detectably different averages for the same measurements of the same parts

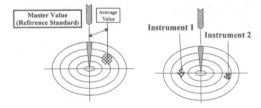

Linearity

A measure of the difference in *Accuracy* or *Precision* over the range of instrument capability.

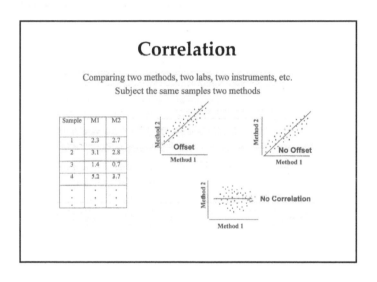

Correlation

Comparing two methods, two labs, two instruments, etc.
Subject the same samples two methods

Stability

A Measure of the Difference in Precision and Accuracy Over Time

- The distribution of measurements remains constant and predictable over time for both mean and standard deviation

- No drifts, sudden shifts, cycles, etc…

- Evaluated using a trend chart

- Ensured through a regular calibration, R&R and SPC program

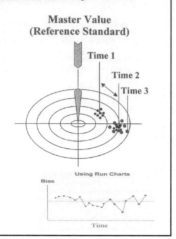

Index

A

Abbreviations, 552–553
Abilities–needs fit model, 73
ACC, *see* Air Combat Command (ACC)
Acceleration formulas, 576
Accounting methods, 182–184
Accuracy
  bias, 684
  precision, *vs.*, 682–683
Accurate online support vector regression (AOSVR), 170
Acoustic fiber optic, 154
Acquisition disaggregation, 405
Acquisition Management and Integration Center
    (AMIC), 377
Acquisition module, 306
Acquisition Program Baseline (APB), 530
Acquisition, Technology, and Logistics (USD, *see*
    Under Secretary of Defense (Acquisition,
    Technology, and Logistics (USD)
Action stage, 46
Active light work, 57
Activity attributes, 479
Activity scheduler, 473, 479, 488; *see also* Resource
    mapper
  activity prioritizing weight, 481
  amount of depleted slack, 480
  coefficients of activity scheduling objective
      function, 481
  constraints formulation, 484–485
  decision maker's supplied weight, 481, 483–484
  Gantt chart and resource loading graphs, 482–483
  LFT, 480
  number of activity successors, 480
  objective function for activity scheduling, 485–486
  peak flag, 484
  resource-driven activity attribute, 479
Activity successors, number of, 480
Actual relative % completion, 586
Acute Exposure Guideline Levels (AEGLs), 91
ADAMS, *see* Anomaly Detection at Multiple Scales
    (ADAMS)
Adaptive online support vector regression
    (AOLSVR), 170
  predicting power grid frequency, 175–176

ADC, *see* Analog-to-digital converter (ADC)
Additive interactive dependency, 475–476
Additive manufacturing (AM), 354
Adenosine diphosphate (ADP), 56
Adenosine triphosphate (ATP), 56
ADJF, *see* Australian Deployable Joint Force (ADJF)
ADP, *see* Adenosine diphosphate (ADP)
Advanced closed-loop control systems, 153
AEDC, *see* Arnold Engineering Development Center
    (AEDC)
AEGLs, *see* Acute Exposure Guideline Levels (AEGLs)
AET, *see* Arbeitswissenschaftliches
    Erhebungsverfahren zur Tatigkeitsanalyse
    (AET)
AETC, *see* Air Education and Training Command
    (AETC)
AFDDs, *see* Air Force Doctrine Documents (AFDDs)
AFIT, *see* Air Force Institute of Technology (AFIT)
Africa, ancient measurement systems in, 12
  Gebet'a game, 13–14
  Ishango bone, 13
  Lebombo bone, 12–13
Agency for Toxic Substances and Disease Registry
    (ATSDR), 83
Agency Theory, 382–383
AIHA, *see* American Industrial Hygiene Association
    (AIHA)
Air Combat Command (ACC), 377
Aircraft service contracts, 538
Air Education and Training Command (AETC), 377
Air Force Doctrine Documents (AFDDs), 230
Air Force Institute of Technology (AFIT), 545
  RCS measurement range, 546, 547
Air Mobility Command (AMC), 538
Algebraic expansions, 575
Algebraic factoring, 575
Allowable quality level (AQL), 538
α-particles, 192
AM, *see* Additive manufacturing (AM)
AMC, *see* Air Mobility Command (AMC)
American Industrial Hygiene Association (AIHA), 91
American Society of Heating, Refrigeration and Air
    Conditioning Engineers (ASHRAE), 141
AMIC, *see* Acquisition Management and Integration
    Center (AMIC)